Affinity and Immunoaffinity
Purification Techniques

BioTechniques Molecular Laboratory Methods Series

1. *Affinity and Immunoaffinity Purification Techniques*
   T.M. Phillips and B.F. Dickens

2. *Apoptosis Detection and Assay Methods*
   Edited by L. Zhu and J. Chun

3. *Gene Cloning and Analysis by RT-PCR*
   Edited by P.D. Siebert and J.W. Larrick

4. *Immunological Reagents and Solutions: A Laboratory Handbook*
   B. Damaj

5. *Protein Staining and Identification Techniques*
   R.C. Allen and B. Budowle

6. *Yeast Hybrid Methods*
   Edited by L. Zhu and G. Hannon

BioTechniques Update Series

The PCR Technique: DNA Sequencing
Edited by J. Ellingboe and U. Gyllensten

The PCR Technique: DNA Sequencing II
Edited by U. Gyllensten and J. Ellingboe

The PCR Technique: Quantitative PCR
Edited by J.W. Larrick

The PCR Technique: RT-PCR
Edited by P.D. Siebert

Expression Genetics: Differential Display
Edited by A.B. Pardee and M. McClelland

Expression Genetics: Accelerated and High-Throughput Methods
Edited by M. McClelland and A.B. Pardee

Polymorphism Detection and Analysis
Edited by J. Burczak and E. Mardis

Affinity and Immunoaffinity Purification Techniques

Terry M. Phillips, PhD, DSc
The George Washington University Medical Center
Washington, DC, USA

Benjamin F. Dickens, PhD
The George Washington University Medical Center
Washington, DC, USA

A BioTechniques® Books publication
Eaton Publishing

Terry M. Phillips
Immunochemistry Laboratory
The George Washington University Medical Center
2300 Eye Street NW
Washington, DC 20067

Benjamin F. Dickens
Immunochemistry Laboratory
The George Washington University Medical Center
2300 Eye Street NW
Washington, DC 20067

Library of Congress Cataloging-in-Publication Data

Phillips, Terry M.
   Affinity and immunoaffinity purification techniques / Terry M. Phillips, Benjamin F. Dickens.
     p. cm. -- (BioTechniques molecular laboratory methods series ; 5)
   Includes bibliographical references and index.
   ISBN 1-881299-22-8
    1. Affinity chromatography--Laboratory manuals. 2. Immunoadsorption--Laboratory
manuals.   3. Immunoassay--Laboratory manuals. 4.
Biomolecules--Separation--Laboratory manuals. I. Dickens, Benjamin F., 1951-II.
Title. III. Series.

QP519.9.A35 P48 2000
572'.36--dc21

00-021411

ISBN 1-881299-22-8

Printed in the United States of America

9 8 7 6 5 4 3 2 1

Eaton Publishing
BioTechniques Books Division
154 E. Central Street
Natick, MA 01760
www.BioTechniques.com

Francis W. Eaton: *Publisher and President*
Stephen Weaver: *Director and Editor-in-Chief*
Christine McAndrews: *Managing Editor*
Sandy Lamont: *Production Manager*
Ken Strom: *Cover Designer*

The cover illustration shows a synthetic ligand (large forward molecule) capturing two Peptide T molecules (small molecules). The graphic was designed by T.M. Phillips using Chemsite v 3.01 PRO software from Pyramid Learning.

DEDICATION

We would like to dedicate this book to our wives, Jennifer and Bonnie, who have given us their devoted support even when we were tired, miserable, and totally separated from reality.

CONTENTS

PREFACE

Separation science has become a major part of modern biotechnology and involves procedures that must be attempted by most experimentalists. The application of affinity and immunoaffinity ligands as aids in the selective isolation of specific analytes is becoming an essential and often convenient aspect of biological separation science. Although reasonably straightforward, these techniques have nuances that hinder successful isolation of the desired material. These nuances, such as ligand orientation, ligand stability, and bound analyte recovery, can play major roles in the separation process and result in suboptimal to nonrecovery of the selected analyte. One of the major reasons for lack of success in affinity or immunoaffinity separations is that separation science in some ways approaches an art form and that many of the "tricks" essential for success are either lost in the text or missing from published articles. Therefore, our intention in this book is to provide a guide to the principles and practice of affinity and immunoaffinity separation techniques. We provide not only an overview of the different aspects of affinity and immunoaffinity separation technology, but also illustrate each chapter with practical, "cookbook" style procedures, thus enabling even a novice to perform successful separations.

The book is divided into two parts: a laboratory-oriented approach to constructing affinity and immunoaffinity matrices, followed by discussions of the most common application of these techniques. Additionally, step-by-step procedures will accompany each chapter in order to help the reader apply this technology to their own experimental design. Initially, in Chapter 1, the text takes the reader through the general principles of affinity and immunoaffinity separation technology. The second chapter provides an overview of the different types of solid-phase matrices available and a review of the procedures used to prepare these matrices. Chapter 3 discusses the variations that are responsible for determining the success of any affinity or immunoaffinity separation, and Chapter 4 discusses the attachment of affinity ligands to the support matrix. Additionally, the chemistries involved in immobilizing these ligands are discussed and examples of practical approaches to the different coupling chemistries given. The use of immunological reagents, such as antibodies, and their attachment to solid-phase matrices is discussed in Chapter 5, along with a practical approach to the preparation and manufacture of immunoaffinity matrices.

Chapters 6 through 9 describe and discuss the applications of affinity and immunoaffinity separation procedures. In Chapter 6, the procedures involved in batch techniques—useful in bulk separation and often applied in process biotechnology—are described. These procedures include the use of affinity and immunoaffinity techniques for precipitation, phase separation technology, and magnetic bead separations. Batch techniques are followed by column chromatography, frequently the more popular separation technology. In this chapter, a review of the types of column chromatography and approaches to using this form of separation to the isolation of biological molecules is discussed. The application of affinity and immunoaffinity techniques to the isolation of cells is reviewed in Chapter 8 and these procedures are extended to the isolation of intact cell organelles in the following chapter. The last chapter, Chapter 10, introduces the reader to a series of specialized separation techniques. Procedures reviewed in this chapter include affinity and immunoaffinity capillary electrophoresis, immunosensors, multi-analyte batch techniques, and recycling affinity and immunoaffinity techniques.

Additionally, three appendices have been included to help the reader in a number of ways. The first describes protein measurement techniques, which are essential for ligand immobilization and estimation of bound ligand. Second, techniques for analyte detection and labeling are included. These procedures will enable the reader to use a number of important labels in order to refine the sensitivity of the detection procedure, because selection of a suitable label is essential for detection of analytes in the picogram and femtogram concentrations. The third appendix is a list of suppliers, for which contact information has been provided. It is felt that this latter information will help in the often tedious chore of finding sources for reagents and materials essential for building separation matrices.

Finally, we hope that the information and techniques outlined in this book will provide a sound guide for investigators and students wishing to use affinity and immunoaffinity separation in their experimental designs. The diversity of affinity ligands coupled with the wider diversity of specificities afforded by antibodies make these techniques applicable to a variety of scientific disciplines, including immunochemistry, biochemistry, virology, molecular biology, and biotechnology.

The authors would like to thank Drs. Karin Nelson and James Dambrosia of the Institute for Neurological Disorders and Stroke, and Dr. Douglas Brenneman, Institute of Child Health and Human Development, National Institutes of Health, for their support and help. Additionally, appreciation is given to Dr. John Stobaugh of the University of Kansas, Dr. Thomas Karnes of the Medical College of Virginia, and James Babashak of Corning Glass for hlepful discussions and useful information. The authors would also like to extend their thanks to the editorial staff at BioTechniques Books, Eaton Publishing, especially Steve Weaver and Christine McAndrews. We would also like to thank Dan Picard for his fine editorial work.

Terry M. Phillips, PhD, DSc
Benjamin F. Dickens, PhD
February 2000

1 Affinity and Immunoaffinity Separation

Many molecules exhibit the ability to selectively associate with a specific complementary molecule, often referred to as their substrate or ligand. In many instances, this ligand specificity can be extremely selective. Exploitation of this specificity has led to the development of a series of isolation techniques known as affinity separation. This unique form of separation science is one of the few procedures that enables biological molecules to be isolated and purified on the basis of their biological activity or function. For this reason, affinity separation techniques are one of the most powerful and useful tools in not only the analytical sciences, but also in other areas such as cell and molecular biology.

Affinity separations were first described in 1910 by Starkenstein (62), who used insoluble starch for the isolation of α-amylase. The technique was enhanced by the introduction of new beaded chromatography supports based on dextran (55) and agarose backbones (30). Such supports possess an abundance of hydroxyl groups that are easily modified by cyanogen bromide (CNBr) and this led to the discovery that molecules could be attached to these modified supports by their primary amine groups (6). Although CNBr derivatization greatly promoted the applicability of affinity techniques to the biological sciences, this technique proved to have limited usefulness in the isolation of macromolecules that bind low-molecular-weight ligands (e.g., pharmaceuticals) due to steric interference between the macromolecule and the support. The introduction of "spacer" arms designed to extend the immobilized ligand away from the support matrix surface greatly helped to overcome this steric hindrance and further enhanced the use of this technique (19). Since these early beginnings, affinity separations have rapidly developed and now applications extend into many areas of analytical and biological sciences. One particular procedure, affinity chromatography, includes separation mechanisms that use nonbiological interactions, such as metal chelate (4,71,72), hydrophobic (13), covalent (47), and mimetic dye chromatography (20,64,70).

It was found that affinity separations could be further enhanced by the use of immunological reagents as the capture ligands, thus adding the selectivity and speci-

Affinity and Immunoaffinity Purification Techniques
Terry M. Phillips and Benjamin F. Dickens
© 2000 Eaton Publishing, Natick, MA

ficity of immunological reactions to the separation. Once again, the most common form of such separations is immunoadsorption or immunoaffinity chromatography. This specialized form of affinity chromatography uses an immobilized antibody or antigen as the ligand, and the selective separation comes about through an immunological reaction (48). Immunoaffinity chromatography was first described by Campbell et al. (15), who isolated a specific antibody using an immunoadsorbent prepared by immobilizing a specific antigen on diazotized aminobenzyl cellulose. The efficiency and capacity of the immunoaffinity matrix was improved by Gurvich et al. (25), who precipitated the cellulose as a fine suspension before diazotization. Others have described similar procedures using bromoacetyl-cellulose (58) and ester-derivatized cellulose (7) as the support matrix. Since these early reports, techniques for the attachment of antigens to dextran (35), agarose (18), polyacrylamide (32), plastic beads (17,23), and silica or glass (50,68,75) have been reported. Immobilization chemistry has also evolved to include procedures involving cross-linking of the ligand with glutaraldehyde (5), carbodiimides (24,63), and a large variety of other immobilization chemistries (8,14,28,43,67).

AFFINITY VS. IMMUNOAFFINITY

When considering the application of affinity or immunoaffinity separations, it is wise to take some time to examine the chemical characteristics of the analyte to be isolated in order to select the correct ligand to be used for the separation. Factors such as amount, abundance, and expense might play important parts in the separation plan. Likewise, the chemical characteristics of the ligand must be considered. It would be foolish to use an expensive antibody to isolate a glycoprotein, if a suitable and selective lectin were available. This situation is also true for the isolation of analytes exhibiting group specificity for certain ligands. The biomimetic dyes are able to inexpensively and effectively isolate a variety of different analytes such as human serum albumin (HSA). Dye ligands are not only inexpensive, but also abundant and therefore, suitable to large or preparative scale separations.

Although expensive, antibodies are extremely useful ligands and they can be engineered to react to a variety of chemical, biological, and pharmaceutical agents. Antibodies also provide well-defined sites for attachment to suitable supports. These specific sites can also be manipulated to maintain the correct alignment for the immobilized antibody. Once antibodies are attached to the support, the immunoaffinity column can be reused up to 200 times without loss of activity or specificity (44,59). This reusability makes them attractive as affinity ligands, especially for the isolation of defined analytes.

Selection of the separation procedure also depends on the application. Both affinity and immunoaffinity can be used in batch techniques or as selective depletion agents for the removal of specific molecules. They can both be used in a variety of chromatographic techniques, although immobilized antibodies tend to clog macroporous membranes following interaction with macromolecular analytes. Affinity ligands, especially the biomimetic dyes and metal chelators, have found more applications in analytical chemistry while the immunoaffinity ligands appear to be more applicable to clinical and biological studies.

PRINCIPLES OF AFFINITY SEPARATION

Affinity reactions include a large family of noncovalent binding interactions between molecules that depend on structural relations between the two reactants. Classic examples of this family include substrate binding to a specific enzyme, and the binding of a hormone to its receptor. The forces governing this reaction (Figure 1) are relatively simple and represent the basic forces employed to maintain the structural integrity of macromolecules. The molecule to be isolated (the analyte) binds to the immobilized partner (the ligand; Figure 1A) and is held by a series of forces generated by interactions between the two molecules (Figure 1B). The forces involved in these reactions (Figure 1C) are known as van der Waals, coulombic, ion attractive, and hydrogen bonding forces (37). The van der Waals forces are the most important bonding forces and are generated by interactions between the electron clouds of the two molecules. These interactions cause a temporary electron disturbance, which causes dipoles to form in both molecules. The two dipoles then set up a series of attractive forces between the molecules. As the displaced electrons swing back

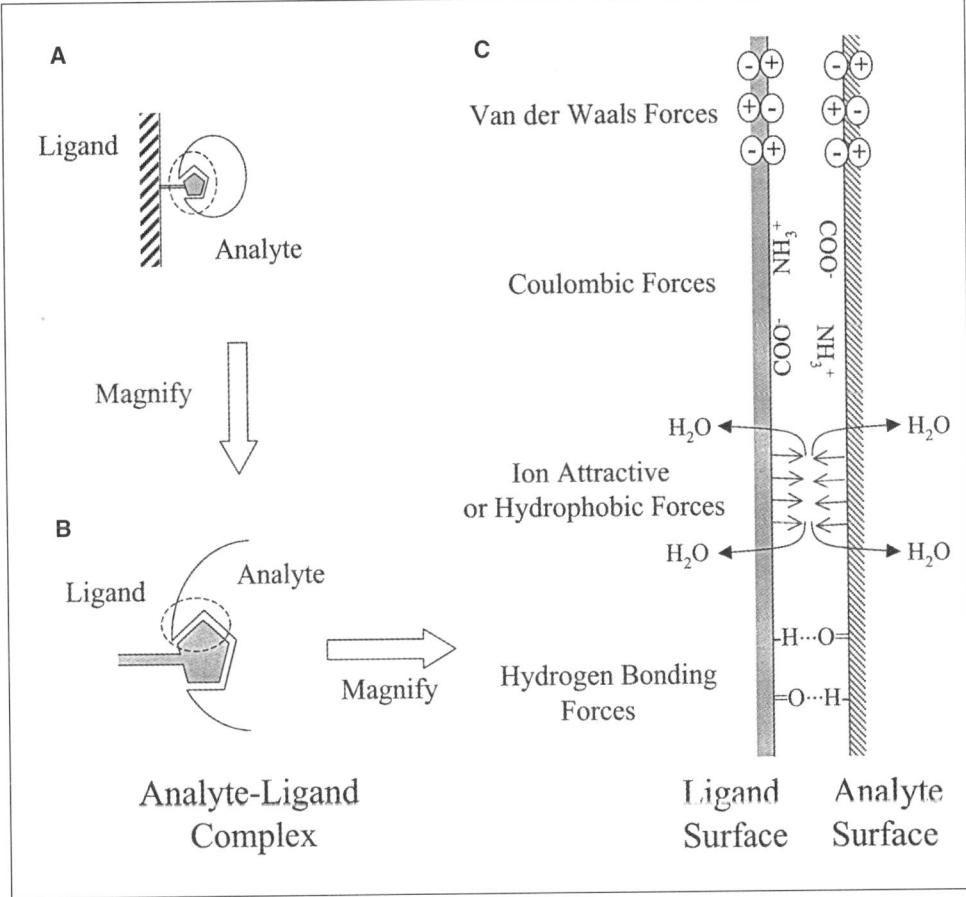

Figure 1. Molecular interactions involved in affinity binding. (A) Binding of an analyte (circle) by an immobilized ligand (solid pentagon). (B) Magnification of the analyte-ligand complex showing the good fit between analyte and ligand surface necessary for intramolecular attractive forces required for strong affinity binding. (C) The four types of noncovalent bonds involved in affinity binding.

through equilibrium, the dipoles oscillate. This generates an attractive force that is inversely proportional to the seventh power of the distance; the force greatly increases as the distance between the two molecules decreases. The coulombic forces that arise from the attraction between two oppositely charged ionic groups are the second most important force. Such groups could be reactive ionized amino (NH_3^+) groups on one molecule attracting ionized carboxyl (COO^-) groups on the other. The attractive force created is inversely proportional to the square of the distance between the opposing charges. Therefore, this type of attractive force also increases as the proximity of the two molecules decreases. Ion attractive forces are also known as hydrophobic forces and again result from the close proximity of one molecule to the other. This coming-together of the molecules forces water from between them, thus reducing the net charge. When two molecules are compatible, they prefer to remain in this low energy state and thus in close proximity to each other. The closeness of the two molecules allows the fourth bonding force, hydrogen bonding, to come into play. This type of bonding forms when the weak attractive forces formed between hydrophilic groups on the two molecules come into contact and interact.

The interactions between the analyte and the ligand can be explored mathematically and any affinity reaction involving the interaction of a single analyte (A) with a single ligand (L) can be expressed by the following:

$$A + L \rightleftharpoons AL$$ [Eq. 1]

This is the simplest reaction involved in affinity or adsorption interactions and results in the formation of the reversible complex AL. In most affinity separations, the ligand is immobilized and the solution containing the analyte is passed over the ligand-coated support (Figure 2, step 1). Upon contact, the analyte will specifically recognize and bind to the ligand (Figure 2, step 2). Once this interaction has occurred, the analyte becomes part of the immobilized analyte-ligand complex (Figure 2, step 3) until the complex is dissociated. This releases the analyte, allowing it to be recovered (Figure 2, step 4). This dissociation usually takes place after the nonreactive material has been removed. In this way, the analyte can be isolated in a pure, active form.

There are several theoretical models for the study of bio-selective reactions (69), although no single model is able to encompass the wide range of interactions that occur in affinity interactions. This is due to the nature of the different sources that comprise the overall interaction. Molecular interactions arise not only from specific reactions between the analyte and its ligand, but also as a result of nonspecific interactions between the analyte and the surface of the support. However, the sum of these forces contributes to the basic equilibrium reaction and is usually examined as a whole. There are two other forces that are major contributors to the stability of the complex. Most affinity complexes are reversible and factors such as the association and dissociation constants must be considered. Therefore, the simple binding equation (equation 1) should correctly be written as:

$$A + L \overset{a}{\underset{d}{\rightleftharpoons}} AL$$ [Eq. 2]

In this modification, "a" represents the association constant while "d" represents the dissociation constant. Equilibrium is reached when the association (a) is equal to the dissociation (d). The equilibrium constant (K) of this reaction can now be expressed as:

$$K = [A][L]/[AL] \qquad\qquad\qquad \text{[Eq. 3]}$$

or

$$K = k_a/k_d = [AL]/[A][L] \qquad\qquad\qquad \text{[Eq. 4]}$$

In equation 4, the concentrations of each component are written in square brackets and the association and dissociation constants are expressed as "k_a" and "k_d", respectively. In this equation, K is expressed as an association constant. When actually using equation 4, the association constant is expressed in liters per mole and can be

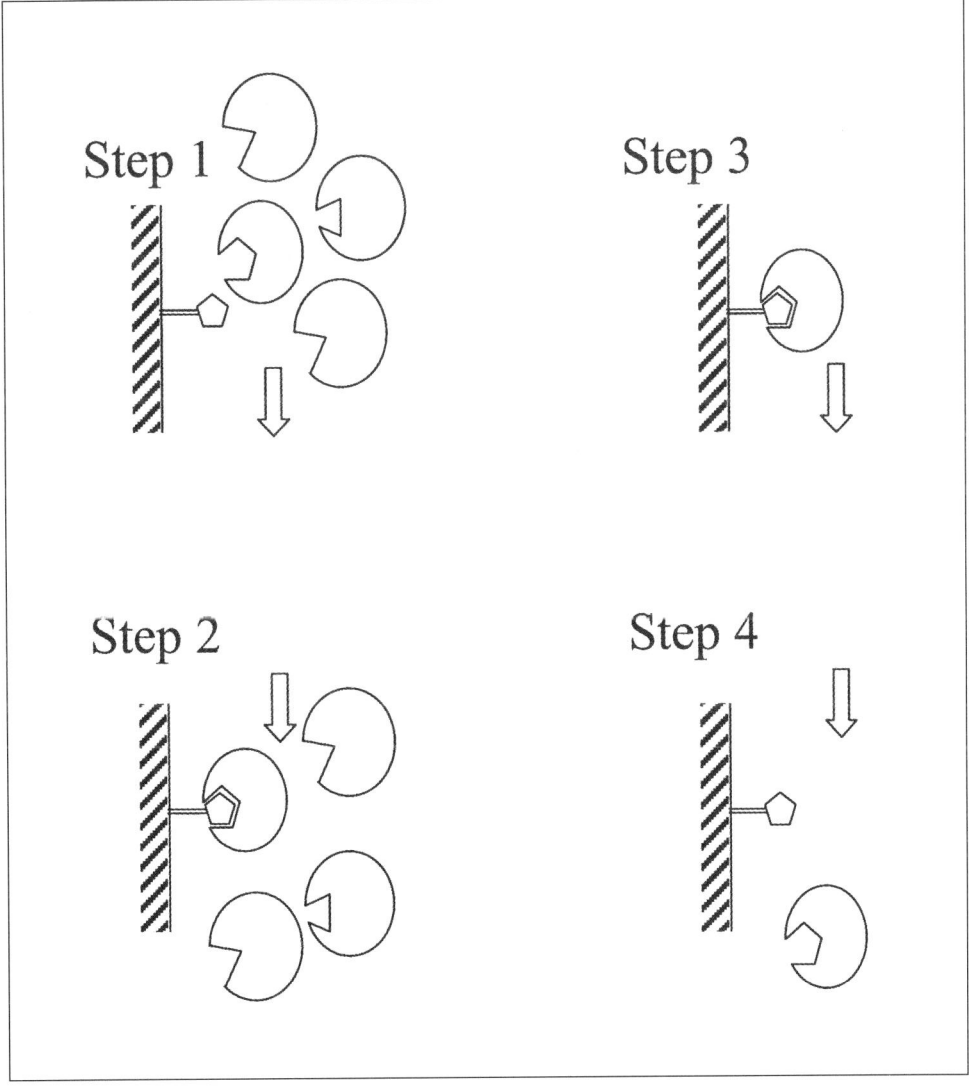

Figure 2. Formation of the reversible analyte-ligand complex. The analyte-containing solution is passed over a ligand-coated support (step 1). Binding occurs between the analyte and the immobilized, specifically recognized ligand (step 2), forming an analyte-ligand complex (step 3). This complex is maintained until all of the nonreactive material is removed. The complex is then dissociated, allowing the analyte to be recovered (step 4).

used as an estimate of binding between the two reactants. The thermodynamic affinity of a reaction can be defined as the negative of the free energy change of the reaction (ΔG^O), as expressed by the following:

$$\Delta G^O = -RT \ln K \qquad \text{[Eq. 5]}$$

where "R" represents the gas constant (8.314 J/mol/K) and "T" the temperature in Kelvin (K). Although one cannot usually calculate the equilibrium concentrations for all of the components of the reaction ([A], [L], and [AL]), it is possible to use equation 3 to calculate values for [AL] by inserting [A] = (A)-[AL], where (A) is the upper limit of A:

$$[AL] = K[L]/1 + K[L] \, (A) \qquad \text{[Eq. 6]}$$

or

$$[AL] = [L]/k_d + [L] \, (A) \qquad \text{[Eq. 7]}$$

Both equations become identical as $K = 1/k_d$. At low concentrations, more ligand will become bound if more is added, and at higher concentrations, saturation will be observed as the complex [AL] approaches (A).

When using affinity chromatography, one can substitute A_o and L_o in equation 7 for the initial concentrations of A and L, respectively, and if $L_o \gg A_o$, then:

$$K = [A_o - AL][L_o - AL]/[AL] \qquad \text{[Eq. 8]}$$
$$= (A_o - L/AL)L_o$$

The chromatographic distribution coefficient can then be defined as:

$$k_d = (AL/A_o - AL) = L_o/K \qquad \text{[Eq. 9]}$$

or in terms of column units:

$$V_e = V_o + k_d V_o \qquad \text{[Eq. 10]}$$

where V_e and V_o represent the elution volume and the void volume, respectively. From this, we can further derive:

$$V_e/V_o = L_o/K + 1 \qquad \text{[Eq. 11]}$$

From equations 9 and 11, it is possible to demonstrate that effective separations can be achieved only when we can fulfill the condition $L_o \gg K$. The upper limit value for L is in the range $10^{-2}–10^{-3}$ M^{-1}. If we assume that $V_e/V_o > 3$ for the separation to be efficient, then the lower limit for K will fall in the range $10^{-3}–10^{-4}$ M^{-1}. It can be further inferred that, using bio-selective complexes with $K < 10^{-4}$ M^{-1}, a minimum ligand concentration is required.

PRINCIPLES OF IMMUNOAFFINITY SEPARATION

A further refinement of affinity separations is the use of antibodies or antigens as the capture ligands. Immunoaffinity separations are similar to those described for affinity, but in this case, the specificity and selectivity of an immunological reagent can broaden the scope of the selective process. Unless the analyte is a specific antibody (in which case immobilized antigens are used as the ligands), it is usually the antibody that is immobilized. The advantage of this approach is that antibodies are versatile reagents that can be manufactured against a variety of different molecules. To understand this property, it is essential to understand the antibody molecule itself. Antibodies are glycoproteins shaped in a "Y" (1). In some ways, they resemble a molecular lobster, with a tail and two arms, with each arm containing a terminal claw or antigen receptor. This terminal area is also known as the hypervariable region. Figure 3 is a diagram of a basic antibody molecule. The molecule is composed of two long or heavy chains, joined by disulfide bonds, which run the entire length of the molecule. At the N terminus, two shorter or light chains are attached by disulfide bonds to the heavy chains. Together, the light and their heavy chain counterparts form the fragment antibody or FAb. The region immediately behind the attachment point for the light chains is extremely flexible, allowing the arms of the molecule to move in a sweeping arc. This movement enables a single antibody molecule to attach to two antigens. The flexible area of an antibody molecule is called the hinge region. Finally, the portion of the heavy chain that extends from the hinge region down to the C terminus is called the fragment crystallizable or Fc. This latter region is responsible for binding to a number of receptors, as well as providing a focal point for the attachment of accessory molecules such as complement (1).

The hypervariable regions of each FAb form receptor cups resembling the electron cloud shape of the antigen and gives the antibody its specificity (Figure 4). Due to the nature of the noncovalent bonds involved in antibody-antigen binding, replication of the electron cloud shape of the antigen is required for the antibody to establish a good conformational fit. When this situation is achieved, then the antibody also maintains its specificity. Shape recognition plays a large part in the ability of any antibody to recognize its specific antigen. However, in many cases, the electron cloud shape of

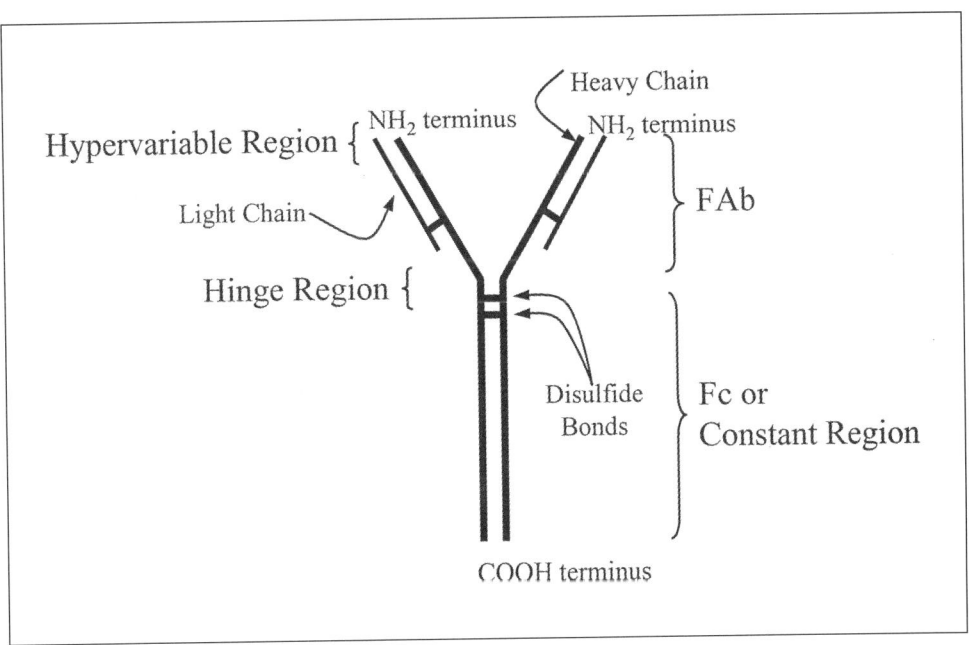

Figure 3. Basic antibody molecule. The basic antibody molecule is composed of two identical heavy chains and two identical light chains, held together by disulfide bonds. The single Fc is at the carboxy-terminus end of the heavy chain molecules, and the two antigen-binding or hypervariable regions are located at the amino-terminus of the light and heavy chains. After one of the variable regions binds an antigen, flexibility in the hinge region aids the binding of a second antigen.

the specific antigen is similar to that of other molecules, thus allowing the reactive antibody to react with a wide variety of related molecules. When this situation arises, a poor conformational fit occurs and the usefulness of the antibody is seriously impaired. These antibodies are called cross-reactive, and, although useless for specific separations, might find application in the isolation of a certain class of molecules, all exhibiting a common electron cloud shape (16).

The unique structure of the basic antibody molecule can be modified by enzymatic cleavage to produce a number of reactive fragments (Figure 5). Cleavage of the molecule with papain produces two FAb fragments plus a single Fc fragment joined by disulfide bonds (56). This type of cleavage essentially produces monovalent antibodies (FAb) that are useful for performing quantitative studies. Digestion of the antibody with pepsin produces different fragments such as a $F(Ab')_2$ fragment and a Fc fragment (65). The former represents a single unit that still possesses the activity of the original molecule and exhibits an antigen-binding capacity of 2. The pepsin cleavage takes place in such a way that the two arms of the antibody plus a single disulfide bond remain intact. Breaking the disulfide bond to give two FAb fragments, each complete with a reactive thiol group, can further reduce this unit. This single reactive thiol group is extremely useful for immobilizing the reduced FAb to free thiol groups present on the support. Like the enzyme-produced FAb fragments, these fragments are useful for quantitative work where single valency reactions take place.

Immunoaffinity separations are performed in three phases (Figure 6): a primary capture or adsorption phase, a holding phase, and finally, a recovery or elution phase. Initially, the analyte to be isolated reacts with the immobilized antibody to form an insoluble complex. The bound analyte is held in this complex until all of the nonreactive materials have been removed. Once this has been achieved, the complex is dissociated during the elution phase, thus releasing the captured analyte. In immuno-

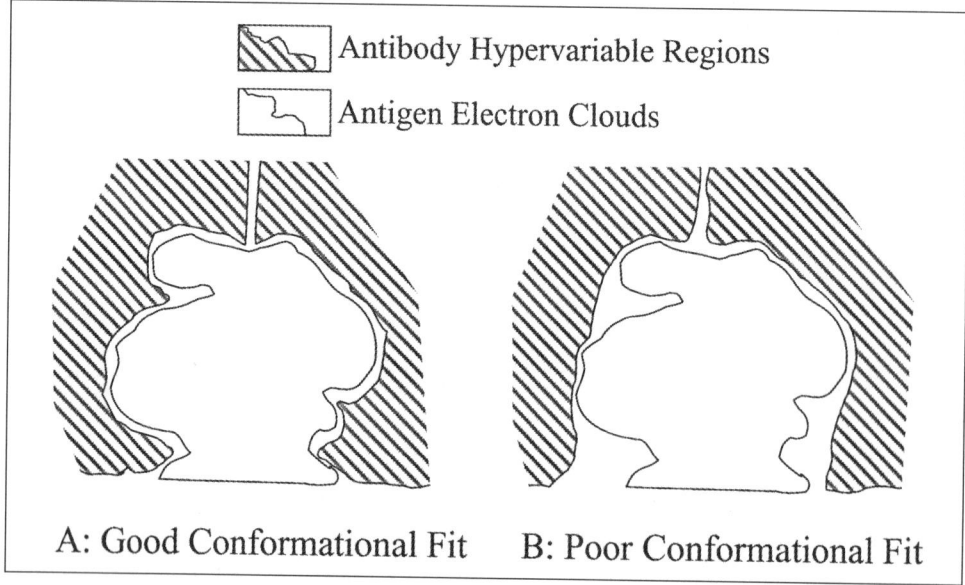

Figure 4. Conformational fit of antibody and antigen. When a good conformational fit occurs between antibody and antigen (A) in the antigen-binding cups of the hypervariable region of the antibody, it maximizes the noncovalent, intra-molecular attractive interactions (see Figure 1) by minimizing the distance between the two surfaces. However, when a poor fit exists (B), repulsive forces associated with electron cloud overlap prevent affinity binding.

affinity chromatography, the initial two phases of the process are recorded as a single, large peak on a chromatogram, as illustrated in the lower aspects of Figure 6. Recovery of the analyte is indicated by the presence of a small, sharp peak appearing later in the chromatogram. This second peak represents the immunoaffinity-purified analyte.

The general reactions governing immunoaffinity separations are:

Immobilized Antibody + Analyte ⇌ Immobilized Antibody-Analyte Complex

Immobilized Antibody-Analyte Complex + Elution Agent ⇌ Immobilized Antibody + Free Analyte

Antibody-analyte complexes are formed in the same way as described for analyte-ligand complexes and are held together by identical noncovalent bonds. The specificity of the reaction is governed by a number of interactive parameters, especially the ability of the combining sites or antigen receptors to structurally conform to the electron cloud shape of the antigen (Figure 4). A good conformational fit will ensure that the antibody will exhibit a high degree of binding. To achieve strong binding between antibodies and their reactive analytes, the electron cloud shapes of both the antibody receptors and the analyte must be complementary. When this situation arises, strong attractive forces and weak repulsive forces exist between the two mol-

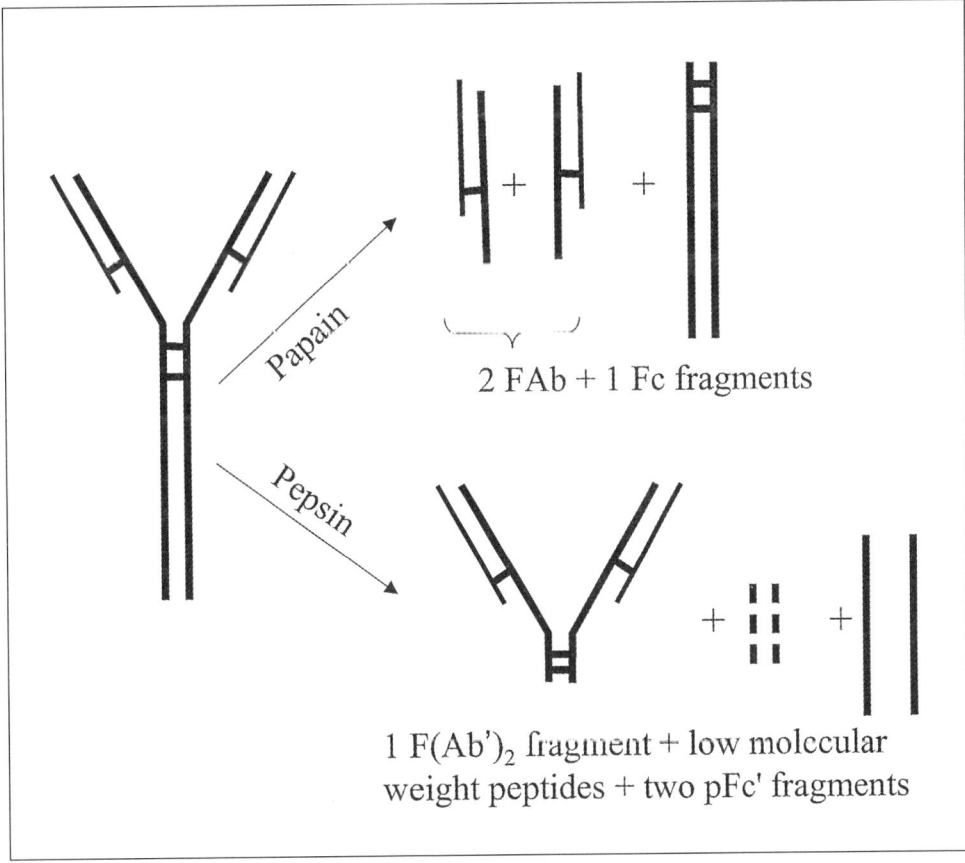

Figure 5. Production of reactive fragments by enzymatic cleavage of an antibody molecule. Digestion with papain creates two FAb fragments and one Fc fragment. Pepsin digestion creates one F(Ab')₂ fragment and one Fc fragment.

ecules (49). Antibody-antigen reactions can be expressed in a similar manner to that used for the expression of generalized affinity reactions. At equilibrium, the sum of both the attractive and repulsive forces can be expressed as:

$$[Ab] + [Ag] \rightleftharpoons [AbAg] \qquad \text{[Eq. 12]}$$

In equation 12, [Ab] represents free antibody, [Ag] is free antigen, and [AbAg] represents the complex. When the law of mass action is applied, the equation can be expressed as:

$$k_a [Ab][Ag] = k_d [AbAg] \qquad \text{[Eq. 13]}$$

and

$$K = [AbAg]/[Ab][Ag] \qquad \text{[Eq. 14]}$$

where k_a and k_d represent the association and dissociation constants, respectively, and $K = k_a/k_d$ represents the affinity constant of the reaction.

Experimentally, immunological reactions can be examined by a number of different techniques. The diversity of antibodies present in antisera makes many of these studies impossible and investigators have turned to simplistic systems or models for

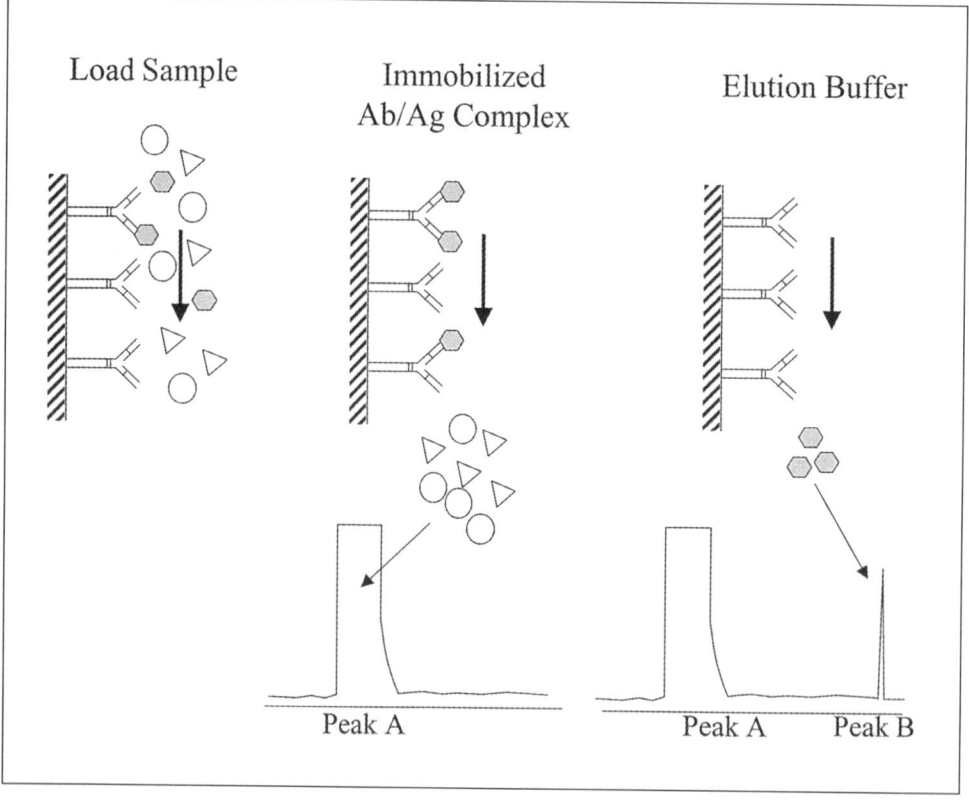

Figure 6. Three phases of immunoaffinity separation. In the adsorption phase, an antigen-containing sample is passed over an antibody-coated support where the antigen (shaded hexagon) reacts with the immobilized antibody (left). All nonreactive materials are removed while the antibody-antigen (Ab/Ag) complex remains (the holding phase; middle). A representative chromatogram (bottom) records these two phases as a large, single peak (peak A). Finally, in the elution phase, an elution buffer is added to release the bound antigen (right). Analyte recovery is recorded as a small, sharp peak in the chromatogram (peak B).

their studies. Examination of the interactions between an antibody and a monovalent analyte or hapten has proven one of the best models. At equilibrium, such a system can be expressed as:

$$[Ab] + [H] \underset{k_d}{\overset{k_a}{\rightleftharpoons}} [AbH] \qquad \text{[Eq. 15]}$$

Here, [Ab] represents free antibody, [H] represents free haptenic analyte, and [AbH] represents the antibody-hapten complex. Once again, the symbols k_a and k_d represent the association and dissociation rate constants of the interaction. Applying the law of mass action to this model gives:

$$K = [AbH]/[Ab][H] \qquad \text{[Eq. 16]}$$

However, for the calculations to be reasonably accurate, both reactants must be pure, homogeneous in their reactivity, and in solution.

The interaction of antibodies with their specific antigens can be described by two separate and different terms; affinity describes the thermodynamic interactions of a single antibody-antigen reaction, while avidity is used to describe the universal interactions. Avidity, therefore, should only be used when additional contributory factors (e.g., antibody and/or antigen valence) and multiple interactions are involved. In an attempt to resolve this situation, Hornick and Karush (31) used the term "intrinsic affinity" to describe the interaction of a monospecific antibody with its haptenic analyte, and "functional affinity" to describe the interaction of an antibody with multiple antigenic sites.

Experimentally, the measurement of antibody-analyte interactions can be achieved by several techniques. The simplest of these is equilibrium dialysis as shown diagrammatically in Figure 7, with the simplest of situations illustrated in Figure 7A. Molecules are placed in a container divided by a semipermeable membrane. As the molecules pass from the initial chamber to the other chamber by diffusion, a concentration gradient is set up until finally, the molecular concentrations in both chambers reach equilibrium. In Figure 7B, labeled analytes plus ligand are placed in the initial chamber and the process repeated. By recording the label in both chambers, experimental evidence of ligand-analyte binding can be obtained. In a similar vein, precipitin curves can also be used to assess the binding capacity of a ligand-analyte system. In this case, a series of tubes are set up in which a constant amount of antibody is placed. Increasing amounts of labeled antigen or analyte are added to each tube and the resulting complex precipitated and measured. From this data, a precipitin curve (Figure 8) can be constructed and the equilibrium zone calculated. Applying the law of mass action to the data allows the following form of the Langmuir adsorption isotherm to be derived:

$$[AbH]/[Ab] = r = nK[H]/1 + K[H] \qquad \text{[Eq. 17]}$$

In this equation, "r" represents the amount (moles) of analyte bound per mole of antibody, [AbH] represents the bound analyte concentration, [Ab] represents the total antibody concentration, [H] represents the free analyte concentration at equilibrium, "n" represents the antibody valence, and K represents the association equilibrium constant. From this, the Scatchard equation (60) can be expressed as:

$$r/[H] = nK - rK = - rK + nK \qquad \text{[Eq. 18]}$$

A Scatchard plot of n/[H] vs. r over a range of free hapten concentrations (Figure 9) allows values of antibody affinity K and antibody valence (n) to be assessed. When a divalent antibody (n = 2) is used, the Scatchard equation can calculate the intrinsic association constant K of the antibody for the given analyte. In situations where one-half the antibody binding sites are bound (r = 1), then:

$$r/[H] = nK - rK \qquad \text{[Eq. 19]}$$

which becomes

$$r/[H] = 2K - K = K \qquad \text{[Eq. 20]}$$

and K is equal to the reciprocal of the free analyte concentration at equilibrium. An alternative equation for calculation of affinity from binding data is:

$$[Ab]/[AbH] = 1 + K[H]/nK[H] \qquad \text{[Eq. 21]}$$

from which

$$1/r = 1/n \cdot 1/[H] \cdot 1/K \cdot 1/n \qquad \text{[Eq. 22]}$$

A plot of 1/r vs. 1/[H] gives the reciprocal Langmuir plot (Figure 10). This plot allows values for affinity (in terms of K), and antibody valence to be calculated. In ideal experiments, both Scatchard and Langmuir equations should result in linear plots.

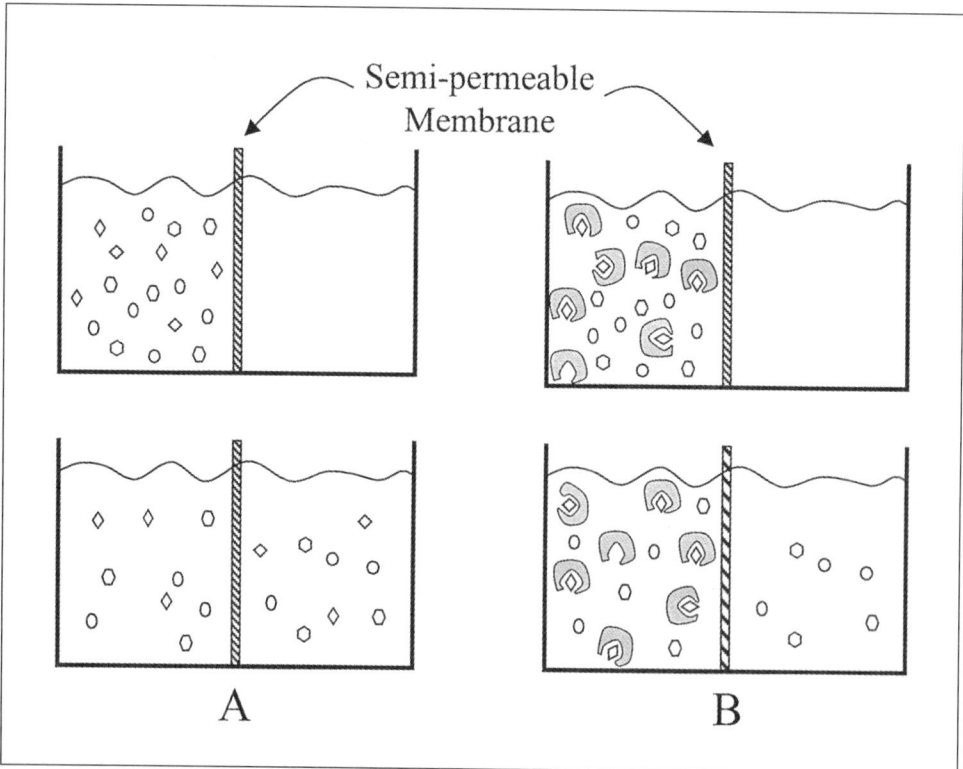

Figure 7. Equilibrium dialysis. In the absence of a binding ligand, low-molecular-weight analyte and other solutes are free to pass through the semipermeable (or dialysis) membrane, and rapidly reach equilibrium in both chambers (panel A). Adding nonpermeable ligands (filled object) that bind their specific analytes (diamonds) alters the analyte distribution between the chambers (panel B). For the Scatchard plot (see Figure 9), the analyte concentration in the right-hand chamber gives the free hapten concentration, while the difference between the concentration on the left side and right side gives the bound-analyte concentration.

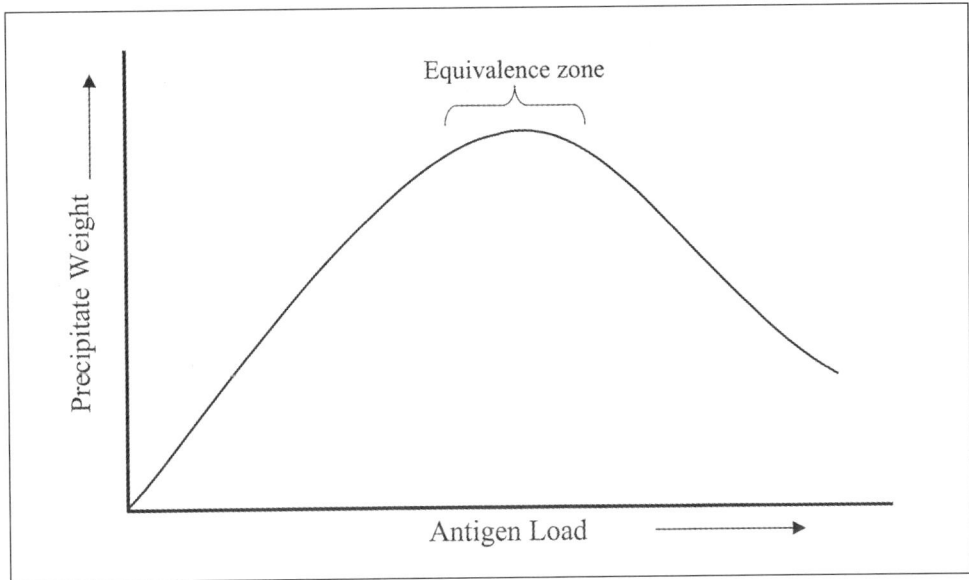

Figure 8. Classic precipitin reaction in vitro. For any constant amount of antibody, as the antigen load is increased, the amount of immune precipitation increases until no free antibody or antigen can be detected in the supernatant (equivalence zone). As antigen load is increased further, the amount of immune precipitate decreases and free antigen can be found in the supernatant.

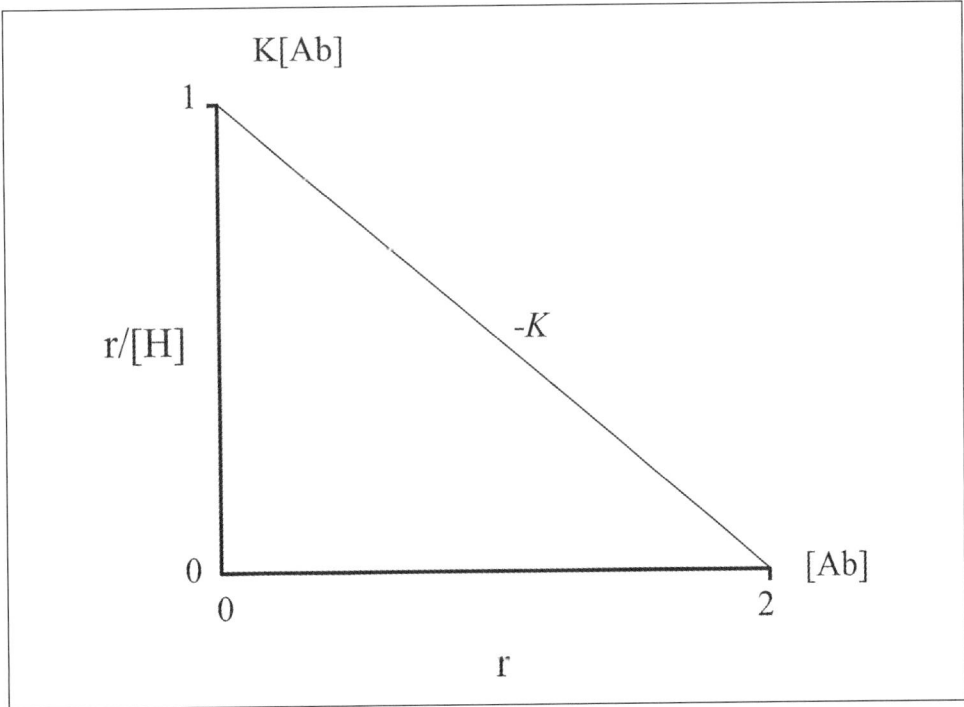

Figure 9. Scatchard plot produced by equilibrium dialysis to assess antibody-binding affinity constant (K) and antigen-binding valence (n). Dialysis is with constant antibody concentration [Ab] and with varying free hapten concentrations [H]. The ratio of bound hapten (r) to free (nonbound) hapten [H] is plotted vs. r. K and n can be determined from the slope and the x-intercept, respectively. While Scatchard plots are still useful to visualize affinity binding data, modern nonlinear regression software has replaced the need to linearize binding data to calculate these values.

However, such situations exist only in systems where pure antibody and analyte exist. It should also be remembered that the majority of mathematical models tend to study ideal or monovalent systems and might not be fully applicable to real-life situations.

OVERVIEW OF APPLICATIONS IN BIOTECHNOLOGY AND THE BIOMEDICAL SCIENCES

The ability to selectively isolate a specific analyte has great application in many fields of analytical science. The rapid expansion of affinity and immunoaffinity techniques has recently been witnessed by the number of special issues of analytical journals devoted to describing affinity applications. Affinity and immunoaffinity techniques have been applied to both basic and clinical sciences. Affinity separations have been the basis for a number of important clinical tests, including the measurement of glycosylated hemoglobin in diabetics. Likewise, the application of immunological separations to clinical assessment is tremendous and has led to the application of numerous diagnostic tests. The discovery of magnetic beads as a support for both affinity and immunoaffinity ligands has led to their use in cell isolation, molecular

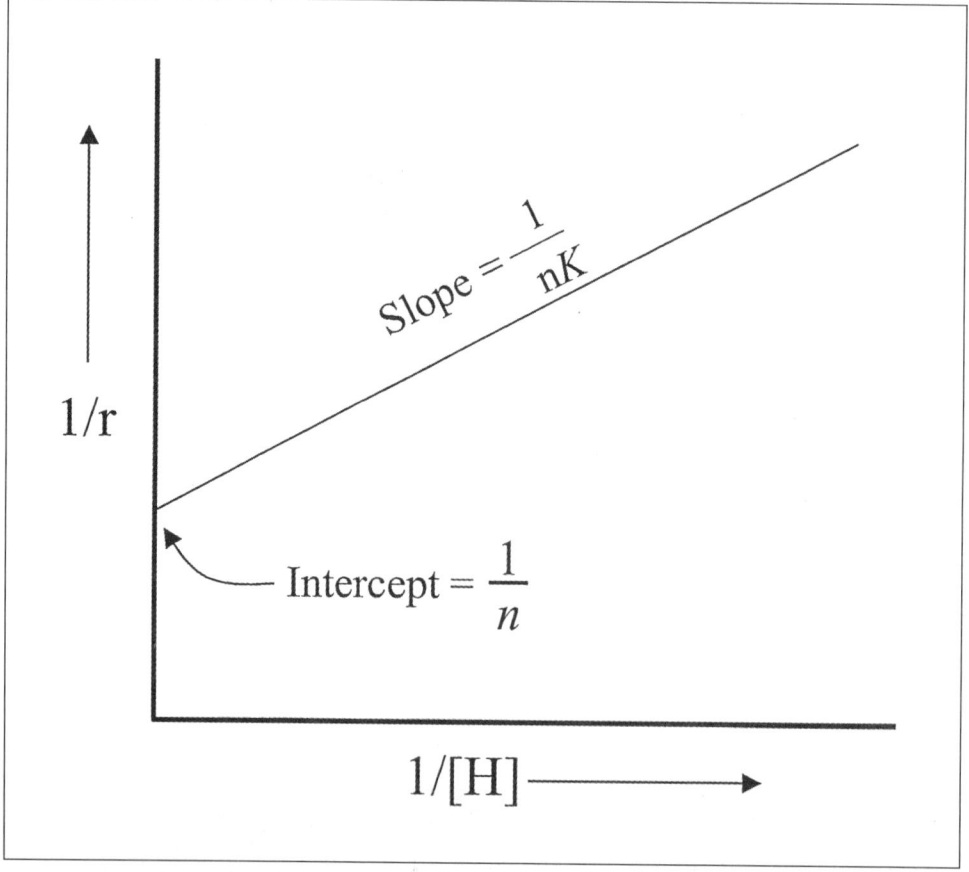

Figure 10. Langmuir plot used to calculate binding affinity constant and antigen-binding valance. Symbols are as described in Figure 9.

biology, and clinical assays. Bruno et al. (12) have recently described an immuno-magnetic technique for the recovery of active chromatin restriction fragments from body fluids in under one hour using antibody-coated agarose magnetic beads.

The development of metal chelates as affinity ligands has gained considerable popularity. In 1989, Margolis et al. (42) described the use of immobilized organic chelators as potential affinity ligands for the isolation of proteins. This application was reviewed by Arnold (4) and the potential applications in protein separations expanded in 1991. Wuenschell et al. (71) further developed the theme of metal chelators being used as selective separators for histidine-rich proteins. Metal chelate supports were also used as models for studying complex behavior in protein displacement systems and the estimation of equilibrium parameters (36).

The application of membrane affinity techniques to the separation sciences has made a large impact. Improved membrane design has made this form of affinity separation available for high performance techniques (33) and selective protein isolations (39). New modeling procedures have produced improved membranes expressed not only in sheets, but also in rods and hollow fibers (66).

Affinity and immunoaffinity chromatographic techniques have been applied to evaluating a number of chemicals, both environmental and pharmaceutical, in the plasma, urine, and tissues of both humans and animals. Immunoaffinity columns have been used to isolate a number of agents, such as environmental drug residues (34), flunitrazepain and its metabolites in urine (21), and aflatoxins (38). Lawrence and Menchard describe an immunoaffinity procedure for the pre-analysis cleanup of beef liver and muscle samples for the detection of clenbuterol by HPLC (40). A similar immunoaffinity pre-analysis preparation was described for the detection of corticosteroids in various animal tissues and fluids, including milk by GC-MS (22). There have also been a number of studies on the binding of clinically relevant drugs to human serum proteins (61,73). Recently, studies have been described in which a mathematical approach to studying the binding of anti-anxiety and inflammatory drugs to HSA was developed based on competitive binding in high-performance affinity chromatography (74).

The application of affinity and immunoaffinity techniques to capillary electrophoresis (CE) has greatly opened the applications available to these techniques. The use of affinity procedures coupled with CE separators has been applied to a number of different separations including the isolation of pharmaceutical agents (2,3), amyloid proteins (29), and DNA restriction fragments (26). Recently, Hage (27) has extensively reviewed the use of proteins as chiral affinity agents in CE analysis. The application of immunoaffinity CE is less advanced but has been applied to the measurement of neuropeptides (51), cytokines (53,54), and a number of pharmaceuticals (45,46,52).

Perhaps one of the most exciting advances in affinity techniques is its application to mass spectrometry. A number of investigators have reported a system called probe-immobilized affinity chromatography/mass spectrometry in which affinity separations are performed directly on the surface of a matrix-assisted laser desorption/ionization mass spectrometer probe (9,10,41). In a similar vein, Bruce et al. (11) developed a system called bio-affinity characterization mass spectrometry (BACMS), a Fourier-transform ion cyclotron resonance mass spectrometry technique allowing analysis of noncovalent molecular complexes in solution. To achieve this, the authors coupled the system to an electrospray ionization detector. The sensitivity achieved by this technique is claimed to be in the order of 1×10^9 M. The application of

15

multidimensional chromatographic techniques and their potential application to mass spectrometry have been reviewed by Regnier and Huang (57).

REFERENCES

1. **Alzari, P.M., M.-B. Lascombe and R.J. Polkak.** 1988. Three-dimensional structure of antibodies. Annu. Rev. Immunol. *6*:555-580.
2. **Amini, A. and D. Westerlund.** 1998. Evaluation of association constants between drug enantiomers and human alpha 1-acid glycoprotein by applying a partial-filling technique in affinity capillary electrophoresis. Anal. Chem. *70*:1425-1430.
3. **Arai, T., M. Ichinose, H. Kuroda, N. Nimura and T. Kinoshita.** 1994. Chiral separation by capillary affinity zone electrophoresis using an albumin-containing support electrolyte. Anal. Biochem. *217*:7-11.
4. **Arnold, F.H.** 1991. Metal-affinity separations: a new dimension in protein processing. Biotechnology *9*:151-156.
5. **Avrameas, S.** 1985. Immobilisation of proteins onto polyacrylamide beads using glutaraldehyde, p. 64-65. *In* P.D.G. Dean, W.S. Johnson, F.A. Middle (Eds.), Affinity Chromatography: A Practical Approach. IRL Press, Oxford.
6. **Axen, R., J. Porath and S. Ernback.** 1967. Chemical coupling of peptides and proteins to polysaccharides by means of cyanogen halides. Nature *214*:1302-1304.
7. **Boeden, H.F., K. Pommerening, M. Becker, C. Rupprich, M. Holtzhauer, F. Loth, R. Muller and D. Bertram.** 1991. Bead cellulose derivatives as supports for immobilization and chromatographic purification of proteins. J. Chromatogr. *552*:389-414.
8. **Brinkley, M.** 1992. A brief survey of methods for preparing protein conjugates with dyes, haptens, and cross-linking agents. Bioconjug. Chem. *3*:2-13.
9. **Brockman, A.H. and R. Orlando.** 1995. Probe-immobilized affinity chromatography/mass spectrometry. Anal. Chem. *67*:4581-4585.
10. **Brockman, A.H. and R. Orlando.** 1996. New immobilization chemistry for probe affinity mass spectrometry. Rapid Commun. Mass Spectrom. *10*:1688-1692.
11. **Bruce, J.E., G.A. Anderson, R. Chen, X. Cheng, D.C. Gale, S.A. Hofstadler, B.L. Schwartz and R.D. Smith.** 1995. Bio-affinity characterization mass spectrometry. Rapid Commun. Mass Spectrom. *9*:644-650.
12. **Bruno, J.G., H. Yu, J.P. Kilian and A.A. Moore.** 1996. Development of an immunomagnetic assay system for rapid detection of bacteria and leukocytes in body fluids. J. Mol. Recognit. *9*:474-479.
13. **Builder, S.E.** 1993. Hydrophobic Interaction Chromatography: Principles and Methods. Pharmacia, Uppsala.
14. **Burton, S.C., N.W. Haggarty and D.R.K. Harding.** 1991. Efficient substitution of 1,1'-carbonyldiimidazole activated cellulose and sepharose matrices with amino acyl spacer arms. J. Chromatogr. *587*:271-275.
15. **Campbell, D.H., E. Luescher and L.S. Lerman.** 1951. Immunologic immunoadsorbents - I. Isolation of antibody by means of a cellulose-protein antigen. Proc. Natl. Acad. Sci. USA *7*:575-582.
16. **Clark, W.R.** 1991. Structure-function relationships in antibody molecules, p. 136-143. The Experimental Foundation of Modern Immunology. John Wiley & Sons, New York.
17. **Coupek, J., J. Labsky, J. Kalal, J. Turkova and O. Valentova.** 1977. Reactive carriers of immobilized compounds. Biochim. Biophys. Acta *481*:289-296.
18. **Cuatrecasas, P.** 1970. Protein purification by affinity chromatography. Derivatizations of agarose and polyacrylamide beads. J. Biol. Chem. *745*:3059-3065.
19. **Cuatrecasas, P., M. Wilchek and C.B. Anfinsen.** 1968. Selective enzyme purification by affinity chromatography. Proc. Natl. Acad. Sci. USA *61*:636-643.
20. **Dean, P.D.G. and D.H. Watson.** 1979. Protein purification using immobilised triazine dyes. J. Chromatogr. *165*:301-319.
21. **Deinl, I., L. Angermaier, C. Franzelius and G. Machbert.** 1997. Simple high-performance liquid chromatographic column-switching technique for the on-line immunoaffinity extraction and analysis of flunitrazepam and its main metabolites in urine. J. Chromatogr. B Biomed. Sci. Appl. *704*:251-258.
22. **Delahaut, P., P. Jacquemin, Y. Colemonts, M. Dubois, J. De Graeve and H. Deluyker.** 1997. Quantitative determination of several synthetic corticosteroids by gas chromatography-mass spectrometry after purification by immunoaffinity chromatography. J. Chromatogr. B Biomed. Sci. Appl. *696*:203-215.
23. **Fulton, S.P., M. Meys, L. Varady, R. Jansen and N.B. Afeyan.** 1991. Antibody quantitation in seconds using affinity perfusion chromatography. BioTechniques *11*:226-231.
24. **Gilles, M.A., A.Q. Hudson and C.L. Borders.** 1990. Stability of water-soluble carbodiimides in aqueous solutions. Anal. Biochem. *184*:244-248.
25. **Gurvich, A.E., O.B. Kuzovlena and A.E. Tumanova.** 1961. Production of protein-cellulose complexes (immunoadsorbents) in the form of suspensions able to bind great amounts of antibodies. Biokhimiia *26*:803-817.

26. **Guttman, A. and N. Cooke.** 1991. Capillary gel affinity electrophoresis of DNA fragments. Anal. Chem. *63*:2038-2042.
27. **Hage, D.S.** 1997. Chiral separations in capillary electrophoresis using proteins as stereoselective binding agents. Electrophoresis *18*:2311-2321.
28. **Hearn, M.T.W., G.S. Bethall, J.S. Ayers and W.S. Hancock.** 1979. Application of 1,1'-carbonyldiimidazole-activated agarose for the purification of proteins. II. The use of activated matrix devoid of additional charged groups for the purification of thyroid proteins. J. Chromatogr. *185*:463-470.
29. **Heegaard, N.H., H.D. Mortensen and P. Roepstorff.** 1995. Demonstration of a heparin-binding site in serum amyloid P component using affinity capillary electrophoresis as an adjunct technique. J. Chromatogr. *717*:83-90.
30. **Hjerten, S.** 1962. Chromatographic separation according to size of macromolecules and cell particles on columns of agarose suspensions. Arch. Biochem. Biophys. *99*:466-475.
31. **Hornick, C.L. and F. Karush.** 1972. Antibody affinity. 3. The role of multivalance. Immunochemistry *9*:325–340.
32. **Inman, J.K. and H.M. Dintzis.** 1969. The derivatization of cross-linked polyacrylamide beads. Controlled introduction of functional groups for the preparation of special-purpose, biochemical adsorbents. Biochemistry *8*:4074-4085.
33. **Josic, D., J. Reusch, K. Loster, O. Baum and W. Reutter.** 1992. High-performance membrane chromatography of serum and plasma membrane proteins. J. Chromatogr. *590*:59-76.
34. **Katz, S.E. and M. Siewierski.** 1992. Drug residue analysis using immunoaffinity chromatography. J. Chromatogr. *624*:403-409.
35. **Kennedy, J.F.** 1974. Chemically reactive derivatives of polysaccharides. Adv. Carbohydr. Chem. Biochem. *29*:305-405.
36. **Kim, Y.L.** 1995. Prediction of protein displacement by simplified immobilized metal ion affinity chromatographic model. Bioseparation *5*:295-306.
37. **Kuby, J.** 1992. Antigen-antibody interactions, p. 121-125. Immunology. W.H. Freeman and Co., New York.
38. **Kussak, A., B. Andersson, K. Andersson and C.A. Nilsson.** 1998. Determination of aflatoxicol in human urine by immunoaffinity column clean-up and liquid chromatography. Chemosphere *36*:1841-1848.
39. **Langlotz, P. and K.H. Kroner.** 1992. Surface-modified membranes as a matrix for protein purification. J. Chromatogr. *591*:107-113.
40. **Lawrence, J.F. and C. Menard.** 1997. Determination of clenbuterol in beef liver and muscle tissue using immunoaffinity chromatographic cleanup and liquid chromatography with ultraviolet absorbance detection. J. Chromatogr. B Biomed. Sci. Appl. *696*:291-297.
41. **Liang, X., D.M. Lubman, D.T. Rossi, G.D. Nordblom and C.M. Barksdale.** 1998. On-probe immunoaffinity extraction by matrix-assisted laser desorption/ionization mass spectrometry. Anal. Chem. *70*:498-503.
42. **Margolis, S.A, A.J. Fatiadi, L. Alexander and J.J. Edwards.** 1989. Chromatographic separations of serum proteins on immobilized metal ion stationary phases. Anal. Biochem. *183*:108-121.
43. **Marriott, G. and J. Ottl.** 1998. Synthesis and applications of heterobifunctional photocleavable cross-linking reagents. Methods Enzymol. *291*:155-175.
44. **McConnell, J.P. and D.J. Anderson.** 1993. Determination of fibrinogen in plasma by high-performance immunoaffinity chromatography. J. Chromatogr. *615*:67-75.
45. **McDonnell, P.A., G.W. Caldwell and J.A. Masucci.** 1998. Using capillary electrophoresis/frontal analysis to screen drugs interacting with human serum proteins. Electrophoresis *19*:448-454.
46. **Ohara, T., A. Shibukawa and T. Nakagawa.** 1995. Capillary electrophoresis/frontal analysis for microanalysis of enantioselective protein binding of a basic drug. Anal. Chem. *67*:3520-3525.
47. **Oscarsson, S. and J. Porath.** 1993. Covalent Chromatography, p. 43-52. *In* T.T. Ngo (Ed.), Molecular Interactions in Bioseparations. Plenum Press, New York.
48. **Phillips, T.M.** 1985. High performance immunoaffinity chromatography: an introduction. LC·GC *3*:962-972.
49. **Phillips, T.M.** 1989. High-performance immunoaffinity chromatography. Adv. Chromatogr. *29*:134-173.
50. **Phillips, T.M.** 1994. Immunoaffinity measurement of recombinant granulocyte colony stimulating factor in patients with chemotherapy-induced neutropenia. J. Chromatogr. B Biomed. Appl. *662*:307-313.
51. **Phillips, T.M.** 1998. Determination of in situ neuropeptides by capillary immunoelectrophoresis. Anal. Chim. Acta. *372*:209-218.
52. **Phillips, T.M. and J.J. Chmielinska.** 1994. Immunoaffinity capillary electrophoretic analysis of cyclosporin in tears. Biomed. Chromatogr. *8*:242-246.
53. **Phillips, T.M. and B.F. Dickens.** 1998. Analysis of recombinant cytokines in human body fluids by immunoaffinity capillary electrophoresis. Electrophoresis *19*:2991-2996.
54. **Phillips, T.M. and P.L. Kimmel.** 1994. High-performance capillary electrophoretic analysis of inflammatory cytokines in human biopsies. J. Chromatogr. B Biomed. Appl. *656*:259-266.
55. **Porath, J. and P. Flodin.** 1959. Gel filtration: a method for desalting and group separation. Nature *183*:1557-1559.

56. **Porter, R.R.** 1959. The hydrolysis of rabbit gamma-globulin and antibodies with crystalline papain. Biochem. J. *73*:119-126.

57. **Regnier, F.E. and G. Huang.** 1996. Future potential of targeted component analysis by multidimensional liquid chromatography-mass spectrometry. J. Chromatogr. *750*:3-10.

58. **Robbins, J.B., J. Haimovich and M. Sela.** 1967. Purification of antibodies with immunoadsorbents prepared using bromoacetyl cellulose. Immunochemistry *4*:11-16.

59. **Ruhn, P.F., J.D. Taylor and D.S. Hage.** 1994. Determination of urinary albumin using high-performance immunoaffinity chromatography and flow injection analysis. Anal. Chem. *66*:4265-4271.

60. **Scatchard, G.** 1949. The attraction of proteins for small molecules and ions. Ann. NY Acad. Sci. *51*:660-672.

61. **Sebille, B., R. Zini, C.V. Madjar, N. Thuaud and J.P. Tillement.** 1990. Separation procedures used to reveal and follow drug-protein binding. J. Chromatogr. *531*:51-77.

62. **Starkenstein, E.** 1910. Uber fermentwirkung und deren beeinflussung durch neutralsalze. Biochem. Z. *24*:210-218.

63. **Staros, J.V., R.W. Wright and D.M. Swingle.** 1986. Enhancement by N-hydroxysulfosuccinimide of water-soluble carbodiimide-mediated coupling reactions. Anal. Biochem. *156*:220-222.

64. **Stead, C.V.** 1991. The use of dyes in protein purification. Bioseparation *2*:129-136.

65. **Stein, S.R., J.L. Palmer and A. Nisonoff.** 1964. Re-formation of interchain bonds linking half-molecules of rabbit gamma-globulin. J. Biol. Chem. *239*:2872-2877.

66. **Svec, F. and J.M. Frechet.** 1996. New formats of polymeric stationary phases for HPLC separations: molded macroporous disks and rods. J. Mol. Recognit. *9*:326-334.

67. **Turkova, J., S. Vohnik, S. Helusova, M.J. Benes and M. Tichia.** 1992. Galactosylation as a tool for the stabilization and immobilization of proteins. J. Chromatogr. *597*:19-25.

68. **Weetall, H.H. and A.M. Filbert.** 1974. Porous glass for affinity chromatography applications. Methods Enzymol. *34*:59-64.

69. **Winzor, D.J.** 1997. Quantitative affinity chromatography, p. 39-60. *In* P. Matejtschuk (Ed.), Affinity Separations. A Practical Approach. IRL Press, Oxford.

70. **Worrall, D.M.** 1996. Dye-ligand affinity chromatography. Methods Mol. Biol. *59*:169-176.

71. **Wuenschell, G.E., E. Wen, R. Todd, D. Shnek and F.H. Arnold.** 1991. Chiral copper-chelate complexes alter selectivities in metal affinity protein partitioning. J. Chromatogr. *543*:345-354.

72. **Yang, Q. and J.W. DePierre.** 1998. Rapid one-step isolation of mouse liver catalase by immobilized metal ion affinity chromatography. Protein Expr. Purif. *12*:277-283.

73. **Yvon, M. and J.M. Wal.** 1991. Tandem immunoaffinity and reversed-phase high-performance liquid chromatography for the identification of the specific binding sites of a hapten on a proteic carrier. J. Chromatogr. *539*:363-371.

74. **Zhivkova, Z. and V. Russeva.** 1998. New mathematical approach for the evaluation of drug binding to human serum albumin by high-performance liquid affinity chromatography. J. Chromatogr. B Biomed. Sci. Appl. *707*:143-149.

75. **Zhou, F.L., D. Muller, X. Santarelli and J. Jozefonvicz.** 1989. Coated silica supports for high-performance affinity chromatography of proteins. J. Chromatogr. *476*:195-203.

2 | Preparation of Solid-Phase Affinity Matrices

Affinity and immunoaffinity separations can be performed in solution, although this usually involves further steps to recover the bound or precipitated analyte. In cases where affinity separations are used to remove specific molecules from the test solution, recovery of the isolated material is not required. However, this can be an expensive approach and the majority of affinity and immunoaffinity procedures are applied for the isolation and recovery of a specific analyte. To simplify the handling of affinity ligands, most procedures use solid-phase reagents, composed of the affinity ligand immobilized to a bead or particle. Solid supports facilitate both the easy recovery of the ligand-analyte complex and ligand removal following dissociation of the complex. Many different materials have the potential to be affinity supports, and a variety of such materials are commercially available. For use in the biological sciences, affinity supports must possess a series of important properties. Chief among these is a mechanical, strong, inert surface that can be chemically modified to attach the desired ligand. This property assures that minimal nonspecific interaction will occur between the support and the materials being separated. Additionally, it is also important that the inert support surface does not chemically interact with and activate biological pathways. A classic example of such an interaction is the nonspecific adsorption of proteins to silanol groups present on silica supports (11).

Great care must be taken in selecting the support materials for affinity and immunoaffinity separations. Although many affinity supports are commercially available, frequently they will not meet the specific requirements of the investigator. In these cases, it is necessary to construct suitable supports in the laboratory. For this reason, we have included procedures for the design and construction of a variety of affinity supports. Beaded or particulate supports are the most popular and are the easiest to handle. This is especially true when column chromatographic techniques are employed.

Beaded or particulate materials can be divided into those constructed from naturally occurring materials or from synthetic materials, as well as a variety of rigid supports. Additionally, there has been a recent increase in the use of membrane-based supports, which preclude the necessity of buying chromatographic equipment. With advances in modern technology, newer supports are being forwarded as potential candidates for use in both affinity and immunoaffinity separations.

Affinity and Immunoaffinity Purification Techniques
Terry M. Phillips and Benjamin F. Dickens
© 2000 Eaton Publishing, Natick, MA

ACTIVATED NATURAL SUPPORTS

Many natural materials can be successfully formed into affinity supports. Long-chain polysaccharides, such as dextrans, agaroses, and cellulose (Figure 1), are naturally occurring polymers that can be chemically altered to form beaded supports suitable for both affinity and immunoaffinity separations. Natural materials also frequently possess functional moieties such as hydroxyl, thiol, and amino groups that can be modified to allow the attachment of suitable reactive side chains. Table 1 lists some of the commercially available sources of these supports.

Dextrans

These materials are microbial polysaccharides obtained from the *Leuconostoc*

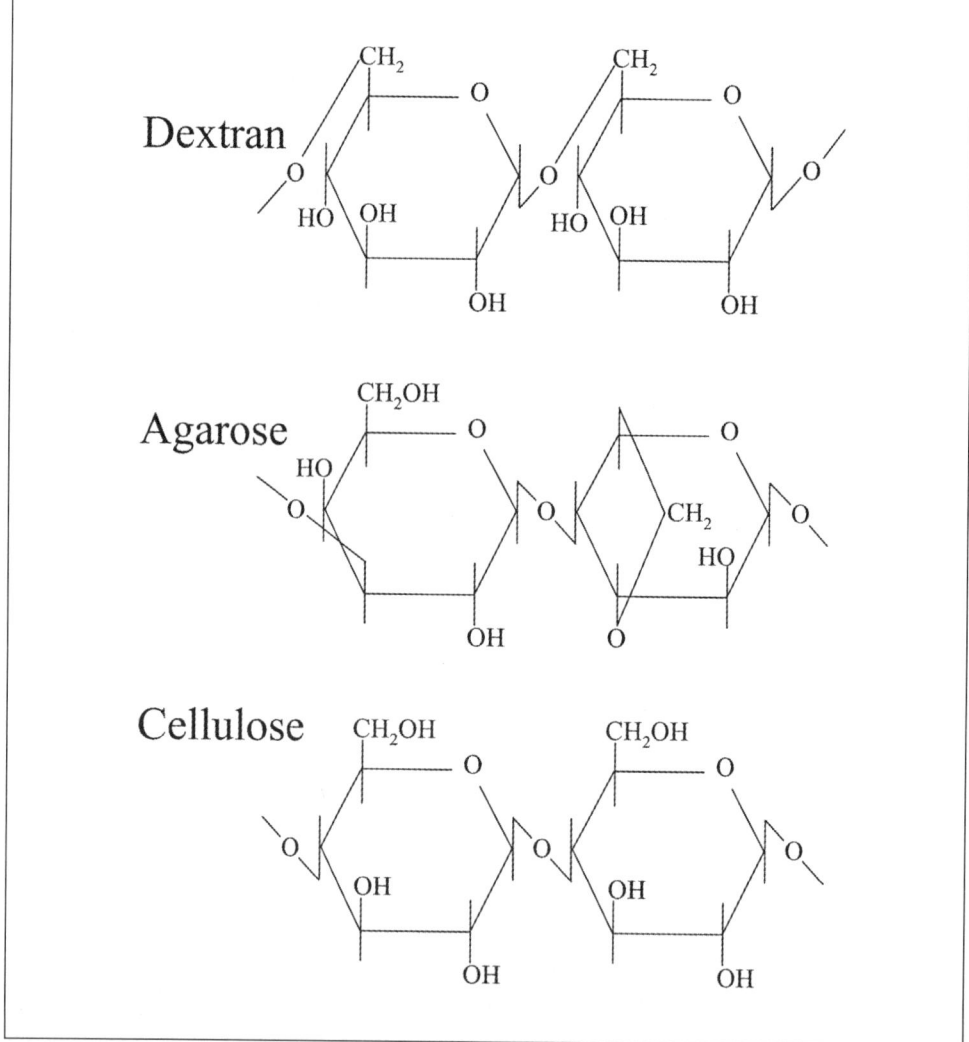

Figure 1. Basic repeating units of the three long-chain polysaccharides most frequently used as affinity ligand supports.

Table 1. Commercial Sources of Dextran, Agarose, and Cellulose Supports

Support	Material	Vendor
Affinica®	Agarose	Schleicher & Schuell
Affi-Gel®	Agarose	Bio-Rad Laboratories
Bio-Gel®	Agarose	Bio-Rad Laboratories
CC31®	Cellulose	Whatman Inc.
Divicell®	Cellulose	Leipziger Arzneimittelwerk GmbH
Reacti-Gel®	Agarose	Pierce Chemical
Sephadex®	Dextran	Amersham Pharmacia Biotech
Sepharose®	Agarose	Amersham Pharmacia Biotech

species. Dextran polymers are composed of α-1,6-linked glucose units that can possess 1,2; 1,3; and 1,4 side branches. The polymers can be cross-linked in alkaline medium with epichlorohydrin to produce a beaded gel, which has been universally used as the support material for size exclusion chromatography (38). Dextran beads are highly hydrophilic due to the abundance of hydroxyl groups that can be used as attachment sites for ligands (19). Although these beads are stable in 0.1 M HCl and 0.25 M NaOH, their structural integrity becomes damaged when they are placed in strong acid, alkaline, or oxidant solutions for prolonged periods of time.

Agaroses

Agarose is a purified, linear, water-soluble polysaccharide derived from seaweed and composed of alternating units of 1,3-linked β-D-galactose and 1,4-linked 3,6-anhydro-α-L-galactose. Several different types of agarose are available and only the purest form should be used to make beads. It should also be remembered that agar and agarose are not the same, the latter being a purified form of the former. This polysaccharide was first introduced as an affinity support by Cuatrecasas et al. (12). Since its introduction, beaded agarose has become one of the most popular materials in affinity chromatography. The useful range of commercially available beaded agaroses offered by a number of companies has enhanced this popularity. It can easily be modified by the abundant hydroxyl groups found throughout the entire matrix (see Figure 1), and, under the majority of chromatographic conditions, exhibits minimal nonspecific binding properties. However, agaroses can show a tendency to bind proteins, peptides, and nucleic acids nonspecifically, especially when run under high salt conditions (52) such as high molarity salt gradients and chaotropic elution.

Magnetic Agaroses

Agarose beads can be further modified through the inclusion of metals (such as iron) into the bead matrix. Magnetic beads are popular for batch separation techniques, since the affinity ligand can easily be recovered with a magnet. These supports were originally made by incorporating 7% (wt/vol) Fe_3O_4 into 4% (wt/vol) epichlorohydrin-cross-linked agarose beads (29). Ligand coupling is performed by any of the techniques to be described (in Chapter 4 of this book) for the activation of agarose supports, although magnetic agarose beads are better suited to batch purifications and immunoassays.

Celluloses

This natural plant product was the first material to be used for attachment of affinity ligands to isolate a variety of enzymes (2,21,48). The material is usually obtained as long strands, although microcrystalline forms are also available. Cellulose is composed of linear units of β-1,4-linked D-glucose with occasional 1,6 bonds, which helps to make the structure strong, fibrous, and uniform. The fibrous structure does not promote fast chromatographic flow rates (23), but its low cost, high tensile strength, and chemical inertness make cellulose very useful for large-scale batch separations. Divicell is a macroporous beaded cellulose that exhibits high flow rates with excellent flow parameters. Activation of Divicell can be achieved with sodium periodate, epichlorohydrin, and 5-norbornene-2,3-dicarboximido carbonochloridate for the attachment of amines, diamines, amino acids, carbohydrates, and proteins. Additionally, Divicell is available with Cibacron Blue F3GA covalently bound to its surface for the selective isolation of HSA (7).

Nitrocellulose paper has been described as a potential solid-phase matrix for protein immobilization. Small segments of paper were frozen in liquid nitrogen and then ground to a powder, and fractionated according to size. The material was used in both batch techniques and column chromatography for the isolation of anti-hapten antibodies and for the removal of anti-carrier proteins in the anti-hapten antiserum. The former was achieved by reacting the antiserum against hapten-immobilized support, while the latter was achieved with carrier-immobilized support (15).

Derivatization of Natural Gels

Polysaccharide-based materials have been extensively used as affinity supports since the introduction of this technology and consequently, a wide variety of derivatization procedures have been applied to these supports. One of the prominent features of these supports is their abundance of hydroxyl groups, which can be chemically derivatized to form attachment sites. However, many of the reactions utilize highly reactive chemistries that require great caution when performed. Although most of these reactions can safely be performed at a standard laboratory bench, it is advisable to have a chemical hood available. Many of the routine manipulations involving organic solvents should be performed under a hood, if only to protect the worker from continually inhaling the chemical fumes. It should also be pointed out that some of the reactions described are volatile, so eye and face protection must be used. The derivatizations typically can be performed using standard laboratory equipment, although one will need access to several types of mixer. Most reactions are performed in a overhead mixer (Figure 2A), but there are times when a rotary mixer (Figure 2B) is also useful. This latter mixer is especially gentle and can be used for attaching delicate ligands to activated supports (see Chapter 4). The investigator will also need access to standard magnetic stirrers and to a low energy sonicator (such as those used by electron microscopists or jewellers).

CNBr derivatization of polysaccharide supports introduces one of the most common reactive groups in affinity chromatography (Figure 3). This modification was introduced in 1967 by Axen et al. (4) and is still one of the basic chemistries used for the attachment of affinity ligands, especially for several commercially available supports. CNBr reacts with hydroxyl groups to produce a reactive support that can be further modified by the attachment of a spacer arm or ligand containing primary amino groups (20). During the reaction, *N*-substituted isourea is formed, positively charging the amide and imparting anionic exchange properties to the support under physiological conditions. Although relatively stable, the isourea derivative is susceptible to nucleophilic attack, which causes slow hydrolysis at extreme pH. This hydrolysis results in ligand leakage during elution under acidic or alkaline conditions. Thus, care must be exercised in choosing a suitable elution agent, such as a chaotropic ion (see Chapter 7). Although polysaccharide derivatization with CNBr is simple,

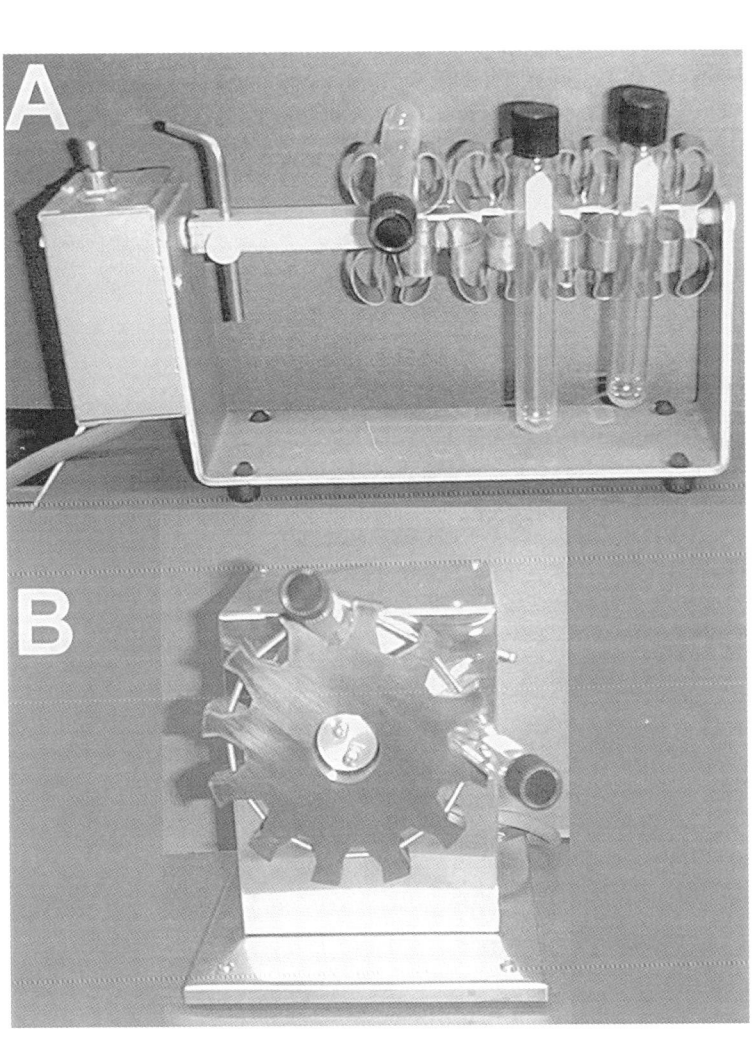

Figure 2. Laboratory mixers used to perform derivatizations. (A) Overhead mixer. (B) Rotary mixer.

extremely toxic cyanide gas is created during acidification. For this reason, it is recommended that only the most experienced chemists perform the CNBr derivatization; the majority of investigators therefore should use commercially available supports.

Organic sulfonyl halides readily react with hydroxyl groups to form sulfonyl esters (34) and are a convenient route for activating polysaccharide-based supports (Figure 4). Two forms of sulfonyl chloride exist, the p-toluene (or tosyl) form, and the trifluoroethyl (or tresyl) form. Both are easily introduced by reactions performed in organic solvents, the resulting support being stable for extended periods of time under anhydrous conditions. Protein amino and thiol groups efficiently and rapidly displace the sulfonate, thus easily producing a stable ligand-coupled support. Tresyl-activated supports are more reactive than their tosyl counterparts and both types react more readily with thiols than with amide groups. Ligand attachment to tosyl-activated supports is performed at pH 9.0–10.5 in bicarbonate buffers, while attachment to tresyl-activated gels is performed at pH 7.0–7.5 in phosphate buffers. The reactions are performed at different temperatures. The tosyl-activated supports require attachment to be performed for 2–3 h at 40°C, while the tresyl-activated supports require overnight incubation at 4°C. Apart from the relatively short time required for attachment, the tosyl-activated supports have an added advantage in that tosyl displacement can be

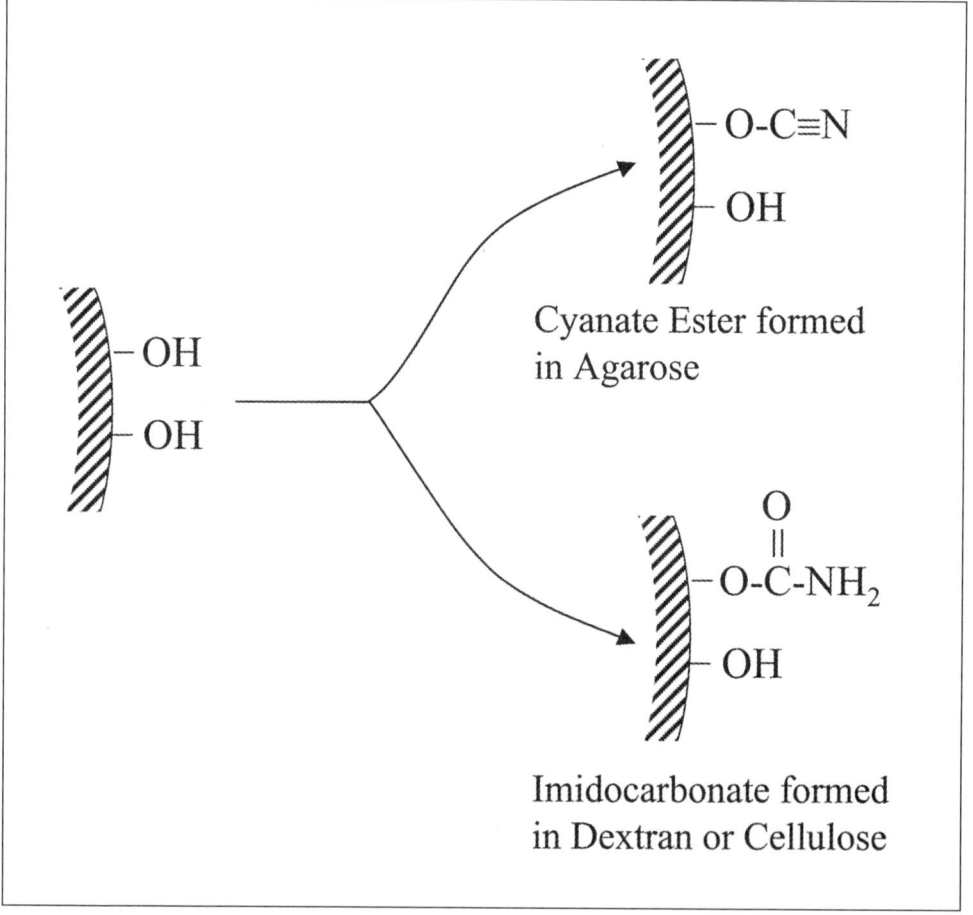

Figure 3. CNBr derivatization of polysaccharide supports.

directly monitored spectrophotometrically.

A recent advance in the modification of polysaccharide hydroxyl groups is activation chemistry using 2-fluoro-1-methylpyridinium toluene-4-sulfonate (FMP) in polar organic solvents (33). This reaction is performed in the presence of a tertiary amine, such as triethylamine, for a relatively short period of time (10 min) at room temperature (Figure 5). This introduces 2-alkoxy-1-methylpyridinium salts, which can react at pH 8.0–9.0 with amino- or sulfhydryl-containing ligands in either aqueous or polar organic solvents. An advantage of this procedure is that both activation and coupling can be followed by spectrophotometry because the leaving group (a cleavage product from a derivatizing agent "leaving" the agent following modification of the ligand), 1-methyl-2-pyridine, absorbs at 297 nm. Further advantages are that FMP is inexpensive, nontoxic, and forms stable thioether or secondary amide linkages at high degrees of substitution. BioProbe International is a commercial source for FMP-activated agarose.

Oxidation of polysaccharide supports is another important chemistry for the attachment of protein ligands. Vicinal *cis*-hydroxyl groups of polysaccharides can be oxidized by sodium m-periodate ($NaIO_4$) to generate aldehyde functions (46) (Figure

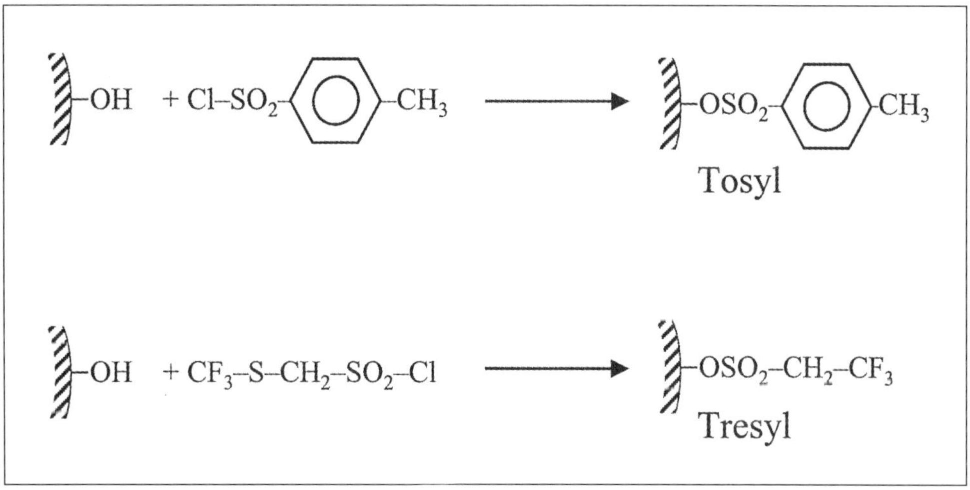

Figure 4. Activation of polysaccharide-based supports by organic halides. Two forms exist; the tosyl (p-toluene) form and the tresyl (trifluoroethyl) form.

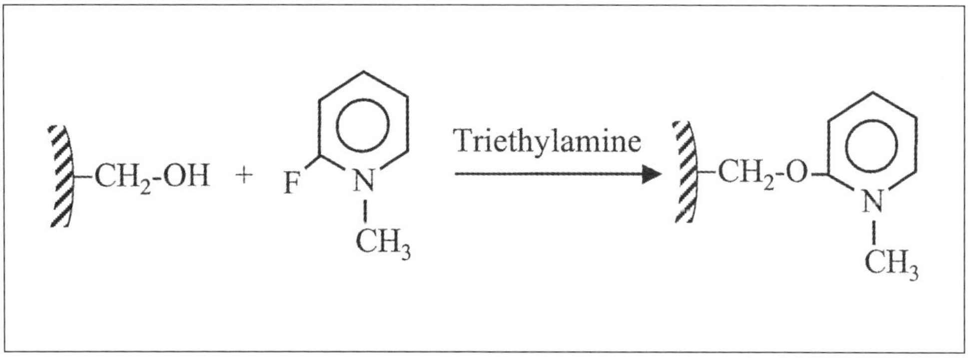

Figure 5. FMP activation in polar organic solvents in the presence of triethylamine.

6). These aldehydes can be further converted into either secondary amides by reductive amination or into hydrazides by reaction with dihydrazides (Figure 7). One advantage of using $NaIO_4$ is that it is extremely soluble in water and can be easily removed from the activated support by extensive washing. The activated support is reasonably stable and can be stored at 4°C for several days. Ligands are attached to the activated support by primary amino groups, which react with the aldehydes at pH 4.0–6.0 to form Schiff's base. This reaction can then be stabilized by reduction with sodium borohydride. Periodate oxidation is a convenient procedure for producing stable ligand supports, especially when the ligands to be attached are sensitive to high pH. Additionally, the reagents used in this procedure are nontoxic, as is the alkylamine product.

Reaction of bis-oxiranes, such as 1,4-butanediol glycidyl ether, with polysaccharide hydroxyl groups at alkaline pH produces a support with terminal epoxide or oxirane groups (Figure 8). These groups can then be reacted with ligands to produce affinity matrices. An advantage of this procedure is that a long-chain, hydrophilic spacer arm is automatically introduced during the activation. The reaction can be enhanced by performing it at temperatures above 40°C. However, a degree of cross-linking of the support occurs during the activation procedure. Although this reduces the effective pore size of the support, it does increase its rigidity, thus producing an extremely stable support. Epoxide chemistry can be used to couple ligands by amino, thiol, and hydroxyl (including phenyl) groups in pH conditions ranging from pH 9.0 to pH 13.0 at room temperature. Changing the pH conditions can influence the type of ligand linkage. Conditions at pH 7.5–8.5 favor linkage through thiol groups, while amino

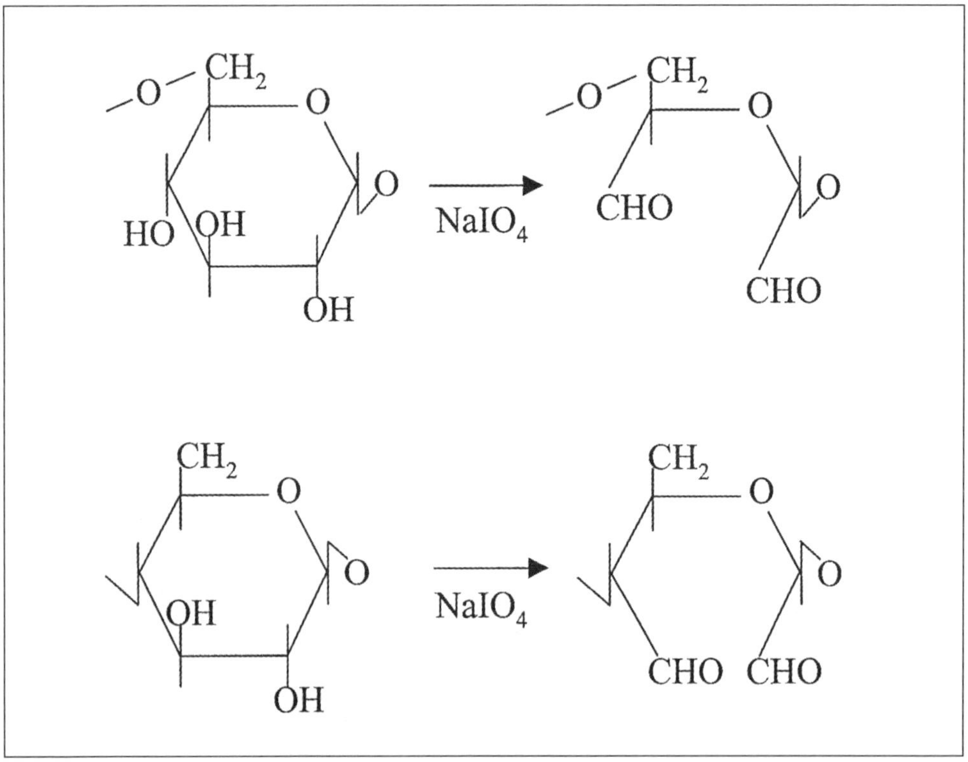

Figure 6. Oxidation of polysaccharide *cis*-hydroxyl groups by $NaIO_4$.

groups are usually coupled at pH 8.5–9.5, and hydroxyl groups at pH 11.0–13.0.

Carbodiimides, such as *N,N'*-dicyclohexylcarbodiimide, can activate polysaccharide supports under anhydrous conditions, introducing reactive groups to which

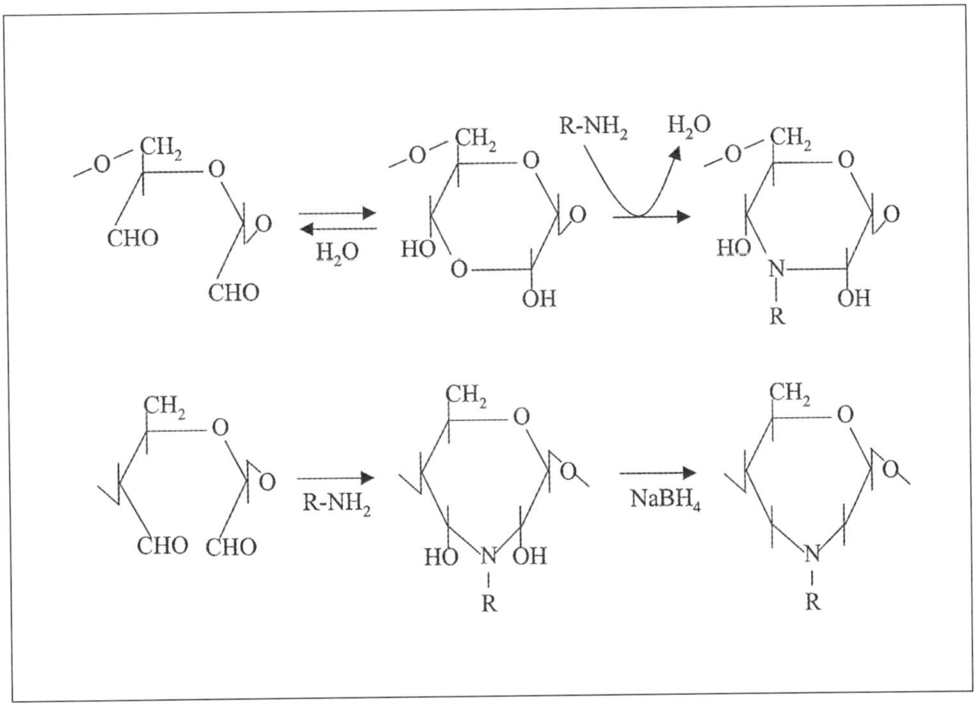

Figure 7. Conversion of aldehydes by reductive amination or dihydrazide reaction.

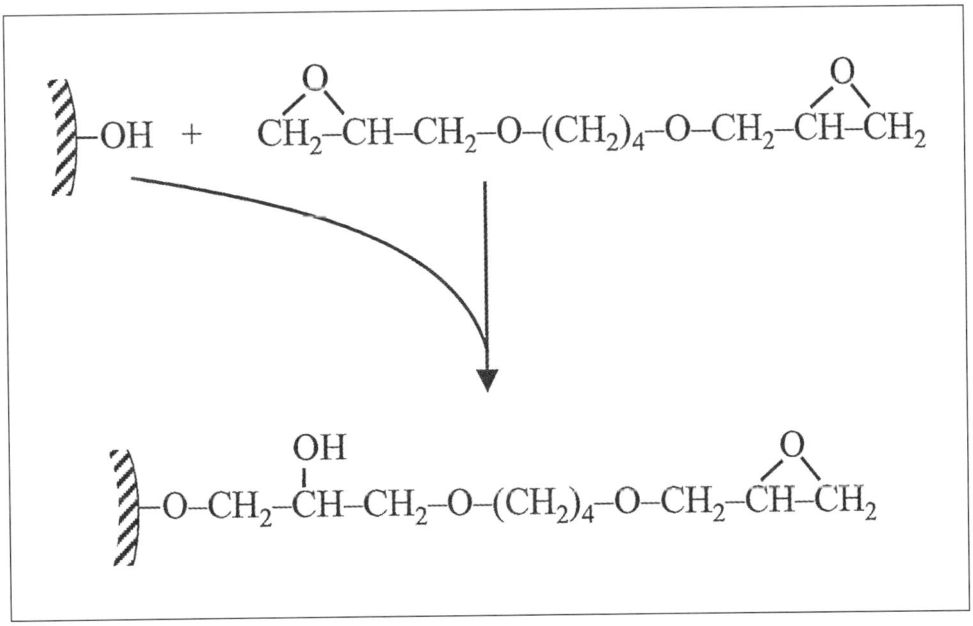

Figure 8. Reaction scheme for covalent bonding of bis-oxiranes (this example uses 1,4-butanediol glycidyl ether) to polysaccharide supports to produce a hydrophilic function terminated with a reactive oxirane.

carbonyldiimidazole (CDI) groups can be attached (6) (Figure 9). At alkaline pH, these groups react with primary amino groups on the ligand to give stable carbonate derivatives. The advantages of this procedure are that CDI is nontoxic, gives high levels of activation, and does not introduce ion-exchange groups into the matrix. Potential disadvantages are that CDI-activated supports are more stable in anhydrous conditions than in aqueous ones, and the activated support must either be used immediately or stored sealed under nitrogen. Coupling is usually performed at pH 8.5–9.5 in 0.1 M borate at room temperature.

Activation of supports with amino or amide functions can be performed using a fresh solution of 25% glutaraldehyde in 0.5 M sodium phosphate, pH 7.5, at 20°–40°C (3,49) (Figure 10). The reaction is usually allowed to proceed overnight and the support is washed extensively to remove the excess aldehyde. The ligand is then coupled by primary amines in 0.5 M phosphate buffer at 4°C for 16–24 h. The method is simple, inexpensive, and suitable for attachment of ligands that are unstable or damaged at alkaline pH. The linkage is stable and has the advantage that a spacer arm is simultaneously introduced during the reaction. One should note that glutaraldehyde is moderately toxic and unstable; it can polymerize on storage, thus making it unreactive.

Triazine derivatization is used to introduce reactive groups suitable for immobilizing proteins and peptides onto polysaccharide surfaces (32). The technique involves activation of the matrix with 2-amino-4,6-trichloro-s-triazine, during which a chlorine is replaced by solubilization groups (Figure 11). The ligand is covalently bound by a primary amino group during a secondary step in which another chlorine under-

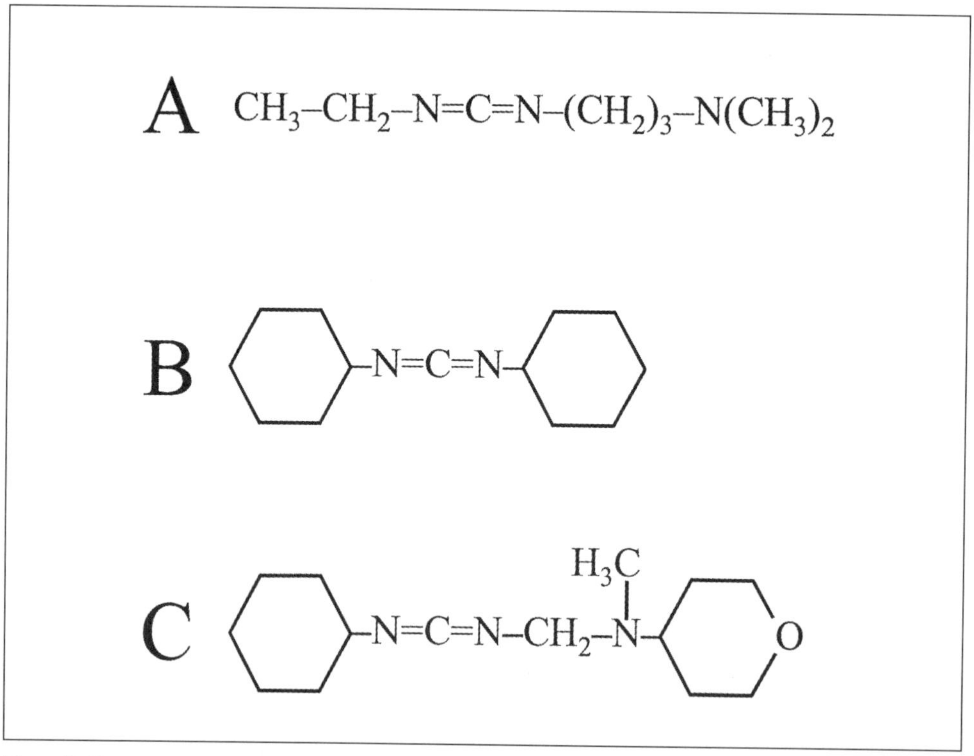

A $CH_3-CH_2-N=C=N-(CH_2)_3-N(CH_3)_2$

B

C

Figure 9. Common carbodiimide derivatizing agents. (A) 1-ethyl-3-(3-dimethylaminopropyl)carbodiimide, (B) dicyclohexyl carbodiimide, (C) 1-cyclohexyl-3-(2-morpholinoethyl)-carbodiimide.

goes nucleophilic substitution.

Natural supports can also be modified by chemically introducing thiol groups onto the surface of the beads (Figure 12). This type of support is extremely useful for immobilizing ligands that possess free thiol groups, such as enzyme-digested antibody fragments. Sulfhydryl groups can easily be derivatized into FMP-activated polysaccharide supports by reacting the support with β-mercaptoethanol or dithiothreitol (DTT). Ligand attachment proceeds by first performing a thiol-disulfide interchange followed by covalent linkage by a disulfide bridge.

PROTOCOLS

Preparation of Agarose Beads
(US Standard Sphere Size: 170–300)

1. Dissolve 10 g of agarose in 100 mL of distilled H_2O by autoclaving in a covered beaker at 30–40 psi.
2. Transfer the hot agarose solution to a round-bottom flask (Figure 13).
3. Add a stirrer bar attached to a variable speed motor.
4. Dissolve 15 g of polyethylene sorbitan monostearate in a solvent solution containing 440 mL of toluene and 160 mL of anhydrous CCl_4.
5. Heat the solvent solution to 50°C.
6. Set the stirrer to 1700 rpm and carefully add the hot solvent solution to the agarose.

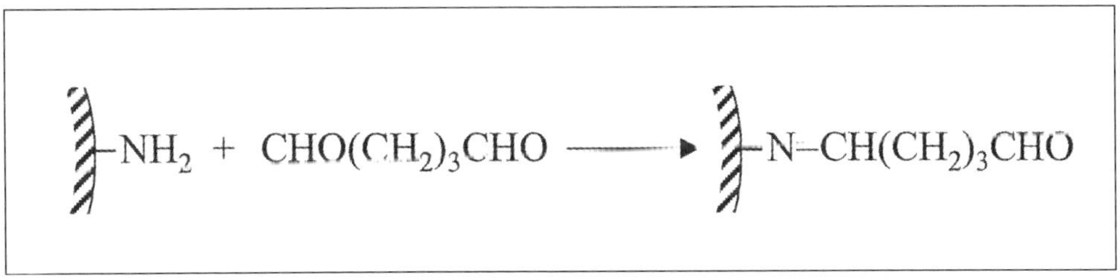

Figure 10. Activation of supports with amino or amide functions with glutaraldehyde.

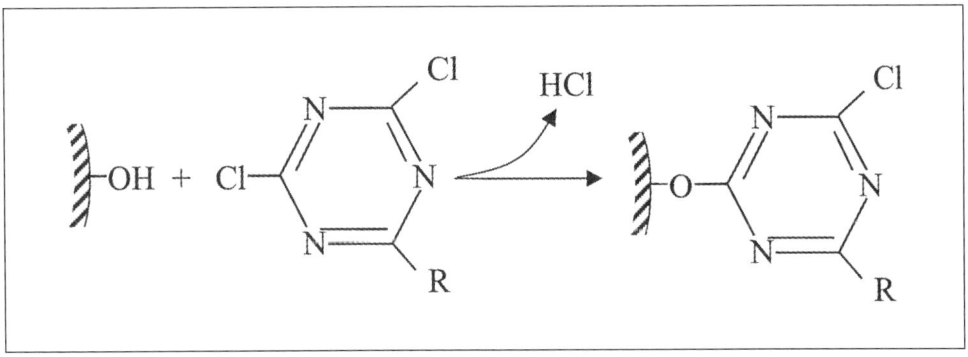

Figure 11. Activation of polysaccharide supports by triazine derivatization. Typical functional groups in the 2-position (R) include amino (2-amino-4,6-trichlorostriazine) and Cl (2,4,6-trichloros-triazine).

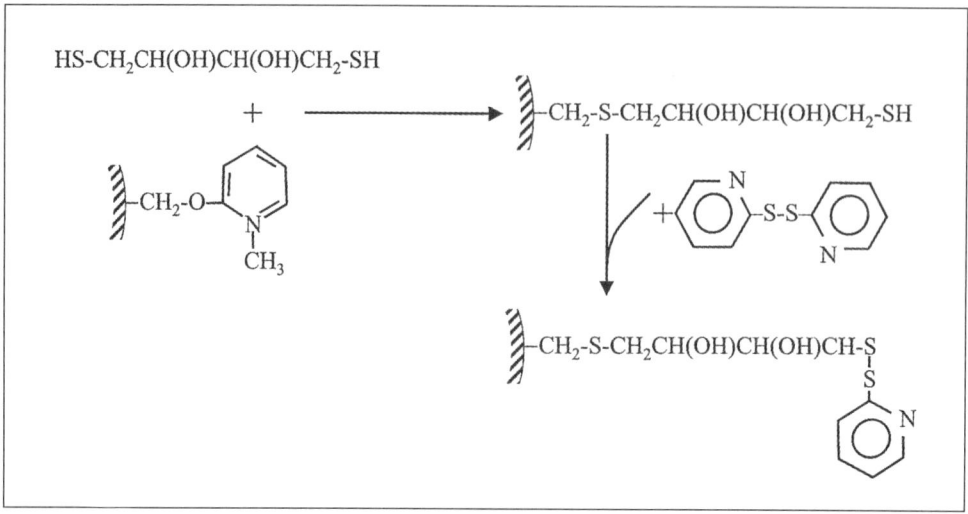

Figure 12. Introduction of a reactive thiol group to FMP-activated supports. Sequential treatment of the FMP-activated support with DTT (also known as Cleland's Reagent) and 2,2'-dipyridyl disulfide creates a support with a reactive disulfide bridge.

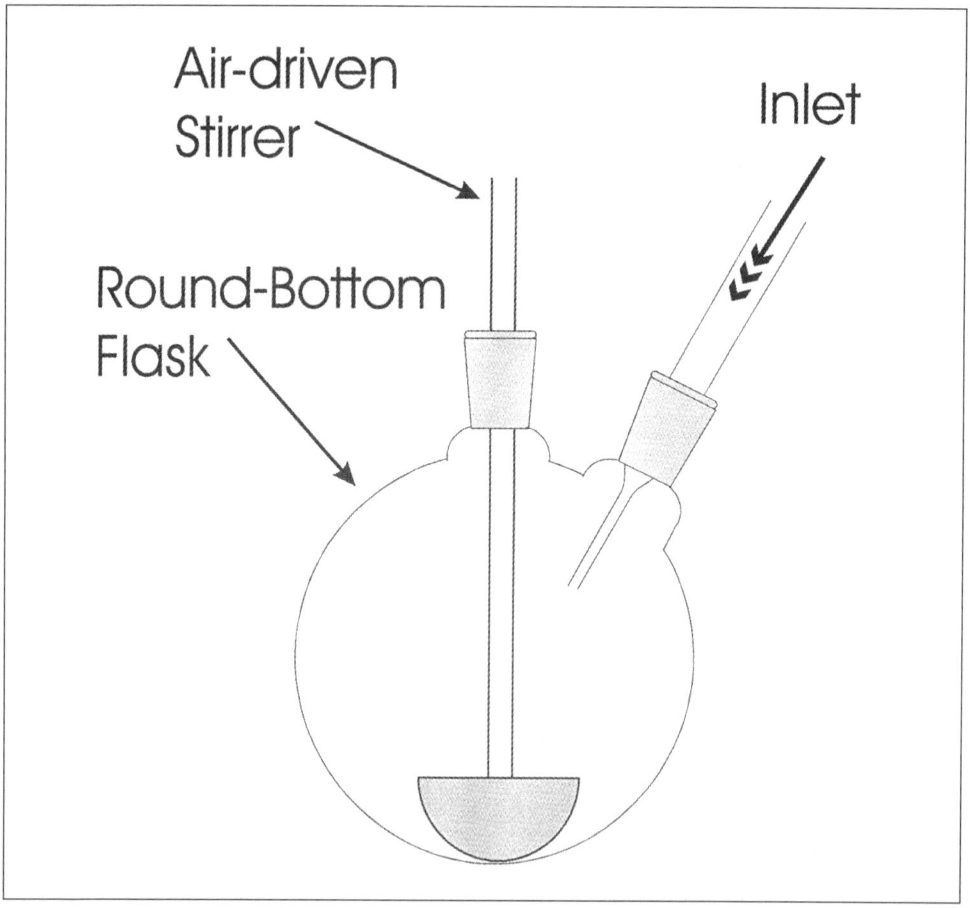

Figure 13. Bead derivatization vessel. Beads are introduced by the inlet port and constantly stirred during the reaction by the air-driven stirrer. This setup ensures minimal damage to the beads during the derivatization process.

7. Allow the mixture to stir for 1 min.
8. Cool by surrounding the flask with cold H_2O.
9. Allow the mixture to cool for approximately 5 min and then transfer the beads to a sintered glass filter.
10. Wash the beads three times (using 200 mL of ether in each wash) to remove the solvent solution.
11. Place the bead-ether mixture in a large glass beaker and add 1000 mL of distilled H_2O.
12. Carefully decant the ether layer.
13. Place the beads in a Buchner funnel and wash five times in 200 mL of H_2O (using 200 mL of distilled H_2O in each wash) to remove the residual ether.

CNBr Derivatization

Caution: Extremely toxic cyanide gas is released during this procedure. Therefore, this procedure should only be performed by highly experienced chemists with appropriate safeguards.

1. Swell 10 g of beads by placing them overnight in 400 mL of 0.1 M sodium phosphate, pH 7.2.
2. Dry the beads in a sintered glass filter and add the semi-dry beads to a 50-mL flask containing 8 mL of 0.1 M sodium phosphate.
3. Cool the flask in an ice bath to 4°C.
4. Place the ice bath in a fume hood and add 8 mL of cold aqueous CNBr.
5. Stopper the flask and gently shake for 10 min.
6. Pour the slurry into a sintered glass filter and wash with 400 mL of ice-cold distilled H_2O.
7. Store at 4°C.

Succinimide Esters

1. Add 10 g of the derivatized beads to 10 mL of a 10% (wt/vol) solution of succinyl chloride (dissolved in dry chloroform) in a 50-mL round-bottom flask.
2. Add 10 mL of 10% triethylamine (dissolved in dry chloroform) and reflux for 30 min.
3. Wash the beads five times (1–3 min/wash) in chloroform and dry at 100°C.
4. Resuspend the beads in 20 mL of anhydrous dioxane containing 0.1 M N-hydroxysuccinimide and 0.1 M N,N'-dicyclohexylcarbodiimide.
5. Stir the mixture for 2 h at room temperature.
6. Collect and wash the beads in 200 mL of anhydrous dioxane, followed by two washes in 95% methanol (1–2 min in 200 mL/wash), to remove the accumulated dicyclohexylurea.
7. Resuspend the beads in enough dioxane to cover their surface and store at

room temperature in light-tight containers.

Note: This preparation is stable for 1–2 months.

Introduction of Triazine Groups

1. Swell 10 g of beads by placing them overnight in 400 mL of distilled H_2O.
2. Resuspend the beads by "hand-swirling" and pour the slurry into a sintered glass filter.
3. Wash the beads with 200 mL of distilled H_2O, followed by 200 mL of 12% (wt/vol) NaOH.
4. Allow the NaOH to drain by gravity and move the beads to a beaker.
5. Resuspend the beads in 50 mL of 12% (wt/vol) NaOH.
6. Cool to 4°C.
7. Prepare a 0.5 M solution of 2-amino-4,6-trichloro-s-triazine in acetone.
8. Cool the triazine solution to 4°C.
9. Carefully add 100 mL of the triazine solution to the beads by running the solution gently down the side of the container. A bilayer should form immediately.
10. Add a magnetic stirrer bar and slowly mix for 2 h at 4°C, during which time the bilayer will disappear.
11. Remove the supernatant, and wash the activated beads in 200 mL of acetone by sedimentation.
12. Wash in 200 mL of distilled H_2O by sedimentation.
13. Immediately react with the ligand.

Periodate Oxidation

1. Swell 10 g of beads by placing them overnight in 400 mL of distilled H_2O.
2. Resuspend the beads and pour the slurry into a sintered glass funnel.
3. Semi-dry by applying a vacuum.
4. Move the beads to a round-bottom flask and add to 20 mL of 0.05 M aqueous sodium periodate.
5. Swirl the mixture for 45 min at room temperature.
6. Add 10 mL of 2 M ethylene glycol and rotate for a further 30 min.
7. Return the beads to the glass filter and wash in 200 mL of distilled H_2O followed by 200 mL of 0.01 M $NaCO_3$, pH 9.5.
8. Store at 4°C.

Introduction of Hydrazide Groups

1. Place 10 g of succinimide-activated support into a sintered glass filter.
2. Wash five times in 10 mL of 0.1 M 2-(*N*-morpholino)ethanesulfonic acid (MES) dissolved in 0.9% NaCl, pH 4.7.
3. Dry the support by applying a vacuum and move the support to a round-bottom flask.

4. Add a volume of adipic dihydrazide solution equal to the bead displacement volume, followed by 0.5 g of 1 M 1-ethyl-3-(3-dimethylaminopropyl)carbodiimide.

5. Place an overhead stirrer into the mixture and agitate constantly for 3 h at room temperature.

6. Transfer to a sintered glass filter and wash 5 times with distilled H_2O.

7. Wash 5 times with 1 M NaCl.

8. Wash 5 times in distilled H_2O.

9. Wash 5 times in 0.1 M sodium phosphate, pH 7.2.

10. Dry the support by applying a vacuum and react with ligand.

Epoxide Derivatization

1. Place 10 g of pre-swollen agarose or silanized beads in a sintered glass filter and wash with 200 mL of double-distilled H_2O.

2. Dry by applying a vacuum.

3. Resuspend the beads in 10 mL of 1 M NaOH containing 2 mg/mL $NaBH_4$.

4. Slowly add 10 mL of 1,4-butanediol glycidyl ether and place in a capped 50-mL tube.

5. Place on an overhead mixer at room temperature for 12–16 h.

6. Place the activated beads back in the glass filter and wash with 400 mL of double-distilled H_2O.

7. Store at 4°C for up to 1 week.

Diazo Derivatization

Caution: Extremely toxic cyanide gas is released during this procedure. Therefore, this procedure should only be performed by highly experienced chemists with appropriate safeguards.

1. Swell 10 g of beads by placing them overnight in 400 mL of 0.1 M sodium phosphate, pH 7.2.

2. Dry the beads in a sintered glass filter by applying a vacuum and add the semi-dry beads to a 50-mL round-bottom flask containing 8 mL of 0.1 sodium phosphate.

3. Cool in an ice bath to 4°C.

4. Place the beads in an ice bath in a fume hood and add 8 mL of cold aqueous CNBr.

5. Stopper the flask and gently shake for 10 min.

6. Pour the slurry into a sintered glass filter and wash with 400 mL of ice-cold distilled H_2O.

7. Dry the beads in a sintered glass filter and chill to 4°C.

8. Resuspend the cold beads in 150 mL of 1.0 M benzamidine, pH 10.0, and stir

with an overhead stirrer for 16 h at 4°C.

9. Dry the beads in a sintered glass filter and wash with 200 mL of distilled H_2O.

10. Resuspend the bead-cake in 3 vol of chilled 0.5 M HCl to which 0.2 M $NaNO_2$ has previously been added.

11. Stir at room temperature for 15 min.

12. Place the beads in a sintered glass filter and wash with 500 mL of cold distilled H_2O.

13. Wash with 50 mL of 1% (vol/vol) sulfamic acid.

14. Wash with 200 mL of distilled H_2O and use immediately.

FMP Derivatization

1. Place the support in a sintered glass filter and wash with distilled H_2O.

2. Wash the support twice each sequentially in 25:75, 50:50, and 75:25 acetone:water.

3. Wash in 100% acetone.

4. Resuspend the equivalent of 10 mL of packed support in acetonitrile.

5. Place in a heat-dried beaker containing 1 vol of freshly prepared dry acetonitrile plus 2% (vol/vol) dry triethylamine.

6. Insert an overhead stirrer and stir continuously during mixing.

7. Prepare 600 mg of FMP in 20 mL of the acetonitrile/triethylamine solution.

8. Slowly add to the support mixture and stir for 10 min at room temperature.

9. Transfer to a sintered glass filter.

10. Wash with acetonitrile.

11. Wash twice with 50:50 acetone:water.

12. Wash twice with 2 mM HCl.

13. Calculate the FMP concentration by determining the 1-methyl-2-pyrrolidone concentration in the pooled washes at 297 nm using a molar extinction coefficient of 5900.

14. Store the support for up to 4 months at 4°C in 2 mM HCl.

Preparation of Thiol-Derivatized Supports

1. Wash FMP-activated support with distilled water in a sintered glass filter.

2. Dry by applying a vacuum.

3. Resuspend the support in 1 vol of 0.2 M $NaHCO_3$ containing 1 M DTT.

4. Place an overhead mixer into the mixture and stir for 5 h at room temperature.

5. Transfer to a sintered glass filter and wash in 10 vol of 0.2 M $NaHCO_3$.

6. Wash 5 times in distilled H_2O.

7. Wash twice in 2 mM HCl.

8. Wash twice in distilled H_2O.

9. Transfer to a glass tube containing 50:50 acetone in distilled H_2O.

10. To 1 mL of the support, add 100 mg of 2,2′-dipyridyl disulfide dissolved in 50% (vol/vol) acetone.

11. Stir constantly for 30 min at room temperature.
12. Wash with 50% (vol/vol) acetone to remove the disulfide (filtrate will lose its yellow color).
13. Wash in 1 mM EDTA.
14. Store in 1 mM EDTA until required.

Activation of Supports with Glutaraldehyde

1. Place 10 g of support in a sintered glass filter.
2. Wash three times in distilled H_2O.
3. Resuspend the support in 25 mL of 0.1 M sodium phosphate, pH 7.4.
4. Add 10 mL of 25% (vol/vol) glutaraldehyde.
5. Incubate at 37°C for 12–16 h.
6. Return the support to the sintered glass filter.
7. Wash five times in 200 mL of distilled H_2O.
8. Store at 4°C.

ACTIVATED SYNTHETIC POLYMERS

In addition to beaded supports made from natural products, there are a number of synthetic materials such as polyacrylamide, polysaccharide-acrylamide blends, and plastics commonly used as affinity supports. Table 2 lists some of the commercially available sources of synthetic supports.

Polyacrylamide Beads

This material is composed of a cross-linked hydrocarbon matrix to which carboxyamide groups are attached (Figure 14). Polyacrylamide is made by the copolymerization of acrylamide with bis-acrylamide. When produced in a beaded form, the matrix contains a hydrocarbon structural framework with abundant carboxamide side chains. In their classic paper, Inman and Dintzis (17) describe several chemical pathways for the modification of polyacrylamide beads to make matrices suitable for affinity ligand attachment. A pre-activated form of polyacrylamide called Enzacryl® (Fluka) is available for protein immobilization. The support is available as Enzacryl AH, which contains reactive hydrazide side chains, and Enzacryl AA, which contains aromatic acid residues that bind proteins by their amino groups.

Polysaccharide-Acrylamide Composites

The combination of polysaccharides and acrylamide polymers possesses the tensile strength of polyacrylamide and the high-molecular-weight sieving qualities of the polysaccharide gels (28). Sephacryl HR (Amersham Pharmacia Biotech) is a commercially available form of these gels and possesses both the amide groups of the polyacrylamide and the hydroxyl groups of allyl dextran for ligand attachment. A magnetic agarose-polyacrylamide gel, Magnogel AcA 44, is available from Serva. Another magnetic bead is composed of iron oxide particles coated with Enzacryl that

Table 2. Sources of Synthetic Supports

Support	Material	Vendor
Bio-gel P	Polyacrylamide	Bio-Rad Laboratories
Trisacryl®	Polyacrylamide variant	BioSepra
Ultrogel®	Agarose/polyacrylamide	BioSepra
Sephacryl®	Dextran/polyacrylamide	Amersham Pharmacia Biotech
EUPERGIT®	Methacrylate	Alltech Associates
POROS®	Polystyrene	PE Applied Biosystems
HyperD®	Polystyrene/hydrogel	BioSepra

is used in stirred batch reactors or fluidized beds, where it can be easily recovered after the purification is completed.

Methacrylate

These plastic beads are prepared by copolymerization of hydroxyethyl methacrylate and ethylene dimethyacrylates. The rigidity of the beads maintains a high flow rate that enables them to be used in a variety of chromatographic techniques. Ligand coupling is by hydroxyl groups found on the surface of the beads. Although these beads exhibit a high degree of nonspecific binding, they can effectively be used for hydrophobic interaction separations of proteins and peptides (43). Oxirane or epoxide side groups are the most popular attachment groups for the coupling of proteins to plastic affinity matrices.

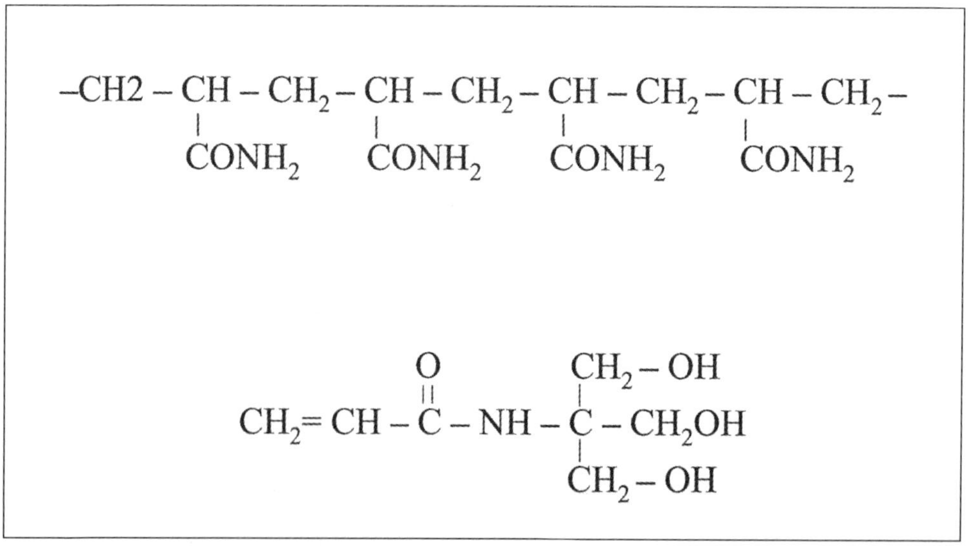

Figure 14. The backbone of a linear polyacrylamide chain (upper) and the basic monomer used in making triacyl gels (lower).

Acrylic Beads

Another plastic support can be made from acrylic beads, which are produced by the copolymerization of methacrylamide, methylene bis-methacrylamide, and either glycidyl methacrylate or allyl glycidyl ether. Commercially available forms of these neutrally charged and hydrophilic beads can be obtained from Sigma Chemical or from Serva (EUPERGIT C). Ligand immobilization is achieved through activation of oxirane groups on the bead matrix over a wide pH range (45).

Trisacryl Beads

Trisacyl is a synthetic gel that can be used for a number of chromatographic techniques including affinity chromatography. The plastic can be formed into a bead according to the procedure outlined in the French Patent No. 7,702,391. During the process, defined-size beads are formed by exothermic polymerization. The beads are made from the polymerization of the unique plastic, N-acryloyl-2-amino-2-hydroxymethyl-1,3-propanediol. The beads formed during the polymerization bear three hydroxymethyl groups plus one alkyamide group per repeating unit. This structure makes the polymer very hydrophilic and especially suitable for the separation of proteins, peptides, and cells. The gels are nonbiodegradable, thermally stable, and resistant to denaturing agents. Although they are quite stable in acidic pH, slow hydrolysis of the amide linkage occurs in alkaline conditions. Trisacryl beads are commercially available from Amersham Pharmacia Biotech.

Derivatization of Synthetic Supports

Polyacrylamide beads can be modified by converting the carboxamide groups to acyl azides by reaction with hydrazine hydrate (Figure 15). *Handle hydrazine with care* and only in a chemical hood, as hydrazine can cause lung and skin damage. Once acyl azides have been produced, the azide can be further derivatized by a series of different reactive groups, thus allowing attachment of ligands through interactions with primary amino groups. Many commercially available synthetic supports have oxirane groups on their surfaces, thus making ligand attachment simple. Synthetic supports can also be activated by glutaraldehyde using the protocol described for natural material supports.

PROTOCOLS

Preparation of Trisacryl Beads

1. Prepare 1000 mL of polymer solution by adding 330 g of N-acryloyl-2-amino-2-hydroxymethyl-1,3-propanediol with 40 g of N,N'-diallyltartardiamide (the cross-linking agent) to distilled H_2O.
2. Heat this solution to 55°C in a water bath.
3. Add 120 mg of ammonium persulfate plus 1.6 mL of 6.6 M N,N,N',N'-tetramethylethylenediamine.
4. Emulsify the solution by thoroughly mixing with 2 L of paraffin oil.

5. Stir the mixture continuously.

6. Recover the formed beads by allowing them to sediment and decant the fluid.

7. Transfer the beads to a Buchner funnel.

8. Wash three times in 1% (vol/vol) aqueous Triton® X-100 detergent in distilled H_2O.

9. Sieve the beads to obtain the 40–80-μ diameter fraction.

10. Return the beads to the Buchner funnel and wash five times in 1 M NaCl.

11. Finally, wash three times in double-distilled H_2O.

Acyl Azide Derivatization

1. Swell 5 g of polyacrylamide beads by suspending the beads in 200 mL of distilled H_2O.

2. Leave overnight at room temperature.

3. Decant the excess water and resuspend the beads in an equal volume of distilled H_2O.

4. Mix and place the slurry in a stoppered glass container.

5. Add an equal volume of 99% (vol/vol) hydrazine hydrate.

6. Tightly stopper the container, then mix the contents by vigorous swirling.

7. Place in a 50°C water bath for 8–10 h.

8. Agitate by hand-swirling every 30 min.

9. Remove the container from the water bath, and *under a chemical hood*, very carefully remove the stopper.

    Note: Use precautions to avoid inhaling the fumes that escape.

10. Place the gel in a sintered glass funnel and repeatedly wash with 0.1 M NaCl until the filtrate becomes clear when tested. The filtrate is considered clear when it no longer gives a pale-violet coloration when mixed with the test reagent (3% (vol/vol) sodium 2,4,6-tri-nitrobenzene sulfate in distilled H_2O).

    a. To test the filtrate, add 200 μL of the test reagent to 5 mL of the filtrate and mix.

    b. Add 1 mL of saturated $Na_2B_4O_7$, mix again, and read after 5 min.

11. Once the filtrate is clear, the gel can be stored cold in 0.1 M NaCl.

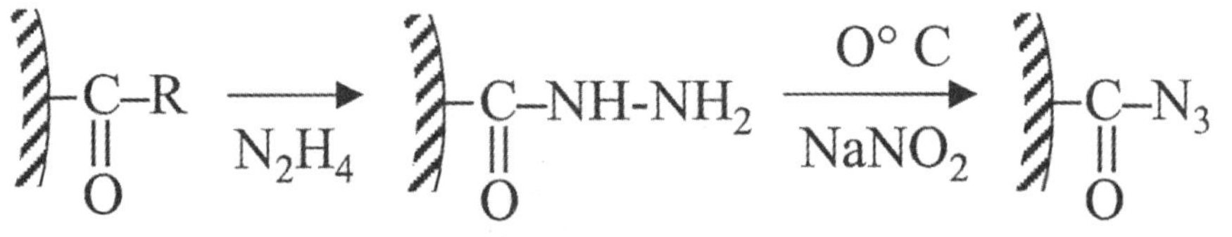

Figure 15. Hydrazine activation of carboxamide groups on polyacrylamide supports to produce activated acyl azides.

SILICA AND GLASS

Silica possesses many of the properties essential for a good affinity support, and in a gel form, it is the most abundantly used packing material for high-performance liquid chromatography (HPLC). Together with glass, this family of supports has enabled affinity and immunoaffinity chromatography to be performed at high flow rates and reasonably high pressures. Table 3 lists some of the commercially available sources of pre-activated silica and glass supports.

Silica

This material has gained popularity as a support for high performance affinity applications, especially in its microparticulate form. Silica is chemically modified to produce silane groups to which reactive side chains are attached. The most commonly used side chains are epoxy or thiol groups that can provide attachment sites for the ligand to be immobilized. A disadvantage of silica is its loss of stability in extreme acidic and alkaline conditions. However, it is possible to chemically modify the surface with a thin coating of hydrophilic polymers to form both a protective coat and a spacer arm for ligand attachment. Diol-bonded silica has superior ligand-binding capacities and is commercially available from Supelco, although the material can easily be made by the technique developed by Walters and colleagues (47,51).

Glass

Glass is a versatile material and can be used in many forms, especially solid beads, controlled-pore beads, glass-coated plastic beads, and glass capillaries (the latter being employed to make micro-affinity and immunoaffinity columns). Glass beads and glass-coated plastic beads, suitable for affinity chromatography, are available in sizes ranging from 0.25–1 mm in diameter (Sigma Chemical). Unfortunately, solid glass beads are not available with reactive side chains, which therefore have to be chemically added to the bead surface by the investigator. This process involves silanization of the bead surface followed by chemical modification of the silanol groups to attach the reactive side chains. The advantage of these beads is that high flow rates can be achieved without increasing the running pressures, and the beads are excellent for constructing preparative columns.

Capillary Tubes

The need for ultra-microanalytical techniques has led to the development of capillaries as small affinity and immunoaffinity columns. The interior surfaces of fused silica capillaries can easily be modified to design suitable attachment for a variety of ligands.

Derivatization of Silica and Glass

Both silica and glass can easily be derivatized with a number of reactive side groups once the surface has been treated with silanol (Figure 16). Carbonylation with 1,1′-carbonyldiimidazole can be performed on both glass (5,35) and silica supports.

Table 3. Commercially Available Silica and Glass Supports

Support	Material	Vendor
Carboxymethyl, CGP	Glass beads	Sigma
Lichrospher	Silica	Merck KGaA
Ultrasphere®	Silica	Beckman Coulter
Spherisorb®	Silica	Waters Corporation
ZORBAX®	Silica	Agilent Technologies
HiPAC	Silica	ChromatoChem
Protein-PAK™	Silica	Waters Corporation
Ultraffinity™-EP	Silica	Beckman Coulter

Following activation, reactive CDI groups can easily be attached to hydroxyl groups on the support by the technique described below (Figure 17). The introduction of thiol groups is perhaps one of the most useful side-group additions for immunoaffinity chromatography. This creates attachment points for enzyme-digested antibodies and other free thiol-containing ligands. Thiol addition is especially useful in small columns or capillaries where these groups can then be used to attach FAb fragments of reactive antibodies directly to the silica or glass surface by the formation of a disulfide bond (36,37). A popular and useful reactive group is N-hydroxysuccinimide ester. This reactive group can be attached to modified silica and glass following attachment of succinyl chloride to silanized surfaces (5). Successful attachment of this group coats the silica or glass with a succinimide ester that will readily react with primary amino groups on protein ligands.

PROTOCOLS

Preparation of Diol-Bonded Silica

1. Place 1 g of silica in a 25-mL round-bottom flask.
2. Add 15 mL of 0.1 M $C_2H_3NaO_2$, pH 5.5.
3. Sonicate for 5 min to remove trapped air.
4. Dry by applying a vacuum for 5 min.

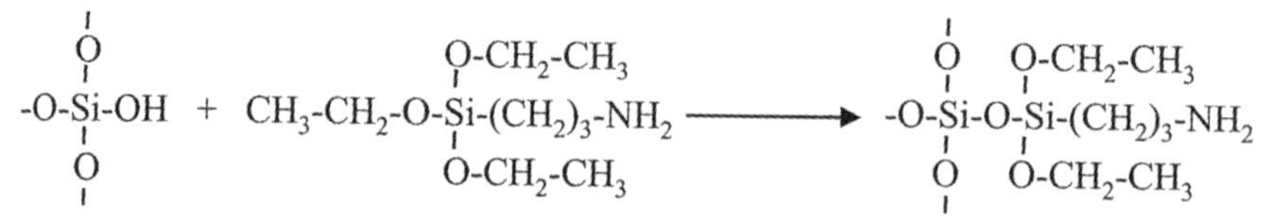

Figure 16. Derivatization of silica or glass by treatment with silanol.

5. Resuspend the silica in 740 µL of 10% (3-glycidoxypropyl) trimethoxysilane by gentle shaking.
6. Stopper the flask.
7. Incubate for 5 h at 90°C in a shaking water bath.
8. Pour the silanized silica into a Buchner funnel.
9. Wash in 200 mL of distilled H_2O.
10. Repeat the washing with 200 mL of distilled H_2O, adjusted to pH 3.0 with 0.5 M H_2SO_4.
11. Place the silica into a round-bottom flask that contains 50 mL of 1 M H_2SO_4, pH 3.0.
12. Reflux for 1 h.
13. Recover the silica and wash in 200 mL of distilled H_2O.
14. Wash 3 times by sedimentation in 50 mL of methanol.
15. Dry overnight at 70°C and store in a desiccator.

Silanization of Silica or Glass Beads

1. Place 1 g of glass beads into a 50-mL graduated cylinder, filled with distilled H_2O.
2. Shake and allow the beads to sediment.
3. Replace the H_2O and repeat steps 1 and 2.
4. Repeat this washing process five times to remove manufacturing impurities.
5. Following the final wash, recover the beads, blot them on clean filter paper, and place them in 10 mL of 1 *N* HCl.
6. Place the beads in a low energy sonicator.
7. Sonicate for 25 min.

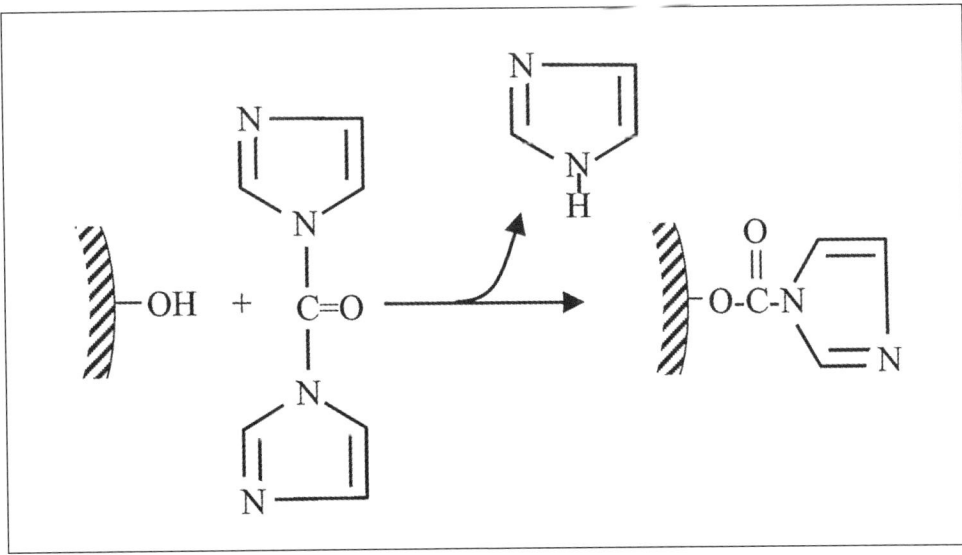

Figure 17. Polysaccharide activation by covalent attachment of 1,1′-carbonyldiimidazole.

8. Wash the beads (as described in steps 1 and 2) in 100 mL of 1 N HCl until the supernatant is clear.
9. Place the clean beads in a 25-mL round-bottom flask (Figure 18).
10. Reflux the beads in 25 mL of 1 N HNO$_3$ for 30 min with slow boiling.
11. Recover and blot-dry the beads on filter paper.
12. Place the beads in a clean round-bottom flask.
13. Add 10 mL of a 10% (vol/vol) solution of 3-aminopropyltriethoxysilane dissolved in toluene.
14. Reflux for 16 h with slow boiling.
15. Recover the beads and wash five times (for 5 min by sedimentation) in 95% methanol.
16. Place the beads in a clean 25-mL round-bottom flask.
17. Reflux for 10 min in 95% methanol.
18. Remove the supernatant and replace with 50 mL of fresh 95% methanol.
19. Reflux for a further 10 min to remove the excess silane.
20. Wash the beads three times (for 5 min by sedimentation) in distilled H$_2$O and blot dry.

CDI Activation of Silica and Glass

1. Add 1 g of silanized beads to 10 mL of dioxane in a 15-mL capped glass tube.
2. Add 100 mg of N,N'-carbonyldiimidazole and mix well.
3. Place on an overhead mixer.
4. Incubate for 6–8 h at room temperature.
5. Recover the beads.
6. Wash extensively (10–20 times for 5 min each by sedimentation) in dioxane.
7. Air-dry and use immediately for protein coupling.

CDI-Derivatized Capillaries

1. Pipet 500 nL of 10% (vol/vol) aqueous 3-aminopropyltriethoxysilane onto a 10-cm square piece of Parafilm.
2. Allow the capillary to take up the liquid by capillary action (Figure 19A).
3. Seal the ends of the capillary with heat-resistant tape.
4. Incubate at 100°C for 60 min.
5. Remove the tape.
6. Attach one end of the capillary to a 3-mL syringe. A capillary purge adaptor available from Supelco can easily complete this attachment (Figure 19B).
7. Purge the capillary using air.
8. Repeat steps 1–7 four times.
9. Following the last purge, fill the capillary with 500 nL of 10 mM HCl.
10. Reseal the ends of the capillary with heat-resistant tape.
11. Incubate at 100°C for 60 min.

12. Purge the capillary with 1 mL of distilled H_2O.
13. Flush the capillary with 200 µL of dioxane.
14. Pipet 500 nL of a 1-mg/mL solution of CDI dissolved in dioxane onto the Parafilm sheet.
15. Allow the capillary to take up the liquid by capillary action.
16. Seal the ends of the capillary.
17. Incubate at room temperature for 6–8 h.
18. Remove the tape and flush the capillary ten times with dioxane.
19. Air-dry and use immediately for protein coupling.

Thiol-Derivatized Capillaries

1. Follow steps 1–8 of the procedure for CDI-derivatized capillaries to successfully derivatize silanol groups onto the interior surface of the capillary.
2. Pipet 500 nL of a 1-mg/mL solution of sulfosuccinimidyl 4-(*N*-maleimidomethyl)-cyclohexane-1-carboxylate dissolved in 50 mM $Na_2B_4O_7$, pH 7.6, onto a Parafilm sheet.
3. Allow the capillary to take up the liquid by capillary action.
4. Seal the ends of the capillary with heat-resistant tape.
5. Incubate at 30°C for 60 min.
6. Flush the capillary three times with 50 mM $Na_2B_4O_7$.
7. Seal the ends of the capillary.
8. Store at room temperature.
   Note: Avoid direct contact with overhead fluorescent lights.

Attachment of N-Succinimide Esters to Glass Beads

1. Dry 1 g of silanized beads by applying a vacuum.
2. Place in a 25-mL round-bottom flask.
3. Add 5 mL of 10% (vol/vol) succinyl chloride dissolved in anhydrous chloroform.
4. Add 5 mL of 10% (vol/vol) triethylamine dissolved in anhydrous chloroform.
5. Gently reflux the mixture for 30 min.
6. Recover the beads and wash 10 times in chloroform for 5 min by sedimentation.
7. Spread the beads on Whatman 3MM filter paper and dry at 100°C.
8. Resuspend the beads in 20 mL of anhydrous dioxane containing 0.1 M *N*-hydroxysuccinimide plus 0.1 M *N,N'*,-dicyclohexylcarbodiimide.
9. Add a small magnetic stirrer bar and stir gently for 2 h at room temperature.
10. Recover the beads and wash twice in dioxane.
11. Wash twice for 5 min each in 100% methanol.
12. Re-wash in anhydrous dioxane.
13. Resuspend the beads in dioxane.
14. Store in a light-tight container for up to 4 months.

MEMBRANES

Membrane supports are becoming popular because of the ease with which they can be used and because of their commercial availability (8). They are constructed of a variety of different materials ranging from nitrocellulose to silica-plastic hybrids and glass. One of the main attractions of membrane supports is the extensive pore

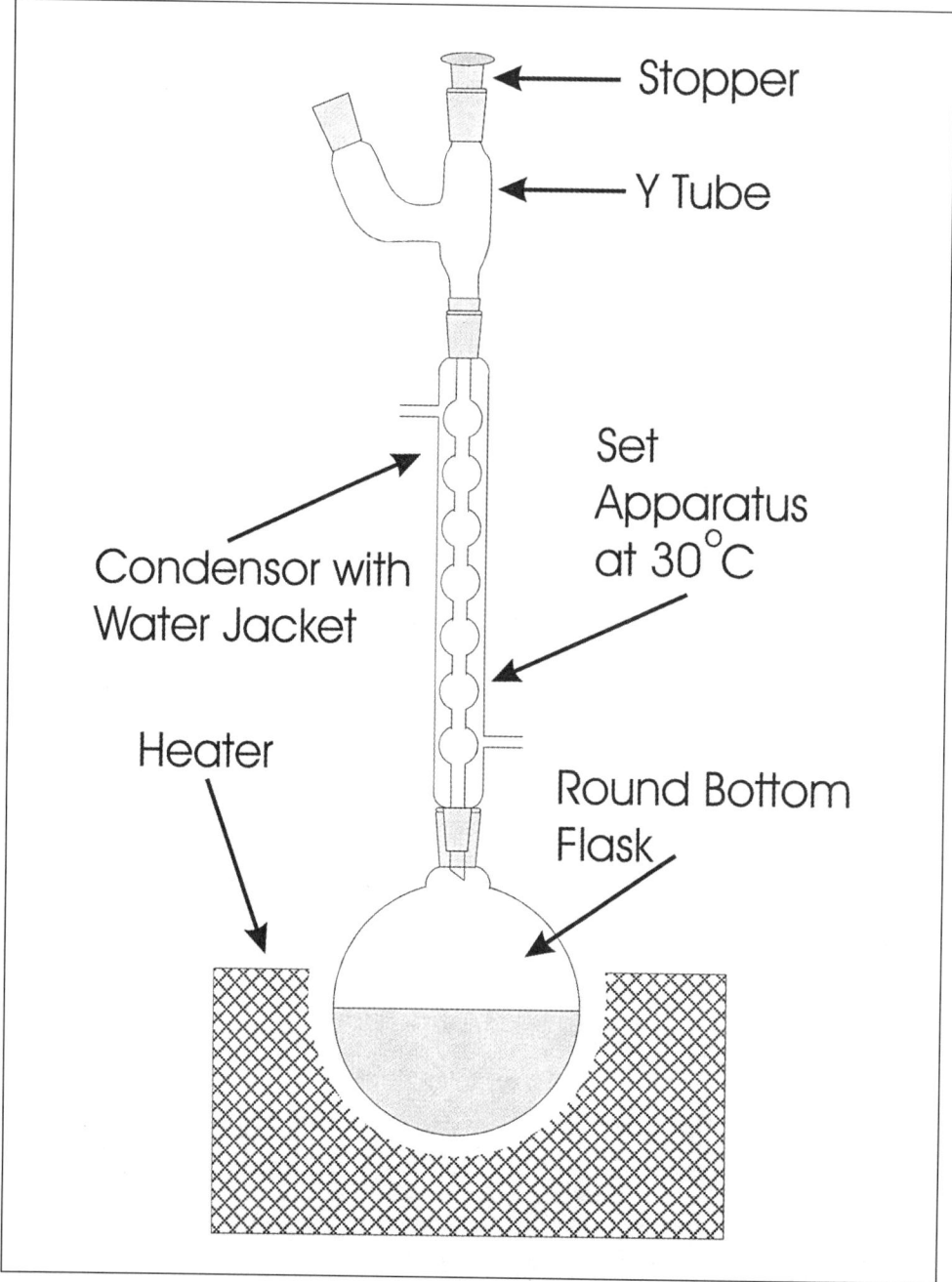

Figure 18. Reflux apparatus used in the derivatization of silica or glass beads.

system that runs through the membrane matrix (Figure 20). These pores allow for high flow rates and an extensive area for ligand immobilization. The technique utilizes immobilized ligands on stacked sheets or disks to provide an immense cross-sectional area within a reasonably thin (ca. 100-μm) package. In this configuration, the membranes overcome two of the major problems associated with conventional supports, namely diffusion effects and mass transfers limitations. Affinity membranes exhibit faster processing times with higher throughput, mainly due to the reduction in diffusion limitations (30). It is also claimed that membrane systems can process nonclarified solutions, which makes them increasingly attractive for the selective isolation of analytes from large volumes of media, such as feedstock in bio-processing

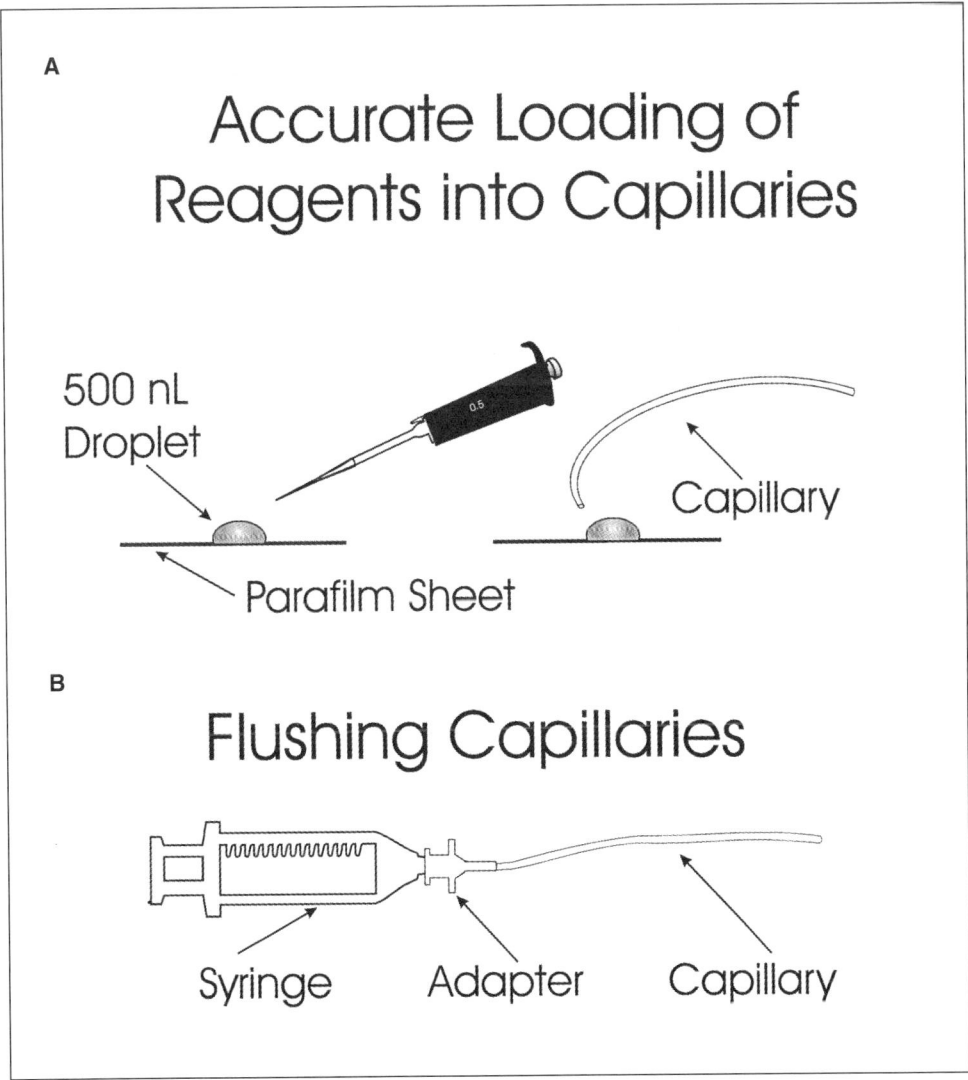

Figure 19. Addition of derivatization agents to a capillary column. A 500-nL microdroplet of the appropriate derivatization agent is placed on a Parafilm sheet (step A, left). The microdroplet is taken into the column by capillary action (step A, right). Supelco provides a capillary adapter and screw-type plunger adapter (not shown) for capillary flushing (step B).

or pharmaceutical plants. The relative open architecture of the membranes allows greater utilization of the immobilized ligand, which, in certain cases where expensive ligands are used, can be advantageous.

Despite the relative advantages of membrane separations, there are a few pitfalls that prevent this technology from becoming the ideal procedure for both affinity and immunoaffinity separations. In many instances, the membrane pores can become clogged, especially when processing viscous fluids such as plasma. Serum proteins can easily become attached to membrane surfaces, thus forming a film over the surface and potentially causing a masking of the membrane pores. This situation can also occur when analytes are being isolated from crude bacterial growth medium. Another area where potential pitfalls arise is in the design of the membrane housing. The containers often have "dead spaces" that can seriously interfere with the efficiency of the separation process. These spaces arise from the channels required to deliver the materials to be analyzed to the membrane itself. The overall effect of such spaces is to increase the size of the housing and diminish the efficacy of the fractionation or final concentration of the analyte during the recovery phase (26).

Techniques for immobilizing ligands to the surfaces of pre-activated membranes are the same as those described for other supports (see Chapter 4). However, in membrane systems, passing the reagents through the membrane performs the immobilization chemistries. Washing is also achieved by passing solutions through the membrane. Several membranes are commercially available with activated groups already attached to their surfaces. Such membranes usually possess succinimide esters and oxirane groups for attachment of ligands by primary amino groups. Membranes are also available with affinity ligands such as protein A, protein G, mimetic dyes, and metal chelators already attached to their surfaces. The ACTI-DISK cartridges available from Whatman are a convenient way to isolate proteins. The membranes are based on a microporous plastic-silica composite sheet sealed in a polypropylene housing. The membranes are available with recombinant protein A covalently attached to the membrane. Other ligands such as protein G and heparin are also avail-

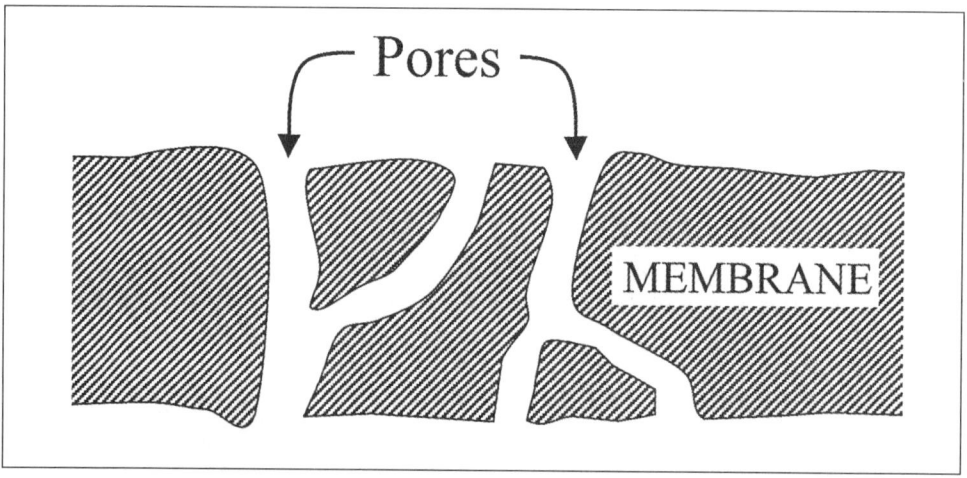

Figure 20. The effect of spacer arm additions to immobilized ligands. The immobilization of a ligand (triangle) on a solid support often results in a steric hindrance that prevents the binding of the complementary analyte (A). The addition of an appropriate spacer arm can restore the binding affinity (B). However, inappropriate selection of spacer arm molecules can lead to nonspecific binding of contaminating proteins or even exhibit intra-molecular bonding that can lead to steric hindrance (C).

able, as well as membranes with glutaraldehyde and hydrazide reactive groups for custom ligand attachment. A series of recent publications have focused on nonchromatographic affinity systems using more robust and less expensive group-specific ligands such as metal chelates and reactive dyes (1,9,18,41). Table 4 lists some of the commercially available sources of pre-activated membrane supports.

OTHER SURFACES

In recent years, a number of new developments have taken place in liquid chromatography, especially the introduction of new phases for chromatographic separations and the introduction of microseparation techniques. In affinity separation technology, two new supports are available, including the POROS support introduced by PE Applied Biosystems and the HyperD support introduced by BioSepra. These exciting new supports have already shown great potential and have contributed greatly to improvements in the speed and efficiency of affinity separations (especially the POROS system).

Other supports have also been reported, although many are purely experimental. The use of zirconium-1 and fluorocarbon-based supports are worthy of mention as these both appear to be potentially useful for the attachment of a number of ligands, including biomimetic dyes.

POROS

PerSeptive Biosystems has developed a unique polymeric particle called POROS, made of stable cross-linked polystyrene, for perfusion chromatography (13). These particles contain two distinct classes of pores: "through pores" that penetrate the particle and are large enough to allow some convective flow through the particles, and smaller diffusive pores that line the through pores and provide a large, adsorptive surface area. During a chromatographic run above a critical flow rate, convection dominates over diffusion in the through pores, allowing efficient access to the high surface-area diffusive pores of the particle. The combination of rapid perfusive transport within the through pores and the ultra-short (<1-μm) diffusion path-lengths of the diffusive pores make both the resolution and capacity essentially independent of flow rate. The chemical and physical nature of the support is exceptionally robust. It is resistant to extremes of pH (1.0–14.0) and can withstand 0.5 M NaOH and 1 M HCl for cleaning and sterilizing purposes. The maximum pressure limit is 3000 psi, which is higher than the HyperD support.

According to Afeyan et al. (1), the high throughput of perfusive particle columns allows great flexibility. Upon process applications, for example, a small column can be used in rapid cycles to produce a given throughput or amount of material in a specified time. The authors compared the effect of cycling mass throughput using soft gel, HPLC, and perfusion chromatography. They claim that these studies showed that the POROS particles allowed the separations to be performed at 10–100 times the flow rate of the other systems without loss in resolution or binding capacity.

POROS material is available in two forms: 50-μm beads for low-pressure operation, and 20-μm particles for analytical-scale high-performance affinity chromatography. Matrices are available as aldehyde-, epoxy-, hydroxyl-, and hydrazide-activated beads, as well as ready-to-use adsorbents with protein A, protein G, metal chelate, and heparin immobilized as ligands.

Table 4. Commercial Sources of Pre-Activated Membrane Supports

Support	Material	Vendor
ACTI-DISK®	Silica/PVC	Whatman
Bioran-M	Glass fiber	Schott Glass Technologies
Immobilon-AV	PVDF	Millipore
MAC Protein A	Glass fiber	Millipore
MAC Protein G	Glass fiber	Millipore
Sartobind™ dye	Cellulose	Sartorius North America
Sartobind metal	Cellulose	Sartorius North America
UltraBind™	Polyethersulfone	Pall Gelman Laboratory

HyperD

Composite particles called HyperD beads are commercially produced by BioSepra. The HyperD particles are based on 35-μm diameter beads composed of a polystyrene composite shell filled with a soft gel. This combination gives the beads rigidity (in the plastic shell) plus the chromatographic performance of the soft gel. The manufacturers claim that the beads are able to withstand pressures of up to 2000 psi with linear velocities of greater than 5000 cm/h. The unique design of the beads suggests a theoretically high capacity since the bead pores are filled with the gel. This allows binding sites to be located throughout the entire pore, not just confined to the surface. The size and composition of the beads will also allow for high throughput, which, together with the improved capacity, should provide a marked improvement over conventional supports. HyperD beads are commercially available with protein A, heparin, or basilene blue dye as affinity ligands. At present, this support does not appear to be available in activated forms suitable for the attachment of custom ligands.

Zirconium Dioxide

Zirconium dioxide (zirconia) has been used as a chromatographic support, but usually in the capacity of an ion exchanger (10). The presence of hard Lewis acids on the surface of the support can be modified with ethylenediamine-N,N'-tetramethylphosphonic acid (EDTPA), a phosphonate analog of EDTA, to make the support more biocompatible (39). Such modifications have now advanced porous zirconia into the growing list of potential affinity supports. Zirconia particles exhibit a high chemical stability over a broad pH range (1.0–14.0) and have physical strength equal or superior to that of silica (50). Wirth and Hearn (50) have described the first applications of the material for affinity chromatography. They modified the surface of the particles by silanization and immobilized concanavalin A (con A) and iminodiacetic acid-Cu(II) for the adsorption of horseradish peroxidase and myoglobin, respectively. The surface can be modified in many ways; dynamically, by addition of competing Lewis bases to the mobile phase, or permanently, by covering its surface with polymers or by depositing carbon (44). Affinity columns containing immobilized con A have been described, wherein the ligand was made by in situ adsorption onto a microparticulate zirconium dioxide support and used to resolve monosaccharides (14). Enzymes (trypsin, chymotrypsin, papain, and pepsin) have been covalently attached to the surface of porous zirconia. The immobilization efficiency was found to be comparable to that of silica, although the solid-phase enzymes exhibited different characteristics (16).

Fluorocarbon-Based Supports

Perfluorocarbons hold the potential to be excellent affinity supports as their structure makes them chemically and biologically inert. The material is highly dense, and unaffected by both aqueous and organic solvents, as well as being thermally stable. Perfluorocarbons possess highly hydrophobic surfaces that can be wetted by water only in the presence of fluorosurfactants. The application of stabilized modifications of these materials to affinity separations has been described by Sii and Sadana (42), who coated the solid support with a perfluoro-octanoyl-polyvinyl alcohol layer prior to coupling a ligand. Lowe et al. (22) described the potential for incorporating triazine biomimetic dyes into the support for the isolation of proteins of potential pharmaceutical interest. The idea was further developed and resulted in a new support system composed of a perfluorocarbon emulsion (24,25). This support was made by homogenizing a saturated solution of perfluorocarbon oil with a polymeric fluorosurfactant, based on polyvinyl alcohol, that had previously been derivatized with the ligand of interest. The final emulsion was cross-linked in situ and used in a fluidized bed for the continuous extraction of HSA from blood plasma.

SPACER ARMS

Problems often arise when small molecules such as enzyme substrates, receptor-binding ligands, pharmaceutical drugs, or chemical antigens are used as the immobilized capture molecule in either affinity or immunoaffinity procedures. In such cases, steric hindrance often occurs between the ligand support and the molecule to be isolated, thus causing reduced reactivity or nonreactivity between the analyte and the affinity ligand (Figure 21). This further leads to either inefficient or an absence of recovery of the analyte. The generally accepted approach to this situation is to employ an accessory molecule or spacer arm to extend the ligand away from the support surface. Such molecules already attached to suitable affinity supports are commercially available and are listed in Table 5 and the structures chosen examples shown in Figure 22. The addition of a spacer arm ensures that the analyte will have adequate access to the immobilized ligand. Spacer arms are attached either directly by modifying the derivatized support surface or by attaching a preconstructed spacer arm to activated groups on the support surface by a secondary reaction. In either case, the ligand is chemically attached to the reactive group on the end of the spacer arm that is furthest from the support.

A spacer molecule may also be employed if the ligand is immobilized through a site near to the analyte-binding site that may interfere with protein binding. The length of the spacer arm is crucial and must be determined empirically, with six to eight methylene groups most often being successful (27). The spacer should be hydrophilic and not itself bind proteins, either because of its hydrophobicity or its charge. Some activation chemistries will automatically insert a suitable spacer between the support and the ligand (for example, bis-oxirane as illustrated in Figure

49

22). Supports with spacer arms already attached are commercially available.

The simplest technique for introducing a reasonable spacer arm onto the surface of any support is to directly derivatize a long-chain carbon onto the surface itself. This can be performed with relative ease. Many supports can be modified to accept a long carbon chain, located between the support and the reactive end-group. A convenient reaction that enables the attachment of any molecular structure, containing a free carboxyl group, to amino groups on the support surface is the use of carbodiimides (40). During the coupling stage, the carbodiimide reacts with carboxyl groups to form *o*-cylisourea at pH 4.0–5.0. The activated carboxyl group then condenses with an amino group on the silanized support surface to form a peptide bond plus the urea, which needs to be removed from the reaction mixture. Two water-soluble carbodiimides are available; 1-ethyl-3-(3-ethylamino-propyl)carbodiimide (EDC) or 1-cyclohexyl-3-(2-morpholinoethyl)carbodiimide metho-p-toluene sulfonate (CMC). Both of these agents have been successfully used to couple spacer arms to any support that has free amino groups.

Table 5. Commercially Available Support Spacer Arm Molecules

Hexamethylenediamine
Aminocaproic acid
N-hydroxysuccinimide
Arylsulfonate
Bis-oxirane
3,3′-diaminodipropylamine
Long-chain hydrazine biotin

PROTOCOLS

Attachment of Diaminodipropylamine Spacer Chains

1. Dissolve 20% (wt/vol) diaminodipropylamine in acetone.
2. Place in a 15-mL stoppered glass tube.

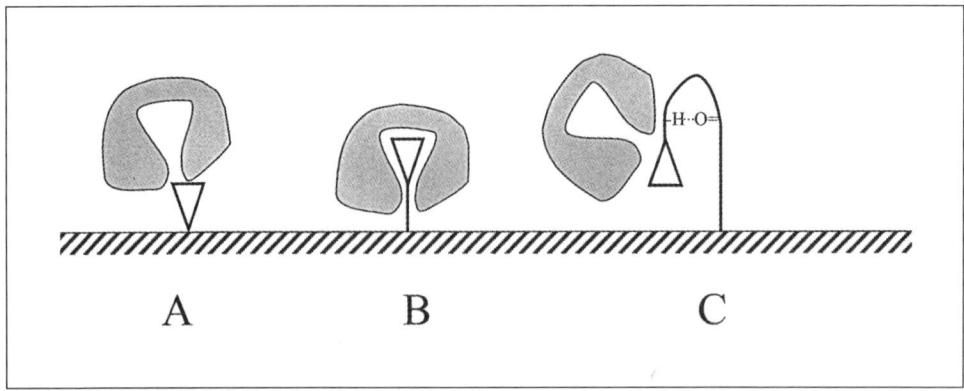

Figure 21. The effect of spacer arm additions to immobilized ligands. The immobilization of a ligand (triangle) on a solid support often results in a steric hindrance that prevents the binding of the complementary analyte (A). The addition of an appropriate spacer arm can restore the binding affinity (B). However, inappropriate selection of spacer arm molecules can lead to nonspecific binding of contaminating proteins or even exhibit intra-molecular bonding that can lead to steric hindrance (C).

3. Transfer 1 g of freshly prepared CDI-activated support to a sintered glass filter.

4. Dry by applying a vacuum.

5. Add the dried support to the diaminodipropylamine.

6. Place on an overhead stirrer.

7. Continuously stir the mixture for 4–6 h at room temperature.

8. Return the support to the sintered glass filter.

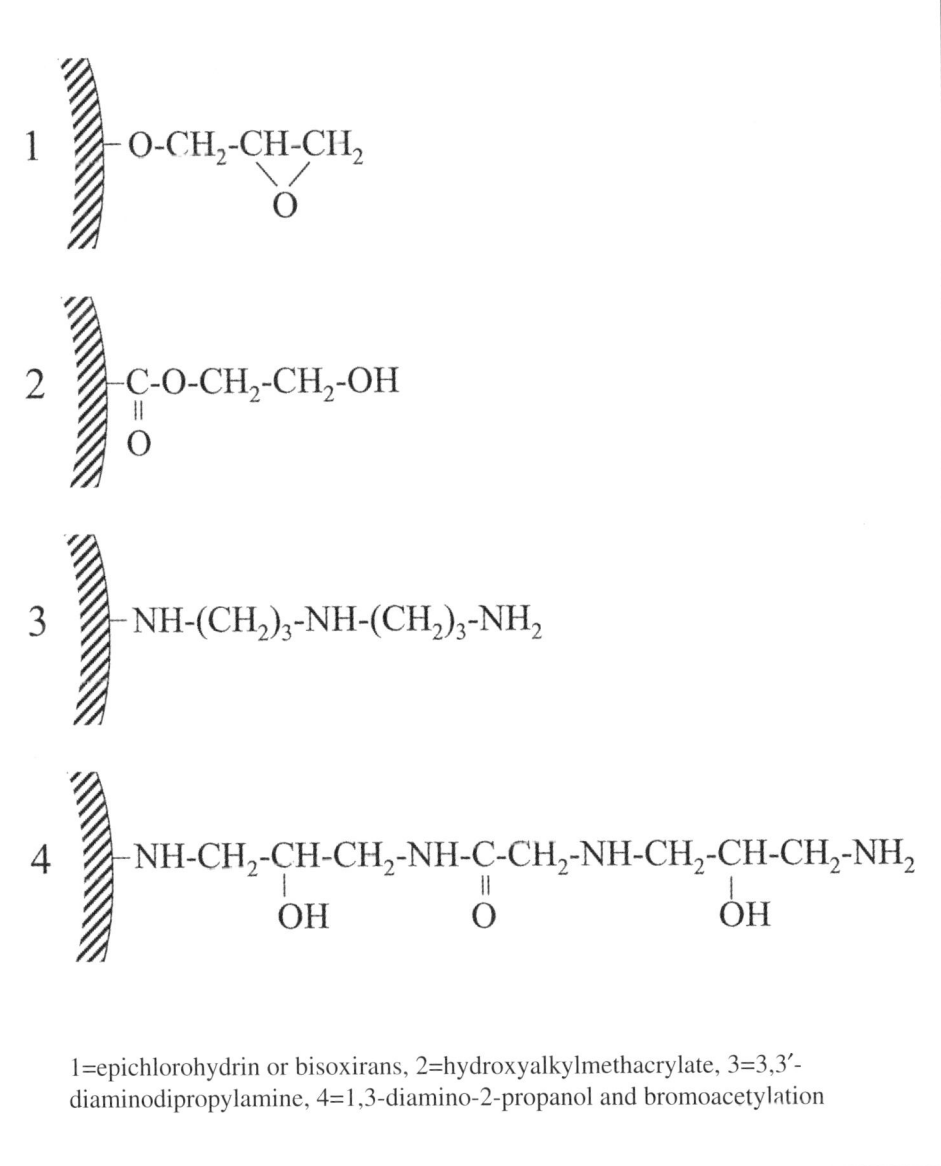

1=epichlorohydrin or bisoxirans, 2=hydroxyalkylmethacrylate, 3=3,3′-diaminodipropylamine, 4=1,3-diamino-2-propanol and bromoacetylation

Figure 22. Examples of spacers used in affinity chromatography. These spacers are formed by reaction of various supports with: 1 = Chloromethyloxirane; 2 = hydroxy bis-oxiranes, 2 = hydroxy alkyl methacrylate, 3 = 3,3′-diaminodipopylamine, and 4 = 1,3-diamino-2-propanol and bromoacetylation.

9. Wash five times in acetone.

10. Wash five times in 50:50 acetone:distilled H_2O.

11. Wash five times in distilled H_2O.

12. Wash twice in 1 M NaCl.

13. Wash in distilled H_2O.

14. Resuspend the support in double-distilled H_2O.

15. Store at room temperature.

Carbodiimide Attachment

1. Place 1 g of the beads in a sintered glass filter.

2. Wash in 200 mL of 0.5 M NaCl.

3. Dissolve 1 mg of the spacer in 1 mL of 0.1 M sodium phosphate, pH 4.5.

4. Dissolve 10 mg of carbodiimide in 1 mL of distilled H_2O.

5. Adjust the carbodiimide solution to pH 4.5 with either 0.1 N HCl or 0.1 N NaOH (depending on the initial pH of the solution).

6. Place the beads in a 15-mL capped tube.

7. Add 1 mL of spacer plus 1 mL of carbodiimide.

8. Cap the tube tightly.

9. Incubate overnight on an overhead mixer at room temperature.

10. Return the beads to the glass filter.

11. Wash with 500 mL of 0.1 M sodium phosphate, pH 7.2.

12. Store at 4°C.

REFERENCES

1. **Afeyan, N.B., S.P. Fulton and F.E. Regnier.** 1991. High-throughput chromatography using perfusive supports. LC·GC Internat. *4*:14.
2. **Arsenis, C. and D.B. McCormick.** 1966. Purification of flavin mononucleotide-dependent enzymes by column chromatography on flavin phosphate cellulose compounds. J. Biol. Chem. *241*:330-334.
3. **Avrameas, S. and T. Ternynck.** 1969. The cross-linking of proteins with glutaraldehyde and its use for the preparation of immunoadsorbents. Immunochemistry *6*:53-66.
4. **Axen, R., J. Porath and S. Ernback.** 1967. Chemical coupling of peptides and proteins to polysaccharides by means of cyanogen halides. Nature *214*:1302-1304.
5. **Babashak, J.V. and T.M. Phillips.** 1989. Isolation of a specific membrane protein by immunoaffinity chromatography with biotinylated antibodies immobilized on avidin-coated glass beads. J. Chromatogr. *476*:187-194.
6. **Bethall, G.S., J.S. Ayeers, W.S. Hancock and M.T.W. Hearn.** 1979. A novel method of activation of cross-linked agaroses with 1,1′-carbonyldiimidazole which gives a matrix for affinity chromatography devoid of additional charged groups. J. Biol. Chem. *254*:2572.
7. **Boeden, H.F., K. Pommerening, M. Becker, C. Rupprich, M. Holtzhauer, F. Loth, R. Muller and D. Bertram.** 1991. Bead cellulose derivatives as supports for immobilization and chromatographic purification of proteins. J. Chromatogr. *552*:389-414.
8. **Brandt, S., R.A. Goffe, S.B. Kessler, J.L. O'Connor and S.E. Zale.** 1988. BioTechnology *6*:779.
9. **Champluvier, B. and M.R. Kula.** 1992. Sequential membrane-based purification of proteins, applying the concept of multidimensional liquid chromatography (MDLC). Bioseparation *2*:343-351.
10. **Clausen, A.M. and P.W. Carr.** 1998. Chromatographic characterization of phosphonate analog EDTA-modified zirconia support for biochromatographic applications. Anal. Chem. *70*:378-385.
11. **Cuatrecasas, P.** 1971. Affinity chromatography. Ann. Rev. Biochem. *40*:259-278.
12. **Cuatrecasas, P., M. Wilchek and S. Ernback.** 1968. Selective enzyme purification by affinity chro-

matography. Proc. Natl. Acad. Sci. USA *61*:636-643.

13. **Fulton, S.P., M. Meys, L. Várady, R. Jansen and N.B. Afeyan.** 1991. Antibody quantitation in seconds using affinity perfusion chromatography. BioTechniques *11*:226-231.

14. **Glavanovich, M.H. and P.W. Carr.** 1994. Easily regenerable affinity chromatographic zirconia-based support with concanavalin A as a model ligand. Anal. Chem. *66*:2584-2589.

15. **Hammerl, P., A. Hartl and J. Thalhamer.** 1993. Particulate nitrocellulose as a solid phase for protein immobilization in immuno-affinity chromatography. J. Immunol. Methods *165*:59-66.

16. **Huckel, M., H.J. Wirth and M.T.W. Hearn.** 1996. Porous zirconia: a new support material for enzyme immobilization. J. Biochem. Biophys. Methods *31*:165-179.

17. **Inman, J.K. and H.M. Dintzis.** 1969. The derivatization of cross-linked polyacrylamide beads. Controlled introduction of functional groups for the preparation of special-purpose, biochemical adsorbents. Biochemistry *8*:4074-4082.

18. **Iwata, H., K. Saito, S. Furasaki, T. Sugo and J. Okamoto.** 1991. Adsorption characteristics of an immobilized metal affinity membrane. Biotechnol. Prog. *7*:412-418.

19. **Kennedy, J.F.** 1974. Chemically reactive derivatives of polysaccharides. Adv. Carbohydr. Chem. Biochem. *29*:305-405.

20. **Kohn, J. and M. Wilchek.** 1982. A new approach (cyano-transfer) for cyanogen bromide activation of Sepharose at neutral pH, which yields activated resins, free of interfering nitrogen derivatives. Biochem. Biophys. Res. Comm. *107*:878-884.

21. **Lerman, L.S.** 1953. A biochemical specific method for enzyme isolation. Proc. Natl. Acad. Sci. USA *39*:232-236.

22. **Lowe, C.R., S.J. Burton, N. Burton, D.J. Stewart, D.R. Purvis, I. Pitfield and S. Eapen.** 1990. New developments in affinity chromatography. J. Mol. Recognit. *3*:117-122.

23. **Madden, J.K. and D. Thom.** 1982. Properties and interaction of polysaccharides–their use as chromatographic supports, p. 113. *In* T.C.J. Gribnau, J. Visser and R.J.F. Nivard (Eds.), Affinity Chromatography and Related Techniques. Elsevier, Amsterdam.

24. **McCreath, G.E., H.A. Chase, D.R. Purvis and C.R. Lowe.** 1992. Novel affinity separations based on perfluorocarbon emulsions. Use of a perfluorocarbon affinity emulsion for the purification of human serum albumin from blood plasma in a fluidised bed. J. Chromatogr. *597*:189-196.

25. **McCreath, G.E., H.A. Chase, D.R. Purvis and C.R. Lowe.** 1993. Novel affinity separations based on perfluorocarbon emulsions. Development of a perfluorocarbon emulsion reactor for continuous affinity separations and its application in the purification of human serum albumin from blood plasma. J. Chromatogr. *629*:201-213.

26. **Mohan, S.B. and A. Lyddiatt.** 1997. Recent developments in affinity separations, p. 10-11. *In* P. Matejtschuk (Ed.), Affinity Separations: A Practical Approach. IRL Press, Oxford.

27. **Mohr, P. and K. Pommerening.** 1985. General considerations of the adsorption and elution step, p. 89-95. *In* J. Cazes (Ed.), Affinity Chromatography: Practical and Theoretical Aspects. Marcel Dekker, New York,

28. **Monsigny, M., M. Cornet, R. Tixier, M. Corgier and P. Girot.** 1978. p. 57. *In* R. Epton (Ed.), Chromatography of Synthetic and Biopolymers. Vol. 1, Column Packings, G.P.C., G.F., and Gradient Elution. Ellis Horwood, Chichester.

29. **Mosbach, K. and L. Andersson.** 1977. Magnetic ferrofluids for the preparation of magnetic polymers and their application in affinity chromatography. Nature *270*:259-261.

30. **Najarian, S. and B.J. Bellhouse.** 1997. Effect of oscillatory flow on the performance of a novel cross-flow affinity membrane device. Biotechnol. Prog. *13*:113-116.

31. **Nawrocki, J., M.P. Rigney, A. McCormick and P.W. Carr.** 1993. Chemistry of zirconia and its use in chromatography. J. Chromatogr. *657*:229-282.

32. **Neame, P.J. and J. Parikh.** 1982. Sepharose-immobilized triazine dyes as adsorbents for human lymphoblastoid interferon purification. Appl. Biochem. Biotechnol. *7*:295-305.

33. **Ngo, T.T.** 1986. BioTechnology *4*:134.

34. **Nilsson, K. and K. Mosbach.** 1980. p-Toluenesulfonyl chloride as an activating agent of agarose for the preparation of immobilized affinity ligands and proteins. Eur. J. Biochem. *112*:397-402.

35. **Phillips, T.M.** 1989. High-performance immunoaffinity chromatography. Adv. Chromatogr. *29*:133-174.

36. **Phillips, T.M.** 1998. Determination of in-situ tissue neuropeptides by capillary electrophoresis. Anal. Chim. Acta *372*:209-218.

37. **Phillips, T.M. and J.M. Krum.** 1998. Recycling immunoaffinity chromatography for multiple analyte analysis in biological samples. J. Chromatogr. B Biomed. Sci. Appl. *11*:55-63.

38. **Porath, J.** 1979. Molecular sieving and non-ionic adsorption in polysaccharide gels. Biochem. Soc. Trans. *7*:1197-1222.

39. **Rigney, M.P., E.F. Funkenbusch and P.W. Carr.** 1990. Physical and chemical characterization of microporous zirconia. J. Chromatogr. *499*:291-304.

40. **Scouten, W.H.** 1981. Synthetic methods, p. 42. *In* Affinity Chromatography. John Wiley & Sons, New York.

41. **Serifica, G.C., J. Pimbley and G. Belfort.** 1994. Protein fractionation using fast flow immobilized metal

chelate affinity membranes. Biotechnol. Bioeng. *43*:21-36.

42. **Sii, D. and A. Sadana.** 1991. Bioseparation using affinity techniques. J. Biotechnol. *19*:83-98.

43. **Strop, P., F. Mikes and Z. Chytilova.** 1978. Hydrophobic interaction chromatography of proteins and peptides on Spheron 300. J. Chromatogr. *156*:239-254.

44. **Sun, L., A.V. McCormick and P.W. Carr.** 1994. Study of the irreversible adsorption of proteins on polybutadiene-coated zirconia. J. Chromatogr. *658*:465-473.

45. **Turkova, J., K. Blaha, M. Malanikova, D. Vancurova, F. Svec and J. Kalal.** 1978. Methacrylate gels with epoxide groups as supports for immobilization of enzymes in pH range 3-12. Biochim. Biophys. Acta *524*:162-169.

46. **Turkova, J., L. Petkov, J. Sajdok, J. Kas and M.J. Benes.** 1990. Carbohydrates as a tool for oriented immobilization of antigens and antibodies. J. Chromatogr. *500*:585-593.

47. **Walters, R.R.** 1982. High-performance affinity chromatography: pore-size effects. J. Chromatogr. *249*:19-28.

48. **Wellicky, N. and H.H. Weetall.** 1965. The chemistry and use of cellulose derivatives for the study of biological systems. Immunochemistry *2*:293-322

49. **Weston, R.D. and S. Avrameas.** 1971. Proteins coupled to polyacrylamide beads using glutaraldehyde. Biochem. Biophys. Res. Commun. *45*:1574-1580.

50. **Wirth, H.J. and M.T.W. Hearn.** 1993. High-performance liquid chromatography of amino acids, peptides and proteins. CXXX. Modified porous zirconia as sorbents in affinity chromatography. J. Chromatogr. *646*:143-151.

51. **Wu, D. and R.R. Walters.** 1988. Protein immobilization on silica supports: a ligand density study. J. Chromatogr. *458*:169-174.

52. **Zeichner, M. and R. Stern.** 1977. Resolution of ribonucleic acids by Sepharose 4B column chromatography. Biochemistry *16*:1378-1382.

3 Factors Influencing Affinity and Immunoaffinity Separation

Like all analytical techniques, affinity and immunoaffinity procedures rely on a number of external factors for their success. All too often, investigators proceed with a technique without considering all of the factors involved. This is certainly true for affinity separations. Often, failure of the procedure is caused by an external or environmental factor not examined during the planning of the study. One must always remember that a number of different parameters dictate the success of any separation process. Elements like heat, ionic balance, analyte and/or ligand density, and the style in which the separation is performed (in batches vs. column chromatography) are influential in optimizing any separation. This chapter gives an overview of the major factors affecting affinity and immunoaffinity separations.

ENVIRONMENTAL CONDITIONS

The successful affinity or immunoaffinity separation of any analyte is dependent upon ligand and support stability, pH, temperature, ionic strength of the buffer, and incubation time. Additionally, when column chromatography is employed, other factors such as column size, flow rate, and pressure also influence the efficiency of the separation process (31,38,82,94). Temperature and pH are perhaps the most influential parameters and can become extremely important when biological ligands, such as enzymes, antibodies, binding proteins, and receptors, are employed. Likewise, affinity separations utilizing ionic interactions as their primary binding parameter (hydrophobic interaction chromatography) are subject to both pH and temperature effects.

Temperature Effects

Many biological systems require a reasonably narrow temperature range for optimal activity. The most noted examples of this temperature-dependence are enzymes, many of which are also pH-dependent. In such cases, immobilized enzymes or enzyme substrates have defined environmental requirements that must be met before optimal activity can be achieved. High temperature can inactivate (72) or even

Affinity and Immunoaffinity Purification Techniques
Terry M. Phillips and Benjamin F. Dickens
© 2000 Eaton Publishing, Natick, MA

destroy the structural integrity of enzymes (55,83), thus often rendering them useless as affinity ligands. In addition, all proteins are susceptible to heat degradation and many non-enzymatic protein ligands can become irreversibly damaged if subjected to prolonged exposure to high temperatures. Temperature is also known to affect hydrophobic interactions (the higher the temperature, the stronger the hydrophobic bonding), thus requiring more severe elution parameters. Under chromatographic conditions, this results in increased retention time with associated changes in the chromatogram. For example, at 15°C, it has been reported that affinity elution of cytochrome *c* produces a sharp peak, whereas peak broadening occurs as the temperature increases up to 35°C (30). This situation is reversed when the temperature is increased to 50°C, when the peak again becomes sharp.

The temperature used in immunoaffinity separations greatly affects both the efficiency of the separation and the effective life of the immunoaffinity support. Temperatures above 60°C can denature antibodies, especially immobilized ones, thus seriously impairing and shortening their activity and the life of the activated support (61,64). Even room temperature (generally considered to be 21°–24°C), which does not cause immediate structural damage to proteinaceous ligands, can cause immobilized antibodies to denature over time. These temperature effects also hold true for a number of other temperature-sensitive ligands used in both affinity and immunoaffinity separation techniques. Work in our laboratory has shown that the most effective temperature for immunoaffinity separations is 4°C (61,64). We have also found that this temperature can easily be achieved by performing the separations in a cold room or, in the case of column chromatography, by pumping ice-cold water through a water jacket surrounding the column. Table 1 summarizes our findings on the effects of temperature on immunoaffinity supports.

pH Effects

The pH of any environment is dependent on the relative concentration of free hydrogen ions in solution. This concentration makes up the pH scale and is very important in regulating interactions between analytes and their respective ligands. The effects of excess free hydrogen and, to a certain extent, free hydroxyl ions on analyte-ligand interactions are well-documented. These agents have been widely used in the recovery of bound analytes from affinity columns (49,52,61,64,76). Early studies on the binding of enzymes to affinity supports examined the effects of pH on the ability of analytes to bind to immobilized ligands (46,47). The authors of these studies demonstrated that the interactions between the enzyme lactate dehydrogenase and the affinity support was unaffected below pH 8.0. However, the amount of enzyme bound to the immobilized ligand decreased as the pH increased above this level.

While the pH of the system is a significant factor in affinity separation, it can become especially important in the affinity separation of proteins when using mimetic dyes as ligands. It has been shown that plasma proteins exhibit different affinities for immobilized Cibacron Blue 3GA at different pH levels; while the majority of the proteins bound at pH 5.0, only 40% bound when the pH was raised to 9.0 (2). Tejedor et al. (81) have shown that the recovery of phosphofructokinase from rat erythrocytes by affinity partitioning to Cibacron Blue F3GA covalently bound to polyethylene glycol is highly pH-dependent. As the pH was increased from 6.0 to 8.0, the recovery of phosphofructokinase increased. pH can also interfere with the effectiveness of metal chelates as affinity ligands for isolating serum proteins. Patwardhan

Table 1. Effect of Temperature on Immunoaffinity Separations

Temperature	Effect
4°C	Well-defined peak resolution
	Column life extended to 200–250 repetitive elution cycles
10°C	Less defined but reasonable peak resolution
	Column life between 150–200 repetitive elution cycles
15°C	Slight peak broadening, but still acceptable definition
	Column life approximately 100 repetitive elution cycles
20°C	Notable peak broadening
	Column life between 50–100 repetitive elution cycles
Above 25°C	Poorly defined peaks with diffuse peak broadening
	Column life between 10–20 repetitive elution cycles

and Ataai (60) demonstrated that pH can affect the binding of bovine serum albumin (BSA) to a copper ion column. They showed that pH-dependent binding of the protein to the chelator could introduce binding artifacts that seriously affected the estimated saturation capacity of the affinity support.

One of the most effective ways to introduce the desired level of hydrogen and hydroxyl ions into separation techniques is with buffers. The majority of published procedures contain formulas for buffers that require some degree of manipulation to achieve the recommended pH. In such cases, it is easy for the investigator to introduce an excess of either hydrogen or hydroxyl ions during adjustment of the final buffer pH. The presence of such an excess of these ions can seriously affect the efficiency of the separation, causing inhibition or partial inhibition of the bonding between the analyte and its ligand, with generally unsatisfactory results. This situation may cause complete failure to recover the desired analyte, or perhaps just incomplete analyte recovery. While most biological systems are optimal within the range of pH 6.5 to 8.0, some enzymes require more severe conditions for optimal activity. Therefore, it is important to optimize conditions for optimal activity. Table 2 illustrates this point by giving the optimal conditions for a number of enzymes.

Incubation Times and Flow Rate

An important parameter often overlooked in both affinity and immunoaffinity separations is the time required for an analyte to locate and interact with the immobilized ligand. This period can become extremely important, especially when the interaction is performed in columns under high-flow conditions. The binding efficiency of any ligand is dependent upon the affinity it displays for the analyte, which in turn dictates the minimal time required for such reactions to occur. Therefore, it is important to consider the relative selectivity and strength of the interaction between the analyte and the immobilized ligand before selecting these parameters for affinity chromatographic separations. In general, weaker affinity interaction between ligand and analyte requires longer incubation times to allow the analyte to adequately bind to the immobilized ligand. This period may range from seconds to several minutes, depend-

Table 2. pH and Temperature Requirements for Selected Enzymes

Enzyme	pH range	Temperature (°C)
Collagenase type I and II	7.0–8.0	37
Lactate dehydrogenase	7.0	25
Papain	6.0–7.0	25
Pepsin	2.0–4.0	37
Trypsin	8.5–8.8	25
Hyaluronidase	4.5–6.0	37
β-glucosidase	5.0–5.3	37
β-1,4-galactosidase	6.0–6.5	37
Endoprotease	8.0–8.5	25
α-chymotrypsin	7.0	25
Carboxypeptidase Y	5.5–6.5	25
Alkaline phosphatase	8.0–10.5	25
Subtilisin A	7.0–8.0	25

ing on the strength of the bonding required for an adequate separation. Additionally, affinity column chromatography involving weak avidity between ligand and analyte may require slower flow rates or longer columns than higher avidity combinations to achieve longer retention times. The incubation time becomes an important factor when immunoaffinity separations are used. Immobilized antibodies, although relatively rapid in their binding kinetics, cannot always compete with the speed at which the analyte passes through the column. This situation is common when extremely short, narrow-bore columns are employed.

Flow rate dictates the time allowed for the analyte to reside in the body of the column; the faster the flow rate, the quicker the analyte passes through the column, thus leaving less time for the analyte to successfully interact with the ligand. Ideal rates for affinity chromatography have been reported as optimal between 0.4 and 1.5 mL per min (15,61,64,86). High flow rates can be responsible for inadequate separation between the nonreactive primary peak and the elution of the specific analyte. In our laboratory, we have found that flow rates around 1 mL per min are optimal for most immunoaffinity separations (Figure 1).

Ligand Orientation

The majority of the procedures described in Chapter 2 bind ligands in a random manner wherever primary amino groups are available. Although this orientation may be adequate for many ligands (especially those that do not require specific molecular interactions), it lessens the efficiency of specialized ligands such as enzymes, receptors, and antibodies. These specialized molecules possess defined molecular recognition sites to which their corresponding substrates or antigens must bind. With these specialized ligands, random binding to primary amino groups causes poor orientation that often results in partial or complete loss of binding activity. Fortunately, the incorporation of a suitable spacer arm can sometimes overcome this problem, but generally, specialized attachment must be employed. The correct orientation of antibodies can easily be achieved by binding to bacterial protein-coated or avidin supports as described in Chapter 5. Antibodies can also be correctly orientated by oxidative modification of the carbohydrates present in their tail structure (58,87). Orientation of enzymes, receptors, and recombinant proteins is not so easy and one must rely on attachment by carboxyl groups, biotinylation (5), or by the use of recombinant molecules containing structurally placed attachment tags (95).

Pressure Effects

In chromatographic isolation techniques, pressure is produced as a combination of flow and resistance to flow caused by the density of the packing medium. Many affinity supports cannot withstand high operational pressures created during chromatographic separation, and therefore collapse or compress. Similarly, membrane affinity systems can foul or even rupture when excessive pressure buildup occurs, thus causing failure of the procedure. The sheer forces generated by high pressure can cause stripping of the immobilized ligand, leading to contamination of the isolated analyte with fragments of denatured ligand, ligand-analyte complexes, or a mixture of both entities (64). Figure 2 illustrates the effects of increasing pressure on ligand stability. In these experiments, protein A leakage from CDI-activated glass beads increased dramatically as the running pressure in the chromatography system increased. Less than 5% loss was detected at system pressures of 200–500 psi, but the loss increased to 25% when the pressure was raised to 700 psi. At 1000 psi, over 40%

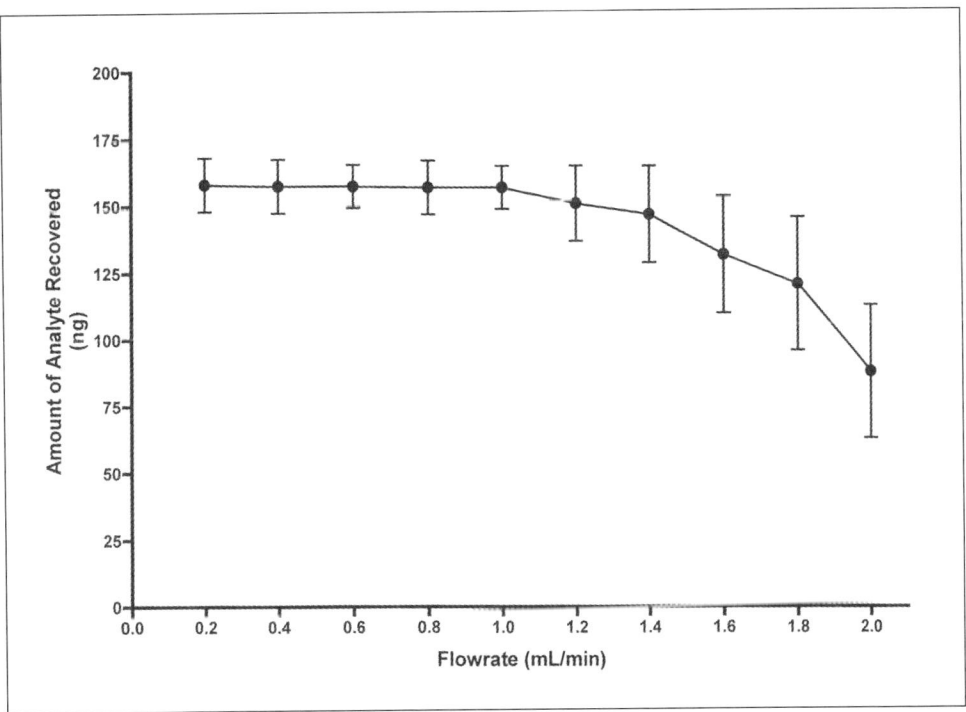

Figure 1. Effect of column flow rate on immunoaffinity separation. While optimal flow rates have to be determined empirically, experience suggests that flow rates near 1.0 mL/min are usually optimal.

of the ligand was stripped from the support. Therefore, it appears important to carefully monitor operational pressures during any affinity or immunoaffinity separation to ensure that system pressures do not become greater than 400 psi.

Effect of Column Dimensions

The modern trend in affinity chromatography is to employ extremely short, narrow-bore columns. Walters has written an excellent account of how to construct such columns suitable for affinity chromatography (88). Other procedures have subsequently been described for making affinity columns and cartridges in a variety of different containers ranging from short, biocompatible plastic cartridges to small-bore, in-line microcolumns (12,59,65). However, the use of such columns imparts their own set of potential problems. Short columns mean reduced retention or incubation times as well as the amount of ligand-activated support available for capturing the analyte of interest. Investigators must therefore carefully plan separations such that the amount of ligand is suitable for the sample size and analyte concentration. It would be impossible for a short, narrow-bore column to adequately isolate high concentrations of analyte from a large sample. In such a case, the analyte would easily saturate the available ligand. Narrow-bore columns can present an additional problem, namely increased running pressures, especially if packed with dense or microparticulate packing. Often, it is necessary to run such columns at relatively slow flow rates and under controlled pressure. Although this may increase the time required to perform the separation, it could greatly increase its efficiency, and the investigator has to decide which parameter (speed or increased recovery) is the most advantageous to the studies at hand.

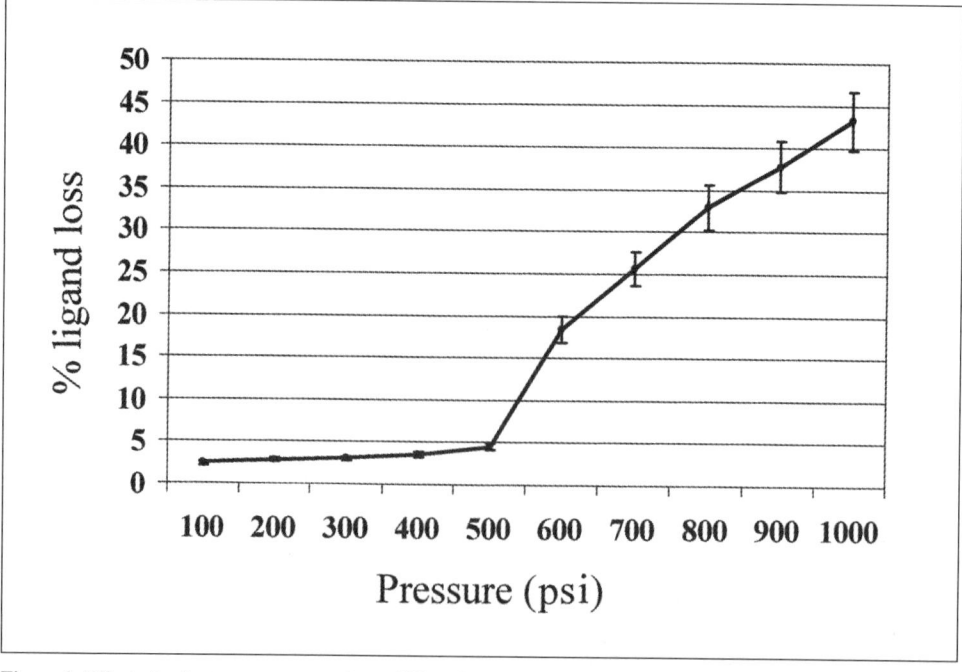

Figure 2. Effect of column pressure on the stability of Protein A immobilized on glass beads.

SUPPORT AND LIGAND STABILITY

One of the major areas in affinity separation where problems arise is in ligand and/or support stability. It is always necessary to check that the support, whether "laboratory-made" or commercially obtained, contains adequate active attachment sites. Additionally, one of the most important factors to consider is ligand density and its effect on the separation process. This factor alone will "make or break" any separation and there are a number of approaches to defining the amount of ligand present in an affinity support.

Support Characteristics

The first area in which potential problems with support characteristics can arise is with the activated support itself. The integrity of the support structure and the number of potential attachment sites all contribute to the success of the separation (34,53,56). Microscopic examination of the support can often quickly provide evidence on its structural integrity. Such examination can be particularly useful when soft supports such as agarose and dextran are used. However, such visual or physical examination of glass and plastic beads is also advantageous as these supports often become damaged due to pressure, chemical attack, or extreme physical handling. It is also wise to check the activity and availability of the support's reactive attachment groups prior to ligand attachment. There are several approaches to resolving this issue. Chemical procedures are available for checking the available concentration of amine, thiol, and imidazole groups in affinity supports prior to attachment of a ligand.

Ligand Density Effects

Insufficient ligand will obviously lead to incomplete recovery of the desired analyte. However, other factors such as the distribution of the ligand and its density have to be considered. Early studies on ligand concentration demonstrated that a sigmoidal curve could be produced when differing concentrations of immobilized ligand were used to isolate reactive enzymes (28). This work suggested that the concentration of bound ligand could affect the ability to recover analytes by increasing the strength of the binding interactions. Similar effects have been reported by Labrou and Clonis (42), who determined that an optimal concentration of 2.2 µM of mimetic dye per gram of support was required before efficient isolation and recovery of bovine heart L-lactate dehydrogenase could be achieved. Ligand density also exerts a direct effect on metal chelate affinity supports. Wirth et al. (92) have reported that ligand density had differing effects according to the type of support. They found that the specific binding capacity per unit area increased continuously with ligand density when nonporous supports were used, but peaked at approximately 50% of the maximum ligand coverage when porous supports were used. It has also been shown that retention of recombinant human growth factor depends on copper ligand density and that regulation of such can be used to optimize both adsorption and recovery of the analyte (44).

Antibody density on immunoaffinity supports greatly affects the efficiency of the separation. Sparse ligand can become saturated by excessive analyte, while densely populated supports can exhibit steric hindrance with an associated loss of activity. This situation can be exaggerated by the molecular size of the analyte itself. Antigen-binding capacities of polyclonal antibodies immobilized onto silica-based supports have been shown to exhibit a decrease in specific activity as ligand density increases. This situation changes according to analyte molecular mass, suggesting a relation between immobilized antibody density and epitope availability (89).

Determination of Ligand Concentration

It is important to have a reasonably accurate estimate of the amount of ligand available prior to performing any affinity separation. The concentration of bound ligand can be measured by a number of different techniques, the simplest of which is measurement of the ligand concentration pre- and post-immobilization. The next easiest approach is to directly measure the ligand in situ. However, this is not always as simple as it sounds and in the majority of cases, the ligand has to be stripped from the support prior to analysis. Many supports, such as agarose, can be solubilized, therefore negating the need to strip the ligand. Finally, more sophisticated approaches (elemental analysis, mass spectrometry) can be applied for determining ligand concentration.

Pre- and Post-Immobilization Analysis

Pre- and post-immobilization analysis simply requires a reasonably accurate measurement of the ligand prior to attachment to the affinity support, followed by another analysis after the procedure and extensive washing (the amount of bound ligand being derived from the difference in the two readings). Although this is not the most accurate approach, in most cases, the results obtained will be sufficient. Measurement of the ligand can be performed by a variety of techniques including spectrophotometry, fluorimetry, and immunoassay. The latter is of great use when analyzing ligands that do not possess an easily identifiable spectral absorbance or fluorescent groups. Additionally, immunoassays can be modified to be extremely sensitive, especially when a chemiluminescence detection system is applied (16,80). When available, further analysis can be performed using sophisticated techniques, but these are usually not required except in cases when quantitative measurements of affinity are being performed (3,91).

Direct Spectroscopy

Direct spectroscopy can be applied when the ligand exhibits an absorbance at wavelengths above 250 nm. In such cases, it is often possible to measure the bound ligand in situ. It is important to thoroughly wash the support prior to analysis and to suspend the support in an optically clear solution. Additionally, prior knowledge of the absorption spectra of the ligand is essential. It is also important to make all of the spectrophotometric measurements at two wavelengths; the first reading is required to measure the ligand itself, while the second is made at a wavelength of approximately 600 nm to subtract the background absorption caused by the support matrix itself.

Solubilization of Activated Supports

Only a few supports can be fully solubilized, the majority of which belong in the natural material category. The most amenable to such treatment is agarose.

Agarose beads or gels can be reduced to a semiliquid state by heating to approximately 75°C with or without the addition of other chemicals. However, the addition of acid, alkali, or oxidizing agents can greatly enhance the solubilization process. Following partial or complete solubilization of the support, quantitative analysis of the ligand can be achieved. It is also possible to hydrolyze the bonds holding the ligand in place, thus releasing the bound ligand for further analysis. Complete hydrolysis is often destructive not only to the support itself but also to the ligand. Under such conditions, the recovered ligand is suitable for analysis by biochemical or chemical analysis, such as saccharide content of a polysaccharide ligand or amino acid analysis for proteins and peptides.

Labeled Ligand Techniques

A more complicated procedure for assessing ligand density is the use of labeled ligands. A simple approach is to employ radiolabeled ligands and to count the activity of the support pre- and post-attachment. This can often be the most accurate approach, providing that suitable activity counts can be obtained and correlated to known amounts of free, labeled ligand. For investigators that do not wish to use radioactivity, the ligand can be labeled with a fluorochrome and the measurements made by fluorimetry. In either case, it is important to ensure that the label does not interfere with groups used for attachment of the ligand to the support. Unfortunately, this is often hard, due to the requirements of both the label and the support. Many labels utilize primary amino groups for their reactive sites; the same sites are used to attach ligands to many supports. In such cases, the investigator has to determine all available reactive sites on the ligand and plan to use different chemistries for the label and the attachment. This can be achieved by using amino groups for the label and carboxyl or thiol groups for attachment. An advantage of using labeled ligands as indicators of ligand content is that not only can the support itself be analyzed for bound ligand, but these measurements can be checked by pre- and post-attachment analysis.

Ligand Leakage

One of the major problems associated with affinity chromatography is loss or leakage of the immobilized ligand. This can be due to inadequate attachment to the activated support or through chemical and biological denaturation of the ligand. Protein ligands, for instance, are susceptible to protease activity, while antibodies and receptors are often damaged by chaotropic agents during elution. The chemistry used to attach the ligand and the degree of attachment play important roles (monovalently attached ligands being more susceptible to breakdown than polyvalent ones). Time and storage of the support is another important issue that can seriously impair efficiency. It should be remembered that all affinity supports, especially immunoaffinity supports, denature over time (31). Degradation can also be caused by bacterial contamination of the medium in which the support is stored. This is especially true of the soft natural supports such as dextran and agarose. Therefore, care should be exercised to ensure that correct storage conditions are used and efficiency checks are performed on supports that have been stored for greater than 1 month.

Fuglistaller (21) tested eight different protein A columns for ligand leakage under different pH elution conditions. He found that considerable leakage (up to 88 parts per million [ppm] [wt/wt]) occurred at pH 4.0, the point at which immunoglobulins elute. Similar studies were performed by Godfrey et al. (24), who investigated the effects of salt concentration, pH, and the presence of antibody analyte on ligand leakage, using stirred-tank and flow-through systems. Ligand leakage in the absence of analyte was lowest when a flow-through mode was employed. In the presence of the analyte, residual protein A contamination was lowest when an adsorption/washing buffer (150 mM NaCl and 1 M glycine) was used. Two supports, Prosep A "high capacity" (Bioprocessing Ltd.) and RepliGen (Repligen Corp.), were found to give the most acceptable results, demonstrating approximately 12 ppm contaminating ligand per 12.8 and 4.3 mg of purified antibody.

Immunoaffinity supports are highly susceptible to ligand leakage and a number of studies have attempted to circumvent this problem. Our laboratory generally employs multiple attachment-point immobilization through the biotin-avidin system (see Chapter 5) and controlled temperature for both storage and operation. However, the addition of protease inhibitors (see *Prevention of Analyte Degradation* in this chapter) is also useful and can prolong the life of protein ligand supports. We have employed a solution of water-soluble carbodiimides to cross-link antibodies to protein A-coated glass beads to reduce ligand leakage during elution (64). Likewise, Sisson and Castor (79) reported increased efficiency in immunoaffinity separations when the antibodies were cross-linked to protein A by dimethyl pimelimidate. Goldberg et al. (25) have introduced an intriguing concept to overcome antibody denaturation and leakage from immunoaffinity supports. They reduced the interchain disulfide bonds and then cross-linked the free thiol groups with a bifunctional reagent, thus changing the four-chain antibody molecule into a single chain one. The investigators claim that such a modification helps to protect the immobilized antibody from reduction by elution agents. However, despite a number of innovative approaches to resolve it, the problem of ligand loss is paramount and one should always check the isolated analyte for possible ligand contamination. A convenient solution to this problem is to screen the isolated product with a ligand-directed immunoassay (6).

PROTOCOLS

Colorimetric Test for Primary Amine Content in Activated Supports

1. Place 10 mg of support in a sintered glass filter.
2. Dry the support by applying a vacuum and place into a glass tube.
3. Resuspend the support in 1 mL of distilled H_2O.
4. Add 1 mL of ninhydrin reagent and heat in a boiling water bath for 15 min.
5. Cool to room temperature and measure the absorbance of the support at 570 nm.
6. Calculate the amine content using a molar extinction coefficient of 8750 for the amine-ninhydrin complex.

2,4,6-Trinitrobenzenesulfonic Acid (TNBS) Determination of Amine Concentration in Activated Supports

1. Suspend 1 mg of support in 1 mL of distilled H_2O.

2. Add 25 μL of 0.03 M TNBS.

3. Mix and place in a quartz cuvette.

4. Allow the mixture to stand for 30 min at room temperature.

5. Prepare a control containing 0.03% (vol/vol) TNBS in 1 mL of borate.

6. Read the absorbance of both the support and the control at 420 nm.

Determination of Thiol Content in Activated Supports

1. Place 10 mg of thiol-activated support in a small Buchner funnel.

2. Wash in 5–10 vol of 0.1 M sodium phosphate containing 1 mM EDTA, pH 8.0.

3. Dry by applying a vacuum.

4. Place the dried support in a 1.5-mL Beckman centrifuge tube.

5. Add 1 mL of 0.1 M sodium phosphate containing 1 mM EDTA, pH 8.0.

6. Add 100 μL of a 4-mg/mL solution of Ellman's reagent [5,5′-dithio-bis(2-nitrobenzoic acid) dissolved in 0.1 M sodium phosphate, 1 mM EDTA, pH 8.0].

7. Place on an overhead mixer and mix at room temperature for 15 min.

8. Centrifuge at $10\,000\times g$ for 5 min.

9. Measure the absorbance at 412 nm.

10. Calculate the thiol content using a molar extinction coefficient of 13 600.

Determination of Imidazole Content in Activated Supports

1. Place 20 mg of CDI-activated support in a sintered glass filter.

2. Dry by applying a vacuum.

3. Transfer to a 15-mL glass tube.

4. Hydrolyze in 2 mL of 0.2 M NaOH for 16–20 h at room temperature.

5. Centrifuge at $2000\times g$ for 5 min.

6. Collect the supernatant and titrate to pH 7.0 with 0.2 M HCl.

7. Adjust the supernatant volume to 1 mL with 0.2 M NaOH (titrated to pH 7.0 with 0.2 M HCl).

8. Add 100 μL of 1 M $NaHCO_3$ to the supernatant.

9. Add 1 mL of 0.06 M p-nitrobenzoyl chloride dissolved in acetone.

10. Shake vigorously for 30 s.

11. Let stand at room temperature for 10 min.

12. Add 2 mL of 1 M NaOH.

13. Incubate at room temperature for 30 min.

14. Read the absorbance at 415 nm.

15. Calculate the imidazole concentration from a standard curve prepared by treating 0–2 μM/mL of imidazole standard by the same technique.

Determination of Bound Ligand Concentration by Difference Analysis Using Direct Spectroscopy

1. Prepare the ligand solution for attachment to the support.
2. Set the spectrophotometer to the specific absorbance wavelength for the ligand.
3. Fill a cuvette with the ligand solution.
4. Measure the absorbance and calculate the ligand concentration (this can easily be done by comparison to a standard curve).
5. Couple the ligand to the support.
6. Collect the solution containing the unbound ligand.
7. Remeasure the absorbance and calculate the new concentration.
8. Subtract the second concentration from the first to measure the amount of bound ligand.

Determination of Bound Ligand Concentration by Difference Analysis Using a Chemiluminescence Immunoassay

1. Prepare a solution containing 1 µg/mL of unlabeled antibody (directed against the ligand) in 0.05 M $NaHCO_3$, pH 9.0.
2. Add 100 µL of antibody solution to every well of a 8-well microplate strip (Nunc–Immuno™ BreakApart™; Nalge Nunc).
3. Incubate overnight at 4°C.
4. Wash the wells three times in 0.01 M sodium phosphate, pH 7.4, to which 0.1% (vol/vol) Tween® 20 has previously been added.
5. Block the electrostatic sites by adding 150 µL of 1% (vol/vol) BSA to each well.
6. Incubate for 3 h at 4°C.
7. Couple the ligand to the support and save 200 µL of ligand for analysis.
8. Collect the ligand solution following immobilization.
9. Add 100 µL of pre-immobilization ligand to wells 1 and 2 of the strip.
10. Add 100 µL of post-immobilization ligand to wells 3 and 4.
11. Place 100 µL of the 0.01 M sodium phosphate/Tween 20 into wells 5 and 6.
12. Place 100 µL of a known concentration of ligand in wells 7 and 8.
13. Cover with Parafilm and incubate the plate overnight at 4°C.
14. Wash the plate five times in 0.01 M sodium phosphate/Tween 20.
15. Make a 1:1000 dilution (in 0.01 M sodium phosphate) of an alkaline phosphatase-labeled antibody directed against the ligand.
16. Add 200 µL of the antibody solution to all of the wells.
17. Incubate for 3 h at room temperature.
18. Wash the wells five times in 0.01 M sodium phosphate/Tween 20.
19. Prepare the chemiluminescence substrate solution (CSPD®) by dissolving 100 µL of enhancer (Tropix) per 1 mL of 0.1 M diethanolamine, 1 mM $MgCl_2$, pH 10.0. Immediately before use, add 17 µL of CSPD and mix well.

20. Add 200 µL of the substrate solution to each well.
21. Incubate for 10–15 min at room temperature.
22. Read in a luminometer at 463 nm.
23. Subtract the post-immobilization measurement from the pre-immobilization measurement to calculate the concentration of bound ligand.

> *Note*: If a luminometer is not available, Tropix makes a camera luminometer (Model No. ICL901) that can be used to obtain a photographic image of the reaction. This image can then be quantified by scanning densitometry. If such an instrument is used, then a standard curve constructed from known amounts of the ligand must be included to calculate ligand concentration.

Determination of Bound Ligand Concentration by Direct Spectrophotometry

1. Place 1 g of activated support in a sintered glass filter.
2. Apply a gentle vacuum and wash the support in 100 mL of distilled H_2O.
3. Dry the support by applying a vacuum.
4. Suspend the support in an optically clear solution of glycerol.
5. Place in a cuvette and read at the appropriate wavelength.
6. Repeat steps 1–5 with an equal amount of unactivated support.

Solubilization of Agarose Supports

1. Place 1 g of derivatized agarose support into a glass tube.
2. Add an equal vol of 0.5 M HCl.
3. Place the tube into a 100°C water bath for 20–30 min or until the support becomes liquid.
4. Cool to 75°C and read the absorbance of the liberated ligand at the appropriate wavelength.

Determination of Bound Ligand Concentration by Use of Fluorochrome-Labeled Ligands

1. Label the ligand to be immobilized (see Appendix II for labeling procedures).
2. Prepare 1 g of support according to any of the procedures given in Chapter 2 or as per the manufacturer's instructions.
3. Add the labeled ligand and perform the immobilization procedure as described in any of the procedures described in Chapter 4 or according to the manufacturer's instructions.
4. Wash the support in 100 mL of 0.01 M sodium phosphate, pH 7.4.
5. Measure the amount of labeled ligand either by direct fluorimetry or by hydrolyzing the support and measuring the liberated ligand by fluorimetry.
6. Calculate the amount of immobilized ligand by comparing the reading to a standard curve constructed by measuring the fluorescence of a series of known fluorescent ligand standards.

RECOVERY OF ISOLATED ANALYTES

Recovery of the ligand-bound material can be achieved by several different techniques. Changes in pH, ionic strength, temperature, or the addition of disruptive (chaotropic) agents are all used to dissociate the ligand-analyte complex and effectively recover bound analytes. While any of these procedures can be generally applied to most affinity separations, there are situations that require a more specialized approach. For instance, lectins require competition elution with a sugar similar to that found on the analyte, while many immobilized receptors require elution with an analog of the analyte.

However, one must always ensure that the elution agent does minimal damage to the structural integrity of the immobilized ligand. Structural damage during elution occurs most frequently when using receptors or antibodies as the immobilized ligand (36). One must always bear in mind that the forces responsible for maintaining the interactive binding between the antibody and its antigen are similar, if not the same, as those that hold the tertiary structure of the antibody in place. Therefore, the dissociation agent must be gentle enough to maintain the antibody structure while enabling dissociation of the complex. A variety of dissociation treatments can be employed, ranging from changes in pH to the introduction of chaotropic agents, thus changing the ionic strength of the medium (49,52,64,76). The effect of different elution buffers on the structural integrity of polyclonal antibodies eluted from an immobilized antigen column was investigated by Narhi et al. (54). They found that antibodies sustain minor structural changes when eluted in 0.1 M glycine, pH 2.9; however, 7 M urea and 50 mM sodium acetate at pH 4.0 caused partial unfolding of the antibody. There was a complete loss of secondary and tertiary structure when the molecules were eluted in 6 M guanidine HCl and 50 mM sodium acetate, pH 4.0. It was found that dialysis against phosphate-buffered saline (PBS) could restore integrity to antibodies eluted in glycine or urea, while the guanidine/HCl-denatured antibodies were refolded by dialysis into 7 M urea, pH 4.0, followed by dialysis into PBS. However, despite the appearance of structure damage, the activity of the refolded antibodies appeared to be unimpaired, as demonstrated by their ability to form antigen-antibody complexes with the appropriate antigen.

pH Dissociation

In affinity chromatography, the most popular technique for dissociating ligand-analyte complexes is through pH manipulation. The standard approach is to lower the pH of the running buffer into the acidic range, with reports of elution pH ranges of 1.0–1.5. However, it must be remembered that some supports, such as silica, are very sensitive to sharp changes in pH (especially the acidic end of the pH range) with prolonged exposure under such conditions causing structural degradation. When acid elution is used, the addition of a gradient of citric acid, formic acid, acetic acid, or a combination of Tris/HCl can all effectively dissociate ligand-analyte complexes (64). In all cases, these agents were added to the running buffer in such a way as to create a linear gradient between pH 7.2 and pH 1.5. Bueno et al. (10) used a discontinuous pH gradient to isolate human IgG from unseparated serum on immobilized L-histidine hollow-fiber membranes. They also report that IgM could be partially separated from IgG by pH gradient elution using the same system. Decreasing linear pH gradients have been used to isolate immunoglobulins on immobilized metal ion affinity supports (8) and to recover antibodies from thiophilic ligand columns (74). IgG subclasses have been recovered from protein A Superose® columns using a unique linear pH gradient created by mixing disodium phosphate with sodium acetate, sodium chloride, and glycine. This system was reported to be able to separate IgG_1 from IgG_2 on the basis of their polysaccharide content (43).

Extreme alkaline conditions have been reported to cause unfolding and denaturation of proteins. Bai et al. have reported that lactate dehydrogenase can become fully unfolded at pH 13.5, but still retained some degree of ordered structure (4). Similarly, at pH 11.5, it has been shown that catalase dissociates from its native tetrameric form to monomers with the complete loss of activity (68). However, despite the potential effects on protein structure and function, alkaline treatment has been used for elution in some affinity techniques. Kamihira et al. (36) have reported recombinant α-amylase could be eluted from monoclonal immunoaffinity columns at pH 12.3–12.5; however, they do conclude that elution could be due to partial denaturation of the antigen, immobilized antibody, or both. Ascending pH gradients have been reported to successfully elute egg yolk antibodies from an Fe_3^+ metal ion affinity column (26) and immobilized antigen columns (41). Our experience using immobilized mammalian antibodies has indicated that alkaline elution has a denaturing effect on the immobilized antibody.

Ionic Strength Alterations

It has been known for some time that increasing the ionic strength of the running buffer can greatly affect the efficiency of affinity chromatography. The potential economic use of this effect and potential approaches have been well described by Robinson et al. (71). They describe the use of neutral salt manipulation to change electrostatic conditions for the successful isolation of analytes from affinity supports. This effect is also true for antibody binding to specific epitopes on immobilized protein ligands. Kamata et al. (35) demonstrated that ionic strength played an important part in determining the efficiency of a number of monoclonal antibodies bound to immobilized bovine β-lactoglobulin. One particular epitope, designated 42–56 (after its amino acid sequence), was shown to be electrostatic interaction-dependent and that the ionic strength of the separation conditions greatly affected the binding of these antibodies to that epitope. Taking this further, Burton and Harding (11) have developed hydrophobic charge induction chromatography. This technique is based on the ability of analytes to bind to uncharged weak acid and base ligands (elution being achieved by altering the pH and ionic conditions, thus weakening the electrostatic potential of the ligand-analyte complex).

Chaotropic Dissociation

Chaotropic dissociation appears to give the most effective elution results without many of the denaturing effects observed when using pH manipulation as the elution technique. There are several salts exhibiting chaotropic characteristics and they follow a scale in their effectiveness as dissociation agents for immobilized ligand-substance complexes (64,86). This scale is as follows:

$$CCl_3COO^- > SCN^- > CF_3COO^- > ClO_4^- > I^- > Cl^-$$

These salts are used in concentrations between 1.5 M and 8.0 M, which have been shown to be effective in the dissociation of high-affinity ligand-substrate complexes. The most widely used chaotropic ion is sodium thiocyanate, which is used in concentrations up to 3 M. It must be remembered that the use of high concentration chaotropic ions will also cause some degree of denaturation to both protein ligands and substrates. Chaotropic salts are effective at high concentrations for dissociating antibody-antigen complexes (9,48) and for elution recovery of receptors from affinity columns (33,62). Chaotropic ions have been used as effective elution agents in a number of different affinity and immunoaffinity separations. Potassium thiocyanate has been reported to be an effective elution agent for hydrophobic interaction chromatography of mixtures of acidic and basic proteins (19). In these studies, it was reported that gradient elution with a combination of two or three salts was superior to the results obtained from single-salt gradients. Potassium thiocyanate has also been used for the elution of bovine seminal plasma proteins from affinity matrices containing active groups analogous to phosphorylcholine (18). Sodium thiocyanate and potassium thiocyanate have been used as effective elution agents in a number of immunoaffinity separations. These agents have been used for the recovery of lymphocyte membrane antigen and complement receptors (63), specific antibodies to filarial antigens (39), and specific IgE antibodies for *Dermatophagoides pteronyssinus* allergens (67). Ammonium thiocyanate has been reported to be useful in the recovery of poliovirus from immobilized antibody columns (85) and in assessing the affinity distribution of polyclonal antibodies directed against malarial parasitic antigens (20).

Sodium chloride is a weak chaotropic ion that can be used as an effective elution agent. Singh and Tiwary (78) have reported that phenylalanyl-tRNA synthetase could be successfully recovered from an immobilized phenylalanine column using 0.8 M NaCl. Likewise, NaCl concentrations ranging from 0.3 M to 1 M have been reported to be effective in dissociating naturally occurring immune complexes (37,50).

Competition Elution

The use of free ligand or ligand analog to displace the adsorbed material from the matrix-immobilized ligand is a highly specific technique for affinity elution that is without risk of denaturing the ligand or its analyte. The disadvantage to this approach is that, in many cases, the simple addition of free ligand to the running buffer may prove to be extremely expensive. Often, extremely high concentrations of free ligand are required to neutralize the strong interaction between the substance of interest and the immobilized ligand. The use of a co-substrate (such as a sugar in lectin affinity chromatography) has proven to be the most efficient and inexpensive form of this type of elution (93). Human factor VIII has been isolated from human plasma by lentil lectin affinity chromatography and recovered using the relatively simple sugar, methyl α-D-mannopyranoside (96). Investigators have also found that the use of other saccharides such as D-glucose, methyl α-D-glucopyranoside, D-mannose, and D-galactose could result in the release of factor VIII, but to a lesser degree. Methyl α-D-mannopyranoside has also been used to release high-mannose-type oligosaccharides (17) and invertase (51) from conconavalin A columns. However, in the latter study, the investigators reported that the efficiency of the process was incubation time-dependent. Sugars can also be used as competition agents in immunoaffinity, especially when polysaccharide antigens are used as the affinity ligand. Anastase et al. (1) reported that, using dextran and functionalized dextran columns, they could distinguish two different anti-dextran antibodies present in human plasma samples.

They recovered these antibodies using a combination of sugar and acid elution (oligo-dextran on the native dextran column and acid on the other).

Substrates or their analogs can successfully be used to recover enzymes from affinity columns. Glatz et al. (23) have shown that reduced glutathione in Tris-HCl buffer can be used to recover glutathione S-transferase from thiopropyl Sepharose. α-Amylase has been successfully recovered from immobilized antibody columns by adding low concentrations of the C-terminal region of the enzyme (40). In a similar way, dipalmitoylphosphatidylcholine has been used as a displacement ligand for the recovery of aldolase B from artificial membrane receptors (45).

Electrophoretic Dissociation

Electrophoretic dissociation has been extensively used to recover separated proteins following gel electrophoretic separation. It has also been reported that electroelution is a mild approach for the recovery of analytes from affinity and immunoaffinity supports (14,90). An advantage of electrophoretic elution is that it can produce high yields of recovered analyte. It can be applied to any irreversible complex, such as recovery of antigens from immobilized antibody supports (73), glycoprotein antigens from a lectin support (70), HSA bound to a mimetic dye support (2), apolipoprotein H (22), and recently, analytes in immunoaffinity CE (66) (see the section on affinity and immunoaffinity capillary electrophoresis in Chapter 10). Techniques for performing this type of elution involve placing the analyte-ligand complex on either a modified flatbed polyacrylamide gel electrophoresis apparatus or a flatbed isoelectric focusing gel (27) to separate the two components of the sample complex. A commercial apparatus suitable for electrophoretically dissociating analyte-ligand complexes called an Elutrap® is available from Schleicher & Schuell and is designed for the isolation and concentration of proteins and nucleic acids by trapping them in an electrical field (32). The trap is formed between a dense matrix membrane and a microporous membrane. The former allows ions and materials less than 5 kDa to pass, while the latter filter prevents unwanted particles from entering the trap. Molecules are electrically driven into the trap, where they remain after the electrical driving force has been removed. The Elutrap is capable of isolating proteins greater than 5 kDa and nucleic acid fragments between 14 and 150 000 base pairs (bp). Enzymatic cleaning of the apparatus can be performed by carrying out a short electrophoretic run using Tris/glycine buffer plus DNase or Pronase, followed by thoroughly washing the entire apparatus in distilled water. The Elutrap can also be autoclaved at 120°C to inactivate traces of remaining enzymes or biological contaminants. This simple apparatus is shown in Figure 3.

Another useful dissociation apparatus is the Electro-Eluter insert available from Bio-Rad Laboratories. This apparatus is a type of tube gel that fits into the miniature vertical gel electrophoresis tank (also available from Bio-Rad). The apparatus is designed to elute separated proteins from polyacrylamide gel slices, with the eluted proteins being captured in a membrane "cup" (Figure 4) (7). This apparatus can be converted to perform electrophoretic elution of ligand-coated supports by packing the support into the tubes prior to elution.

Other Dissociation Agents

In cases where the affinity ligand-analyte interactions are ionic, increasing the ionic strength of the medium can cause dissociation of the complex. The opposite is true for interactions that are hydrophobic in nature (i.e., ligand-protein interactions). Temperature has also been used to dissociate affinity ligand-analyte interactions but this technique has disadvantages, such as degradation of the analyte and affinity matrix. However, it has been shown that a linear temperature gradient could be used to elute a number of enzymes from a 5′-adenosine monophosphate (AMP)-activated support (47). Temperature can also be successfully applied as an aid to other elution procedures (such as ligand concentrations and ionic gradients) for the recovery of both low- and high-affinity analytes from their supports (77). Manipulation of temperature has been reported to be useful in the recovery of weakly bound murine IgG antibodies from protein A columns (84).

The introduction of agents that can change the buffer polarity (such as methanol, dioxane, and ethylene glycol) has also been applied to the recovery of analytes from affinity complexes. Dissociation of antibody-antigen complexes can be achieved by any solution that contains polarity-reducing agents. Ikegawa et al. (29) added 90% (vol/vol) aqueous methanol to an elution buffer to recover equilin and its metabolites. Ethanol in combination with NaCl has been used for the recovery of recombinant proteins expressed in a mammalian cell system (69). While agents such as dioxane and ethylene glycol are not popular as affinity elution agents, their mode of action is noteworthy. They act by reducing the polarity of the solution surrounding the antibody and the antigenic determinant. This reduction neutralizes the hydrophobic attractive forces responsible for holding the antigen in close proximity to the antibody's antigen receptors.

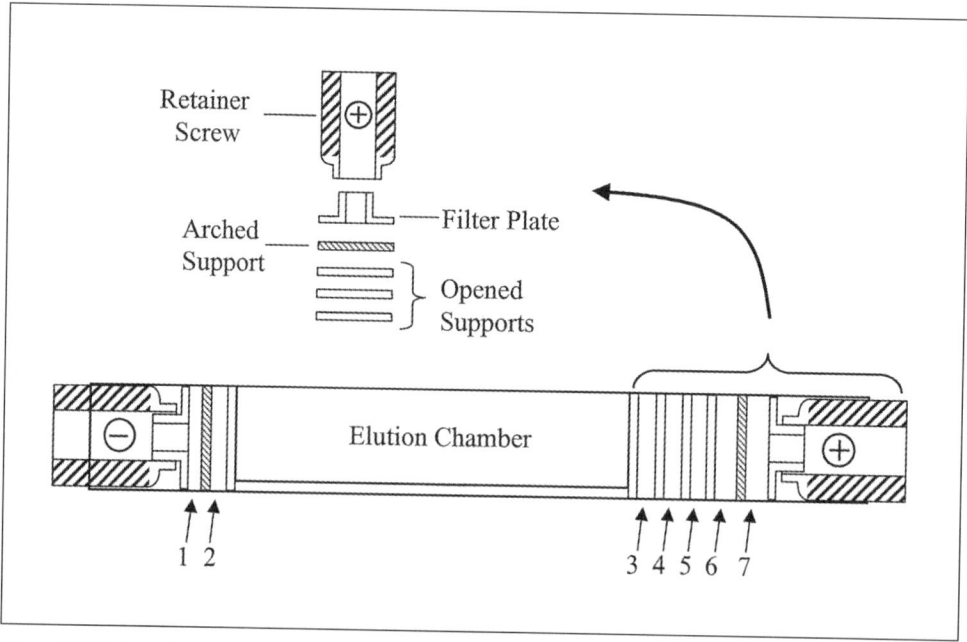

Figure 3. Elutrap electrophoresis chamber allows for rapid (4–8 h for protein, 1–8 h for DNA) isolation and purification of macromolecular samples from fractionating gels or solutions. The macromolecules are driven by electric current through selective membrane filters, yielding highly purified products.

PROTOCOLS

Electro-Elution Using the Elutrap

1. Examine the assembled instrument and note the positive (+) and negative (-) signs etched into the top surface (Figure 3). Also note the two arrows, which indicate the position of the retaining plates.
2. Apply a thin film of silicone grease to the inner threads of the two retainer screws.
3. Apply a very thin film of silicone grease to the outer edges of the arched and open supports (do not grease the inner edges).
4. Insert the two retaining plates into the apparatus.
5. Secure with the two retainer screws.
6. Place the arched supports at points 1 and 7 (the notched side should face into the elution chamber).
7. Place the other supports in front of point 7. Make sure that the notched sides face into the elution chamber.
8. Place a dense membrane between the filter plate and the open supports at points 1 and 7 (the position of these two membranes is marked by engraved arrows on the top of the instrument).
9. Tighten the left-hand retainer screw (Figure 3).
10. Place a microporous membrane at the following points for different sized traps:

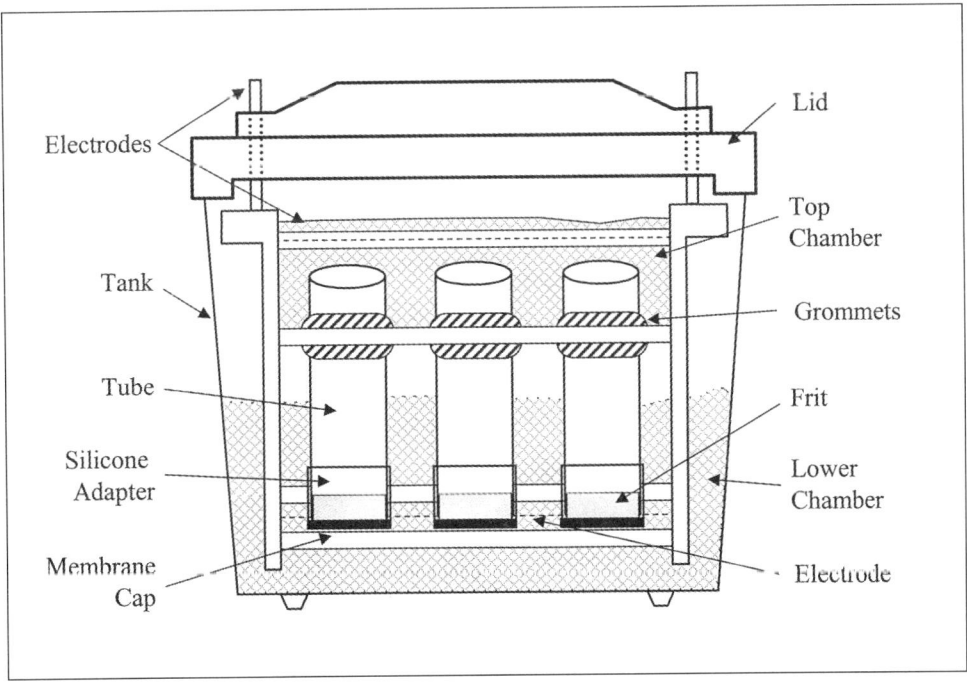

Figure 4. Bio-Rad Laboratories Electro-Eluter insert for a miniature vertical gel electrophoresis tank. This device is used for the elution of proteins from agarose or acrylamide gels.

    a. Point 3 for a 800-μL trap

    b. Point 4 for a 600-μL trap

    c. Point 5 for a 400-μL trap

    d. Point 6 for a 200-μL trap.

    A smaller trap can be constructed by placing a filter at points 6 and 3, 4, or 5.

11. Tighten the right-hand retainer screw.

12. Place the apparatus into a standard, horizontal electrophoresis tank.

13. Fill the electrophoresis buffer tanks and the apparatus with 25 mM Tris, 192 mM glycine buffer, pH 8.3.

14. Place the analyte-ligand complex or analyte-ligand/support complex near the trap (2–3 cm).

15. Apply a 80–100 V current for 12–16 h at 4°C.

16. Raise the voltage to 200 V and reverse the polarity for 20–30 s to remove materials from the dense membrane.

17. Remove the sample by gently aspirating the fluid in the trap.

Electro-elution Using the Bio-Rad Electro-Eluter

1. Soak a membrane cap for 1 h at 60°C in either:

    a.   25 mM Tris base, 192 mM glycine buffer for protein elution

    b.   40 mM Tris, 20 mM acetic acid, 1 mM EDTA buffer for DNA elution.

    Note: Add 0.1% sodium dodecyl sulfate (SDS) to both buffers before use.

2. Take the glass tube and carefully place a frit filter in the bottom. Make sure that the frit is flush with the bottom of the tube.

3. Place the tube in the apparatus stand by pushing it through the rubber grommet.

4. Place rubber stoppers in the open grommets (there are six in the stand).

5. Place the pre-wetted membrane cap in the bottom of a silicone adapter.

6. Fill the sleeve with the buffer used to wet the membrane cap.

7. Slide the silicone sleeves over the bottom of the tube until the cap cannot be moved further.

8. Fill the tube with the analyte-ligand support to within 2 cm of the top.

9. Cover with elution buffer to within 1 cm of the top.

10. Place the entire apparatus into the lower chamber of the electrophoresis apparatus.

11. Fill the tank with elution buffer.

12. Fill the upper chamber with elution buffer.

13. Place a small magnetic stirrer bar in the lower chamber and stir vigorously throughout the run.

14. Connect the electrodes to the power supply and set the current to:

    a.   8–10 mA/tube constant current for protein elution

    b.   10–12 mA/tube constant current for DNA elution.

15. Run the apparatus for:

a. 3–5 h for protein elution

b. 15–60 min for DNA elution.

16. After the elution is complete, reverse the polarity for 1 min before switching the unit off.

17. Remove the stand from the tank.

18. Gently remove each silicone sleeve.

19. With a transfer pipet, carefully remove the contents of the membrane cap.

20. Wash the interior of the cap with 200 µL of fresh elution buffer.

21. Collect the wash and add to the original fluid.

22. Discard the membrane cap. The manufacturer states that membrane caps can be reused but we have found that this practice leads to contamination of future samples.

PREVENTION OF ANALYTE DEGRADATION

An important issue surrounding many biological separations is maintenance of the integrity and biological function of the isolated materials. Denaturation of many molecules, especially proteins, can occur both in the preparative pre-separation phase and during recovery from the separation medium (75). Prevention of such damage can be achieved by a number of techniques ranging from the inclusion of protective agents (enzyme inhibitors) during the pre-separation phase to removal of the dissociation (elution) agent following recovery of the molecule of interest.

Inclusion of Protective Agents

During the process of initially isolating the analyte, there are many situations where additional cell products may also be released. The recovery of materials from cell lysates usually results in the release of a number of enzymes, including those present in lysosomal vesicles. These enzymes can act on the analyte of interest, causing partial or complete denaturation. There are several ways to overcome this situation; however, many of these involve undesirable changes in important parameters such as pH, temperature, or ionic strength. One can also add denaturing agents such as β-mercaptoethanol and DTT, but the addition of such agents can seriously affect the immobilized ligand. In affinity separations, one is left with only one viable solution, namely the addition of enzyme inhibitors. These agents are chemicals or biochemicals that have been shown to inhibit the action of certain proteolytic enzymes. Although a reasonable solution, the addition of such inhibitors does not always solve the problem. There are a number of different categories of proteolytic enzymes including the serine, thiol, and acid proteases, the metalloproteinases, and the carboxypeptidases (75). The broad-spectrum protease inhibitor 4-(2-aminoethyl)-benzenesulfonyl fluoride (AEBSF) has been used successfully to inhibit serine proteases in specific cases [the inhibition of β-secretase in human neural cell lines (57) and dipeptidyl peptidase IV isolated from porcine seminal plasma (13)]. However, in most cases, combinations of several inhibitors are required.

Although it is impossible to inhibit all of these agents, there are commercially available cocktails (Table 3) that are useful. Calbiochem-Novabiochem and Sigma

Chemical both offer cocktails that contain five or six protease inhibitors designed for different situations. Both companies offer a general cocktail, a cocktail for specific use with bacterial cell extracts, and a cocktail for specific use with mammalian cell and tissue extracts. Additionally, Sigma offers a cocktail for use with yeast and fungal extracts. All of these cocktails are designed to inhibit serine proteases, cysteine proteases, aspartic proteases, and metalloproteases with additional coverage of aminopeptidases in the bacterial and mammalian extract cocktails. For the investigator who has a detailed working knowledge of the contaminating enzymes present in the crude extract, Calbiochem offers an impressive array of individual protease inhibitors, which can then be incorporated into a "tailor-made" cocktail.

Removal of the Dissociation Agent

The easiest technique for the removal of unwanted buffers from protein solutions involves precipitation of the protein in saturated sulfate solutions and reconstitution of the precipitate in a more suitable buffer. Elution buffers can also be removed by dialysis using either conventional dialysis tubing or specialized dialysis units. The former is straightforward but requires some degree of skill in preparing and sealing the dialysis tubing. However, these problems can be overcome by employing commercially available dialysis units. Pierce Chemical makes a convenient unit (the Slide-A-Lyzer; Figure 5) that is easy to operate and produces good results. This unit has a rigid plastic frame into which the sample is introduced by a syringe. Once filled, the unit is placed into a beaker containing dialysis buffer and left for a specified period. The clean sample is then recovered by aspirating the unit with a syringe. Although the manufacturer claims that the resealable inner ring prevents sample loss, it has been our experience that sample loss (up to 10%) can occur through an inability to completely recover the sample through the syringe port. The unit is available in three sizes suitable for samples from 100 µL up to 15 mL. Spectrum Laboratories has also produced a dialysis unit that is easy to use and can accommodate very small volumes. The unit is shaped like a mushroom with a large, flat top that enables the unit to float on the surface of the dialysis buffer (Figure 6). A "finger-like" projection extends into the buffer and contains the dialysis membrane. The internal cavity of this project is the sample chamber, which allows the sample to exchange ions through the membrane. An advantage of this unit is that it is available in sizes that can accommodate samples 25 µL and above.

Recovered analytes can also be cleaned by passing them through a short gel filtration or size exclusion column. These columns can be purchased pre-packed from several manufacturers (Amersham Pharmacia Biotech, Bio-Rad, Pierce Chemical), or they can easily be built using packing materials such as Sephadex G-25 or P-10 (64). Generally, cleanup is affected by retardation of the low-molecular-weight elution agents within the smaller pores of the packing matrix while the larger analyte molecules pass through the column. This approach is particularly efficient when large-molecular-weight analytes are being isolated, but can be less efficient when low-molecular-weight analytes are to be cleaned. However, this latter situation also exists for dialysis so the investigator may have to employ a combination of techniques (i.e., salt precipitation followed by column chromatography) to finally obtain a clean preparation.

Table 3. Commercially Available Protease Inhibitor Cocktails

Commercial source	Inhibitors	Target
Calbiochem set I[a]	AEBSF, E-64, aprotinin, leupeptin, EDTA	Serine, cysteine, aspartic, and metalloproteases
Calbiochem set II[b]	AEBSF, E-64, bestatin, pepstatin, EDTA	Serine, cysteine, aspartic, metalloproteases, and aminopeptidases
Calbiochem set III[c]	AEBSF, E-64, aprotinin, bestatin, leupeptin, pepstatin	Serine, cysteine, aspartic, metalloproteases, and aminopeptidases
Sigma general set[a]	AEBSF, E-64, aprotinin, leupeptin, bestatin, EDTA	Serine, cysteine, aspartic, and metalloproteases
Sigma bacterial[b]	AEBSF, E-64, pepstatin, bestatin, EDTA	Serine, cysteine, aspartic, metalloproteases, and aminopeptidases
Sigma mammalian[c]	AEBSF, E-64, aprotinin, bestatin, leupeptin, pepstatin	Serine, cysteine, aspartic, metalloproteases, and aminopeptidases
Sigma yeast[d]	AEBSF, E-64, pepstatin 1,10-phenanthroline	Serine, cysteine, aspartic, and metalloproteases

[a]for general use
[b]for use with bacterial cell extracts
[c]for use with mammalian cell/tissue extracts
[d]for use with yeast and fungal cell extracts

PROTOCOLS

Salt Precipitation of Proteins

1. Prepare a saturated solution of NH_2SO_4 by adding solid NH_2SO_4 to 100 mL of distilled H_2O until the solution is unable to dissolve any more solid. Heat to boiling and cool to room temperature (saturation is achieved when crystals form on the bottom of the container).
2. Cool to 4°C in an ice bath.
3. Place the recovered protein into a glass tube.
4. Cool to 4°C in an ice bath.
5. Slowly add an equal volume of the saturated salt solution to the protein.
6. Mix and let stand at 4°C for 2–3 h.
7. Centrifuge the mixture at $10\,000\times g$ for 15 min at 4°C to sediment the precipitate.
8. Carefully remove the supernatant.
9. Resuspend the precipitate in a fresh volume of the saturated salt solution.

10. Repeat steps 7–10.
11. Place the mixture in dialysis tubing (as described in the protocol for the dialysis of the isolated analyte to remove dissociation agents, below).
12. Dialyze against 0.01 M sodium phosphate, pH 7.2, for 18 h at 4°C.
13. Recover the sample.

Dialysis of the Isolated Analyte to Remove Dissociation Agents

1. Cut a length of dialysis tubing sufficient to hold the volume of the sample plus 1½–2 inches extra at either end.

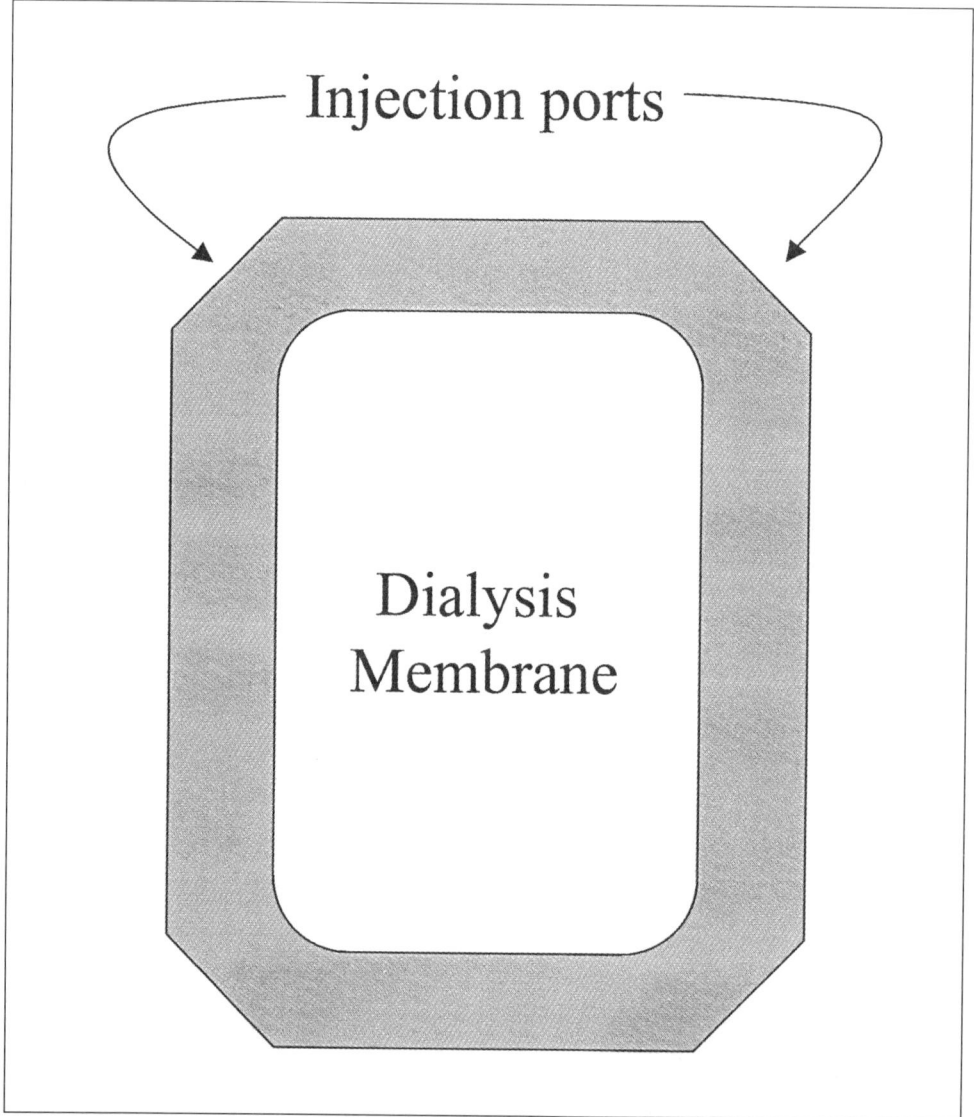

Figure 5. Slide-A-Lyzer membrane dialysis unit for quick and easy removal of elution buffer. Pierce Chemical also makes a Slide-A-Lyzer MINI dialysis unit specifically to handle volumes of 10–100 μL.

2. With gloved hands, wet the cut tubing in distilled H_2O.
3. Dry with a tissue.
4. Tie a knot in one end of the tubing and fill with H_2O (special clips are commercially available for sealing dialysis tubing).
5. Carefully close the open end of the tubing by twisting it.
6. Squeeze the closed tubing to ensure that there are no puncture holes.
7. Carefully drain out the water and fill the tubing with the sample.
8. Seal the open end by tying another knot.
9. Attach a magnetic stirrer bar to the bottom end of the tubing (a rubber band is an excellent way to do this).
10. Place the tubing in a beaker containing 10 vol of 0.01 M PO_4 buffer, pH 7.2.
11. Place on a magnetic stirrer and leave for 6–12 h at 4°C with constant stirring.
12. Retrieve the sample by carefully cutting off the top of the tubing (below the knot) and squeezing the sample into clean beaker or tube.

Microdialysis Using a Pierce Slide-A-Lyzer Unit

1. Examine the edge and locate one of the injection ports (a small hole located in the center of the cut-off corner of the square frame).
2. Fill a syringe with the sample and attach a 18-gauge needle to the syringe.
3. Insert the needle carefully into the port and slowly inject the sample (take great care not to puncture the membrane).
4. Once the sample has been injected, slowly extract the air in the chamber by slowly drawing back on the syringe.
5. Place a flotation cap on the dialyzer unit.

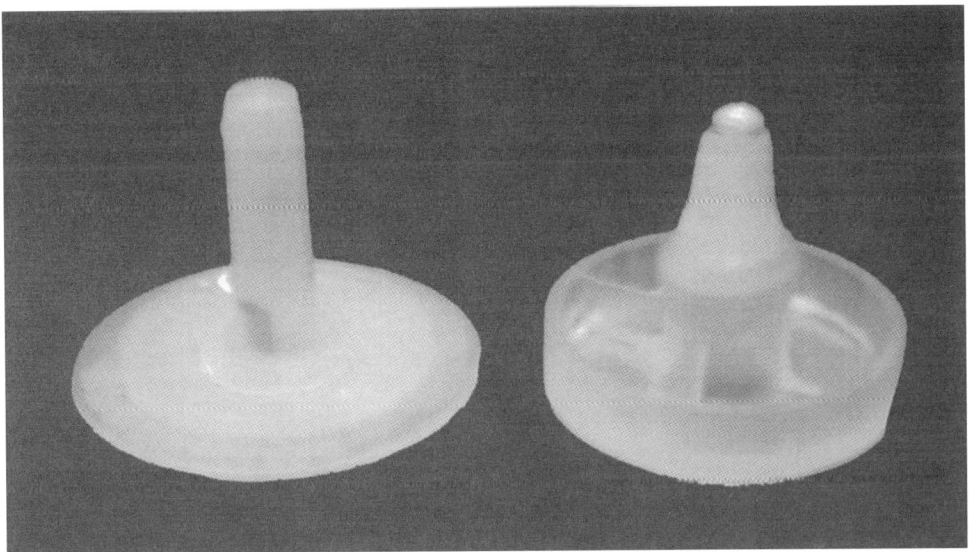

Figure 6. Spectra/Por Micro DispoDialyzer (Spectrum Laboratories) for use in dialysis of sub-milliliter volumes (10–100 μL). These devices come ready to use in a number of molecular weight cut-offs. Larger Spectra/Por units are also available.

6. Submerge the unit in a beaker filled with 0.01 M sodium phosphate, pH 7.4.

7. Incubate at 4°C for 1–2 h.

8. Recover the dialyzed sample by inserting a syringe into the injection port and expanding the membrane chamber with air.

9. Slowly invert the dialyzer so that the syringe hangs downward and withdraw the sample.

Microdialysis Using a Spectrum Micro DispoDialyzer® Unit

1. Fill a small (100 mL) beaker with 0.01 M sodium phosphate.

2. Remove the cap from the dialyzer unit.

3. Carefully float the unit on the surface of the buffer. The finger projection should face downward into the buffer.

4. Fill the finger cavity with 0.1 M sodium phosphate.

5. Allow the unit to equilibrate for 30 min.

6. Carefully remove the buffer from the finger cavity with a pipet.

7. Place the analyte solution in the cavity (maximum volume according to the type of microdialyzer used).

8. Place the beaker at 4°C for 1–2 h.

9. Recover the analyte from the dialyzer with a pipet.

Removal of Dissociation Agents Using Laboratory-Built Desalting Columns

1. Swell 10 g of Sephadex G-25 beads by placing them in 200 mL of 0.01 M sodium phosphate, pH 7.2. Leave overnight at room temperature.

2. Remove the small manufacturing impurities that have floated to the top of the suspension.

3. De-gas using gentle vacuum.

4. Attach a 0.9- × 15-cm column to a stand and attach a flow valve to the bottom. Make sure that the column is vertical.

5. Fill one-third of the column with 0.1 sodium phosphate.

6. With a transfer pipet, gently fill the column with the bead suspension.

7. Open the flow valve of the column and allow the fluid to drip out.

8. Continue to add the bead suspension until the packed bead bed is 1 cm from the top of the column (make sure that the beads are kept in suspension during the filling process).

9. Connect the column to a running buffer reservoir.

10. Fill the reservoir with 0.01 M PO_4 (the running buffer).

11. Allow 20–30 mL of running buffer to flow through the column, then stop the column flow.

12. With a transfer pipet, carefully apply the sample to the top of the column and restart the column.

13. Collect 0.5-mL fractions.

14. Analyze fractions in a spectrophotometer at 280 nm.

Removal of Dissociation Agents Using Commercial Columns

1. Clamp the column in a vertical position.

2. Remove the top and bottom caps.

3. Regenerate the column bed by adding 1 mL of distilled H_2O followed by 1 mL of 100% ethanol.

4. Add 2 mL of butanol, 1 mL of ethanol, and 4 mL of distilled H_2O.

5. Allow the column to run until the flow stops.

6. Add 2 mL of 0.01 M sodium phosphate, pH 7.2.

7. Add the sample and run until it fully enters the column bed.

8. Add 1.25 mL of 0.1 M sodium phosphate.

9. Collect the column effluent, which contains the cleaned sample.

REFERENCES

1. **Anastase, S., D. Letourneur and J. Jozefonvicz.** 1996. Affinity chromatography of human anti-dextran antibodies. Isolation of two distinct populations. J. Chromatogr. B Biomed. Appl. *686*:141-150.
2. **Angal, S. and P.D. Dean.** 1977. The effect of matrix on the binding of albumin to immobilized Cibacron Blue. Biochem. J. *167*:301-303.
3. **Azimzadeh, A. and M.H. Van Regenmortel.** 1990. Antibody affinity measurements. J. Mol. Recognit. *3*:108-116.
4. **Bai, J.H., H.J. Wang and H.M. Zhou.** 1998. Alkaline-induced unfolding and salt-induced folding of pig heart lactate dehydrogenase under high pH conditions. Int. J. Biol. Macromol. *23*:127-133.
5. **Bayer, E.A. and M. Wilchek.** 1990. Avidin column as a highly efficient and stable alternative for immobilization of ligands for affinity chromatography. J. Mol. Recognit. *3*:102-107.
6. **Beer, D.J., A.M. Yates, S.C. Randles and G.W. Jack.** 1995. A comparison of the leakage of a monoclonal antibody from various immunoaffinity chromatography matrices. Bioseparation *5*:241-247.
7. **Bhown, A.S., J.E. Mole, F. Hunter and J.C. Bennett.** 1980. High-sensitivity sequence detection of proteins quantitatively recovered from sodium dodecyl sulfate gels using an improved electrodialysis procedure. Anal. Biochem. *103*:184-190.
8. **Boden, V., J.J. Winzerling, M.A. Vijayalakshmi and J. Porath.** 1995. Rapid one-step purification of goat immunoglobulins by immobilized metal ion affinity chromatography. J. Immunol. Methods *181*:225-232.
9. **Brahimi, K., J.L. Perignon, M. Bossus, H. Gras, A. Tartar and P. Druilhe.** 1993. Fast immunopurification of small amounts of specific antibodies on peptides bound to ELISA plates. J. Immunol. Methods *162*:69-75.
10. **Bueno, S.M., K. Haupt and M.A. Vijayalakshmi.** 1995. Separation of immunoglobulin G from human serum by pseudobioaffinity chromatography using immobilized L-histidine in hollow fibre membranes. J. Chromatogr. B Biomed. Appl. *667*:57-67.
11. **Burton, S.C. and D.R. Harding.** 1998. Hydrophobic charge induction chromatography: salt independent protein adsorption and facile elution with aqueous buffers. J. Chromogr. *814*:71-81.
12. **Cai, J. and J. Henion.** 1997. Quantitative multi-residue determination of beta-agonists in bovine urine using on-line immunoaffinity extraction-coupled column packed capillary liquid chromatography-tandem mass spectroscopy. J. Chromatogr. B Biomed. Sci. Appl. *691*:357-370.
13. **Citron, M., T.S. Diehl, A. Capell, C. Haass, D.B. Teplow and D.J. Selkoe.** 1996. Inhibition of amyloid beta-protein production in neural cells by the serin protease inhibitor AEBSF. Neuron *17*:171-179.
14. **Dean, P.D.G., P. Brown, M.J. Leyland, D.H. Watson, S. Angal and M.J. Harvey.** 1977. Electrophoretic desorption of affinity adsorbents. Biochem. Soc. Trans. *5*:1111-1113.
15. **De Angelis, C., M. Riscazzi, R. Salvini, A. Piccoli, C. Ferri and A. Sanucci.** 1997. Isolation and characterization of a digoxin-like immunoreactive substance from human urine by affinity chromatography. Clin. Chem. *43*:1416-1420.
16. **Deaver, D.R.** 1995. A new non-isotoptic detection system for immunoassays. Nature *377*:758-760.
17. **Deras, I.L., N. Kawasaki and Y.C. Lee.** 1998. Quantitative recovery of Man9GlcNAc2Asn derivatives from concanavalin A. Carbohydr. Res. *306*:469-471.
18. **Desnoyers, L. and P. Manjunath.** 1993. Interaction of a novel class of phospholipid-binding proteins of bovine seminal fluid with different affinity matrices. Arch. Biochem. Biophys. *305*:341-349.
19. **el Rassi, Z., L.F. De Ocampo and M.D. Bacolod.** 1990. Binary and ternary salt gradients in hydrophobic-

interaction chromatography of proteins. J. Chromatogr. *499*:141-152.

20.**Ferreira, M.U. and A.M. Katzin.** 1995. The assessment of antibody affinity distribution by thiocyanate elution: a simple dose-response approach. J. Immunol. Methods *187*:297-305.

21.**Fuglistaller, P.** 1989. Comparison of immunoglobulin binding capacities and ligand leakage using eight different protein A affinity chromatography matrices. J. Immunol. Methods *124*:171-177.

22.**Gambino, R., G. Ruin, M. Cassader and G. Pagano.** 1996. Apolipoprotein H: a two-step isolation method. J. Lipid Res. *37*:902-904.

23.**Glatz, Z., J. Psotova, O. Janiczek, K. Chroust and T. Jowet.** 1997. Use of thiopropyl sepharose for the synthesis of an adsorbent for the affinity chromatography of glutathione S-transferase. J. Chromatogr. B Biomed. Sci. Appl. *688*:239-243.

24.**Godfrey, M.A., P. Kwasowski, R. Clift and V. Marks.** 1993. Assessment of the suitability of commercially available SpA affinity solid phases for the purification of murine monoclonal antibodies at process scale. J. Immunol. Methods *160*:97-105.

25.**Goldberg, M., K.L. Knudsen, D. Platt, F. Kohen, E.A. Bayer and M. Wilchek.** 1991. Specific interchain cross-linking of antibodies using bismaleimides. Repression of ligand leakage in immunoaffinity chromatography. Bioconjug. Chem. *2*:275-280.

26.**Greene, C.R. and P.S. Holt.** 1997. An improved chromatographic method for the separation of egg yolk IgG into subpopulations utilizing immobilized metal ion (Fe^{3+}) affinity chromatography. J. Immunol. Methods *209*:155-164.

27.**Haff, L.A., M. Lasky and A. Manrique.** 1979. A new technique for desorbing substances tightly bound to affinity gels: flat-bed electrophoretic desorption in Sephadex via isoelectric focusing (FEDS-IEF). J. Biochem. Biophys. Methods *1*:275-286.

28.**Harvey, M.J., C.R. Lowe and P.D.G. Dean.** 1974. Affinity chromatography on immobilized adenosine 5′-monophosphate. 2. Some parameters relating to the selection and concentration of the immobilized ligand. Eur. J. Biochem. *41*:347-352.

29.**Ikegawa, S., M. Itoh, N. Murao, H. Kijima, M. Suzuki, T. Fujiyama and J. Goto.** 1996. Immunoaffinity extraction for liquid chromatographic determination of equilin and its metabolites in plasma. Biomed. Chromatogr. *10*:73-77.

30.**Ingraham, R.H.** 1991. Hydrophobic interaction chromatography of proteins, p. 424-435. *In* C.T. Mant and R.S. Hodges (Eds.), High-Performance Liquid Chromatography of Peptides and Proteins. CRC Press, Boca Raton.

31.**Jack, C.W.** 1994. Immunoaffinity chromatography. Mol. Biotechnol. *1*:59-86.

32.**Jacobs, E. and A. Clad.** 1986. Electroelution of fixed and stained membrane proteins from preparative sodium dodecyl sulfate-polyacrylamide gels into a membrane trap. Anal. Biochem. *154*:583-589.

33.**Janusz, M., M. Niezgodka, Z. Wieczorek and J. Lisowski.** 1986. A mild method for the purification of guinea pig peritoneal macrophage Fc gamma receptors. Affinity chromatography and elution of the receptor with reducing agents. J. Immunol. Methods *86*:119-124.

34.**Johnson, R.D. and F.H. Arnold.** 1995. The Tempkin isotherm describes heterogeneous protein adsorption. Biochim. Biophys. Acta. *1247*:293-297.

35.**Kamata, N., A. Enomoto, S. Ishida, K. Nakamura, J. Kurisaki and S. Kaminogawa.** 1996. Comparison of pH and ionic strength dependence of interactions between monoclonal antibodies and bovine beta-lactoglobulin. Biosci. Biotechnol. Biochem. *60*:25-29.

36.**Kamihira, M., S. Iijima and T. Kobayashi.** 1992. Stabilities of antigen and antibody under elution conditions in immunoaffinity chromatography using monoclonal antibody. Bioseparation *3*:185-188.

37.**Kanai, Y. and T. Kubota.** 1989. A novel trait of naturally occurring anti-DNA antibodies: dissociation from immune complexes in neutral 0.3-0.5 M NaCl. Immunol. Lett. *22*:293-299.

38.**Kang, K.A. and D.D. Ryu.** 1991. Studies on scale-up parameters of an immunoglobulin separation system using protein A affinity chromatography. Biotechnol. Prog. *7*:205-212.

39.**Kannan, K., P. Lalitha, K.V. Rao, R.B. Narayanan and P. Kaliraj.** 1997. Optimisation of immunoaffinity purification of Wuchereria bancrofti specific antibodies from human sera. Indian J. Exp. Biol. *35*:1076-1079.

40.**Katoh, S., M. Terashima and K. Miyaoku.** 1997. Purification of alpha-amylase by specific elution from anti-peptide antibodies. Appl. Microbiol. Biotechnol. *47*:521-524.

41.**Kuronen, I., H. Kokko, I. Mononen and M. Parviainen.** 1997. Hen egg yolk antibodies purified by antigen affinity under highly alkaline conditions provide new tools for diagnostics. Human intact parathyrin as a model antigen. Eur. J. Clin. Chem. Clin. Biochem. *35*:435-440.

42.**Labrou, N.E. and Y.D. Clonis.** 1995. Biomimetic dye affinity chromatography for the purification of bovine heart lactate dehydrogenase. J. Chromatogr. *718*:35-44.

43.**Leibl, H., W. Erber, M.M. Eibl and J.W. Mannhalter.** 1993. Separation of polysaccharide-specific human immunoglobulin G subclasses using a protein A superose column with a pH gradient elution system. J. Chromatogr. *639*:51-56.

44.**Liesiene, J., K. Racaityte, M. Morkeviciene, P. Valancius and V. Bumelis.** 1997. Immobilized metal affinity chromatography of human growth hormone. Effect of ligand density. J. Chromatogr. *764*:27-33.

45.**Liu, H., D.E. Cohen and C. Pidgeon.** 1997. Single step purification of rat liver aldolase using immobilized artificial membrane chromatography. J. Chromatogr. B Biomed. Sci. Appl. *703*:53-62.

46.**Lowe, C.R., M.J. Harvey and P.D. Dean.** 1974. Affinity chromatography on immobilised adenosine 5′-monophosphate. 4. Variation of the binding of dehydrogenases and kinases with pH. Eur. J. Biochem. *41*:347-351.

47.**Lowe, C.R., M.J. Harvey and P.D. Dean.** 1974. Affinity chromatography on immobilised adenosine 5′-monophosphate. 5. Some applications of the influence of temperature on the binding of dehydrogenases and kinases. Eur. J. Biochem. *41*:353-357.

48.**Macdonald, R.A., C.S. Hosking and C.L. Jones.** 1988. The measurement of relative antibody affinity by ELISA using thiocyanate elution. J. Immunol. Methods *106*:191-194.

49.**Matejtschuk, P. (Ed.)** 1997. Affinity Separations. IRL Press, Oxford.

50.**Mintz, K.P. and S. Brimijoin.** 1985. Monoclonal antibodies to rabbit brain acetylcholinesterase: selective enzyme inhibition, differential affinity for enzyme forms, and cross-reactivity with other mammalian cholinesterases. J. Neurochem. *45*:284-292.

51.**Mislovicova, D., M. Chudinova, P. Gemeiner and P. Docolomansky.** 1995. Affinity chromatography of invertase on concanavalin A-bead cellulose matrix: the case of an extraordinary strong binding glycoenzyme. J. Chromatogr. B Biomed. Appl. *664*:145-153.

52.**Mohr, P. and K. Pommerening.** 1985. Affinity Chromatography. Practical and Theoretical Aspects. Marcel Dekker, NewYork.

53.**Narayanan, S.R. and L.J. Crane.** 1990. Affinity chromatography supports: a look at performance requirements. Trends Biotechnol. *8*:12-16.

54.**Narhi, L.O., D.J. Caughey, T. Horan, Y. Kita, D. Chang and T. Arakawa.** 1997. Effect of three elution buffers on the recovery and structure of monoclonal antibodies. Anal. Biochem. *253*:236-245.

55.**Nosworthy, N.J. and A. Ginsberg.** 1997. Thermal unfolding of dodecameric glutamine synthetase: inhibition of aggregation by urea. Protein Sci. *6*:2617-2623.

56.**Oates, M.R., W. Clarke, E.M. Marsh and D.S. Hage.** 1998. Kinetic studies on the immobilization of antibodies to high-performance liquid chromatographic supports. Bioconjug. Chem. *9*:459-465.

57.**Ohkubo, I., K. Huang, Y. Ochiai, M. Takagaki and K. Kani.** 1994. Dipeptidyl peptidase IV from porcine seminal plasma: purification, characterization, and N-terminal amino acid sequence. J. Biochem. *116*:1182-1186.

58.**O'Shannessy, D.J.** 1990. Hydrazido-derivatized supports in affinity chromatography. J. Chromatogr. *510*:13-21.

59.**Ouyang, S., Y. Xu and Y.H. Chen.** 1998. Selective determination of a group of organic compounds in complex sample matrixes by LC/MIMS with on-line immunoaffinity extraction. Anal. Chem. *70*:931-935.

60.**Patwardhan, A.V. and M.M. Ataai.** 1997. Site accessibility and the pH dependence of the saturation capacity of a highly cross-linked matrix. Immobilized metal affinity chromatography of bovine serum albumin on chelating Superose. J. Chromatogr. *767*:11-23.

61.**Phillips, T.M.** 1989. High-performance immunoaffinity chromatography. Adv. Chromatogr. *29*:133-173.

62.**Phillips, T.M.** 1991. Isolation of an interleukin 2-binding receptor from activated lymphocytes by high-performance immunoaffinity chromatography. J. Chromatogr. *550*:741-749.

63.**Phillips, T.M.** 1991. Theory and practical aspects of high-performance immunoaffinity chromatography, p. 507-515. *In* C.T. Mant and R.S. Hodges (Eds.), High-Performance Liquid Chromatography of Peptides and Proteins. CRC Press, Boca Raton.

64.**Phillips, T.M.** 1992. Analytical Techniques in Immunochemistry. Marcel Dekker, New York.

65.**Phillips, T.M.** 1994. Immunoaffinity measurement of recombinant granulocyte colony stimulating factor in patients with chemotherapy-induced neutropenia. J. Chromatogr. B Biomed. Appl. *662*:307-313.

66.**Phillips, T.M., L.M. Kennedy and E.C. De Fabo.** 1997. Microdialysis-immunoaffinity capillary electrophoresis studies on neuropeptide-induced lymphocyte secretion. J. Chromatogr. B Biomed. Sci. Appl. *697*:101-109.

67.**Poirier, M.P., S. Ahlstedt, J. Ford and W.K. Dolen.** 1997. Use of thiocyanate elution to estimate relative avidity of allergen specific IgE antibodies. Allergy Asthma Proc. *18*:359-362.

68.**Prajapati, S., V. Bhakuni, K.R. Babu and S.K. Jain.** 1998. Alkaline unfolding and salt-induced folding of bovine liver catalase at high pH. Eur. J. Biochem. *255*:178-184.

69.**Reifsnyder, D.H., C.V. Olson, T. Etcheverry, H. Prashad and S.E. Builder.** 1996. Purification of insulin-like growth factor-I and related proteins using underivatized silica. J. Chromatogr. *753*:73-80.

70.**Reinwald, E., P. Rautenberg and H.J. Risse.** 1981. Purification of the variant antigens of Trypanosoma congolense: a new approach to the isolation of glycoproteins. Biochim. Biophys. Acta *668*:119-131.

71.**Robinson, J.B., J.M. Strottmann and E. Stellwagen.** 1981. Prediction of neutral salt elution profiles for affinity chromatography. Proc. Natl. Acad. Sci. USA *78*:2287-2291.

72.**Schokker, E.P. and M.A. van Boekel.** 1998. Mechanism and kinetics of inactivation at 40-70 degrees C of the extracellular proteinase from Pseudomonas fluorescens 22F. J. Dairy Res. *65*:261-272.

73.**Schulze-Osthoff, K., E. Michels, B. Overwein and C. Sorg.** 1989. Electroelution of antigens immobilized on antibody-linked affinity matrices. Anal. Biochem. *177*:314-317.

74. **Schwarz, A., F. Kohen and M. Wilchek.** 1995. Novel heterocyclic ligands for the thiophilic purification of antibodies. J. Chromatogr. B Biomed. Appl. *664*:83-88.
75. **Scopes, R.K.** 1994. Optimization of procedures; final steps. Section 12.2. Stabilizing factors for enzymes and other proteins, p. 317-324. Protein Purification. Principles and Practice. Springer-Verlag, New York.
76. **Scouten, W.H.** 1981. Affinity Chromatography. John Wiley & Sons, New York.
77. **Scouten, W.H.** 1991. Affinity chromatography for protein isolation. Curr. Opin. Biotcchnol. *2*:37-43.
78. **Singh, N.K. and B.N. Tiwary.** 1996. Fast dissociation of phe-tRNA synthetase from Aspergillus nidulans immobilized on sepharose-6B columns by NaCl. J. Basic Microbiol. *36*:59-62.
79. **Sisson, T.H. and C.W. Castor.** 1990. An improved method for immobilizing IgG antibodies on protein A-agarose. J. Immunol. Methods *127*:215-220.
80. **Stott, R.A.** 1998. Enhanced chemiluminescence immunoassay. Methods Mol. Biol. *80*:197-205.
81. **Tejedor, M.C., C. Delgado, M. Grupeli and J. Luque.** 1992. Affinity partitioning of erythrocytic phosphofructokinase in aqueous two-phase systems containing poly(ethylene glycol)-bound cibacron blue. Influence of pH, ionic strength and substrates/effectors. J. Chromatogr. *589*:127-134.
82. **Tharakan, J., F. Highsmith, D. Clark and W. Drohan.** 1992. Physical and biochemical characterization of five commercial resins for immunoaffinity purification of factor IX. J. Chromatogr. *595*:103-111.
83. **Thomas, T.M. and R.K. Scopes.** 1998. The effects of temperature on the kinetics and stability of mesophilic and thermophilic 3-phosphoglycerate kinases. Biochem. J. *330*:1087-1095.
84. **Tu, Y.Y., F.J. Primus and D.M. Goldenberg.** 1988. Temperature affects binding of murine monoclonal IgG antibodies to protein A. J. Immunol. Methods *109*:43-47.
85. **van der Marel, P., A.L. van Wezel, A.G. Hazendonk and A. Kooistra.** 1980. Concentration and purification of poliovirus by immune adsorption on immobilized antibodies. Dev. Biol. Stand. *46*:267-273.
86. **van Eyk, J.E., C.T. Mant and R.S. Hodges.** 1991. High-performance affinity chromatography of peptides and proteins, p. 479-491. *In* C.T. Mant and R.S. Hodges (Eds.), High-Performance Liquid Chromatography of Peptides and Proteins. CRC Press, Boca Raton.
87. **Vankova, R., A. Gaudinova, H. Sussenbekova, P. Dobrev, M. Strnad, J. Holik and J. Lenfeld.** 1998. Comparison of oriented and random antibody immobilization in immunoaffinity chromatography of cytokinins. J. Chromatogr. *811*:77-84.
88. **Walters, R.R.** 1985. Design and use of high performance affinity columns, p. 94-96. *In* P.D.G. Dean, W.S. Johnson and F.A. Middle (Eds.), Affinity Chromatography. A Practical Approach. IRL Press, Washington, DC.
89. **Wheatley, J.B.** 1991. Effect of antigen size on optimal ligand density of immobilized antibodies for a high-performance liquid chromatographic support. J. Chromatogr. *548*:243-253.
90. **Williamson, K.C., P.E. Duffy and D.C. Kaslow.** 1992. Immunoaffinity chromatography using electroelution. Anal. Biochem. *206*:359-362.
91. **Winzor, D.J.** 1997. Quantitative affinity chromatography, p. 39-60. *In* P. Matejtschuk (Ed.), Affinity Separations. IRL Press, Oxford.
92. **Wirth, H.J., K.K. Unger and M.T.W. Hearn.** 1993. Influence of ligand density on the properties of metal-chelate affinity supports. Anal. Biochem. *208*:16-25.
93. **Yamamoto, K., T. Tsuji and T. Osawa.** 1995. Analysis of asparagine-linked oligosaccharides by sequential lectin affinity chromatography. Mol. Biotechnol. *3*:25-36.
94. **Yarmush, M.L., K.P. Antonsen, S. Sundaram and D.M. Yarmush.** 1992. Immunoadsorption: strategies for antigen elution and production of reusable adsorbents. Biotechnol. Prog. *8*:168-178.
95. **Zahn, R., C. von Schroetter and K. Wuthrich.** 1997. Human prion proteins expressed in Escherichia coli and purified by high-affinity column refolding. FEBS Lett. *417*:400-404.
96. **Zhou, F.L., M. Burnouf-Radosevich and T. Burnouf.** 1994. Purification of factor VIII/von Willebrand factor from human plasma on immobilized lentil lectin. Protein Expr. Purif. *5*:138-143.

4 | Ligands and their Attachment to Affinity Supports

The materials used as ligands in affinity separations are quite varied, with selectivity ranging from individually specific through group-specific to nonspecific. Although both affinity and immunoaffinity ligands can interact with their selected analytes in free solution, they are usually attached to a carrier or support to simplify the isolation of the complex. The most widely used ligands are proteins and peptides, a large group that includes enzymes, transporter molecules, and the main reactive molecules of the immune system, namely antibodies. In addition, there are carbohydrate ligands that are capable of reacting with their target analytes through group-specific interactions, as do lectins and mimetic dye ligands. These group-specific ligands have helped to extend the usefulness of affinity partitioning into applications such as preparative- and production-scale separations. Ligands can also be nucleic acids, nucleotides, pharmaceutical drugs, membrane or recombinant receptors, metal ions, or any other material exhibiting molecular selectivity for either a defined analyte or analyte group.

PROTEINS AND PEPTIDES

Proteins and peptides are among the most popular affinity ligands and range from immobilized enzymes to antibodies and antigens. Recently, enzymes have been immobilized onto a number of different supports including hydroxyapatite (29), methacrylate microspheres (3), glass beads (89), polyacrylamide and calcium alginate beads (23), and grafted polymer surfaces (57). An interesting application for immunochemistry is the use of immobilized papain and pepsin for the affinity isolation and cleavage of IgG antibodies. This cleavage takes place in a column, thus allowing the easy isolation of the final antibody fragments (22). Immobilized pepsin and immobilized papain are both commercially available from Pierce Chemical. Pierce also provides immobilized ficin, which is a thiol protease that is especially useful for making $F(Ab')_2$ fragments from mouse monoclonal antibodies. The Pierce Catalog and Handbook contains complete details on the use of these three immobilized enzymes. Although the majority of investigators used relatively conventional approaches for the immobilization process, Kozulic et al. (56) described a technique

Affinity and Immunoaffinity Purification Techniques
Terry M. Phillips and Benjamin F. Dickens
© 2000 Eaton Publishing, Natick, MA

for immobilizing glyco-enzymes by periodate oxidation of their carbohydrate chains.

Immobilized enzymes attached to membranes have been extensively used in the construction of biosensors and chemical sensors (80,92). Campanella et al. (18) used L-aspartase for analyzing aspartate and aspartame in pharmaceutical preparations and artificial sweeteners. α-Amylase has been incorporated into a nonwoven polyester support filled with an acrylamide–methacrylate hydrogel and used as a sensor for starch hydrolysis (20). Carbon electrodes have been coated with acetylcholinesterase, choline oxidase, and osmium poly(vinylpyridine)-based redox polymer containing horseradish peroxidase to form an on-line acetylcholine sensor for the determination of extracellular acetylcholine without interference from choline (76). Osborne et al. (79) developed a sensitive, enzymatic glucose electrode that was coupled with a microdialysis sampling technique to enable the continuous, on-line measurement of dialysate glucose. The electrode was constructed by immobilizing glucose oxidase onto the surface of an osmium poly(vinylpyridine) horseradish peroxidase gel cast-coated onto a glassy carbon electrode.

A large array of other proteins have been used as affinity ligands. Apomyoglobin, covalently linked to CNBr-activated agarose, was used as a heme acceptor for investigating heme transfer reactions from hemoproteins (34). Lactose, when coupled to divinyl sulfone-activated agarose (30), was used to isolate galectin-1, a polypeptidic factor that can have major effects on cell growth. Immobilized pro-metalloproteinase-9 has been used to study the binding characteristics of the α2(IV) chain of collagen IV (78), and ecotin, a unique protease inhibitor derived from *E. coli*, has been used to purify trypsinogen (60). Immobilized crotoxin, a phospholipase A_2 toxin isolated from the venom of the South American rattlesnake, has been used to isolate 87-, 65-, and 50-kDa polypeptides from a synaptic membrane fraction of guinea pig brain (48). Heparin has been used for a variety of different affinity applications, recently as an affinity ligand for the isolation of a 30-kDa laminin-binding protein from bovine peripheral nerve membranes (94).

Immobilized proteins and amino acids have also been used as affinity removal agents. Removal of endotoxin from large-scale *E. coli* recombinant protein preparations was attempted by passing the medium over immobilized polymyxin B or histidine. Although reasonably effective, this approach did not achieve the same degree of efficiency as phase extraction with Triton X-114 (64). In a similar vein, protein A immobilized onto polymethyl-methacrylate uniform microbeads was used to remove cholesterol and IgG from human plasma obtained from a hypercholesterolemic patient (90). Although cholesterol does not appear to interact with protein A in aqueous solution, these studies demonstrated a significant degree of adsorption when applying the patient's serum.

Immobilized hormones have been used to isolate specific receptors (21), just as peptide antigens have been used to isolate specific antibodies. Proteins with specific binding properties, such as protein A, protein G, avidin, and streptavidin, have specialized uses in immunochemistry. These proteins have been used to isolate monoclonal antibodies from cell cultures and for the isolation of biotin-labeled and biotin-containing molecules. A more complete discussion of these proteins is provided in Chapter 5.

CARBOHYDRATES

The use of carbohydrates as affinity ligands has been applied to many different scientific disciplines. Immobilized sugars (*N*-acetylglucosamine, mannose, and maltose) have been employed for the isolation and recovery of ficolin, a TGFβ1 binding protein with an overall structure similar to that of complement C1q and the collectins (58). Likewise, immobilized oligosaccharides have been used to study the binding kinetics of lectins to simple sugars. It was found that both the apparent association and dissociation rate constants showed a decrease as the oligosaccharide density increased, indicating a nonhomogeneous state limited by the mass transport effect. In such cases, the effect of rebinding becomes so large that it cannot be disregarded, indicating the importance of the mass transport effect in modulating the affinity of lectin for oligosaccharides on a solid-phase surface (98). Perhaps the most popular carbohydrate ligand is heparin. This molecule exhibits a number of useful binding characteristics that can be exploited in affinity separations. Immobilized heparin has been used to enrich the proteins of *Haemophilus influenzae* prior to analysis by two-dimensional electrophoresis (31). It has also been used to study interactions and reactivities with xanthine oxidase (91). Heparin-Sepharose has been investigated as a potential group-specific ligand for the recovery of antithrombin III in an industrial bioprocessing unit (53), and as a ligand in affinity CE for analyzing synthetic heparin-binding peptides (109).

An interesting and potentially novel approach to studying the affinity of D-glucose and the transport inhibitor cytochalasin B for the glucose transporter Glut1 has been described (67) using immobilized membranes as the carbohydrate-containing ligand. Similarly, carbohydrate affinity chromatography has been performed utilizing human red blood cells immobilized in a gel bed. This system was also used to analyze Glut1 activities in a cell membrane. The results from these studies suggest that chromatography on immobilized cells is a potentially useful tool for studies of cellular membrane functions (121).

A novel approach to the immobilization of antibodies by their carbohydrate moieties has been developed by Matson and Little (70). They showed that immunoaffinity supports prepared by immobilizing IgG by carbohydrate linkages demonstrated dramatic increases in antigen capacity over those prepared by coupling by primary amino groups. Recently, a similar approach has been applied to the isolation of plant hormones by carbohydrate-oriented immunoaffinity chromatography (110).

NUCLEIC ACIDS

The bio-selective interactions between nucleic acids and their binding proteins have made the use of immobilized nucleic acid ligands a great success. Interactions between immobilized DNA and a number of intercalators and DNA-binding molecules were monitored by reflectometric interference spectroscopy, thus providing a nonlabel, non-isotopic technique for studying such reactions (87). A column containing native DNA complexed with chitosan and celite has been shown to effectively concentrate carcinogenic heterocyclic amines (46). Capillary affinity gel electrophoresis has been applied to sequence-specific and base composition-specific recognition of oligodeoxynucleotides, utilizing the formation of heteroduplexes between a nucleic acid analog immobilized into the capillary gel and soluble oligodeoxynucleotides with different sequences (7). RNA can be immobilized in a similar fashion for the isolation of specific RNA-binding nucleotides and proteins. Columns containing specific polynucleotides have been used to isolate complementary DNA and RNA (77). An oligonucleotide [(dT)$_{12}$]-immobilized capillary was prepared for the affinity CE of DNA (81). Synthetic oligonucleotides, possessing a recognition sequence for the transcription factor NF-κB, have been immobilized onto epoxide-activated methacrylate and used to isolate p50, the DNA-binding element of NF-κB (115). Biotinylated double-stranded oligonucleotides containing thyroid hormone response elements (TRE) have been immobilized to streptavidin-coated scintillating microplates and used to study the binding of two thyroid hormone receptors (19). Both DNA and various synthetic oligonucleotides have been incorporated into biosensor chips by immobilizing them onto micro-fabricated thick-film carbon transducers. These chips have been used for hybridization detection of nucleic acid sequences, for determination of small molecules based upon their collection into the double-stranded DNA layer or by monitoring their effect upon the intrinsic DNA oxidation signal, and for the direct ultra-trace measurements of nucleic acids using constant-current stripping chronopotentiometry (112).

LECTINS

Lectins are specialized proteins that exhibit specificity for sugar groups on a variety of biological molecules (Table 1). Con A is one of the most popular lectins and has been used extensively for affinity separations. Examples of such affinity applications include the isolation of erythropoietin (101), yeast cells (71), submitochondrial particles prepared from beef liver (40), apolipoprotein H (33), and HIV-1 gp120 glycoprotein (2). Other lectins of potential importance include the red kidney bean lectin *Phaseolus vulgaris*, the mistletoe lectin (26,32), soybean agglutinin for the isolation of cells (96), the lentil lectin (88,122), wheat germ agglutinin (61), and jacalin (51,105) (the latter three lectins being used for the isolation of important serum proteins).

New lectins are being described from both plant and animal sources. The latest lectins from plant sources with potential for use in affinity studies have been isolated from crocus bulbs (73), the rhizomes of hedge bindweed (83), and the lotus (119). Newly described animal-derived lectins have been isolated from horseshoe crabs (95), the albumin gland of land snails (15), bovine plasma (47), white shrimp (111), and sea urchins (43).

METALS AND OTHER CHEMICALS

Immobilized metal ions have gained popularity as affinity ligands (1,8,104,118, 120) and have been used to isolate a variety of different analytes. The most popular ion for use in affinity separations appears to be zinc, although copper and nickel have also been widely used. Zinc metal affinity has been applied to the isolation of cystic fibrosis antigen from human saliva (24), bovine α-lactalbumin from whey protein concentrate (14), and trout pituitary gonadotropins (38). Copper has been used for the isolation of hen egg-white lysozyme (50), perforin and granzymes from cytotoxic lymphocytes and natural killer cells (116), BSA (82), and somatotropin (63). Copper ions have also been

Table 1. A Selection of Lectins and their Reactive Sugar Moeities

Common name	Latin name	Reactive sugar residues
Castor bean RCA$_{120}$	*Ricinus communis*	β-D-galactosyl
Chickpea	*Cicer arietinum*	fetuin
Fava bean	*Vicia faba*	D-mannose
		D-glucose
Gorse UEA I	*Ulex europaeus*	α-L-fucose
UEA II		N,N′-diacetylchitobiose
Jacalin	*Artocarpus integrifolia*	α-D-galactosyl
		β-(1,3) N-acetylgalactosamine
Jack bean[1]	*Canavalia ensiformis*	α-D-mannosyl
		α-D-glucosyl
Jequirity bean	*Abrus precatorius*	α-D-galactose
Lentil	*Lens culinaris*	α-D-mannosyl
		α-D-glucosyl
Mistletoe	*Viscum album*	β-D-galactosyl
Mung bean	*Vigna radiata*	α-D-galactosyl
Osage orange	*Maclura pomifera*	α-D-galactosyl
		N-acetyl-D-galactosaminyl
Pea	*Pisum sativum*	α-D-glucosyl
		α-D-mannosyl
Peanut	*Arachis hypogaea*	β-D-galactosyl
Pokeweed	*Phytolacca americana*	N-acetyl-β-D-glucosamine oligomers
Red kidney bean[2]	*Phaseolus vulgaris*	oligosaccharide
Snowdrop	*Galanthus nivalis*	nonreducing terminal end of α-D-mannosyl
Soybean	*Glycine max*	N-acetyl-D-galactosamine
Wheatgerm	*Triticum vulgaris*	N-acetyl-β-D-glucosaminyl
		N-acetyl-β-D-glucosamine oligomers
Green marine algae	*Codium fragile*	N-acetyl-D-galactosamine
Red marine algae	*Ptilota plumosa*	α-D-galactosyl
Edible snail	*Helix pomatia*	N-acetyl-α-D-galactosaminyl
Eel (fresh water)	*Anguilla anguilla*	α-L-fucosyl
Garden snail	*Helix asperia*	N-acetyl-α-D-galactosaminyl
Horseshoe crab	*Limulus polyphemus*	N-acetylneuraminic acid
		Glucuronic acid
		Phosphorylcholine analogs
Slug	*Limax flavus*	N-glycolylneuramicic acid

[1]concanavalin A
[2]Phytohemagglutinin

89

applied to the separation of a number of proteins by affinity CE (45).

Immobilized nickel ions have successfully been used to isolate a variety of analytes ranging from fusion proteins (55) and nickel-binding proteins (28) to phosphatase isolated from the hyperthermophilic marine archaeon, *Thermococcus pacificus* (10), and retroviral integrases (5). Nickel metal affinity chromatography has also been extensively used to isolate recombinant molecules that have histidine tags (11,39,74), as well as DNA containing 6-histidine tags (72).

MIMETIC DYES

Synthetic textile dyes, by virtue of their reactivity with a wide variety of biological materials, can be used as affinity ligands (6,25). The triazine dye Cibacron Blue is the most popular textile dye for use as an affinity ligand. Cibacron Blue 3GA binds HSA, fibroblast interferon, and nucleotide-binding enzymes. Binding is affected at pH 7.0 but can be enhanced at lower pH, possibly because of the ion-exchange effects of the sulfonate groups on the dye. The most efficient elution is in 1 M KCl or 0.5 M thiocyanate gradients. This triazine dye has been used in the isolation of lipoproteins and serum albumin (59), α-fetoprotein (13), IgG (93), complement components (35), α_2-macroglobulin (17), and human clotting factors (36). It has also been used in the purification of an enzymatically active microsomal semicarbazide amine oxidase from bovine lungs (65). A recent report has even shown that Cibacron Blue 3GA and some other triazine dyes can bind the functional region of HIV-1 Env glycoproteins that play a key role in HIV infection (44). Bilirubin molecules have been shown to interact directly with immobilized Cibacron Blue F3GA molecules that were covalently linked to microbeads (27).

Other triazine-based dyes have led to an interest in dye-ligand chromatography (102,103) (Table 2). Such ligands are often referred to as biomimetic ligands because their structure does not have a direct relation to biological ligands. Exploitation of sophisticated molecular modeling techniques in conjunction with binding and crystallographic studies has permitted the design of new, highly selective biomimetic ligands for target proteins (66). Tucker et al. (106) studied the behavior of a heterogeneous protein mixture and of a series of homogeneous proteins on immobilized tetraiodofluorescein. They found that a 9-atom spacer arm was required to improve the efficiency of the immobilized dye. Using a technique called dye-ligand centrifugal affinity chromatography, it was found that Drimarene Brilliant Blue K-R and Drimarene Rubine R/K-5BL exhibited a high affinity for goat IgG, specially binding to the Fc fragment of the molecule (12). Dye-ligand chromatography can be used to isolate adenine/thymidine-specific and guanine/cytosine-specific nucleic acids. The technique is similar to that described above for the isolation of proteins. Although mimetic dye chromatography has developed into an important method for the large-scale purification of proteins (mainly due to their selectivity, stability, and economy of use), dyes can effectively be used for a variety of other affinity procedures such as membrane separations, cross-flow filtration, and precipitation (16,117).

RECEPTORS

The use of receptors as affinity ligands has received only modest experimental

Table 2. Selected Triazine-Based Mimetic Dyes and their Suppliers

Dye	Supplier
Reactive Blue 4	Sigma Chemical[1]
Reactive Blue 72	Sigma Chemical
Reactive Brown 10	Sigma Chemical
Reactive Green 5	Sigma Chemical, ICN
Reactive Green 19	Sigma Chemical
Reactive Yellow 2	Sigma Chemical, ICN
Reactive Yellow 3	Sigma Chemical
Reactive Yellow 86	Sigma Chemical
Reactive Red 120	Sigma Chemical, ICN, Pharmacia
Dyematrex Blue A	Millipore[1]
Dyematrex Blue B	Millipore
Dyematrex Green A	Millipore
Dyematrex Orange A	Millipore
Dyematrex Red A	Millipore

[1]These suppliers offer screening kits composed of dye selections in bulk sample or pre-packed columns.

interest, perhaps due to the difficulty in orienting the receptor during the immobilization process. Bailon and colleagues, who developed a recombinant system for producing functional forms of the soluble human interleukin-2 receptor (114), have pioneered this field. This soluble receptor was used as an affinity ligand for developing a solid-phase receptor binding assay (41). This group also developed the procedure of receptor-affinity chromatography for the isolation of human interleukin 2 (9,113). Later, they developed a system called membrane-based receptor affinity chromatography (MRAC), based on the immobilization of a soluble form of interleukin-2 receptor chemically bonded to hollow-fiber membranes. This technique was found to be not only practical but also scalable (75). In a similar vein, we have used receptors immobilized on solid glass beads by CDI as affinity ligands to measure bioactive interleukin 2 in extracts of renal biopsies (85). Immobilized tachykinin receptors have also been employed as selective affinity ligands for the isolation of bioactive neuropeptides from extracts of nasal biopsies obtained from patients undergoing allergic reactions (84).

MRAC has now been developed into a multipurpose fluidized-bed receptor-affinity purification system. The affinity support is composed of the receptor immobilized on aldehyde-activated controlled-pore glass beads and the separation system is comprised of a column fitted at the inlet end with a perforated distributor plate and at the outlet with an adjustable piston. This system has been applied to the purification of recombinant human interleukin 2 and related molecules (100).

OTHER LIGANDS

There is a large array of other molecules reported to be useful as affinity ligands and this list is growing yearly. Such applications range from the immobilization of trifluoromethyl ketones as potential ligands for the isolation of juvenille hormone esterase from insects (99) to the development of thiophilic ligands based on mercaptoheterocycles for the purification of antibodies (97). The isolation of high-affinity antibodies to testosterone 17 β-acetate has been reported using the immobilized steroid as the ligand (37), and lactose coupled to divinyl sulfone-activated agarose has been employed to purify galectin-1, a polypeptide cell growth factor (30). Molecular and biochemical engineering has introduced a new facet to affinity separation technology (42,49). Recombinant technology, together with the use of computational chemistry, has also enabled the engineering of low-molecular-weight synthetic molecules capable of mimicking the activity of their native macromolecular counterparts (68). Li et al. (62) have developed a nonpeptidyl mimic for protein A. This mimic is capable of recognizing the Fc portions of IgG molecules and, when immobilized on agarose beads, exhibited an affinity constant between 10^5 and 10^6 M^{-1}. The authors demonstrated the ability of this mimetic ligand to isolate IgG from human plasma, murine ascitic fluid, and fetal calf serum. Recombinant technology has also introduced single-chain antibodies that carry the same specificity as intact molecules. These easily produced, engineered protein fragments (54) have the major advantage of offering a relatively inexpensive means for applying affinity separations to preparative- and production-scale separations.

LIGAND IMMOBILIZATION

Ligands can be attached to supports by a number of interactive linkages. The most common group utilized for ligand attachment is a primary amide or free thiol group. Supports activated with CNBr, succinimide ester, triazine, epoxide, FMP, CDI, periodate, and sulfonyl ester all bind ligands by interactions with a primary amine. However, sulfonyl esters, FMP, and epoxy-activated supports can also bind ligands by free thiol or sulfhydryl groups. Epoxide-activated supports can also utilize hydroxyl groups as reactive groups for ligand attachment. Oxidized carbohydrates can be attached to hydrazide-activated supports and many ligands can be condensed to carboxyl groups on the support by cross-linking agents. Pierce Chemical is a commercial source of one of the largest selections of cross-linkers available in the US and reference to the Pierce catalog can give investigators an insight into the variety of reagents available for cross-linking proteins, peptides, and carbohydrates.

The covalent immobilization of proteins to a number of supports can be achieved by using glutaraldehyde or either water-soluble or organic solvent-soluble carbodiimides as condensing agents. These cross-linking agents produce a reactive bond between carboxyl groups on the support and amino groups on the ligand (52,108). Recently, a new heterobifunctional photocleavable cross-linking reagent has been developed that can be used to cage amino groups in the ligand prior to covalently bonding it to the support by a thiol-reactive oxirane group (69).

BLOCKING FREE ACTIVE GROUPS ON AFFINITY SUPPORTS

Attachment of a ligand to any activated support rarely utilizes all of the active groups present on the support. Thus, any uncoupled side chains will be available for interaction and possibly attachment of analytes or other materials in the sample. In either case, an undesirable situation arises that can easily be avoided. Following every immobilization, it is necessary to inhibit or block any unused, but potentially reactive, side chains. There are a number of simple procedures available to block such sites, ranging from chemical neutralization to binding a nonspecific molecule. Perhaps the easiest technique is to block the free active groups with a chemical such as ethanolamine or glycine, both of which are capable of reacting with a number of different amino-binding groups. In some cases, excessive ligand can be used to completely saturate the reactive groups, but this is usually impractical because of insufficient ligand or the expense involved in performing such treatment. It has also been reported that specialized blocking agents, such as the Fc fragment of human IgG, can be employed to block free Fc receptors on protein A-coated supports (86). However, when such agents are used, they need to be chemically cross-linked to the support by carbodiimides.

EVALUATING THE EFFICIENCY OF THE AFFINITY SUPPORT

Once the ideal conditions for adsorption and elution of the selected analyte have been established, it is desirable to calculate the effective capacity of the matrix. The simplest approach to determine this effective capacity is to saturate the column with excess analyte and measure the difference in analyte content between the starting and final solutions. This, however, is not very accurate and a more scientific approach is to perform a frontal analysis (4,107). Basically, a frontal analysis requires that a test column be built and saturated with excess test analyte. Once the column-binding capacity has been fully saturated, it is washed and the analyte eluted. The recovered analyte is measured and the effective analyte capacity calculated from

$$A_t (V_e - V_0) \qquad \text{[Eq. 1]}$$

where A_t is the concentration of the analyte in the test solution, V_e represents the volume at which the concentration of the protein present in the effluent reaches 50% of the total, and V_0 represents the volume at which the non-adsorbed protein is 50% of the starting solution. This equation assumes that A is an insignificant proportion of the total protein.

PROTOCOLS

Ligand Attachment to CNBr-Activated Supports (Figure 1)

1. Dissolve 1 μg of protein ligand in 1 mL of 0.1 M NaCO₃, pH 8.5.
2. Add to 1 g of CNBr-activated beads.
3. Incubate for 2 h at room temperature on an overhead or rotary mixer.
4. Wash three times in 200 mL of 0.1 M NaCO₃ by sedimentation.

5. Block free reactive side chains (see protocol for blocking free activated groups below)
6. Store at 4°C.

Ligand Attachment to Succinimide Ester-Activated Supports (Figure 2)

1. Place 1 g of the succinimide-activated support in a sintered glass filter.
2. Wash quickly with 50 mL of ice-cold distilled H_2O.
3. Dissolve 100 µg of ligand in 10 mL of 0.05 M $NaCO_3$, pH 8.5, in a 15-mL capped glass tube.
4. Transfer the washed support to the ligand solution.
5. Resuspend the support in the solution.
6. Place on an overhead or rotary mixer for 6 h at 4°C.
7. Add 5 mL of 2 M ethanolamine to terminate the reaction.
8. Incubate for 1 h at room temperature.
9. Wash the beads five times in 0.1 M sodium phosphate, pH 7.4.
10. Store at 4°C.

Ligand Attachment to Triazine-Activated Supports (Figure 3)

1. Dissolve 1 µg of ligand in 1 mL of 0.5 M KH_2PO_4, pH 8.0.
2. Add 1 g of triazine-activated beads.

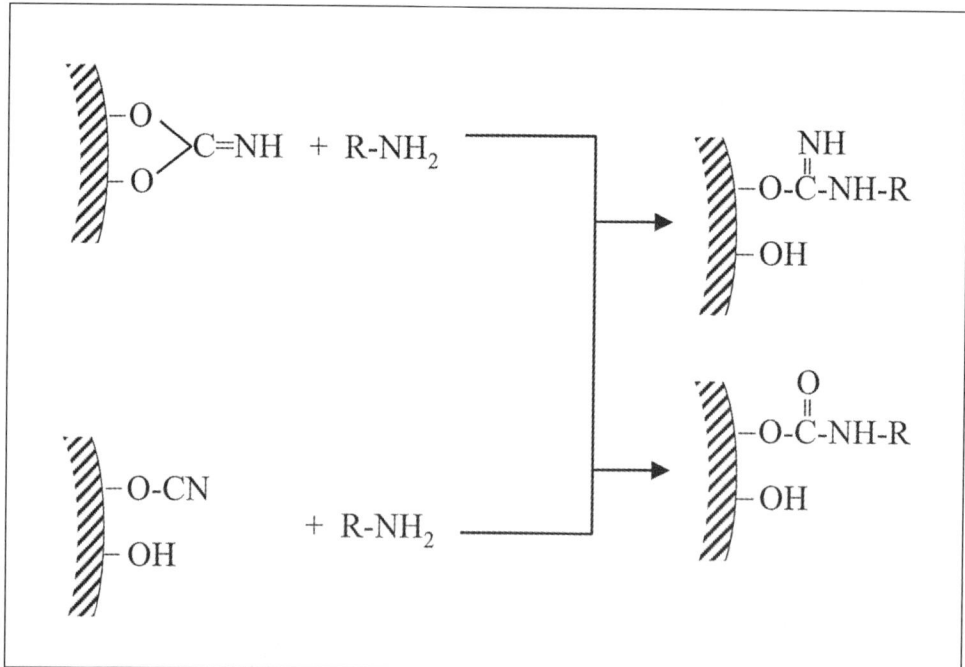

Figure 1. Ligand attachment to CNBr-activated supports by primary aliphatic (or aromatic) amino groups.
Three prominent products are produced, isourea-derivatives, N-substituted carbomates, and N-substituted imidocarbonate (not shown).

3. Place on an overhead or rotary mixer and incubate for 2 h at room temperature.
4. Wash five times in 0.1 M KH_2PO_4, pH 7.2, by sedimentation.
5. Block free reactive side chains (see the protocol for blocking free activated groups, below).

Attachment of Ligands to Periodate-Activated Supports (Figure 4)

1. Dissolve 50 µg of ligand in 1 mL of 0.5 M $NaCO_3$, pH 9.5.
2. Add to 1 g of the periodate-activated beads.
3. Place on an overhead or rotary mixer and mix for 24 h at 4°C.

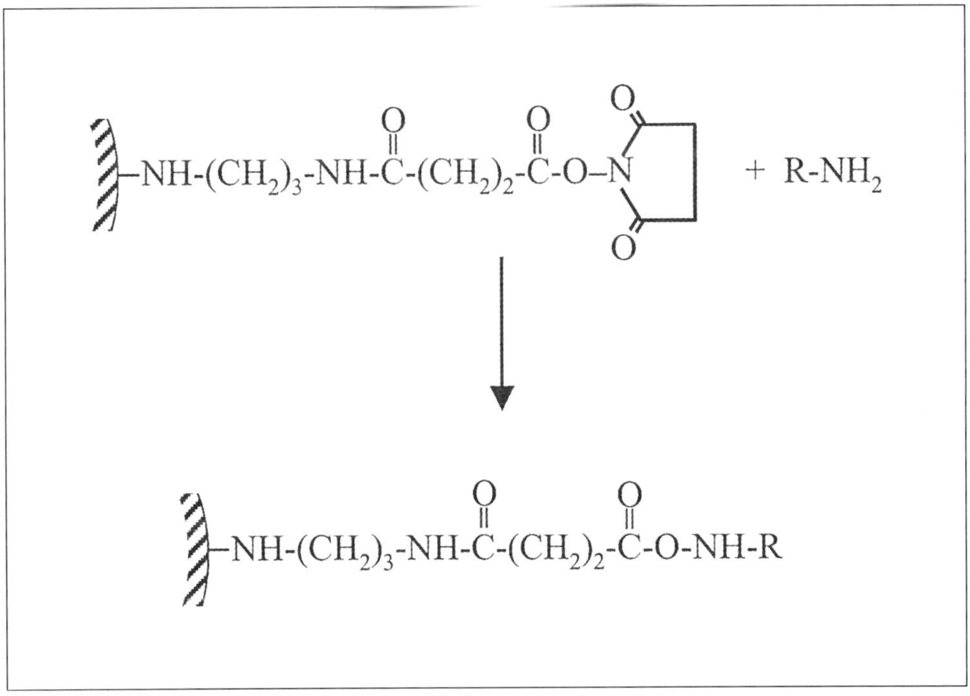

Figure 2. Ligand attachment to *N*-hydroxysuccinimide ester-activated supports by primary aliphatic amino groups.

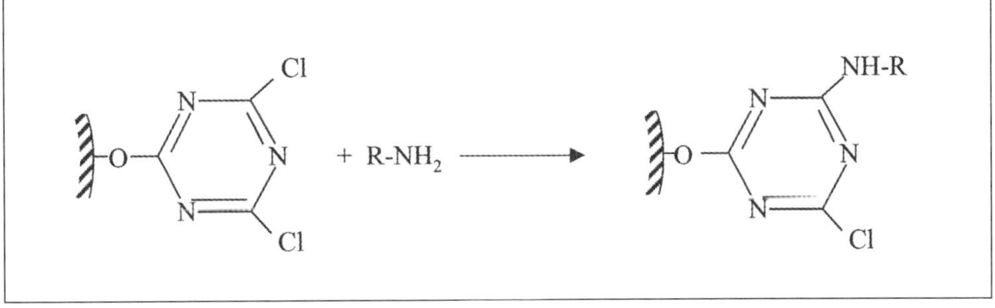

Figure 3. Ligand attachment to triazine-activated supports by primary aliphatic amino groups. The triazine used to activate the support in this example was 2,4,6-trichloro-s-triazine.

95

4. Return the beads to the glass filter and remove the solution containing unbound ligand by applying a vacuum.

5. Place the beads in a 15-mL tube.

6. Resuspend in 2 mL of a cold (4°C) solution of 1 mg/mL $NaBH_4$.

7. Incubate on an overhead mixer for 1 h at 4°C.

8. Wash the beads in 200 mL of 0.01 M sodium phosphate, pH 7.2.

9. Block free reactive side chains (see the protocol for blocking free activated groups, below).

10. Store at 4°C.

Preparation of Ligands for Attachment to Hydrazide-Activated Supports (Figure 5)

1. Dissolve 1 µg of ligand in 1 mL of 0.1 M sodium phosphate, pH 7.2.

2. Transfer to a 15-mL glass tube.

3. Add 1 mL of sodium phosphate containing 10 mg of $NaIO_4$.

4. Mix and place on an overhead or rotary mixer.

5. Incubate for 30 min at room temperature (protect from light by covering the tube with aluminum foil).

6. Pass the oxidized ligand through a desalting column (see the protocol for the removal of dissociation agents using laboratory-built desalting columns in Chapter 3 for details) to remove the periodate.

7. Immediately couple to 1 g of the hydrazide-activated support.

Attachment of Ligands to Epoxide-Activated Supports (Figure 6)

1. Add 1 mL of a 1-µg/mL solution of the ligand dissolved in 0.5 M $NaCO_3$, pH 9.5, to a glass tube.

2. Add 1 g of epoxide-activated support.

3. Place on an overhead or rotary mixer for 16 h at 4°C.

4. Wash in 1 M NaCl.

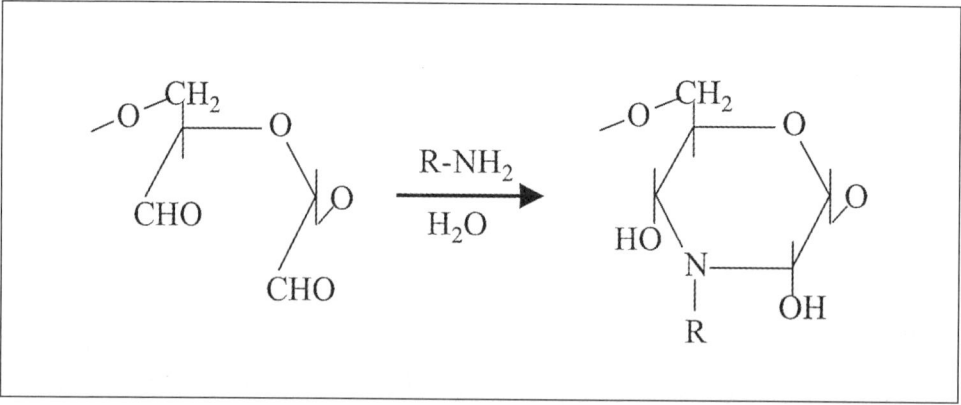

Figure 4. Ligand attachment to periodate-activated supports by primary aliphatic amino groups.

5. Block free reactive side chains (see protocol for blocking free activated groups below) and use within a week.

Attachment of Ligands to Commercial Epoxide-Activated Supports

There are a number of commercially available epoxide-activated supports, although the agarose beads available from Pharmacia are the easiest to use. Coupling of any ligand by primary amino groups can be achieved by the following procedure.

1. Swell 10 g of beads by placing them overnight in 400 mL of distilled H_2O.
2. Resuspend the beads in fresh distilled H_2O.
3. Pour the slurry into a sintered glass funnel.
4. Semi-dry by applying a vacuum.
5. Dissolve the ligand in 0.5 M $NaCO_3$, pH 9.5, at a concentration of 1 mg/mL.

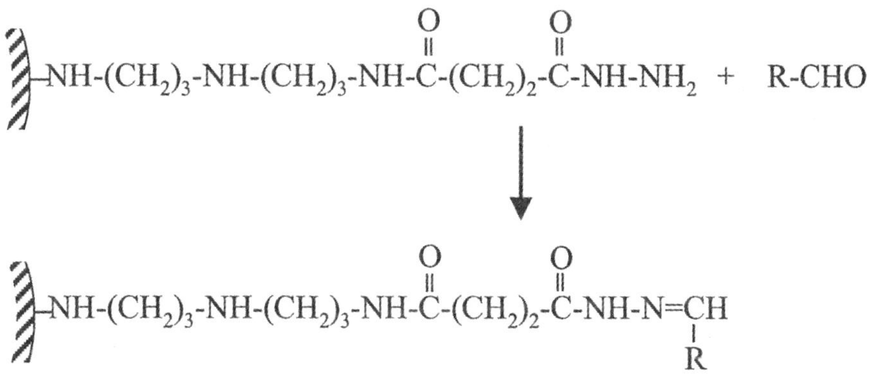

Figure 5. Ligand attachment to hydrazine-activated supports by aldehyde side-chain groups.

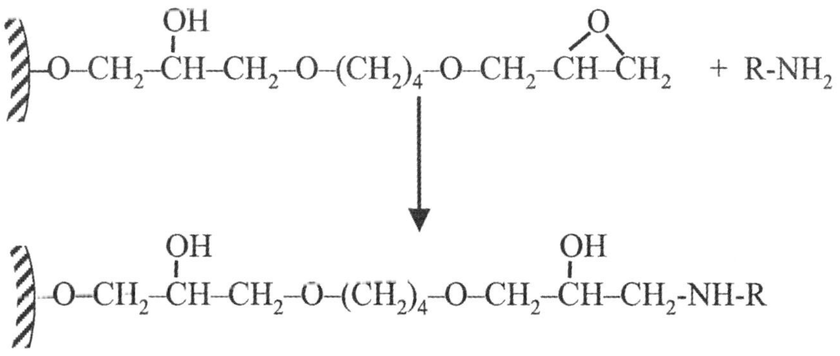

Figure 6. Ligand attachment to epoxide-activated supports by primary aliphatic amino groups.

97

6. Add the ligand to the swollen beads in a ratio of 1:1, in the sintered glass funnel.
7. Transfer to a 15- or 50-mL capped tube.
8. Incubate on an overhead or rotary mixer for 16 h at room temperature.
9. Return the beads to the glass filter.
10. Wash with 50 mL of 0.5 M $NaCO_3$, pH 9.5.
11. Wash with 50 mL of distilled H_2O.
12. Wash with 50 mL of 0.1 M $NaCO_3$, pH 8.5.
13. Wash with 100 mL of 0.1 M $C_2H_3NaO_2$, pH 4.0.
14. Block free reactive side chains (see protocol for blocking free activated groups below) and use within a week.

Attachment of Ligands to Diazo-Activated Supports (Figure 7)

1. Dissolve 1 mg of ligand in 35 mL of 0.2 M NaB_4O_7, pH 8.5.
2. Resuspend 10 g of dry support in this solution.
3. Place the suspension in a 50-mL capped tube.
4. Incubate overnight at 4°C on an overhead or rotary mixer.
5. Recover the beads and wash extensively with 0.2 M NaB_4O_7, pH 8.5.
6. Block free reactive side chains (see protocol for blocking free activated groups below).
7. Store at 4°C.

Ligand Coupling to FMP-Activated Supports (Figure 8)

1. Place 10 g of the support in a sintered glass filter.
2. Wash twice with distilled H_2O.
3. Dry by applying a vacuum.
4. Resuspend the dry support in 1 vol of 0.2 M $NaCO_3$, pH 8.5, containing the ligand (1 mg/mL).
5. Transfer to a 15-mL capped glass tube.
6. Place on an overhead or rotary mixer.
7. Mix overnight at 4°C.
8. Transfer to a sintered glass filter.
9. Wash in 10 vol of 0.2 M $NaCO_3$, pH 9.0.
10. Wash twice in 1 M NaCl.
11. Wash five times in distilled H_2O.
12. Block nonreacted hydroxyl groups by resuspending the support in 1 vol of 0.2 M Tris-HCl, pH 9.0.
13. Place on an overhead or rotary mixer for 2–4 h.
14. Wash twice by sedimentation in 1 M NaCl.
15. Store at 4°C in NaCl.

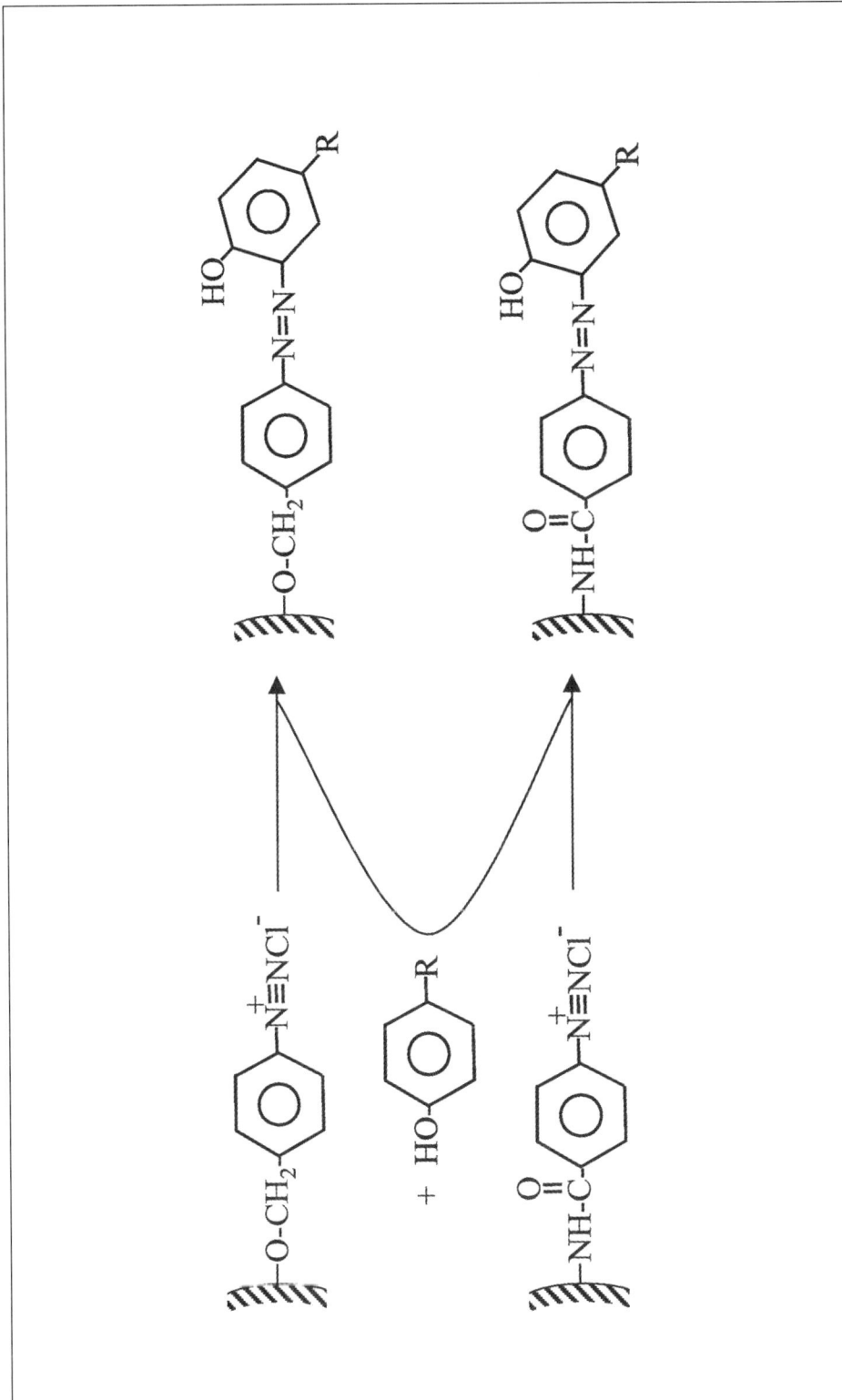

Figure 7. Ligand attachment to diazonium containing activated supports by aromatic amino acids (tyrosine or histidine).

Ligand Attachment to Thiol-Activated Supports (Figure 9)

1. Place 1 g of thiol-activated support in a sintered glass filter.
2. Wash three times in 50 mL of 50 mM $Na_2B_4O_7$, pH 7.6.
3. Transfer the support to a capped glass tube.
4. Add 100 μg of ligand containing a free thiol or sulfhydryl group in 1 mL of 50 mM $Na_2B_4O_7$.
5. Place on a rotary mixer at 4°C for 12–16 h.

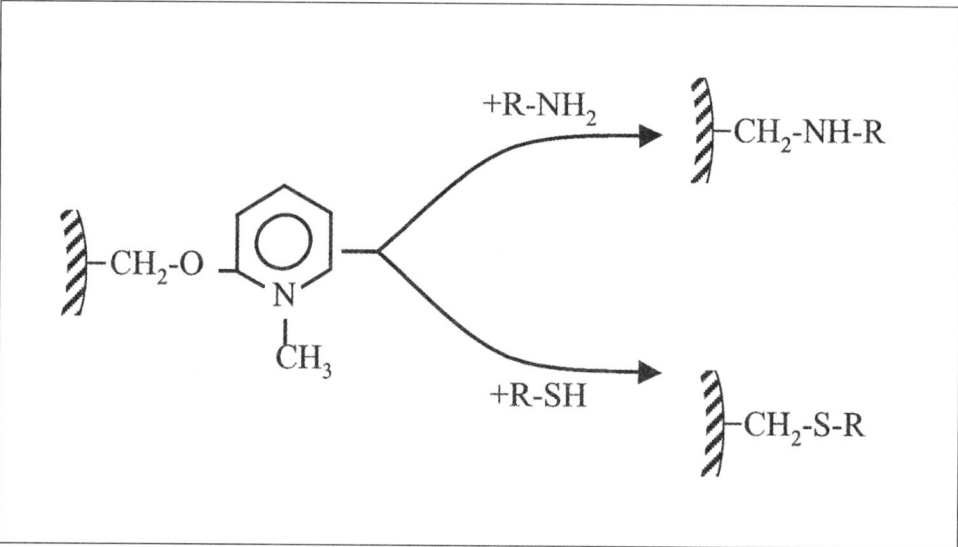

Figure 8. Ligand attachment to FMP-activated supports by either primary aliphatic amino groups or sulfhydryl side chains.

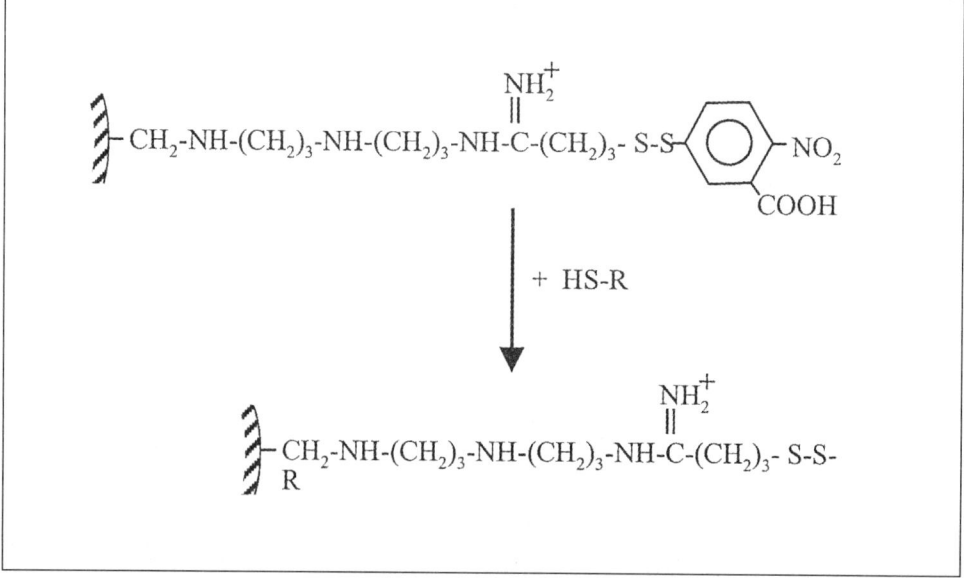

Figure 9. Ligand attachment to thiol-activated supports by sulfydryl side groups.

6. Transfer to a sintered glass filter.

7. Wash five times in 100 mL of 0.1 M sodium phosphate, pH 7.4.

8. Store at 4°C for up to one week.

Ligand Attachment by Glutaraldehyde (Figure 10)

1. Place 10 g of the support in a sintered glass filter.

2. Wash twice in 0.01 M sodium phosphate, pH 7.4.

3. Dissolve 1 mg of ligand in 10 mL of the sodium phosphate.

4. Resuspend the support in the ligand solution.

5. Place in a 15-mL capped glass tube.

6. Place on an overhead or rotary mixer.

7. Incubate for 12–16 h at 4°C.

8. Wash the beads five times in 0.01 M sodium phosphate, pH 7.4.

9. Store at 4°C.

Ligand Attachment to Acyl-Activated Supports (Figure 11)

1. Place 100 mg of the activated support in a small beaker.

2. Add 50 mL of cold 0.3 N HCl.

3. Mix and allow to the support to sediment.

4. Remove the supernatant.

5. Resuspend the support in 15 mL of cold 1 N HCl.

6. Add 1 mL of 1 M NaNO$_2$.

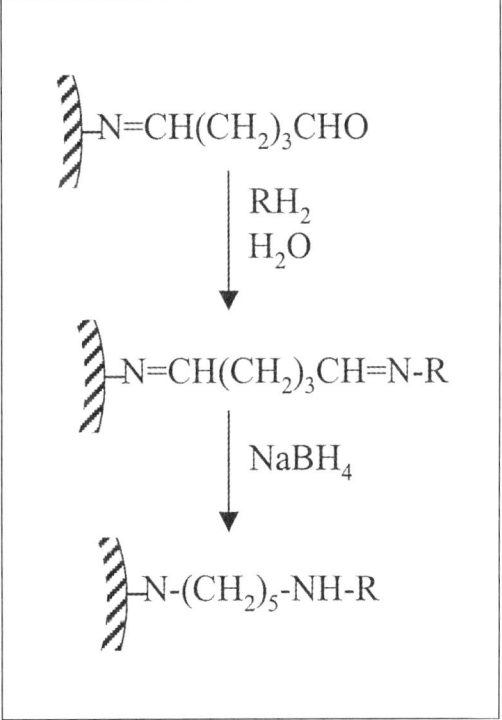

Figure 10. Ligand attachment to glutaraldehyde-activated supports by primary aliphatic amino groups.

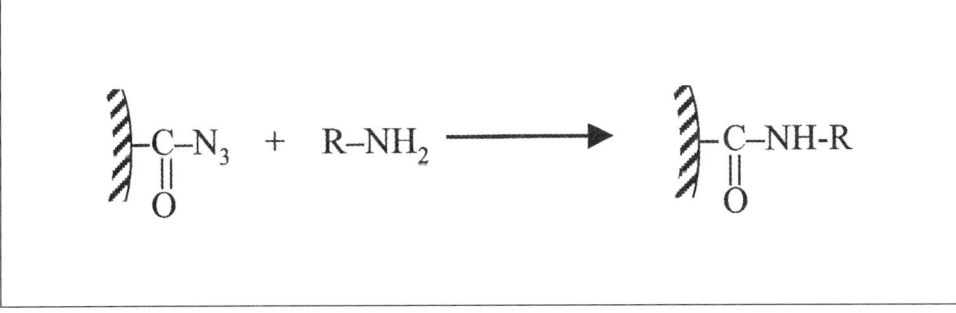

Figure 11. Ligand attachment to acyl azide-activated supports by primary aliphatic amino groups.

7. Incubate for 20 min in an ice bath.

8. Transfer the support to a sintered glass funnel.

9. Wash with 100 mL of 0.3 *N* HCl.

10. Wash with 100 mL of 0.1 M sulfamic acid.

11. Wash with 200 mL of cold distilled H_2O.

12. Dissolve 1 mg of ligand in 15 mL of cold 0.1 M $Na_2B_4O_7$.

13. Resuspend the support in this solution.

14. Place the mixture in a 15-mL capped tube.

15. Incubate for 1 h at 4°C on an overhead or rotary mixer.

16. Block the free acyl groups by adding 4 mL of 1 M NH_4Cl.

17. Mix for 1 h.

18. Wash the gel in 300 mL of 0.2 M NaCl and use within a week.

Attachment of Solvent-Insensitive Ligands to CDI-Activated Supports (Figure 12)

1. Dissolve 1 µg of ligand in 10 mL of formamide.

2. Resuspend 10 g of the support in the ligand solution.

3. Place the mixture in a 15-mL glass tube.

4. Place on an overhead or rotary mixer for 6 h at room temperature.

5. Wash the support 10 times in 10 mL of formamide by sedimentation.

6. Add 10 mL of 0.3 M ethanolamine.

7. Mix for 2 h at room temperature.

8. Wash 10 times with 10 mL formamide.

9. Store in solvent solution.

Attachment of Solvent-Sensitive Ligands to CDI-Activated Supports

1. Place 10 g of support in a sintered glass filter.

2. Remove the storage solution by applying a vacuum.

3. Wash with 20 mL of 70% (vol/vol) formamide.

4. Wash with 20 mL of 30% (vol/vol) formamide.

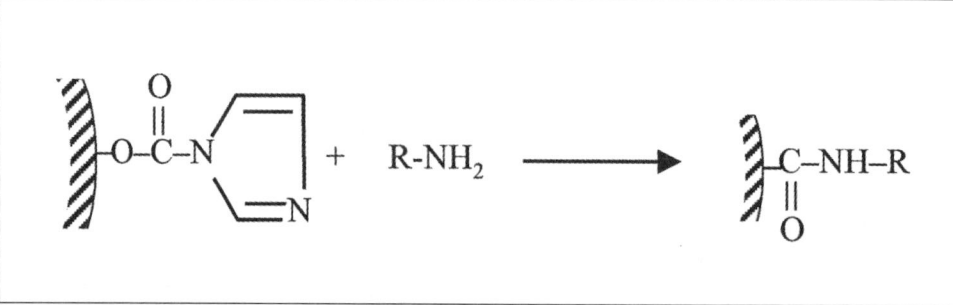

Figure 12. Ligand attachment to carbodiimide-activated supports by primary aliphatic amino groups.

5. Wash with 20 mL of distilled H_2O.
6. Allow the support to sediment.
7. Dissolve 1 μg of ligand in 0.1 M sodium phosphate, pH 7.2.
8. Resuspend the support in the ligand solution.
9. Place in a 15-mL glass tube.
10. Mix on an overhead or rotary mixer for 16–20 h at 4°C.
11. Wash five times in 10 mL of 0.1 M sodium phosphate.
12. Resuspend the support in 10 mL of a 0.2 M solution of Tris-HCl, pH 9.0, to block free CDI side chains.
13. Wash 10 times in 0.1 M sodium phosphate.
14. Store in 0.1 M sodium phosphate at 4°C.

Attachment of Ligands to Thiol-Activated Capillaries

The following technique has successfully been used in our laboratory to coat the internal surfaces of thiol-activated capillaries with antibody FAb fragments. However, the procedure is suitable for attaching any ligand with a free thiol group. Reference to the protocol for CDI-derivatized capillaries in Chapter 2 is helpful.

1. Flush the capillary with 200 μL of 50 mM borate.
2. Flush the capillary with 200 μL of 0.01 M sodium phosphate, pH 7.4.
3. Dissolve 1 μg of ligand in 1 mL of the sodium phosphate.
4. Place 500 nL of the ligand solution onto a Parafilm sheet.
5. Allow the fluid to be drawn into the capillary by capillary action.
6. Incubate overnight at 4°C.
7. Flush the capillary three times with 200 μL of 0.1 M sodium phosphate, pH 7.4,
8. Seal the ends of the capillary with tape.
9. Store at 4°C for up to 6 weeks.

Ligand Attachment by Water-Soluble Carbodiimide Coupling

1. Swell 1 g of the support in distilled H_2O and place in a sintered glass filter.
2. Wash the support in 200 mL of 0.5 M NaCl.
3. Wash the support in 100 mL of 0.1 M sodium phosphate, pH 4.5.
4. Dissolve the ligand at a concentration of 1 mg/mL in 0.1 M sodium phosphate, pH 4.5.
5. Dissolve 10 mg of either EDC or CMC (carbodiimide) in 1 mL of distilled H_2O.
6. Adjust the pH to 4.5 with 0.1 N HCl.
7. Place the support in a 15-mL capped tube.
8. Add 1 mL of ligand solution.
9. Slowly add 1 mL of the carbodiimide (gently shake the tube during the addition of the carbodiimide).
10. Cap the tube tightly.

11. Incubate overnight on an overhead or rotary mixer at room temperature.
12. Return the support to the glass filter.
13. Wash with 500 mL of 0.1 M sodium phosphate, pH 7.2.
14. Store at 4°C.

Ligand Attachment by Organic Solvent-Soluble Carbodiimide Coupling

1. Place 1 g of the support in a sintered glass filter.
2. Wash the support in 200 mL of formamide.
3. Dissolve the ligand at a concentration of 1 mg/mL in formamide.
4. Dissolve 10 mg of dicyclohexylcarbodiimide in 1 mL of formamide.
5. Adjust the pH to 4.5 with 0.1 N HCl.
6. Place the support in a 15-mL capped tube.
7. Add 1 mL of ligand solution.
8. Slowly add 1 mL of the carbodiimide (gently shake the tube during the addition of the carbodiimide).
9. Cap the tube tightly.
10. Incubate overnight on an overhead or rotary mixer at room temperature.
11. Return the support to the glass filter.
12. Wash with 500 mL of formamide.
13. Store at 4°C.

Blocking Free Activated Groups

Effective blocking of free activated groups can also be achieved by substituting 1 M glycine for ethanolamine.

1. Following attachment of the ligand, place the washed support in a capped tube.
2. Add 2 vol of 1 M ethanolamine to the support (a volume is equal to the approximate volume in mL of the gravity-packed support).
3. Incubate on an overhead or rotary mixer for 4 h at room temperature.
4. Wash the support five times in 100 mL of 0.1 M sodium phosphate, pH 7.2, by sedimentation.
5. Store at 4°C.

Frontal Analysis for Estimating the Capacity of an Affinity Matrix

1. Prepare a 1-mL column of fresh matrix.
2. Wash with 8–10 mL of 0.01 M sodium phosphate, pH 7.2.
3. Apply a constant flow of analyte solution.
4. Collect 0.5-mL fractions.
5. Screen the fractions at 280 nm for the presence of protein. If possible, screen the fractions for the presence of the analyte. When the concentration of the analyte in the column effluent equals the concentration of the analyte in the starter solution, the column is saturated. Stop adding the analyte solution.

6. Wash the column with buffer until no analyte or protein is detectable.

7. Recover the analyte by running 0.33 M citric acid, pH 1.5, through the column.

8. Measure the amount of analyte recovered.

REFERENCES

1. **Abudiab, T. and R.R. Beitle.** 1998. Preparation of magnetic immobilized metal affinity separation media and its use in the isolation of proteins. J. Chromatogr. *795*:211-217.

2. **Akashi, M., T. Niikawa, T. Serizawa, T. Hayakawa and M. Baba.** 1998. Capture of HIV-1 gp120 and virions by lectin-immobilized polystyrene nanospheres. Bioconjug. Chem. *9*:50-53.

3. **Aksoy, S., H. Tumturk and N. Hasirci.** 1998. Stability of alpha-amylase immobilized on poly(methyl methacrylate-acrylic acid) microspheres. J. Biotechnol. *60*:37-46.

4. **Angal, S. and P.D.G. Dean.** 1989. Purification by exploitation of activity, p. 245-262. *In* E.L.V. Harris and S. Angal (Eds.), Protein Purification Methods. A Practical Approach. IRL Press, Oxford.

5. **Asante-Appiah, E., G. Merkel and A.M. Skalka.** 1998. Purification of untagged retroviral integrases by immobilized metal ion affinity chromatography. Protein Expr. Purif. *12*:105-110.

6. **Atkinson, T., P.M. Hammond, R.D. Hartwell, P. Hughes, M.D. Scawen, R.F. Sherwood et al.** 1981. Triazine-dye affinity; chromatography. Biochem. Soc. Trans. *9*:290-293.

7. **Baba, Y., T. Sawa, A. Kishida and M. Akashi.** 1998. Base-specific separation of oligodeoxynucleotides by capillary affinity gel electrophoresis. Electrophoresis *19*:433-436.

8. **Baek, W.O., K. Haupt, C. Colin and M.A. Vijayalakshmi.** 1996. Immobilized metal ion affinity gel electrophoresis: quantification of protein affinity to transition metal chelates. Electrophoresis *17*:489-492.

9. **Bailon, P. and D.V. Weber.** 1988. Receptor-affinity chromatography. Nature *335*:839-840.

10. **Bannikova, G.E., V.P. Varlamov, M.L. Miroshnichenko and E.A. Bonch-Osmolovskaya.** 1998. Isolation of thermostable phosphatase from the hyperthermophilic archaeon Thermococcus pacificus by immobilized metal affinity chromatography. Biochem. Mol. Biol. Int. *44*:363-370.

11. **Beers, E.P. and J.J. Callis.** 1993. Utility of polyhistidine-tagged ubiquitin in the purification of ubiquitin-protein conjugates and as an affinity ligand for the purification of ubiquitin-specific hydrolases. Biol. Chem. *268*:21645-21649.

12. **Berg, A. and W.H. Scouten.** 1990. Dye-ligand centrifugal affinity chromatography. Bioseparation *1*:23-31.

13. **Berkenmeier, G., E. Usbeck, L. Saro and G. Kopperschlager.** 1983. Triazine dye binding of human α-fetoprotein and albumin. J. Chromatogr. *265*:27-35.

14. **Blomkalns, A.L. and M.R. Gomez.** 1997. Purification of bovine alpha-lactalbumin by immobilized metal ion affinity chromatography. Prep. Biochem. Biotechnol. *27*:219-226.

15. **Booth, J.R., R. Munks and R.J. Sokol.** 1995. Isolation of IgA1 from human serum by affinity chromatography using an immobilized extract of the albumin gland of Helix pomatia. Transfus. Med. *5*:117-121.

16. **Boyer, P.M. and J.T. Hsu.** 1993. Protein purification by dye-ligand chromatography. Adv. Biochem. Eng. Biotechnol. *49*:1-44.

17. **Bridges, M.A., D.A. Applegarth, J. Johannson, A.G.F. Davidson and L.T.K. Wong.** 1982. Isolation of pure, active α2-macroglobulin from small plasma samples. Clin. Chim. Acta *118*:21-31.

18. **Campanella, L., Z. Aturki, M.P. Sammartino and M. Tomassetti.** 1995. Aspartate analysis in formulations using a new enzyme sensor. J. Pharm. Biomed. Anal. *13*:439-447.

19. **Carlsson, B., H. Ahola and J. Haggblad.** 1997. Application of a novel method for the comparison of DNA binding parameters of the two human thyroid hormone receptor subtypes hTR alpha 1 and hTR beta 1. J. Recept. Signal Transduct. Res. *17*:355-371.

20. **Chen, J.P., Y.M. Sun and D.H. Chu.** 1998. Immobilization of alpha-amylase to a composite temperature-sensitive membrane for starch hydrolysis. Biotechnol. Prog. *14*:473-478.

21. **Conti-Tronconi, B.M. and M. Raftery.** 1982. The nicotinic cholinergic receptor: correlation of molecular structure with functional aspects. Annu. Rev. Biochem. *51*:491-530.

22. **Coulter, A. and R. Harris.** 1983. Simplified preparation of rabbit Fab fragments. J. Immunol. Methods *59*:199-203.

23. **Das, N., A.M. Kayastha and O.O. Malhotra.** 1998. Immobilization of urease from pigeonpea (Cajanus cajan L.) in polyacrylamide gels and calcium alginate beads. Biotechnol. Appl. Biochem. *27*:25-29.

24. **Davey, H.P., G. Embery, J.E. Creeth and D. Cummins.** 1997. Identification of a zinc-binding cystic fibrosis antigen in human saliva by 65Zn probing and N-terminal sequencing. Arch. Oral Biol. *42*:861-867.

25. **Dean, P.D.G. and D.H. Watson.** 1979. Protein purification using immobilised triazine dyes. J. Chromatogr. *165*:301-319.

26. **Debray, H., J. Montreuil and H. Franz.** 1994. Fine sugar specificity of the mistletoe (Viscum album) lectin I. Glycoconj. J. *11*:550-557.

27. **Denizli, A., M. Kocakulak and E. Piskin.** 1998. Bilirubin removal from human plasma in a packed-bed column system with dye-affinity microbeads. J. Chromatogr. B Biomed. Sci. Appl. *707*:25-31.

28. **de Pina, K., C. Navarro, L. McWalter, D.H. Boxer, N.C. Price, S.M. Kelly, M.A. Mandrand-Berthelot and L.F. Wu.** 1995. Purification and characterization of the periplasmic nickel-binding protein NikA of Escherichia coli K12. Eur. J. Biochem. *227*:857-865.

29. **Eis, C., R. Griessler, M. Maier, A. Weinhausel, B. Bock, K.D. Kulbe, D. Haltrich, R. Schinzel and B.J. Nidetzky.** 1997. Efficient downstream processing of maltodextrin phosphorylase from Escherichia coli and stabilization of the enzyme by immobilization onto hydroxyapatite. J. Biotechnol. *58*:157-166.

30. **Fouillit, M., M. Levi-Strauss, V. Giudicelli, D. Lutomski, D. Bladier, M. Caron and R. Joubert-Caron.** 1998. Affinity purification and characterization of recombinant human galectin-1. J. Chromatogr. B Biomed. Sci. Appl. *706*:167-171.

31. **Fountoulakis, M., H. Langen, S. Evers, C. Gray and B. Takacs.** 1997. Two-dimensional map of Haemophilus influenzae following protein enrichment by heparin chromatography. Electrophoresis *18*:1193-1202.

32. **Galanina, O.E., H. Kaltner, L.S. Khraltsova, N.V. Bovin and H.J. Gabius.** 1997. Further refinement of the description of the ligand-binding characteristics for the galactoside-binding mistletoe lectin, a plant agglutinin with immunomodulatory potency. J. Mol. Recognit. *10*:139-147.

33. **Gambino, R., G. Ruiu, G. Pagano and M. Cassader.** 1997. Characterization of the carbohydrate structures of apolipoprotein H through concanavalin A affinity chromatography. J. Lipid Mediat. Cell Signal. *16*:11-21.

34. **Gattoni, M., A. Boffi and E. Chiancone.** 1998. Immobilized apo-myoglobin, a new stable reagent for measuring rates of heme dissociation from hemoglobin. FEBS Lett. *424*:275-278.

35. **Gee, A.P., T. Borsos and M.D.P. Boyle.** 1979. Interaction between components of the human classical complement pathway and immobilized Cibacron Blue F3GA. J. Immunol. Methods *30*:119-126.

36. **Gianazza, E. and P. Arnaud.** 1982. A general method for fractionation of plasma proteins. Dye-ligand affinity chromatography on immobilized Cibacron Blue F3-GA. Biochem. J. *201*:129-136.

37. **Giraudi, G. and C. Baggiani.** 1996. Strategy for fractionating high-affinity antibodies to steroid hormones by affinity chromatography. Analyst *121*:939-944.

38. **Govoroun, M.S., J.C. Huet, J.C. Pernollet and B. Breton.** 1997. Use of immobilized metal ion affinity chromatography and dye-ligand chromatography for the separation and purification of rainbow trout pituitary gonadotropins, GTH I and GTH II. J. Chromatogr. B *698*:35-46.

39. **Grisshammer, R. and J. Tucker.** 1997. Quantitative evaluation of neurotensin receptor purification by immobilized metal affinity chromatography. Protein Expr. Purif. *11*:53-60.

40. **Habibi-Rezaei, M. and M. Nemat-Gorgani.** 1997. Adsorptive immobilization of submitochondrial particles on concanavalin A Sepharose-4B. Appl. Biochem. Biotechnol. *67*:165-181.

41. **Hakimi, J., C. Seals, L.E. Anderson, F.J. Podlaski, P. Lin, W. Danho et al.** 1987. Biochemical and functional analysis of soluble human interleukin-2 receptor produced in rodent cells. Solid-phase reconstitution of a receptor-ligand binding reaction. J. Biol. Chem. *262*:17336-17341.

42. **Harakas, N.K.** 1994. Protein purification process engineering. Biospecific affinity chromatography. Bioprocess Technol. *18*:259-316.

43. **Hatakeyama, T., Y. Miyamoto, H. Nagatomo, I. Sallay and N. Yamasaki.** 1997. Carbohydrate-binding properties of the hemolytic lectin CEL-III from the holothuroidea Cucumaria echinata as analyzed using carbohydrate-coated microplate. J. Biochem. *121*:63-67.

44. **Hattori, T., X. Zhang, C. Weiss, T. Xu, T. Kubo, Y. Sato, S. Nishikawa, H. Sakaida and T. Uchiyama.** 1997. Triazine dyes inhibit HIV-1 entry by binding to envelope glycoproteins. Microbiol. Immunol. *41*:717-724.

45. **Haupt, K., F. Roy and M.A. Vijayalakshmi.** 1996. Immobilized metal ion affinity capillary electrophoresis of proteins—a model for affinity capillary electrophoresis using soluble polymer-supported ligands. Anal. Biochem. *234*:149-154.

46. **Hayatsu, H., Y. Tanaka and K. Negishi.** 1997. Preparation of DNA-chitosan columns and their applications: binding of carcinogens to the column. Nucleic Acids Symp. Ser. *37*:139-140.

47. **Holmskov, U., P.B. Fischer, A. Rothmann and P. Hojrup.** 1996. Affinity and kinetic analysis of the bovine plasma C-type lectin collectin-43 (CL-43) interacting with mannan. FEBS Lett. *393*:314-316.

48. **Hseu, M.J., C.Y. Yen, C.C. Tseng and M.C. Tzeng.** 1997. Purification and partial amino acid sequence of a novel protein of the reticulocalbin family. Biochem. Biophys. Res. Commun. *239*:18-22.

49. **Irving, R., J. Atwell and P. Hudson.** 1993. Protein engineering: the selection of proteins with improved binding affinity using complex expression libraries. Australas. Biotechnol. *3*:86-93.

50. **Jiang, W. and M.T.W. Hearn.** 1996. Protein interaction with immobilized metal ion affinity ligands under high ionic strength conditions. Anal. Biochem. *242*:45-54.

51. **Kabir, S.** 1998. Jacalin: a jackfruit (Artocarpus heterophyllus) seed-derived lectin of versatile applications in immunobiological research. J. Immunol. Methods *212*:193-211.

52. **Kang, I.K., B.K. Kwon, J.H. Lee and H.B. Lee.** 1993. Immobilization of proteins on poly(methyl methacrylate) films. Biomaterials *14*:787-792.

53. **Kennedy, R.M.** 1997. Medium format enhancement: design of an immobilized heparin affinity media for industrial bioprocessing. J. Mol. Recognit. *10*:88-92.

54. **Kipriyanov, S.M., G. Moldenhauer and M. Little.** 1997. High level production of soluble single chain antibodies in small-scale Escherichia coli cultures. J. Immunol. Methods *200*:69-77.

55. **Kondo, K., T. Ozaki, Y. Nakamura and S. Sakiyama.** 1995. DAN gene product has an affinity for Ni2+. Biochem. Biophys. Res. Commun. *216*:209-215.

56. **Kozulic, B., I. Leustek, B. Pavlovic, P. Mildner and S. Barbaric.** 1987. Preparation of the stabilized glycoenzymes by cross-linking their carbohydrate chains. Appl. Biochem. Biotechnol. 15:265-278.

57. **Kulik, E.A., K. Kato, M.I. Ivanchenko and Y. Ikada.** 1993. Trypsin immobilization on to polymer surface through grafted layer and its reaction with inhibitors. Biomaterials *14*:763-769.

58. **Le, Y., S.M. Tan, S.H. Lee, O.L. Kon and J. Lu.** 1997. Purification and binding properties of a human ficolin-like protein. J. Immunol. Methods *204*:43-49.

59. **Leatherbarrow, R.J. and P.D.G. Dean.** 1980. Studies on the mechanism of binding of serum albumin to immobilized Cibacron Blue F3GA. Biochem. J. *189*:27-34.

60. **Lengyel, Z., G. Pal and M. Sahin-Toth.** 1998. Affinity purification of recombinant trypsinogen using immobilized ecotin. Protein Expr. Purif. *12*:291-294.

61. **Lepiku, M. and J. Jarv.** 1994. Optimization of synthesis of agarose-based lectin affinity sorbents. Prep. Biochem. *24*:61-67.

62. **Li, R., V. Dowd, D.J. Stewart, S.J. Burton and C.R. Lowe.** 1998. Design, synthesis, and application of a protein A mimetic. Nat. Biotechnol. *16*:190-195.

63. **Liesiene, J., K. Racaityte, M. Morkeviciene, P. Valancius and V. Bumelis.** 1997. Immobilized metal affinity chromatography of human growth hormone. Effect of ligand density. J. Chromatogr. *764*:27-33.

64. **Liu, S., R. Tobias, S. McClure, G. Styba, Q. Shi and G. Jackowski.** 1997. Removal of endotoxin from recombinant protein preparations. Clin. Biochem. *30*:455-463.

65. **Lizcano, J.M., K.F. Tipton and M. Unzeta.** 1998. Purification and characterization of membrane-bound semicarbazide-sensitive amine oxidase (SSAO) from bovine lung. Biochem. J. *331*:69-78.

66. **Lowe, C.R., S.J. Burton, N.P. Burton, W.K. Alderton, J.M. Pitts and J.A. Thomas.** 1992. Designer dyes: "biomimetic" ligands for the purification of pharmaceutical proteins by affinity chromatography. Trends Biotechnol. *10*:442-448.

67. **Lundqvist, A. and P. Lundahl.** 1997. Glucose affinity for the glucose transporter Glut1 in native or reconstituted lipid bilayers. Temperature-dependence study by biomembrane affinity chromatography. J. Chromatogr. A *776*:87-91.

68. **MacLennan, J.** 1995. Engineering microprotein ligands for large-scale affinity purification. Biotechnology *13*:1180-1183.

69. **Marriott, G. and J. Ottl.** 1998. Synthesis and applications of heterobifunctional photocleavable cross-linking reagents. Methods Enzymol. *291*:155-175.

70. **Matson, R.S. and M.C. Little.** 1988. Strategy for the immobilization of monoclonal antibodies on solid-phase supports. J. Chromatogr. *458*:67-77.

71. **McCreath, G.E. and H.A. Chase.** 1996. Affinity adsorption of Saccharomyces cerevisiae on concanavalin A perflurocarbon emulsions. J. Mol. Recognit. *9*:607-616.

72. **Min, C. and G.L. Verdine.** 1996. Immobilized metal affinity chromatography of DNA. Nucleic Acids Res. *24*:3806-3810.

73. **Misaki, A., M. Kakuta, Y. Meah and I.J. Goldstein.** 1997. Purification and characterization of the alpha-1,3-mannosylmannose-recognizing lectin of Crocus vernus bulbs. J. Biol. Chem. *272*:25455-25461.

74. **Muller, K.M., K.M. Arndt, K. Bauer and A. Pluckthun.** 1998. Tandem immobilized metal-ion affinity chromatography/immunoaffinity purification of His-tagged proteins—evaluation of two anti-His-tag monoclonal antibodies. Anal. Biochem. *259*:54-61.

75. **Nachman, M., A.R. Azad and P. Bailon.** 1992. Membrane-based receptor affinity chromatography. J. Chromatogr. *597*:155-166.

76. **Niwa, O., T. Horiuchi, R. Kurita and K. Torimitsu.** 1998. On-line electrochemical sensor for selective continuous measurement of acetylcholine in cultured brain tissue. Anal. Chem. *70*:1126-1132.

77. **Okamura, S., F. Crane, H.A. Messner and T.W. Mak.** 1978. Purification of terminal deoxynucleotidyl-transferase by oligonucleotide affinity chromatography. J. Biol. Chem. *253*:3765-3767.

78. **Olson, M.W., M. Toth, D.C. Gervasi, Y. Sado, Y. Ninomiya and R. Fridman.** 1998. High affinity binding of latent matrix metalloproteinase-9 to the alpha2(IV) chain of collagen IV. J. Biol. Chem. *273*:10672-10681.

79. **Osborne, P.G., O. Niwa, T. Kato and K.J. Yamamoto.** 1997. On-line, continuous measurement of extracellular striatal glucose using microdialysis sampling and electrochemical detection. J. Neurosci. Methods *77*:143-150.

80. **Osborne, P.G. and K. Yamamoto.** 1998. Disposable, enzymatically modified printed film carbon electrodes for use in the high-performance liquid chromatographic-electrochemical detection of glucose or

hydrogen peroxide from immobilized enzyme reactors. J. Chromatogr. B Biomed. Sci. Appl. *707*:3-8.

81. **Ozaki, Y., T. Ihara, Y. Katayama and M. Maeda.** 1997. Affinity capillary electrophoresis using DNA conjugates. Nucleic Acids Symp. Ser. *37*:235-236.

82. **Patwardhan, A.V. and M.M. Ataai.** 1997. Site accessibility and the pH dependence of the saturation capacity of a highly cross-linked matrix. Immobilized metal affinity chromatography of bovine serum albumin on chelating Superose. J. Chromatogr. *767*:11-23.

83. **Peumans, W.J., H.C. Winter, V. Bemer, F. van Leuven, I.J. Goldstein, P. Truffa-Bachi and E.J. van Damme.** 1997. Isolation of a novel plant lectin with an unusual specificity from Calystegia sepium. Glycoconj. J. *14*:259-265.

84. **Phillips, T.M.** 1996. Measurement of bioactive neuropeptides using a chromatographic immunosensor cartridge. Biomed. Chromatogr. *10*:331-336.

85. **Phillips, T.M.** 1997. Measurement of total and bioactive interleukin-2 in tissue samples by immunoaffinity-receptor affinity chromatography. Biomed. Chromatogr. *11*:200-204.

86. **Phillips, T.M., N.S. More, W.D. Queen and A.M. Thompson.** 1985. Isolation and quantitation of serum IgE levels by high-performance immunoaffinity chromatography. J. Chromatogr. *327*:205-211.

87. **Piehler, J., A. Brecht, G. Gauglitz, M. Zerlin, C. Maul, R. Thiericke and S. Grabley.** 1997. Label-free monitoring of DNA-ligand interactions. Anal. Biochem. *249*:94-102.

88. **Pujol, F.H. and I.M. Cesari.** 1993. Schistosoma mansoni: surface membrane isolation with lectin-coated beads. Membr. Biochem. *10*:155-161.

89. **Pundir, C.S., N.K. Kuchhal and A.K. Bhargava.** 1998. Determination of urinary oxalate with oxalate oxidase and peroxidase immobilized on to glass beads. Biotechnol. Appl. Biochem. *27*:103-107.

90. **Rad, A.Y., H. Ayhan and E. Piskin.** 1997. Protein A carrying PMMA microbeads: adsorption of cholesterol and HIgG from human plasma. Int. J. Artif. Organs *20*:576-579.

91. **Radi, R., H. Rubbo, K. Bush and B.A. Freeman.** 1997. Xanthine oxidase binding to glycosaminoglycans: kinetics and superoxide dismutase interactions of immobilized xanthine oxidase-heparin complexes. Arch. Biochem. Biophys. *339*:125-135.

92. **Rubtsova, M. Yu, G.V. Kovba and A.M. Egorov.** 1998. Chemiluminescent biosensors based on porous supports with immobilized peroxidase. Biosens. Bioelectron. *13*:75-85.

93. **Saint-Blancard, J., J.M. Kirzin, P. Riberon, F. Petit, J. Foucart, P. Girot and E. Boschetti.** 1982. A simple and rapid procedure for large scale preparation of IgGs and albumin from human plasma by ion exchange and affinity chromatography, p. 305. *In* T.C.J. Gribnau, J. Visser and R.J.F. Nivard (Eds.), Affinity Chromatography and Related Techniques. Elsevier Scientific, Amsterdam.

94. **Saito, F., H. Yamada, Y. Sunada, H. Hori, T. Shimizu and K. Matsumura.** 1997. Characterization of a 30-kDa peripheral nerve glycoprotein that binds laminin and heparin. J. Biol. Chem. *272*:26708-26713.

95. **Saito, T., S. Kawabata, M. Hirata and S. Iwanaga.** 1995. A novel type of limulus lectin-L6. Purification, primary structure, and antibacterial activity. J. Biol. Chem. *270*:14493-14499.

96. **Schain, L.R., D. Okrongly, T.B. Okarma and J.S. Lebkowski.** 1994. Separation of lectin-binding cells using polystyrene culture devices with covalently immobilized soybean agglutinin. J. Hematother. *3*:37-46.

97. **Schwarz, A., F. Kohen and M. Wilchek.** 1995. Novel heterocyclic ligands for the thiophilic purification of antibodies. J. Chromatogr. B Biomed. Appl. *664*:83-88.

98. **Shinohara, Y., Y. Hasegawa, H. Kaku and N. Shibuya.** 1997. Elucidation of the mechanism enhancing the avidity of lectin with oligosaccharides on the solid phase surface. Glycobiology *7*:1201-1208.

99. **Shiotsuki, T., T.L. Huang, T. Uematsu, B.C. Bonning, V.K. Ward and B.D. Hammock.** 1994. Juvenile hormone esterase purified by affinity chromatography with 8-mercapto-1,1,1-trifluoro-2-octanone as a rationally designed ligand. Protein Expr. Purif. *5*:296-306.

100. **Spence, C., C.A. Schaffer, S. Kessler and P. Bailon.** 1994. Fluidized-bed receptor-affinity chromatography. Biomed. Chromatogr. *8*:236-241.

101. **Spivak, J.L., L.S. Avedissian, J.H. Pierce, D. Williams, W.D. Hankins and R.A. Jensen.** 1996. Isolation of the full-length murine erythropoietin receptor using a baculovirus expression system. Blood *87*:926-937.

102. **Stead, C.V.** 1991. The use of dyes in protein purification. Bioseparation *2*:129-136.

103. **Su, Z.G., W. Jiang, P.S. Feng and K.Y. Gao.** 1991. Purification of serum albumin with dye-ligand adsorption chromatography. Chin. J. Biotechnol. *7*:161-168.

104. **Tadepalli, S.S., D.F. Bruley, K.A. Kang and W. Drohan.** 1997. Separation of protein C from fraction IV of the Cohn process using immobilized metal affinity chromatography. Adv. Exp. Med. Biol. *428*:639-644.

105. **To, W.Y., J.C. Leung and K.N. Lai.** 1995. Identification and characterization of human serum alpha2-HS glycoprotein as a jacalin-bound protein. Biochim. Biophys. Acta *1249*:58-64.

106. **Tucker, R.F., J. Babul and E. Stellwagen.** 1981. Protein dye affinity chromatography using immobilized tetra-iodofluorescein. J. Biol. Chem. *256*:10993-10998.

107. **Tweed, S.A., B. Loun and D.S. Hage.** 1997. Effects of ligand heterogeneity in the characterization of affinity columns by frontal analysis. Anal. Chem. *69*:4790-4798.

108.**Valuev, I.L., V.V. Chupov and L.I. Valuev.** 1998. Chemical modification of polymers with physiologically active species using water-soluble carbodiimides. Biomaterials *19*:41-43.

109.**VanderNoot, V.A., R.E. Hileman, J.S. Dordick and R.J. Linhardt.** 1998. Affinity capillary electrophoresis employing immobilized glycosaminoglycan to resolve heparin-binding peptides. Electrophoresis *19*:437-441.

110.**Vankova, R., A. Gaudinova, H. Sussenbekova, P. Dobrev, M. Strnad, J. Holik and J. Lenfeld.** 1998. Comparison of oriented and random antibody immobilization in immunoaffinity chromatography of cytokinins. J. Chromatogr. *811*:77-84.

111.**Vargas-Albores, F., F. Jimenez-Vega and G.M. Yepiz-Plascencia.** 1997. Purification and comparison of beta-1,3-glucan binding protein from white shrimp (Penaeus vannamei). Comp. Biochem. Physiol. B Biochem. Mol. Biol. *116*:453-458.

112.**Wang, J., X. Cai, G. Rivas, H. Shiraishi and N. Dontha.** 1997. Nucleic-acid immobilization, recognition and detection at chronopotentiometric DNA chips. Biosens. Bioelectron. *12*:587-599.

113.**Weber, D.V. and P. Bailon.** 1991. Application of receptor-affinity chromatography to bioaffinity purification. J. Chromatogr. *510*:59-69.

114.**Weber, D.V., R.F. Keeney, P.C. Familletti and P. Bailon.** 1988. Medium-scale ligand-affinity purification of two soluble forms of human interleukin-2 receptor. J. Chromatogr. *431*:55-63.

115.**Wheatley, J.B., M.H. Lyttle, M.D. Hocker and D.E. Schmidt.** 1996. Salt-induced immobilizations of DNA oligonucleotides on an epoxide-activated high-performance liquid chromatographic affinity support. J. Chromatogr. *726*:77-90.

116.**Winkler, U., T.M. Pickett and D. Hudig.** 1996. Fractionation of perforin and granzymes by immobilized metal affinity chromatography (IMAC). Immunol. Methods *191*:11-20.

117.**Worrall, D.M.** 1996. Dye-ligand affinity chromatography. Methods Mol. Biol. *59*:169-176.

118.**Xiao, J. and M.E. Meyerhoff.** 1996. Retention behavior of amino acids and peptides on protoporphyrin-silica stationary phases with varying metal ion centers. Anal. Chem. *68*:2818-2825.

119.**Yan, L., P.P. Wilkins, G. Alvarez-Manilla, S.I. Do, D.F. Smith and R.D. Cummings.** 1997. Immobilized Lotus tetragonolobus agglutinin binds oligosaccharides containing the Le(x) determinant. Glycoconj. J. *14*:45-55.

120.**Yip, T.T. and T.W. Hutchens.** 1996. Immobilized metal ion affinity chromatography. Methods Mol. Biol. *59*:197-210.

121.**Zeng, C.M., Y. Zhang, L. Lu, E. Brekkan, A. Lundqvist and P. Lundahl.** 1997. Immobilization of human red cells in gel particles for chromatographic activity studies of the glucose transporter Glut1. Biochim. Biophys. Acta *1325*:91-98.

122.**Zhou, F.L., M. Burnouf-Radosevich and T. Burnouf.** 1994. Purification of parasitic worm antigens and factor VIII/von Willebrand factor from human plasma on immobilized lentil lectin. Protein Expr. Purif. *5*:138-143.

5 Preparation of Immunoaffinity Matrices

Immunoaffinity separations are gaining wide application in a number of biological, biochemical, and medical fields (2,26,63,68,83,84,99,101,111,114). This interest is now spreading to the use of immunoaffinity chromatography as a tool for pre-analytical cleanup (74,93,98,108), drug screening (16,19,31,86), and toxicology (23,61, 95). Recently, Hage has written an excellent overview of the applications of immunoaffinity techniques in the biological and chemical sciences (32). Immuno-affinity separations are achieved by applying the specificity of the antibody-antigen interaction to isolate a specific analyte. Most often, an immobilized antibody is used as the ligand to isolate a very specific analyte (the antigen to which it is specifically directed). However, the technique can be reversed such that an immobilized antigen is used as the ligand to isolate antibodies directed specifically against it (17,50). In both cases, orientation of the immobilized ligand is important, although more care must be exercised when immobilizing antibodies. The immunoglobulin molecule (as described in Chapter 1) requires that its antigen binding sites be oriented away from the support. This outward-pointing orientation allows the immobilized antibodies to encounter and bind their specific antigen. Most of the supports described in Chapter 2 can be used in immunoaffinity separations; however, direct linkage of the immuno-globulin molecule to the surface of such supports usually results in poor orientation of the immunoglobulin and loss of specific binding activity (47). Additionally, attachment through free amino groups can result in the immunoglobulin molecule being attached to the substrate by its FAb arms, thus inhibiting antigen-binding activity (Figure 1A). To facilitate correct attachment of any antibody or reactive antibody fragment to a support, special surface coating procedures are employed. Such procedures include coating the support surface with bacterial or plant materials that bind to the Fc portion of the antibody, thus ensuring correct orientation during attachment to the support (Figure 1B) (79). Likewise, the Fc portion of the antibody itself can be chemically modified to ensure correct attachment to specialized support surfaces. For a number of years, our laboratory has specialized in the development of a series of coatings designed for specific antibody immobilization and orientation. Based on this experience, this chapter will be devoted to describing practical approaches to the building of suitable supports for antibody and antigen immobilization.

Affinity and Immunoaffinity Purification Techniques
Terry M. Phillips and Benjamin F. Dickens
© 2000 Eaton Publishing, Natick, MA

BACTERIAL PROTEIN MATRICES

Natural selection has designed a number of unique molecules that help organisms evade the ravages of the host immune system. For example, bacteria of the *Staphylococcus* species, especially *S. aureus*, have evolved coat proteins that protect the bacteria from immunological attack by binding host immunoglobulins. This coat protein, commonly called protein A, binds a number of mammalian immunoglobulins (especially IgG) by their Fc portions (9,29,75,92,94). Due to this binding activity, protein A, which is available in several different forms, has become a useful tool in immunoaffinity separations. A crude protein A matrix preparation can be produced by simply treating a bacterial culture with a combination of heat and formalin. This procedure produces whole bacteria that can be used for immunoglobulin capture and isolation (22,46). Alternatively, the protein can be used after extensive purification, and recently, recombinant Protein A has become commercially available.

Genetic engineering studies have demonstrated that the protein A molecule possesses five potential immunoglobulin binding sites (69), although only two IgG molecules can effectively bind to each protein A molecule (56). The most common uses of protein A are as a ligand for the affinity chromatographic separation of immunoglobulins and for the clinical removal of antibody-antigen complexes (11,39). However, immobilized protein A is also used as a bridging molecule to bind antibodies to a physical support for immunoaffinity chromatography (Figure 2) (79,91).

Since protein A binds to the Fc portion of immunoglobulins, it leaves the antibody's FAb binding sites free to capture antigens. This appears to make it an ideal

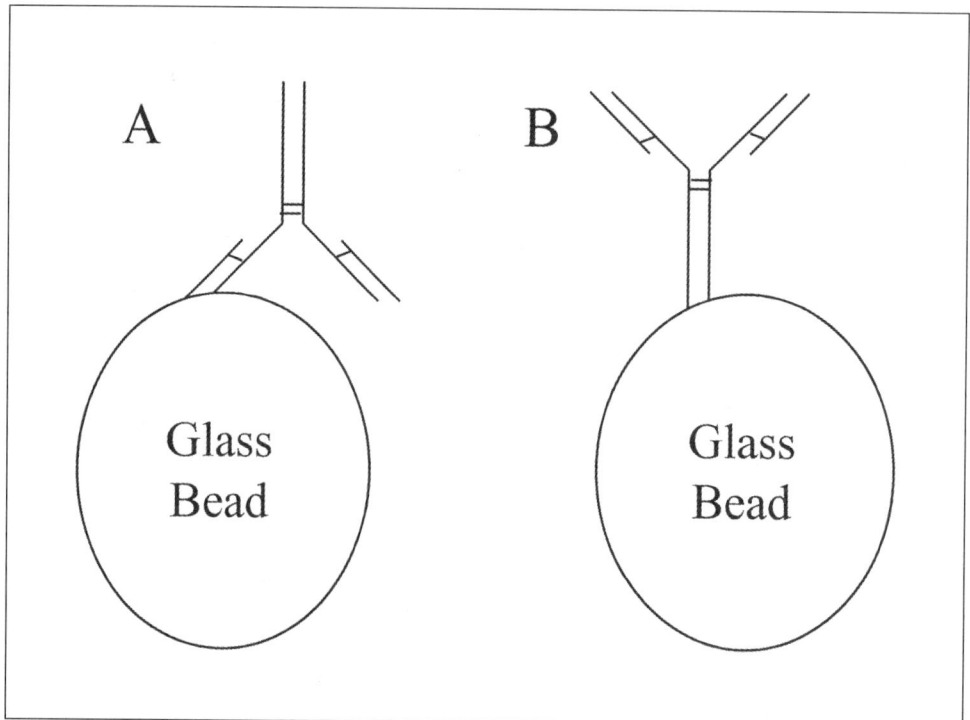

Figure 1. Orientation of immunoglobulins binding to immobilization matrix surfaces. Incorrect orientation involving the amino terminus (A), and correct orientation by the carboxyl terminus (B).

material for the correct orientation of immunoglobulins. However, a serious disadvantage of protein A is that it only binds selective immunoglobulins, and this selectivity extends to immunoglobulins from a variety of species (Table 1). There is evidence that protein A does not readily bind all subclasses of rodent IgG, especially those from mouse and rat (72). This is potentially a huge disadvantage, especially when monoclonal antibodies derived from these animals are to be used. The discovery of another bacterial protein, protein G (which also binds the Fc portion of immunoglobulins), has helped to relieve the problems associated with protein A. Protein G, which is derived from strain G148 of human *Streptococcus* species (30), possesses less selective binding properties and can bind not only rodent immunoglobulins but also immunoglobulins from a wide variety of mammals. Protein G is commercially available as a pure reagent or in combination with protein A. This latter combination ensures successful immunoglobulin binding for a wide variety of species. The binding properties exhibited by protein G are given in Table 1 and compared to those of protein A.

A potential disadvantage to the use of bacterial proteins in immunoaffinity separations is the recent discovery of cross-reactivity to different regions of the immunoglobulin molecule. Protein A has been reported to possess a potential for binding FAb regions as well as Fc regions of human and murine immunoglobulins (44). It was reported that all five domains could bind not only Fc fragments but also polyclonal F(Ab')$_2$ and recombinant Fv fragments. This finding requires further investigation but could indicate that protein A may not be as useful an attachment molecule in immunoaffinity studies as previously believed. Protein G has also been reported to bind to FAb regions of human immunoglobulins. Although this binding

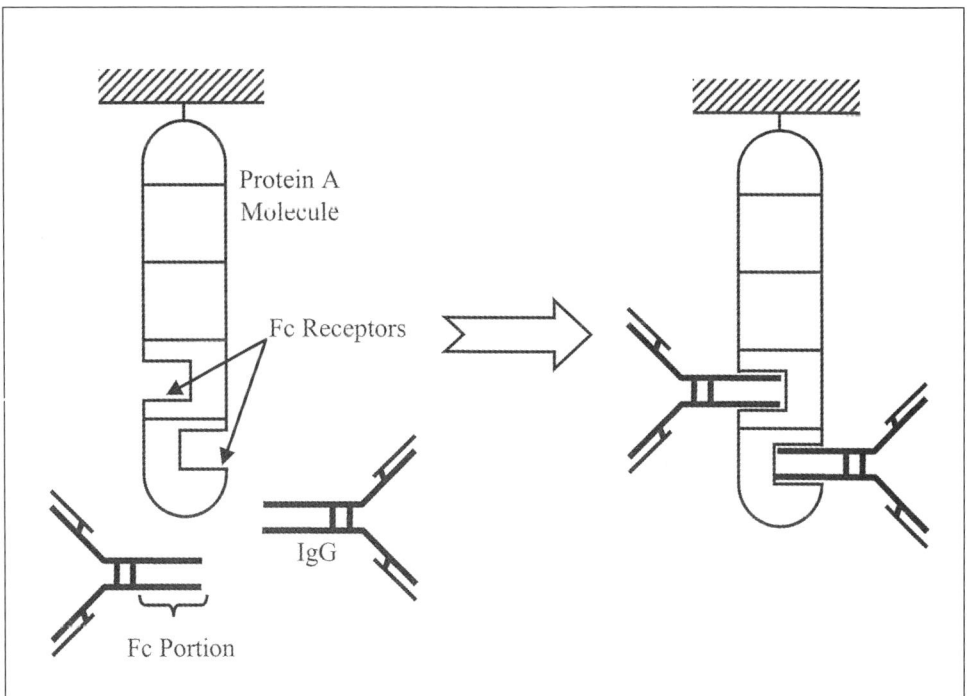

Figure 2. Immunoglobulin immobilization chemistry using bacterial Protein A molecule. Protein A is immobilized to the matrix surface first, followed by Protein A-binding of the Fc portion of the immunoglobulin molecule.

Table 1. Antibody Binding to Bacterial Proteins

Antibody Subclass		Protein A	Protein G
Human	IgG1	Strong binding	Strong binding
	IgG2	Strong binding	Strong binding
	IgG3	Nonbinding	Strong binding
	IgG4	Strong binding	Strong binding
	IgA	Nonbinding	Nonbinding
	IgM	Nonbinding	Nonbinding
Mouse	IgG1	Weak binding	Weak binding
	IgG2a	Strong binding	Strong binding
	IgG2b	Strong binding	Strong binding
	IgG3	Strong binding	Strong binding
	IgM	Nonbinding	Nonbinding
Rat	IgG1	Weak binding	Weak binding
	IgG2a	Nonbinding	Strong binding
	IgG2b	Nonbinding	Weak binding
	IgG2c	Strong binding	Strong binding
	IgM	Nonbinding	Nonbinding
Rabbit	IgG	Strong binding	Strong binding
	IgM	Nonbinding	Nonbinding
Sheep	IgG1	Nonbinding	Strong binding
	IgG2	Strong binding	Strong binding
	IgM	Nonbinding	Nonbinding
Goat	IgG1	Weak binding	Strong binding
	IgG2	Weak binding	Strong binding
	IgM	Nonbinding	Nonbinding
Bovine	IgG1	Nonbinding	Weak binding
	IgG2	Strong binding	Strong binding
Horse	IgGa	Strong binding	Strong binding
	IgGb	Strong binding	Strong binding
	IgGc	Selective binding	Strong binding
	IgG(T)	Selective binding	Strong binding
Dog	IgG	Selective binding	Selective binding
	IgM	Selective binding	Selective binding
	IgE	Selective binding	Strong binding

does not appear to hold for all human IgG subclasses, Perosa et al. (77,78) demonstrated that IgG_2 does not appear to possess binding regions for protein G.

In addition to protein A and G, other bacterial proteins possessing the ability to bind immunoglobulins or their fragments have been described. Zhang et al. (116) have shown that a second gene (*sbi*) in *S. aureus* also encodes for an IgG-binding polypeptide exhibiting similar properties to protein A. Hammerschmidt et al. (35) described a surface protein derived from *Streptococcus pneumoniae* (designated SpsA) capable of specifically binding human secretory IgA with a dissociation constant of 9.3×10^{-9} M. Another bacterial protein called protein L has been described (48) that demonstrates a specificity for IgG κ light chains, but only when complexed to a heavy chain in the $F(Ab')_2$ configuration. This protein has been immobilized to Sepharose to isolate κ light chain-containing IgG molecules and fragments. Additionally, recombinant protein A has been developed not only in the native form but also as isolated domains. A novel approach to the isolation of recombinant Fv fragments has been to produce a fusion product containing the Fv fragment plus a functional domain of protein A (52). This product could then be isolated and purified using an immobilized IgG column. Li et al. (51) have used computer-assisted molecular modeling to develop a low-molecular-weight synthetic molecule or mimetic. This molecule is nonpeptidyl, yet possesses the binding properties of a single protein-A domain. Li et al. claim that one of their mimetic molecules can exhibit IgG binding affinity constants of $10^5–10^6$ M^{-1} when immobilized onto agarose supports.

PROTOCOLS

Preparation of Whole Bacteria for Immunoadsorption

1. Prepare a bacterial suspension containing 1 g of cells (wet weight) in 0.15 M NaCl, 0.04 M Na_2HPO_4, pH 7.4.

   Note: Use *S. aureus* Cowan I strain for protein A and *Streptococcus* species strain G148 for protein G. A preparation of whole, protein A-bearing fixed bacteria is commercially available from Calbiochem-Novabiochem and is sold under the name PANSORBIN®. This preparation provides washed organisms ready for use in immunoprecipitation or antibody capture studies.

2. Sediment the cells by centrifugation at 800× *g* for 10 min.

3. Wash the cell pellet three times by resuspending the cells in 0.15 M NaCl-0.04 M Na_2HPO_4 and by centrifuging at 88× *g* for 10 min.

4. Resuspend the cells in 10 mL of 1.5% (vol/vol) formalin in 0.15 M NaCl-0.04 M Na_2HPO_4.

5. Place the cell suspension on a rotary mixer and mix at room temperature for 90–100 min.

6. Sediment the cells by centrifugation at 800× *g* for 10 min.

7. Wash the cells three times in 0.15 M NaCl-0.04 M Na_2HPO_4.

8. Resuspend the cells in 0.15 M NaCl-0.04 M Na_2HPO_4 and place the suspension in a small Erlenmeyer flask.

9. Heat the cell suspension in a shaking water bath at 80°C for 5–10 min.

10. Rapidly cool the cell suspension by placing the flask in an ice bath.

11. Wash the cells three times in 0.15 M NaCl-0.04 M Na_2HPO_4.

12. Resuspend the cells in 0.15 M NaCl, 0.005 M EDTA, 0.05 M Tris, 0.1% Non-idet® P-40 (a non-ionic detergent used to inhibit cell clumping; Calbiochem-Novabiochem) at pH 7.4.

13. Store at 4°C for up to 4 months. Use a 10% suspension of the cells for affinity isolations.

Preparation of Protein A or Protein G Agarose Supports
Using CNBr-Activated Beads

1. Dissolve 1 mg of purified or recombinant protein A or protein G ligand in 10 mL of 0.1 M $NaCO_3$, pH 8.5.
2. Add to 10 g of CNBr-activated beads.
3. Incubate for 2–4 h at room temperature on an overhead or rotary mixer.
4. Wash three times in 200 mL of 0.1 M $NaCO_3$, pH 8.5, by sedimentation.
5. Resuspend the support in 10 mL of 0.1 M sodium phosphate, pH 7.4, containing 1 M ethanolamine to block free reactive side chains.
6. Incubate on an overhead or rotary mixer for 2 h at room temperature.
7. Wash the support five times in 100 mL of 0.1 M sodium phosphate, pH 7.4, by sedimentation.
8. Store at 4°C.

Preparation of Protein A- or Protein G-Coated Glass Beads
Using CDI-Activated Beads

1. Place 10 g of support in a sintered glass filter.
2. Wash with 20 mL of distilled H_2O.
3. Dissolve 1 mg of protein A or protein G in 10 mL of 0.1 M sodium phosphate, pH 7.4, in a 15-mL glass tube.
4. Resuspend the washed support in the protein solution.
5. Mix on an overhead or rotary mixer for 16–20 h at 4°C.
6. Wash five times in 10 mL of 0.1 M sodium phosphate, pH 7.4.
7. Resuspend the support in 10 mL of 0.2 M Tris-HCl, pH 9.0, to block free CDI side chains.
8. Wash ten times in 0.1 M sodium phosphate, pH 7.4.
9. Store in 0.1 M sodium phosphate at 4°C.

AVIDIN AND STREPTAVIDIN MATRICES

Another protein useful in coupling chemistry to attach antibodies to support materials is avidin, a glycoprotein derived from egg white. This protein gets its name from its ability to form very specific and very strong noncovalent bonds with vitamin H (D-biotin) (55). The three-dimensional structure of avidin has been described as a functional tetramer exhibiting 2-pseudo 22 molecular symmetry. Each monomer is composed of an eight-stranded antiparallel orthogonal β barrel, with extended loop regions that form a deep pocket and define the biotin binding site (55). This site is located at the center of the barrel, displaying both hydrophobic and polar residues for recognition of the bound biotin, which becomes almost completely buried in the protein core. Under ideal conditions, this binding can produce an affinity constant in excess of 10^{-15} M (5). This remarkably high affinity binding, which is several magnitudes higher than most antibodies for their specific antigen, means that the avidin-biotin linkage can withstand harsh elution conditions without dissociation. Since proteins (including antibodies) can be easily biotinylated using one of many commercially available biotin derivatives (Figure 3), avidin is an ideal coating for immunoaffinity applications using biotinylated antibodies.

Streptavidin, a bacterial form of avidin produced by *Streptomyces avidinii* (13), exhibits the same biotin-binding properties but with fewer nonspecific interactions with tissue extracts. As in the case of avidin, streptavidin can be used in coupling chemistry to attach biotinylated antibodies to support materials. Supports can be coated with either streptavidin or avidin and then these beads can be used interchangeably in the procedures below for the attachment of biotinylated antibodies. Recent reports indicate that streptavidin can also be produced in several enhanced forms by genetic engineering (88–90). Variations now available include dimeric and tagged streptavidin useful for a variety of biological applications.

The binding of avidin to biotin is generally irreversible. However, recent advances have led to the development of modified forms of avidin and streptavidin with lower biotin avidity. Nitro-avidin and nitro-streptavidin are examples of such chemical modifications, in which the tyrosine within the biotin-binding site has been nitrated. The reversible attraction of these molecules for biotin has been described (4,70) and used to attach biotinylated ligands to a streptavidin matrix. Another modification of streptavidin that has been developed is a novel chimeric tetramer composed of subunits of both wild-type and genetically engineered streptavidin. The tetramer was constructed by initially inducing guanidine thiocyanate denaturation of an equimolar mixture of wild-type and mutant streptavidin, followed by renaturation and reassociation to form the tetramer. The chimera also exhibits interesting features that have potential applications in affinity separations [i.e., irreversibly binding biotinylated targets at the wild-type unit, while exhibiting reversible separation capabilities at the mutant unit (15)].

Although protein A-, protein G-, avidin-, and streptavidin-coated beads are all commercially available from a number of suppliers (Table 2), it is a very simple and easy process to coat CNBr-, succinimide-, epoxide-, or CDI-activated supports with any of the four proteins. Protocols for making laboratory-prepared supports are provided in Chapters 2 and 4. Alternatively, the three protocols in *Bacterial Protein Matrices* in this chapter are suitable for making avidin or streptavidin supports. A potential disadvantage to using avidin and streptavidin is that, unlike the Fc-binding proteins, they directly link to antibodies at an intermediate molecule, biotin. This not only involves chemical modification of the immunoglobulin but also introduces a serious problem. If the biotinylation involves the antibody-binding site, it can seriously affect the antibody-binding affinity. *Hydrazine Matrices and the Preparation of Antibodies* (below) will describe strategies for assuring that specific Fc fragment iotinylation is achieved to avoid such interference.

LECTIN MATRICES

Glycoproteins and other sugar-containing molecules can be attached to affinity supports bearing suitable immobilized lectins (see *Lectins* in Chapter 4). Conconavalin (Con) A is a popular lectin used for such applications and has been used to immobilize model proteins (45) and glycoenzymes (25) to affinity supports. Immunoglobulins are glycoproteins, with the sugar moieties residing in the Fc fragment (Figure 4). It has been reported that lectins, such as con A, can be used to fractionate and therefore bind immunoglobulins (109). Peng et al. (76) have reported that canine immunoglobulins bind to con A and that, when they applied dog serum to a con A-Sepharose column, 100% of IgE and IgM, 60% of IgG, and 58% of IgA was retained (the different immunoglobulins being recovered by elution with defined sets of oliogsaccharides). Lectins derived from the *Artocarpus* species (36) or the land snail (8) have been shown to bind IgA. A mannose-binding lectin derived from *Artocarpus integer* has been reported to bind to human IgE and IgM antibodies, with a weak binding of IgA_2 (53). However, the authors of the study demonstrated that the lectin also bound to horseradish peroxidase, ovalbumin, porcine thyroglobulin, human α1-acid glycoprotein, transferrin, and α_1-antitrypsin, thus making it less appealing as a selective immunoaffinity attachment ligand.

Care must be exercised to ensure that attachment of certain antigens to the lectin-

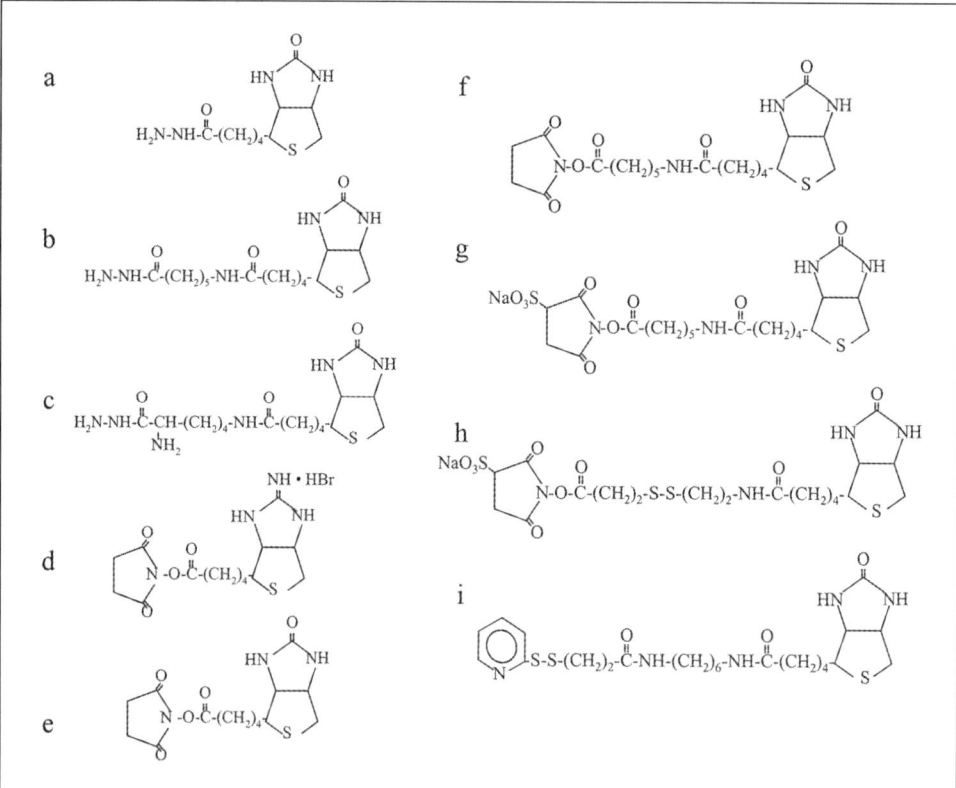

Figure 3. Examples of biotin derivatives available from Pierce Chemical. These agents are: (a) biotin hydrazide; (b) biotin-LC-hydrazide; (c) biocytin hydrazide; (d) NHS-iminobiotin; (e) N-hydrosuccinimide (NHS) biotin; (f) NHS-LC-biotin II; (g) sulfo-NHS-LC-biotin; (h) sulfo-NHS-SS-biotin; and (i) Biotin-HPDP.

coated support does not cause steric hindrance with antibody binding. In many cases, the repeated epitopes of the carbohydrate molecule are the functional sites for antibody binding; in such cases, the immobilized lectin and the antibody to be isolated will compete for identical structures on the antigen surface. This is particularly true of purified bacterial polysaccharide antigens used to isolate or immunoaffinity-purify specific antibodies. As with attachment of antibodies to bacterial proteins, it is wise to cross-link the antigens into place with a suitable cross-linking agent such as a carbodiimide (see *Preparation of Antibodies* in this chapter).

HYDRAZINE MATRICES

The carbohydrate moiety found on the Fc portion of antibodies has become a favored site for antibody immobilization. This is performed by simply oxidizing the antibody's carbohydrate vicinal hydroxyl group with sodium periodate (60) to form a reactive aldehyde function. The aldehyde is then linked to supports coated with hydrazine groups to form stable, covalent hydrazide bonds. Because this linkage is through the Fc portion, it has the same advantage of Fc-binding proteins like protein A (i.e., the antibody immobilization site is well removed from the antigen binding

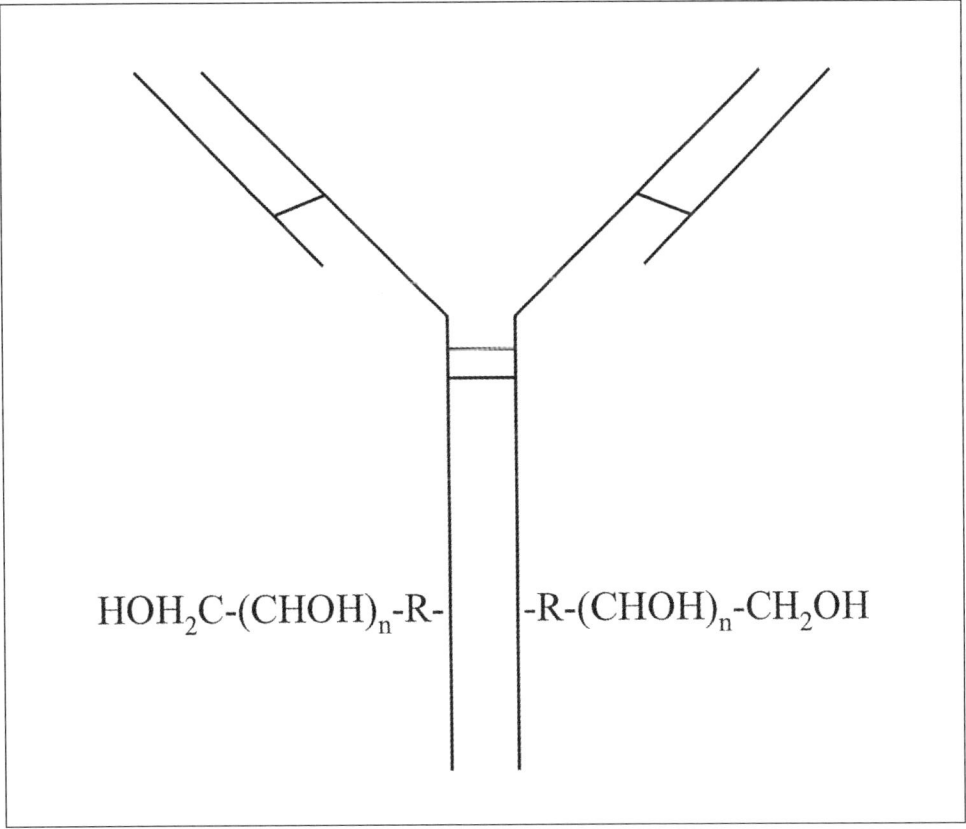

$$HOH_2C\text{-}(CHOH)_n\text{-}R\text{-} \qquad \text{-}R\text{-}(CHOH)_n\text{-}CH_2OH$$

Figure 4. The carbohydrate moieties on the Fc region of IgG are excellent sites for attachment to affinity supports. The carbohydrate moieties on the Fc region of IgG can be altered by reaction with hydrazine or hydrazine-biotin to achieve correct orientation to the surface of the affinity matrix.

119

sites). Another reason this method is so popular is that the oxidation of the carbohydrate portion of the antibody is a very simple technique (33), and commercial sources of hydrazine matrices are readily available. Bio-Rad produces commercial forms of this support called Affi-Gel Hz and Affi-Prep® Hz, while hydrazide beads are also available from Pierce Chemical, as is a similar carbohydrate-binding product called CarboLink® Gel. A protocol for producing laboratory-built hydrazine supports is provided in Chapter 2 of this book.

Table 2. Commercial Sources of Immunoaffinity Supports

Immobilized Ligand	Commercial Source
Protein A	Amersham Pharmacia Biotech Calbiochem-Novabiochem ICN Biomedicals Pierce Chemical Sigma Chemical
Protein G	Amersham Pharmacia Biotech Calbiochem-Novabiochem ICN Biomedicals Pierce Chemical Sigma Chemical
Avidin	Amersham Pharmacia Biotech Calbiochem-Novabiochem ICN Biomedicals Pierce Chemical Sigma Chemical
Streptavidin	Amersham Pharmacia Biotech Calbiochem-Novabiochem ICN Biomedicals Pierce Chemical Sigma Chemical

PREPARATION OF ANTIBODIES

Before immunoaffinity matrices can be constructed, it is important for the investigator to consider the source of the antibody or antigen to be used as the ligand, whether whole or antibody fragments are to be used, and the method of attachment. The following sections will provide an overview of the different procedures that can be applied to achieve the optimal immunoaffinity support.

Antibody Sources

Varieties of different antibody types are available for use in immunoaffinity procedures. These entities range from laboratory-raised polyclonal and monoclonal antibodies to engineered recombinant molecules. This latter category is becoming extremely popular, especially with the introduction of molecular biological techniques. The original procedure for producing antibodies was to inject antigens or analytes into rabbits or other mammals and collect the serum containing the resulting antibodies (57). Such antibodies are called polyclonal because they are the products of multiple responses and can, on occasion, contain multiple specificities. The specificity of these antibodies depends on the purity of the immunizing antigen, the route of administration, and the duration of the immunization program. Polyclonal antibodies arise naturally and can be isolated from a number of sources, including patients with certain autoimmune diseases or from patients sensitized to known allergens. In this latter class, useful antibodies can be obtained to a number of environmental factors, especially chemicals and other analytes known to cause allergic reactions (21,54). Both humans and animals, following prolonged production of antibody responses, can produce another interesting class of antibodies commonly known as anti-antibodies. The most common clinical source of these antibodies are the rheumatoid factors, which are naturally occurring antibodies directed against the Fc portion of other immunoglobulins (117). Such anti-antibodies can be useful for isolating immunoglobulins or specific fragments of immunoglobulins. The most interesting anti-antibody is the anti-idiotypic form. These antibodies are made against the antigen receptors of the target immunoglobulins (112). They can be used to mimic the original antigen and are useful for isolating specific antibodies and their reactive antigens (34).

The use of engineered antibodies has broadened because of the ability to produce such antibodies in genetically modified mice (43). The clinical application of engineered antibodies for therapeutic use has helped to advance the field, especially in the area of humanizing monoclonal antibodies for injection into humans (38,106). Hudson (41) has recently reviewed the application of engineered antibodies to a number of clinical, diagnostic, and bioanalytical fields. The most interesting advance has been in the development of single-chain or Fv antibodies. These molecules possess the specificity of conventional two-chain molecules but can easily be produced by recombinant technology in bacterial (59,71), insect (58), or mammalian cell lines (18), and human phage libraries (62). The stability of these single-chain antibodies has been improved by engineering disulfide linkages between the heavy and light chain variable regions, distant from the areas that conform to the electron cloud shape of the antigen (85). Single-chain antibodies have been used for the production of probes for the measurement of antigen-binding forces by atomic force microscopy (87) and for the construction of optical biosensors (81). The introduction of disulfide linkages has also been applied to the generation of stable, bivalent Fv molecules in which heavy and light chain variable regions are linked together by disulfide bonds and the two molecules are joined by a flexible 15 amino acid bridging molecule (6). Protein engineering has also been used to produce bi-specific antibodies. This was achieved not by chemical cleavage followed by recombination as described earlier (20), but by remodeling the heavy chains and genetically combining them with identical light chains (64).

There are other specialized antibodies (such as catalytic antibodies) that have been shown to drive a number of chemical reactions usually associated with enzyme activity (110). These molecules can be effectively used to induce hydrolysis of a number of target molecules, including the hydrolysis of norleucine phenyl ester with *(S)*-enantioselectivity (10), the hydrolysis of a pro-drug of chloramphenicol monoester to generate the parent drug (49), and the hydrolysis of carbonate esters (100). Monoclonal anti-idiotypic antibodies have also been reported to be a convenient way to produce catalytic molecules exhibiting activities closely resembling those of the original antigen or enzyme, such as carboxypeptidase A (40) and β-lactamase (3). Although the majority of catalytic antibodies are either monoclonal or engineered, the presence of polyclonal catalytic antibodies has been reported (97).

Table 3 gives an overview of these different types of antibodies and their origins.

Antibody Isolation

Prior to immobilization or modification, it is important to isolate either the specific molecules or the general class (i.e., IgG) of the antibody to be immobilized.

Table 3. Types of Antibodies Used in Immunoaffinity Chromatography

Natural antibodies	
Polyclonal	Raised by immunizing animals with selected antigens
Autoantibodies	Obtained from the serum of human patients with autoimmune diseases
Idiotypic	Naturally occurring, highly specific antibodies
Anti-idiotypic	Specialized antibodies that can mimic antigens, hormones, or substrates for cell receptors
Engineered antibodies	
Monoclonal	Raised by fusion of spleen cells from immunized animals with myeloma cell lines
Bifunctional	Biochemical or molecular biological modifications of an antibody producing two FAb fragments with different specificities
Catalytic	Polyclonal, monoclonal, or single-chain antibodies capable of catalyzing chemical reactions
Recombinant antibodies	
Single chain	Single-chain antibodies produced by molecular engineering

Immunoglobulins can be batch-purified from serum, ascitic fluid, or culture medium by precipitation with polyethylene glycol (PEG) or saturated ammonium sulfate. These techniques are well suited to the isolation of monoclonal or recombinant antibodies produced in culture. Antibodies produced by ascites are heavily contaminated with host proteins and must be regarded in the same light as polyclonal antibodies. Salt precipitation is also useful for producing a starting solution for many of the other isolation techniques. IgG can easily be isolated by salt precipitation followed by simple fractionation of the starting material on an ion-exchange resin (102). This process can be performed by either batch techniques or column chromatography. Boden et al. (7) have reported that goat polyclonal antibodies can be easily isolated by immobilized metal ion affinity chromatography using tris(2-aminoethyl)amine chelated with copper. Applying this procedure, they were able to recover an immunoglobulin fraction with greater than 95% homogeneity as assessed by silver-stained native gel electrophoresis and SDS polyacrylamide gel electrophoresis. Likewise, thiophilic adsorption can be used for the isolation of F(Ab')$_2$ fragments following papain digestion of IgG$_1$ monoclonal antibodies to isolate 90% pure fragments (115).

Combinations of size exclusion chromatography and lectin affinity chromatography are useful for isolating IgM and IgA antibodies. IgM antibodies are insoluble in water and this fact can be used to obtain a starting material that is subsequently purified on a size exclusion column (27). However, not all IgM antibodies precipitate in water and in such cases, these entities are recoverable by precipitation in a saturated solution of ammonium sulfate prior to size exclusion chromatography purification. However, the most efficient approach to purifying specific antibodies (especially

Table 4. Extinction Coefficients for Determining Immunoglobulin and Fragment Concentrations

Immunoglobulin	Extinction coefficient
IgG	1.43
IgM	1.18
FAb γ	1.53
F(Ab')$_2$ γ	1.48
FAb μ	1.38
F(Ab')$_2$ μ	1.38

IgG) for use as immunoaffinity ligands is through immobilized antigen immunoaffinity chromatography (28). Although this may appear to be complicated, in principle it is the only way to purify analyte-specific polyclonal antibodies.

Digestion of Antibodies

Antibodies are relatively large molecules exhibiting molecular weights of between 150 000 and 1 000 000 Da. This size can often become a problem when using antibodies in certain situations such as antigen localization within cells and subcellular structures. In 1959, Porter (82) demonstrated that the IgG molecule could be enzymatically digested with papain to smaller (but still biologically active) fragments. Later, it was shown that mercuripapain could also produce functional fragments from IgG molecules (96). The two most commonly used cleavage enzymes, papain and pepsin, both act at the hinge region of the antibody molecule (Figure 5). Papain has been shown to cleave the IgG molecule in the hinge region on the amino side of the disulfide bridges, thus releasing two functional monovalent antibody units. Potential problems can arise if care is not taken to ensure that the correct ratio of immunoglobulin to enzyme is used. This can easily be remedied by carefully measuring the amounts of immunoglobulin added to the reaction. A simple approach to this problem is to perform direct spectrophotometric measurement of the target immunoglobulin solution (see Appendix I). Table 4 gives the extinction coefficients for some immunoglobulins and their fragments. Recently, this procedure has been greatly simplified by the introduction of immobilized papain by Pierce Chemical (73). Mohanty and Rosenthal (67) describe a technique wherein IgG antibodies are first captured on a protein A column and then digested by papain in situ. This procedure eliminates the need to perform further purification to remove the Fc fragments.

Pepsin cleaves the intact IgG molecule at the carboxyl side of the disulfide bridges, producing a bivalent antibody unit (80) (the FAb arms of this molecule are still attached by a disulfide bridge [Figure 5]). This bifunctional unit can be further broken down by reduction of the disulfide bridge to form two FAb units, each with a free thiol group that can be used for attachment. Pepsin is also available in an immobilized form from Pierce Chemical and its use greatly simplifies the production of F(Ab')$_2$ fragments (103). Pepsin can also be used to produce functional F(Ab')$_2$ fragments from IgM (42).

Other enzymes such as clostripain, lysyl endopeptidase, metalloendopeptidase, and V8 protease have been used to produce active fragments from a variety of murine monoclonal antibodies (113). Additionally, molecular biological techniques have introduced the ability to engineer single antibody chains possessing the specificity of the original, intact antibody. Studies on these entities are discussed in *Antibody Sources* in this chapter. Table 5 summarizes the nomenclature, molecular weights, and possible uses for these different immunoglobulin fragments.

Table 5. Reactive Immunoglobulin Fragments

Fragment	Molecular weight	Functional use
FAb	50 000	Smallest digestible fragment that still retains antibody activity
F(Ab′)$_2$	100 000	Bifunctional antibody minus the tail fragment
Fc	50 000	Binds to complement components and Fc receptors
Fv	12 500	Single-chain reactive antibody

Antibody Modification Prior to Attachment

Apart from the digestion that renders antibody molecules into reactive fragments useful as affinity ligands, there are further modifications that can be performed to facilitate attachment of antibodies to the support. Chemical modification of the

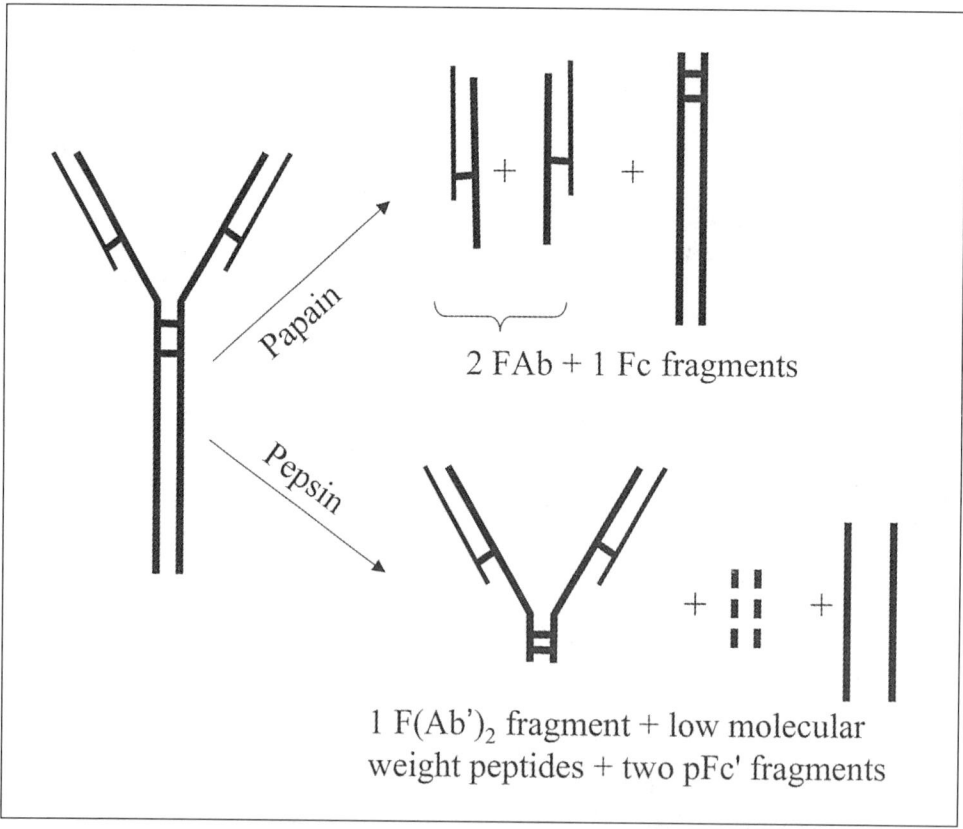

Figure 5. Production of reactive fragments by enzymatic cleavage of an antibody molecule. Digestion with papain creates two reactive FAb fragments and one Fc fragment. Pepsin digestion creates one F(Ab′)$_2$ fragment and one Fc fragment.

saccharide content in the Fc portion has already been described in the *Hydrazine Matrices* section of this chapter. This modification allows the antibody to be correctly oriented on the support. Another popular technique is to attach *N*-hydroxysuccinimidobiotin to the antibody (37). This modification facilitates attachment of the antibody to avidin- or streptavidin-coated supports. However, a disadvantage of this modification is that biotinylation can occur at any free amino group and such random attachment can seriously affect the efficiency of the immobilized antibody. Biotin can become attached to areas directly adjacent to or within the analyte receptors of the antibody. This will result in attachment by the FAb portions, resulting in either impaired analyte capture or complete inhibition. The introduction of commercially available hydrazine-biotin by Pierce Chemical (65) helped overcome the shortcomings of generalized biotinylation agents. As discussed previously, hydrazine effectively attaches to chemically modified carbohydrates in the Fc portion of the IgG molecule. The advantage of this attachment site, like direct attachment to hydrazine–activated supports, ensures that the antibody molecule is biotinylated at the correct site (Figure 6).

Storage of Isolated and Digested Immunoglobulins

Although few studies have been performed on the effects of storage on antibody activity, Virella (107) reported degradation of a monoclonal antibody during storage. Other studies on bioactive proteins have also noted that storage can affect the bioactivity of the recovered molecules (24,66,104). It must be remembered that immunoglobulins are reasonably fragile protein molecules and can easily be denatured during storage. One must always pay attention to how isolated antibodies are

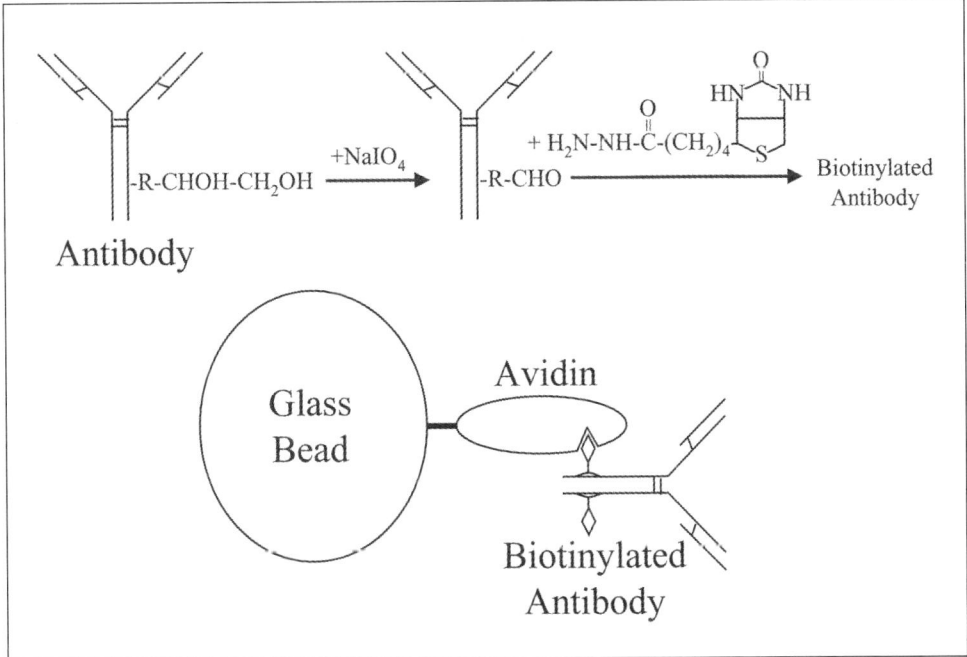

Figure 6. Reaction used to attach biotin by the carbohydrate moieties on an IgG molecule to an avidin-coated support.

stored in order to maintain their activity. This is particularly true for digested fragments, which are already "denatured" and therefore more susceptible to further structural damage. Failure to inactivate or remove residual enzymes will result in slow but continued digestion over time. Additionally, we have noted that a certain degree of digestion always occurs during storage of whole serum and plasma, mainly because of the presence of proteolytic enzymes. This situation can be reduced or overcome by isolation of the antibodies of interest. It must also be remembered that isolated antibodies can undergo dissociation, especially when stored for long periods in protein-free buffers. We have found that there are several approaches to reducing or inhibiting storage-related denaturation. Twenty-percent sucrose or ten-percent polyvinylpyrrolidone (PVP) can be added to solutions stored at 4°C in a laboratory refrigerator. Alternatively, the solutions can be stored in small aliquots at -70°C or below and thawed once (or a maximum of twice). Lyophilization is the most common method used to dry proteins, but this process involves freezing and dehydration, which are both damaging to protein. Additionally, the process often produces aggregates when the materials are reconstituted, thus causing a loss of activity. Studies have also demonstrated a number of problems associated with this form of preservation (1,14). Studies in our laboratory have shown that one of the most common causes of activity loss in lyophilized materials occurs during reconstitution, when "foaming" produced by vigorous reconstitution of the dried materials often causes unfolding and denaturation of the stored product.

PROTOCOLS

Ammonium Sulfate Precipitation of Immunoglobulins

1. Prepare 100 mL of saturated ammonium sulfate solution by dissolving 9 g of $(NH_4)_2SO_4$ in 100 mL of boiling distilled H_2O.
   Note: Allow the solution to cool to room temperature before use.
2. Place 5 mL of clear serum (no fibrous or cell debris) in a glass tube.
3. Slowly add 0.5 mL of the $(NH_4)_2SO_4$ solution until a precipitate forms.
4. Allow the serum to rest for 5 min.
5. Add a further 2 mL of the $(NH_4)_2SO_4$ solution.
6. Cover the tube with Parafilm and incubate overnight at 4°C.
7. Centrifuge the mixture at $10\,000$–$12\,500\times g$ for 10–15 min.
8. Collect the precipitate.
9. Redissolve the precipitate in 2 mL of 0.1 M sodium phosphate, pH 7.4.
10. Place the precipitate solution in a dialysis sac and dialyze overnight at 4°C against 0.1 M sodium phosphate, pH 7.4.
11. Recover the solution from the dialysis sac.
12. Store at 4°C or freeze at -70°C.

Batch Technique for IgG Isolation

1. Suspend 1 g of DEAE-Sepharose gel beads in 0.005 M sodium phosphate, pH 6.5.

2. Dialyze 1 mL of antibody solution or serum against 0.005 M sodium phosphate for 16–24 h at 4°C.

3. Place the gel beads in a 15-mL centrifuge tube.

4. Sediment the gel beads by centrifugation at 300× g for 10 min.

5. Adjust the pH of the gel bead solution to 6.5 with 0.1 N HCl and decant the supernatant.

6. Resuspend the gel beads in the dialyzed antibody solution or serum.

7. Place on an overhead mixer for 1 h at room temperature.

8. Recentrifuge the tube at 300× g for 10 min.

9. Recover the supernatant containing the IgG fraction and recentrifuge at 10 000× g for 20 min to remove all impurities.

10. Filter sterilize, then store at 4°C or freeze at -70°C.

Isolation of IgG by Protein A Ligand Chromatography

1. Clarify the antibody solution by centrifugation at 10 000× g for 30 min at 4°C and save the supernatant.

2. Equilibrate a commercial protein A-Sepharose column by running 10 column volumes of 0.1 M sodium phosphate, pH 7.2 through the column. Allow the buffer to run through the column until the top of the gel bed is barely wet.

3. Stop the flow through the column and carefully load the supernatant (which contains the antibody solution) onto the top of the gel bed.

4. Allow the antibody solution to completely become adsorbed into the gel bed.

5. Add 0.1 M sodium phosphate, pH 7.2, to the top of the gel bed.

6. Pass 5 column volumes of buffer through the column.

7. Check the optical density of the column eluate. (It should be at baseline; if not, then continue to run buffer through the column until the optical density is at baseline.)

8. Recover the IgG by slowly adding 0.1 M citric acid to the running buffer. (This can easily be achieved by a linear gradient; see Chapter 7.)

9. Continue adding the acid until pH 2.5 is reached and continuously collect fractions during the acid elution phase.

10. Pool all fractions containing protein (as measured by increases in the optical density).

11. Dialyze the pooled fractions overnight at 4°C against 0.1 M sodium phosphate, pH 7.4.

12. Store at 4°C or freeze at -70°C.

Isolation of IgM by Size Exclusion Chromatography

1. Place the antibody solution into a dialysis sac and dialyze overnight at 4°C against double-distilled H_2O.

2. Remove the entire contents of the sac (including the insoluble IgM precipitate).

3. Centrifuge the solution at 10 000–12 500× g for 10 min and recover the precipitate.

4. Redissolve the precipitate in 1 mL of 0.05 M $Na_2B_4O_7$, pH 6.4.

5. Pack a column (see Chapter 7) with suitable size exclusion packing (e.g., Sepharose 6B).

6. Equilibrate the column with 10 column volumes of 0.05 M $Na_2B_4O_7$, pH 6.4.

7. Adjust the flow to 1 mL/min.

8. Slowly layer the IgM solution onto the top of the column bed.

9. Monitor the column output with a UV detector set at 280 nm.

10. Collect the first peak (the IgM fraction).

11. Check the protein concentration and adjust to 1 mg/mL with 0.1 M sodium phosphate, pH 7.4.

12. Store at 4°C or freeze at -70°C.

Alternatively, IgM can be isolated from the ammonium sulfate-precipitated fraction and purified by size exclusion chromatography following steps 5–12 in the above protocol.

Isolation of IgA by Lectin Affinity Chromatography

1. Clarify the antibody solution by centrifugation at $10\,000\times g$ for 30 min at 4°C and save the supernatant.

2. Equilibrate a 1-cm × 10-cm column containing Jacalin-Sepharose by running 10 column volumes of 0.1 M sodium phosphate, pH 7.2, through the column. Allow the buffer to run through the column until the top of the gel bed is barely wet.

3. Stop the flow through the column and carefully load the antibody solution onto the top of the gel bed.

4. Allow the antibody solution to become completely adsorbed into the gel bed.

5. Add sodium phosphate to the top of the gel bed and pass 5 column volumes through the column.

6. Check the optical density of the column eluate. (It should be at baseline; if not, then continue to run buffer through the column until the optical density is at baseline.)

7. Recover the IgA by slowly adding 0.1 M α-D-galactoside to the running buffer. (This can easily be achieved by a linear gradient; see Chapter 7.)

   Note: Continuously collect fractions during the elution phase.

8. Pool all fractions containing protein (as measured by increases in the optical density).

9. Dialyze the pooled fractions overnight at 4°C against 0.1 M sodium phosphate, pH 7.4.

10. Store at 4°C or freeze at -70°C.

Antigen Affinity Isolation of Specific Antibodies

1. Place 10 mL of CNBr-activated agarose in a glass tube.

2. Add 1 μg of antigen dissolved in 1 mL of 0.1 M Na_2CO_3, pH 9.0.

3. Mix slowly on an overhead mixer for 24 h at 4°C.

4. Transfer the CNBr-activated agarose to a sintered glass filter.

5. Wash with 200 mL of ice-cold distilled H_2O.

6. Wash with 200 mL of ice-cold 0.1 M Na_2CO_3, pH 9.0.

7. Resuspend CNBr-activated agarose in 0.01 M sodium phosphate, pH 7.2.

8. Pack the CNBr-activated agarose into a suitable column.

9. Slowly add the antibody solution to the column.

10. Run the column in 0.01 M sodium phosphate, pH 7.2.

11. Collect the eluate.

12. Estimate the affinity-purified antibody content by spectrophotometry at 280 nm.

13. Store at 4°C or freeze at -70°C.

Preparation of FAb Fragments

1. Dissolve 1 mg of IgG in 1 mL of digestion buffer (10 mM sodium phosphate, 0.15 M NaCl, 1 mM EDTA, 2 mM cysteine).

2. Add 10 µg of crystalline mercuripapain.

3. Incubate for 18 h at 37°C.

4. Dialyze the solution against double-distilled H_2O for 4 h at 4°C with 10 changes of H_2O.

5. Separate the FAb fragments from the Fc fragments by column chromatography on a carboxymethyl ion-exchange column.

Preparation of F(Ab′)₂ Fragments

1. Dissolve 1 mg of IgG in 10 mL of 0.025 M CH_3 COONa, pH 4.5.

2. Warm the solution to 37°C.

3. Dissolve 100 µg of pepsin in 10 mL of 0.025 M CH_3 COONa, pH 4.5.

4. Warm to 37°C.

5. Mix the IgG and pepsin solutions.

6. Incubate at 37°C for 24 h.

7. Titrate the solution to pH 8.0 with 1 M Tris-HCl.

8. Isolate the $F(Ab′)_2$ fragments on a 2-cm × 25-cm size exclusion column.

Pepsin digestion can be performed on IgM, IgA, and IgE, using a similar protocol to that described above. With these other immunoglobulins, the digested fragments need to be isolated on a Sepharose 6B size exclusion column because the $F(Ab′)_2$ fragments often co-elute with undigested material when applied to size exclusion columns packed with separation medium of a lower resolving range.

Hydroxysuccinimide Biotinylation of Antibodies

1. Place 1 mg of polyclonal antibody or 250 µg of monoclonal antibody in a capped tube.

129

2. Add 1 mg of *N*-hydroxysuccinimidobiotin, dissolved in 1 mL of 0.05 M $CaCl_2$, pH 9.0.

3. Mix well.

   Note: Do not allow the solution to "foam" when mixing.

4. Place the mixture in stoppered glass container.

5. Place on a rotary or overhead mixer and incubate at room temperature for 2 h.

6. Dialyze against 0.01 M sodium phosphate, pH 7.4, overnight at 4°C.

Hydrazine Biotinylation of Antibodies

1. Suspend 1 μg of (monoclonal or polyclonal) antibody in 1 mL of 0.1 M CH_3COONa, pH 5.0.

2. Cool to 4°C in an ice bath.

3. Add 1 mL of ice-cold 10 mM sodium metaperiodate.

4. Mix and place on a rotary or overhead mixer.

5. Incubate for 20 min at 4°C in the dark.

6. Stop the reaction by adding 20 mL of a 5% (vol/vol) solution of ethylene glycol (mol. wt. = 1500).

7. Dialyze the antibody solution for 5 h against 0.01 M sodium phosphate, pH 7.0, at 4°C in the dark.

8. Remove the antibody solution and place in a 15-mL capped glass tube.

9. Dissolve 1 mg of sodium cyanoborate in 1 mL of 0.01 M sodium phosphate, pH 7.4.

10. Add 1 mg of hydrazine-biotin.

11. Add the hydrazine-biotin solution to the antibody solution.

12. Mix and place on a rotary or overhead mixer.

13. Incubate for 1 h at room temperature.

14. Dialyze against 0.01 M sodium phosphate, pH 7.4, overnight at 4°C.

ATTACHMENT OF ANTIBODIES AND ANTIGENS

An increasing interest in immunoaffinity techniques has greatly enhanced the availability of suitable immobilization supports. Although CNBr-activated supports are still commercially available, coupling by succinimide ester, carbonyldiimide, and epoxide groups offer easier ways to immobilize antibody ligands. This direct attachment is often the most convenient way to immobilize antigen ligands. However, in the case of antibodies, these approaches usually do not orient the ligands correctly. Recently, the introduction of hydrazide and thiol supports have helped to overcome this problem. The former, as described in Chapter 2, binds antibodies by oxidized polysaccharides in the Fc portion, while the latter directly attaches FAb fragments by a free thiol group (12,105) [this group being readily available following reduction of the disulfide bridge in $F(Ab')_2$ fragments]. The use of bacterial proteins as attachment surfaces allows for attachment of certain classes of antibodies directly to Fc receptors on the protein coat. Modification of antibodies by biotinylation allows easy immobilization to avidin- or streptavidin-coated supports. The use of such supports is gaining popularity with the availability of numerous commercially available biotinylated antibodies. Biotinylation can also be a useful procedure for preparing antigens for direct immobilization to avidin- or streptavidin-coated supports. Whichever approach is chosen, the immobilization of immunological reagents is quite simple and usually involves mixing the activated support with the molecule to be immobilized. However, as pointed out in Chapter 4, care must be taken to ensure that all of the free attachment sites are blocked before proceeding with the separation.

PROTOCOLS

Direct Ligand Attachment to Derivatized Supports

1. Place 1 g of derivatized beads into a 15-mL capped glass tube.
2. Add 5 mL of 50 mM $NaCO_3$, pH 9.0.
3. Add the antibody dissolved in 5 mL of 50 mM $NaCO_3$, pH 9.0.
4. Place on an overhead mixer.
5. Incubate overnight at 4°C.
6. Remove the tube, and stand it upright in a rack to allow the beads to sediment by gravity.
7. Decant the supernatant.
8. Wash the beads five times in 250 mL of 0.01 M sodium phosphate, pH 7.4. (This is achieved by resuspending the beads and allowing them to settle by gravity.)

Ligand Attachment to Bacterial Protein-Coated Supports

1. Dissolve 1 mg of the antibody in 1 mL of 0.01 M sodium phosphate, pH 7.2.
2. Place in a 15-mL capped glass tube.
3. Add 10 g of protein A- or protein G-coated beads.
4. Place the tube on an overhead mixer.
5. Incubate for 1 h at 4°C.
6. Allow the beads to sediment.
7. Remove the supernatant and store at 4°C (for potential re-use).
8. Add 1 mg of normal rabbit IgG dissolved in 10 mL of 0.01 M sodium phosphate, pH 7.2, to the sedimented beads to block the nonreacted Fc receptors.
9. Mix and incubate for 30 min on an overhead mixer.
10. Allow the beads to sediment.
11. Discard the supernatant.
12. Wash five times in 50 mL of 0.01 M sodium phosphate, pH 7.2.
13. Allow the beads to settle.
14. Resuspend the beads in 10 mL of 50 mM $NaCO_3$, pH 9.0, containing 10 mM CMC to cross-link the antibodies to the bead coat.
15. Place the mixture on the overhead mixer for 30 min.

131

16. Recover the beads by sedimentation.
17. Wash five times in 50 mL 0.01 M sodium phosphate, pH 7.2.

Ligand Attachment to Avidin and Streptavidin Supports

1. Place 1 mL of 0.01 M sodium phosphate, pH 7.2, containing biotinylated antibody (1 mg polyclonal antibody or 250 µg of monoclonal antibody) into a 15-mL glass tube.
2. Add 2 g of avidin-coated (or streptavidin-coated) beads.
3. Mix for 1 h at 4°C in an overhead mixer.
4. Recover the beads and wash three times in 0.01 M sodium phosphate, pH 7.2.
5. Block the nonreactive biotin receptors by incubating the beads for 1 h at 4°C in 5 mL of a solution of 1 mg/mL biotin dissolved in 0.01 M sodium phosphate, pH 7.2.
6. Wash the beads five times in 0.01 M sodium phosphate, pH 7.2.

Ligand Attachment to Lectin Matrices

1. Dissolve 1 µg of antibody in 1 mL of 0.01 M sodium phosphate, pH 7.2.
2. Place in a 15-mL capped glass tube.
3. Add 10 g of con A-coated beads.
4. Incubate for 1 h at 4°C on an overhead mixer.
5. Allow the beads to sediment.
6. Remove the supernatant and store at 4°C (for potential re-use).
7. Block nonreactive lectin by adding 5 mL of the appropriate sugar at a concentration of 1 mg/mL.
8. Mix and incubate for 30 min on an overhead mixer.
9. Allow the beads to sediment.
10. Discard the supernatant.
11. Wash five times in 50 mL of 0.01 M sodium phosphate, pH 7.2.
12. Allow the beads to settle.
13. Resuspend the beads in 10 mL of 50 mM $NaCO_3$, pH 9.0, containing 10 mM CMC to cross-link the antibodies to the bead coat.
14. Place the mixture on the overhead mixer for 30 min.
15. Recover the beads by sedimentation.
16. Wash five times in 50 mL of 0.01 M sodium phosphate, pH 7.2.

Ligand Attachment to Hydrazine Matrices

1. Suspend 1 µg of antibody in 1 mL of 0.1 M sodium acetate, pH 5.0.
2. Cool to 4°C.
3. Add 1 mL of ice-cold 10 mM sodium metaperiodate.
4. Mix well and incubate for 45 min at 4°C in the dark.
5. Stop the reaction by adding 50 µL of glycerol.

6. Continue mixing for a further 20 min.

7. Place in a dialysis sac and dialyze for 5 h against 0.01 M sodium phosphate, pH 7.2, at 4°C in the dark.

8. Remove the antibody solution and place in a 15-mL capped glass tube.

9. Add 1 mL of hydrazine-activated support in 0.01 M sodium phosphate, pH 7.2, plus 1 mg of sodium cyanoborate.

10. Mix well.

11. Incubate for 16 h on an overhead mixer at 4°C.

12. Wash the gel five times in 0.1 M sodium phosphate, pH 7.2, by sedimentation.

REFERENCES

1. **Allison, S.D., T.W. Randolph, M.C. Manning, K. Middleton, A. Davis and J.F. Carpenter.** 1998. Effects of drying methods and additives on structure and function of actin: mechanisms of dehydration-induced damage and its inhibition. Arch. Biochem. Biophys. *358*:171-181.
2. **Attallah, A.M., S.A. El Masry, H. Ismail, H. Attia, M. Abdel Aziz, A.S. Shehatta, A. Tabll, A. Soltan and A. El Wassif.** 1998. Immunochemical purification and characterization of a 74.0-kDa Schistosoma mansoni antigen. J. Parasitol. *84*:301-306.
3. **Avalle, B., D. Thomas and A. Friboulet.** 1998. Functional mimicry: elicitation of a monoclonal anti-idiotypic antibody hydrolizing beta-lactams. FASEB J. *12*:1055-1060.
4. **Balass, M., E. Morag, E.A. Bayer, S. Fuchs, M. Wilchek and E. Katchalski-Katzir.** 1996. Recovery of high-affinity phage from a nitrostreptavidin matrix in phage-display technology. Anal. Biochem. *243*:264-269.
5. **Bayer, E.A. and M. Wilchek.** 1980. The use of the avidin-biotin complex as a tool in molecular biology. Methods Biochem. Anal. *26*:1-45.
6. **Bera, T.K., M. Onda, U. Brinkmann and I. Pastan.** 1998. A bivalent disulfide-stabilized Fv with improved antigen binding to erbB2. J. Mol. Biol. *281*:475-483.
7. **Boden, V., J.J. Winzerling, M. Vijayalakshmi and J. Porath.** 1995. Rapid one-step purification of goat immunoglobulins by immobilized metal ion affinity chromatography. J. Immunol. Methods *181*:225-232.
8. **Booth, J.R., R. Munks and R.J. Sokol.** 1995. Isolation of IgA1 from human serum by affinity chromatography using an immobilized extract of the albumin gland of Helix pomatia. Transfus. Med. *5*:117-121.
9. **Boyle, M.D.P.** 1984. Applications of bacterial Fc receptors in immunotechnology. BioTechniques *2*:334-340.
10. **Buchbinder, J.L., R.C. Stephenson, T.S. Scanlan and R.J. Fletterick.** 1998. A comparison of the crystallographic structures of two catalytic antibodies with esterase activity. J. Mol. Biol. *282*:1033-1041.
11. **Burke, G.W., J. Colona, T. Noto, R. Reik, G. Ciancio, D. Roth et al.** 1997. Removal of preformed cytotoxic antibody using PROSORBA (Staph Protein-A-Silica) column without immunosuppression. Transplant. Proc. *29*:2249-2251.
12. **Catimel, B., M. Nerrie, F.T. Lee, A.M. Scott, G. Ritter, S. Welt, L.J. Old, A.W. Burgess and E.C. Nice.** 1997. Kinetic analysis of the interaction between the monoclonal antibody A33 and its colonic epithelial antigen by the use of an optical biosensor. A comparison of immobilization strategies. J. Chromatogr. *776*:15-30.
13. **Chaiet, L. and F.J. Wolf.** 1964. The properties of streptavidin, a biotin-binding protein produced by *Streptomycetes*. Arch. Biochem. Biophys. *106*:1-5.
14. **Chang, B.S., R.M. Beauvais, A. Dong and J.F. Carpenter.** 1996. Physical factors affecting the storage stability of freeze-dried interleukin-1 receptor antagonist: glass transition and protein conformation. Arch. Biochem. Biophys. *331*:249-258.
15. **Chilkoti, A., B.L. Schwartz, R.D. Smith, C.J. Long and P.S. Stayton.** 1995. Engineered chimeric streptavidin tetramers as novel tools for bioseparations and drug delivery. Biotechnology (NY) *13*:1198-1204.
16. **Creaser, C.S., S.J. Feely, E. Houghton and M. Seymour.** 1998. Immunoaffinity chromatography combined on-line with high-performance liquid chromatography-mass spectrometry for the determination of corticosteroids. J. Chromatogr. *794*:37-43.
17. **de Dios Alche, J. and H. Dickinson.** 1998. Affinity chromatographic purification of antibodies to a biotinylated fusion protein expressed in *Escherichia coli*. Protein Expr. Purif. *12*:138-143.
18. **de Haard, H.J.W., B. Kazemier, M.J.M. Koolen, L.J. Nijholt, R.H. Meloen, B. van Gemen, H.R. Hoogenboom and J.W. Arends.** 1998. Selection of recombinant, library-derived antibody fragments

133

against p24 for human immunodeficiency virus type 1 diagnostics. Clin. Diagn. Lab. Immunol. *5*:636-644.

19. **Deinl, I., L. Angermaier, C. Franzelius and G. Machbert.** 1997. Simple high-performance liquid chromatographic column-switching technique for the on-line immunoaffinity extraction and analysis of flunitrazepam and its main metabolites in urine. J. Chromatogr. B Biomed. Sci. Appl. *704*:251-258.

20. **DeSilva, B.S. and G.S. Wilson.** 1995. Solid phase synthesis of bifunctional antibodies. J. Immunol. Methods *188*:9-19.

21. **Diano, M., A. Le Bivic and M. Hirn.** 1998. Raising polyclonal antibodies using nitrocellulose-bound antigen. Methods Mol. Biol. *80*:5-13.

22. **Doolittle, M.H., D.C. Martin, R.C. Davis, M.A. Reuben and J. Elovson.** 1991. A two-cycle immunoprecipitation procedure for reducing nonspecific protein contamination. Anal. Biochem. *195*:364-368.

23. **Duncan, K., S. Kruger, N. Zabe, B. Kohn and R. Prioli.** 1998. Improved fluorometric and chromatographic methods for the quantification of fumonisins B(1), B(2) and B(3). J. Chromatogr. *815*:41-47.

24. **Fagain, C.O.** 1996. Storage of pure proteins. Methods Mol. Biol. *59*:339-356.

25. **Farooqi, M., M. Saleemuddin, R. Ulber, P. Sosnitza and T. Scheper.** 1997. Bioaffinity layering: a novel strategy for the immobilization of large quantities of glycoenzymes. J. Biotechnol. *55*:171-179.

26. **Frenette, G., R.R. Tremblay and J.Y. Dube.** 1998. Simple purification procedure for human prostatic kallikrein hK2 in its active form. J. Chromatogr. B Biomed. Sci. Appl. *713*:297-300.

27. **Garcia-Gonzalez, M., S. Bettinger, S. Ott, P. Olivier, J. Kadouche and P. Pouletty.** 1988. Purification of murine IgG3 and IgM monoclonal antibodies by euglobulin precipitation. J. Immunol. Methods *111*:17-23.

28. **Girudi, G. and C. Baggiani.** 1996. Strategy for fractionating high-affinity antibodies to steroid hormones by affinity chromatography. Analyst *121*:939-944.

29. **Goding, J.W.** 1978. Use of staphylococcal protein A as an immunological reagent. J. Immunol. Methods *20*:241-253.

30. **Grubb, A., R. Grubb, P. Christensen and C. Schalen.** 1982. Isolation and some properties of an IgG Fc binding protein from group I streptococci type 15. Arch. Allergy Appl. Immunol. *67*:369-376.

31. **Gu, X., M. Meleka-Boules and C.L. Chen.** 1996. Micellar electrokinetic capillary chromatography combined with immunoaffinity chromatography for identification and determination of dexamethasone and flumethasone in equine urine. J. Capillary Electrophor. *3*:43-49.

32. **Hage, D.S.** 1998. Survey of recent advances in analytical applications of immunoaffinity chromatography. J. Chromatogr. B Biomed. Sci. Appl. *715*:3-28.

33. **Hage, D.S., C.E. Wolfe and M.R. Oates.** 1997. Development of a kinetic model to describe the effective rate of antibody oxidation by periodate. Bioconjug. Chem. *8*:914-920.

34. **Hamby, C.V., M. Chinol, C. Manzo and S. Ferrone.** 1997. Purification by affinity chromatography with anti-idiotypic monoclonal antibodies of immunoreactive monoclonal antibodies following labeling with 188Re. Hybridoma *16*:27-31.

35. **Hammerschmidt, S., S.R. Talay, P. Brandtzaeg and G.S. Chhatwal.** 1997. SpsA, a novel pneumococcal surface protein with specific binding to secretory immunoglobulin A and secretory component. Mol. Microbiol. *25*:1113-1124.

36. **Hashim, O.H., C.L. Ng, S. Gendeh and M.I. Nik Jaafar.** 1991. IgA binding lectins isolated from distinct Artocarpus species demonstrate differential specificity. Mol. Immunol. *28*:393-398.

37. **Haugland, R.P. and W.W. You.** 1998. Coupling of antibodies with biotin. Methods Mol. Biol. *80*:173-183.

38. **He, X.Y., Z. Xu, J. Melrose, A. Mullowney, M. Vasquez, C. Queen, V. Vexler, C. Klingbeil, M.S. Co and E.L. Berg.** 1998. Humanization and pharmacokinetics of a monoclonal antibody with specificity for both E- and P-selectin. J. Immunol. *160*:1029-1035.

39. **Hou, K.C., R. Zaniewski and S. Roy.** 1991. Protein A immobilized affinity cartridge for immunoglobulin purification. Biotechnol. Appl. Biochem. *13*:257-268.

40. **Hu, R., G.Y. Xie, X. Zhang, Z.Q. Guo and S. Jin.** 1998. Production and characterization of monoclonal anti-idiotypic antibody exhibiting a catalytic activity similar to carboxypeptidase A. J. Biotechnol. *61*:109-115.

41. **Hudson, P.J.** 1998. Recombinant antibody fragments. Curr. Opin. Biotechnol. *9*:395-402.

42. **Inouye, K. and K. Morimoto.** 1994. Preparation of F(Ab′)$_2$ mu fragments from rat IgM monoclonal antibodies and their application to the enzyme immunoassay of mouse interleukin-6. J. Immunol. Methods *171*:239-244.

43. **Ishida, I., H. Yoshida and K. Tomizuka.** 1998. Production of a diverse repertoire of human antibodies in genetically engineered mice. Microbiol. Immunol. *42*:143-150.

44. **Jansson, B., M. Uhlen and P.A. Nygren.** 1998. All individual domains of staphylococcal protein A show FAb binding. FEMS Immunol. Med. Microbiol. *20*:69-78.

45. **Josic, D., H. Schwinn, A. Strancar, A. Podgornik, M. Barut, Y.P. Lim and M. Vodopivec.** 1998. Use of compact, porous units with immobilized ligands with high molecular masses in affinity chromatography and enzymatic conversion of substrates with high and low molecular masses. J. Chromatogr. *803*:61-71.

46. **Kessler, S.W.** 1975. Rapid isolation of antigens from cells with a staphylococcal protein A-antibody adsorbent: parameters of the interaction of antibody-antigen complexes with protein A. J. Immunol. *115*:1617-1624.

47. **Kortt, A.A., G.W. Oddie, P. Iliades, L.C. Gruen and P.J. Hudson.** 1997. Nonspecific amine immobilization of ligand can be a potential source of error in BIAcore binding experiments and may reduce binding affinities. Anal. Biochem. *253*:103-111.

48. **Kouki, T., T. Inui, H. Okabe, Y. Ochi and Y. Kajita.** 1997. Separation method of IgG fragments using protein L. Immunol. Invest. 26:399-408.

49. **Kristensen, O., D.G. Vassylyev, F. Tanaka, K. Morikawa and I. Fujii.** 1998. A structural basis for transition-state stabilization in antibody-catalyzed hydrolysis: crystal structures of an abzyme at 1.8 A resolution. J. Mol. Biol. *281*:501-511.

50. **Lesley, S.A. and D.J. Groskreutz.** 1997. Simple affinity purification of antibodies using in vivo biotinylation of a fusion protein. J. Immunol. Methods *207*:147-155.

51. **Li, R., V. Dowd, D.J. Stewart, S.J. Burton and C.R. Lowe.** 1998. Design, synthesis, and application of a protein A mimetic. Nat. Biotechnol. *16*:190-195.

52. **Li, Y., W. Cockburn and G.C. Whitelam.** 1998. Filamentous bacteriophage display of a bifunctional protein A:scFv fusion. Mol. Biotechnol. *9*:187-193.

53. **Lim, S.B., C.T. Chua and O.H. Hashim.** 1997. Isolation of a mannose-binding and IgE- and IgM-reactive lectin from the seeds of Artocarpus integer. J. Immunol. Methods *209*:177-186.

54. **Liu, F.T. and D.H. Katz.** 1984. Mouse monoclonal IgE antibodies specific for ragweed pollen antigens. Hybridoma *3*:277-285.

55. **Livnah, O., E.A. Bayer, M. Wilchek and J.L. Sussman.** 1993. Three-dimensional structures of avidin and the avidin-biotin complex. Proc. Natl. Acad. Sci. USA *90*:5076-5080.

56. **Ljungquist, C., B. Jansson, T. Moks and M. Uhlen.** 1989. Thiol-directed immobilization of recombinant IgG-binding receptors. Eur. J. Biochem. *186*:557-561.

57. **Luo, W. and S.H. Lin.** 1997. Generation of moderate amounts of polyclonal antibodies in mice. BioTechniques *23*:630-632.

58. **Mahiouz, D.L., G. Aichinger, D.O. Haskard and A.J. George.** 1998. Expression of recombinant anti-E-selectin single-chain Fv antibody fragments in stably transfected insect cell lines. J. Immunol. Methods *212*:149-160.

59. **Martineau, P., P. Jones and G. Winter.** 1998. Expression of an antibody fragment at high levels in the bacterial cytoplasm. J. Mol. Biol. *280*:117-127.

60. **Matson, R.S. and M.C. Little.** 1988. Strategy for the immobilization of monoclonal antibodies on solid-phase supports. J. Chromatogr. *458*:67-77.

61. **Matsuda, Y., M. Nagao, T. Takatori, H. Niijima, M. Nakajima, H. Iwase, M. Kobayashi and K. Iwadate.** 1998. Detection of the sarin hydrolysis product in formalin-fixed brain tissues of victims of the Tokyo subway terrorist attack. Toxicol. Appl. Pharmacol. *150*:310-320.

62. **McCall, A.M., A.R. Amoroso, C. Sautes, J.D. Marks and L.M. Weiner.** 1998. Characterization of anti-mouse Fc gamma RII single-chain Fv fragments derived from human phage display libraries. Immunotechnology *4*:71-87.

63. **McKercher, G., P.R. Bonneau, L. Lagace, D. Thibeault, M.J. Massariol, R. Krogsrud, C. Lawetz, P.C. McDonald and M.G. Cordingley.** 1997. Improved purification protocol of the HSV-1 protease catalytic domain, using immunoaffinity. Biochem. Cell Biol. *75*:795-801.

64. **Merchant, A.M., Z. Zhu, J.Q. Yuan, A. Goddard, C.W. Adams, L.G. Presta and P. Carter.** 1998. An efficient route to human bispecific IgG. Nat. Biotechnol. *16*:677-681.

65. **Miralles, F., Y. Takeda and M.J. Escribano.** 1991. Comparison of carbohydrate and peptide biotinylation on the immunological activity of IgG1 murine monoclonal antibodies. J. Immunol. Methods *5*:191-196.

66. **Mizutani, T.** 1980. Decreased activity of proteins adsorbed onto glass surfaces with porous glass as a reference. J. Pharm. Sci. *69*:279-282.

67. **Mohanty, J.G. and K.S. Rosenthal.** 1985. A micropreparation of fluorescein conjugates of immunoglobulin G and Fab from serum. Anal. Biochem. *146*:361-365.

68. **Moio, L., C. Marchisano and F. Addeo.** 1998. Isolation of specific oligoclonal antibodies against bovine alpha s1-casein by FPLC tandem immunoaffinity of the polyclonal antibodies. J. Dairy Res. *65*:515-520.

69. **Moks, T., L. Abrahmsen, B. Nilsson, U. Hellman, J. Sjoquist and M. Uhlen.** 1986. Staphylococcal protein A consists of five IgG-binding domains. Eur. J. Biochem. *156*:637-643.

70. **Morag, E., E.A. Bayer and M. Wilchek.** 1996. Immobilized nitro-avidin and nitro-streptavidin as reusable affinity matrices for application in avidin-biotin technology. Anal. Biochem. *243*:257-263.

71. **Muller, K.M., K.M. Arndt and A. Pluckthun.** 1998. A dimeric bispecific miniantibody combines two specificities with avidity. FEBS Lett. *432*:45-49.

72. **Myhre, E.B. and G. Kronvall.** 1980. Binding of murine myeloma proteins of different Ig classes and subclasses to Fc-reactive surface structures in gram-positive cocci. Scand. J. Immunol. *11*:37-46.

73. **Ng, P.C. and Y. Osawa.** 1997. Preparation and characterization of the F(ab)2 fragments of an aromatase activity-suppressing monoclonal antibody. Steroids *62*:776-781.

74. **Ouyang, S., Y. Xu and Y.H. Chen.** 1998. Selective determination of a group of organic compounds in complex sample matrixes by LC/MIMS with on-line immunoaffinity extraction. Anal. Chem. *70*:931-935.

75. **Peng, Z., G. Arthur, E.S. Rector, D. Kierek-Jaszczuk, F.E. Simons and A.B. Becker.** 1997. Hetero-

geneity of polyclonal IgE characterized by differential charge, affinity to protein A, and antigenicity. J. Allergy Clin. Immunol. *100*:87-95.

76. **Peng, Z., G. Arthur, F.E. Simons and A.B. Becker.** 1993. Binding of dog immunoglobulins G, A, M, and E to concanavalin A. Vet. Immunol. Immunopathol. *36*:83-88.

77. **Perosa, F., G. Luccarelli and F. Dammacco.** 1997. Absence of streptococcal protein G (PG)-specific determinant in the FAb region of human IgG2. Clin. Exp. Immunol. *109*:272-278.

78. **Perosa, F., G. Luccarelli, M. Neri and F. Dammacco.** 1997. The FAb region of IgG2 human myeloma proteins does not bear the streptococcal protein G-specific determinant. J. Immunol. Methods *203*:153-155.

79. **Phillips, T.M.** 1985. High-performance immunoaffinity chromatography. LC·GC *3*:962-972.

80. **Phillips, T.M.** 1992. Appendix 3. Antibody digestion techniques, p. 324-328. *In* Analytical Techniques in Immunochemistry. Marcel Dekker, New York.

81. **Piervincenzi, R.T., W.M. Reichert and H.W. Hellinga.** 1998. Genetic engineering of a single-chain antibody fragment for surface immobilization in an optical biosensor. Biosens. Bioelectron. *13*:305-312.

82. **Porter, R.R.** 1959. The hydrolysis of rabbit τ-globulin and antibodies with crystalline papain. Biochem. J. *73*:119-126.

83. **Puri, R.N. and R.W. Colman.** 1997. Immunoaffinity method to identify aggregin, a putative ADP-receptor in human blood platelets. Arch. Biochem. Biophys. *347*:263-270.

84. **Randerath, K., P. Sriram, B. Moorthy, J.P. Aston, R.A. Baan, P.T. van den Berg, E.D. Booth and W.P. Watson.** 1998. Comparison of immunoaffinity chromatography enrichment and nuclease P1 procedures for ^{32}P-postlabelling analysis of PAH-DNA adducts. Chem. Biol. Interact. *110*:85-102.

85. **Reiter, Y., U. Brinkmann, B. Lee and I. Pastan.** 1996. Engineering antibody Fv fragments for cancer detection and therapy: disulfide-stabilized Fv fragments. Nat. Biotechnol. *14*:1239-1245.

86. **Ribeiro Neto, L.M., M.C. Salvadori and H.S. Spinosa.** 1997. Immunoaffinity chromatography in the detection of dexamethasone in equine urine. J. Chromatogr. Sci. *35*:549-551.

87. **Ros, R., F. Schwesinger, D. Anselmetti, M. Kubon, R. Schafer, A. Pluckthun and L. Tiefenauer.** 1998. Antigen binding forces of individually addressed single-chain Fv antibody molecules. Proc. Natl. Acad. Sci. USA *95*:7402-7405.

88. **Sano, T., S. Vajda and C.R. Cantor.** 1998. Genetic engineering of streptavidin, a versatile affinity tag. J. Chromatogr. B Biomed. Sci. Appl. *715*:85-91.

89. **Sano, T., S. Vajda, G.O. Reznik, C.L. Smith and C.R. Cantor.** 1996. Molecular engineering of streptavidin. Ann. NY Acad. Sci. *799*:383-390.

90. **Sano, T., S. Vajda, C.L. Smith and C.R. Cantor.** 1997. Engineering subunit association of multisubunit proteins: a dimeric streptavidin. Proc. Natl. Acad. Sci. USA *94*:6153-6158.

91. **Schramm, W., T. Yang and A.R. Midgley.** 1987. Surface modification with protein A for uniform binding of monoclonal antibodies. Clin. Chem. *33*:1338-1342.

92. **Scott, M.A., J.M. Davis and K.A. Schwartz.** 1997. Staphylococcal protein A binding to canine IgG and IgM. Vet. Immunol. Immunopathol. *59*:205-212.

93. **Shelver, W.L., G.L. Larsen and J.K. Huwe.** 1998. Use of an immunoaffinity column for tetra-chlorodibenzo-p-dioxin serum sample cleanup. J. Chromatogr. B Biomed. Sci. Appl. *705*:261-268.

94. **Sheoran, A.S and M.A. Holmes.** 1996. Separation of equine IgG subclasses (IgGa, IgGb, and IgG(T)) using their differential binding characteristics for staphylococcal protein A and streptococcal protein G. Vet. Immunol. Immunopathol. *55*:33-43.

95. **Solfrizzo, M., G. Avantaggiato and A. Visconti.** 1998. Use of various clean-up procedures for the analysis of ochratoxin A in cereals. J. Chromatogr. *815*:67-73.

96. **Stein, S.R., J.L. Palmer and A. Nisonoff.** 1964. Re-formation of interchain bonds linking half-molecules of rabbit τ-globulin. J. Biol. Chem. *239*:2872-2877.

97. **Stephens, D.B., R.E. Thomas, J.F. Stanton and B.L. Iverson.** 1998. Polyclonal antibody catalytic variability. Biochem. J. *332*:127-134.

98. **Stubbings, G.W., A.D. Cooper, M.J. Shepherd, J.M. Croucher, D. Airs, W.H. Farrington and G. Shearer.** 1998. Determination of 19-nortestosterone and trenbolone in animal tissues by high-performance liquid chromatography with immunoaffinity clean-up. Food Addit. Contam. *15*:293-301.

99. **Suarez, A.M., J.I. Azcona, J.M. Rodriguez, B. Sanz and P.E. Hernandez.** 1997. One-step purification of nisin A by immunoaffinity chromatography. Appl. Environ. Microbiol. *63*:4990-4992.

100. **Suzuki, H., E.B. Mukouyama, C. Wada, Y. Kawamura-Konishi, Y. Wada and M. Ono.** 1998. A catalytic antibody that accelerates the hydrolysis of carbonate esters. Prediction of the binding-site structure of the substrate. J. Protein Chem. *17*:273-278.

101. **Tang, S. and B. Bean.** 1998. A panel of monoclonal antibodies against human sperm. J. Androl. *19*:189-195.

102. **Tishchenko, G.A., M. Bleha, J. Skvor and T. Bostik.** 1998. Effect of salt concentration gradient on separation of different types of specific immunoglobulins by ion-exchange chromatography on DEAE cellulose. J. Chromatogr. B Biomed. Sci. Appl. *706*:157-166.

103. **Tomono, T., T. Suzuki and E. Tokunaga.** 1981. Cleavage of human serum immunoglobulin G by an

immobilized pepsin preparation. Biochim. Biophys. Acta *660*:186-192.

104. **Twomey, C., S. Doonan and A. Giartosio.** 1995. Thermal denaturation as a predictor of stability on long-term storage of a protein. Biochem. Soc. Trans. *23*:369S.

105. **Vankova, R., A. Gaudinova, H. Sussenbekova, P. Dobrev, M. Strnad, J. Holik and J. Lenfeld.** 1998. Comparison of oriented and random antibody immobilization in immunoaffinity chromatography of cytokinins. J. Chromatogr. *811*:77-84.

106. **Vaswani, S.K. and R.G. Hamilton.** 1998. Humanized antibodies as potential therapeutic drugs. Ann. Allergy Asthma Immunol. *81*:105-115.

107. **Virella, G.** 1971. Degradation of IgG3 monoclonal proteins during storage and in the presence of a thiol (DL-penicillamine). Experientia *27*:94-96.

108. **Visconti, A. and M. Pascale.** 1998. Determination of zearalenone in corn by means of immunoaffinity clean-up and high-performance liquid chromatography with fluorescence detection. J. Chromatogr. A *815*:133-140.

109. **Weinstein, Y., D. Givol and P.H. Strausbauch.** 1972. The fractionation of immunoglobulins with insolubilized concanavalin A. J. Immunol. *109*:1402-1404.

110. **Wentworth, P. and K.D. Janda.** 1998. Catalytic antibodies. Curr. Opin. Chem. Biol. *2*:138-144.

111. **Westcott, J.Y., K.M. Maxey, J. MacDonald and S.E. Wenzel.** 1998. Immunoaffinity resin for purification of urinary leukotriene E4. Prostaglandins Other Lipid Mediat. *55*:301-321.

112. **Williams, R.C., C.C. Malone, G. Fry and F. Silvestris.** 1997. Affinity columns containing anti-DNA Id+ human myeloma proteins adsorb human epibodies from intravenous gamma globulin. Arthritis Rheum. *40*:683-693.

113. **Yamaguchi, Y., H. Kim, K. Kato, K. Masuda, I. Shimada and Y. Arata.** 1995. Proteolytic fragmentation with high specificity of mouse immunoglobulin G. Mapping of proteolytic cleavage sites in the hinge region. J. Immunol. Methods *181*:259-267.

114. **Youssoufian, H.** 1998. Immunoaffinity purification of antibodies against GST fusion proteins. BioTechniques *24*:198-200.

115. **Yurov, G.K., G.L. Neugodova, O.A. Verkhovsky and B.S. Naroditsky.** 1994. Thiophilic adsorption: rapid purification of F(Ab)2 and Fc fragments of IgG1 antibodies from murine ascitic fluid. J. Immunol. Methods *177*:29-33.

116. **Zhang, L., K. Jacobsson, J. Vasi, M. Lindberg and L. Frykberg.** 1998. A second IgG-binding protein in Staphylococcus aureus. Microbiology *144*:985-991.

117. **Zhang, M., A. Majid, P. Bardwell, C. Vee and A. Davidson.** 1998. Rheumatoid factor specificity of a VH3-encoded antibody is dependent on the heavy chain CDR3 region and is independent of protein A binding. J. Immunol. *161*:2284-2289.

6 | Batch Separations

Ligand-analyte or antibody-antigen reactions are most often used for the detection and quantification of specific analytes; however, the remarkable specificity of these interactions is also well suited for roles in either affinity extraction or affinity depletion. These two techniques are opposite sides of the same coin. Affinity extraction, or immuno-extraction when using antibodies, is used to obtain the maximum amount of a specific target from a heterogeneous mixture, while affinity depletion is used to remove a single target component from a solution. In the first case, the goal is to obtain the removed target in a purified form; in the latter, it is merely to remove the unwanted target from the mixture. A variety of different affinity and immunoaffinity techniques exist for the removal of targets from heterogeneous mixtures by batch processing. The uses of such batch separation techniques (particularly that of immunoaffinity) are increasing as the advantages of these techniques are recognized and various technical limitations are overcome. This increase has been spurred on by the remarkable selectivity associated with the use of specialized ligands and the marked reduction in the number of processing steps they allow. For clarity, we have divided the techniques involved in batch separations into three separate chapters. In this chapter, we present the general techniques used in affinity-based batch separations for macromolecular targets. Batch affinity separation techniques used for cell isolation are discussed in Chapter 8, and those for cell organelles in Chapter 9.

In general, many of the techniques for immunoaffinity separations discussed in earlier chapters can be adapted for use as a semi-preparative or preparative batch technique. For instance, immunoaffinity chromatography can be used as a batch technique by increasing the column size and the amount of active binding surfaces. One such study involved a single-step immunoaffinity purification of pig liver catechol *O*-methyltransferase by a chromatography system composed of a Sepharose 4B column with appropriate covalently coupled monoclonal antibodies (6). However, as discussed in Chapter 3, a variety of environmental factors can have damaging effects on bound antibodies, thus limiting the usefulness of basic immunoaffinity chromatography for batch separations. One such factor is the development of high back pressure that can be produced during preparative immunoaffinity chromatography. This chapter will discuss four general methods for batch affinity separations that

Affinity and Immunoaffinity Purification Techniques
Terry M. Phillips and Benjamin F. Dickens
© 2000 Eaton Publishing, Natick, MA

utilize affinity reactions for routine batch separations: precipitation reactions, affinity partitioning, affinity membrane separation, and magnetic separation. In addition to these techniques, preparative and semi-preparative immunoaffinity column chromatography can also be routinely performed using HPLC systems in a technique termed high performance immunoaffinity chromatography (5,76), discussed elsewhere (see Chapters 7 and 10).

PRECIPITATION REACTIONS

Isolation of specific analytes from either dilute solutions or complex biological matrices can be achieved by simple precipitation with suitable ligands using a variety of different procedures (35,84). Affinity ligands such as inorganic metals, textile dyes, lectins, and receptor or enzyme substrates have been employed in affinity precipitation approaches. Additionally, the use of specific immunological reagents for the precipitation of analytes from complex biological fluids and cell lysates has been applied for a number of years (25,57).

Affinity Precipitation

Numerous affinity ligands have been used to precipitate specific molecules prior to further purification or to investigate specific molecular interactions. Jackson et al. (37) used tyrosine-phosphorylated peptides of platelet-endothelial cell adhesion molecule (PECAM-1) to precipitate the Src homology 2 (SH2) domains of the protein-tyrosine phosphatase, SHP-2. Using this procedure, they demonstrated that different domains of the enzyme interacted with different phosphopeptides; the amino-terminal domain interacted preferentially with a Tyr-663 PECAM-1 phosphopeptide, while the carboxyl-terminal domain selected a Tyr-686 phosphopeptide. Using the same probes, Hua et al. (34) demonstrated specific binding and precipitation of protein-tyrosine phosphatases SHP-1 and SHP-2. Another example of the application of affinity precipitation to molecular biology is the use of plasmids immobilized on polymer particles to isolate specific endonucleases. Umeno et al. (87) reported that a conjugate of the plasmid pBR322 immobilized on polyacryamide could be used to isolate the restriction endonuclease *Eco*RI.

Utilizing a 39-kDa receptor-associated protein to isolate a low-density lipoprotein (LDL) receptor-related protein resulted in coprecipitation of two isoforms of a G protein from a detergent extract (30). Affinity ligands have also been used to precipitate specific receptors. Li et al. (53) used leptin-coated Sepharose beads to precipitate soluble leptin receptors from mouse plasma. They demonstrated that the beads were able to isolate a receptor molecule of approximately 120 kDa, the nature of which was confirmed from mouse plasma by Western blotting with specific anti-leptin receptor antibodies.

Dyes have also been used to isolate specific analytes. Wu et al. (93) used immobilized Little Rock Orange matrix to coprecipitate four important lectins (namely wheat germ agglutinin, peanut lectin, con A, and red kidney bean lectin). Lilius et al. proposed metal affinity precipitation as a potential technique for the isolation of genetically engineered molecules (55). They used a DNA fragment from *Pseudomonas fluorescens* to encode five histidine residues fused to the 3′-terminal end of the galactose dehydrogenase gene. They expressed this engineered protein in *E. coli* and precipitated it by adding zinc complexed with ethylene glycol-bis(β-aminoethyl ether)-*N,N,N′,N′*-tetraacetic acid. This process isolated the model protein at high purity, and the authors report that the technique is potentially applicable to any molecule containing the histidine tag.

Immunoprecipitation

The use of immunological reagents to bind and precipitate specific analytes has been applied to the isolation of a wide variety of important biological molecules. Antibodies have been applied to the isolation of cellular components (38,59,83), cell membrane receptors (7,19,91), molecules involved in cell regulation (12,49,82), and secreted molecules in culture media (27,31,71). Mandal et al. (59) developed hamster antibodies against murine CD1 molecules and used them to immunoprecipitate both the 52-kDa heavy chain and its associated 12-kDa β_2-microglobulin chain from thymocytes and splenocytes. These studies were used to define the role of these molecules in T-lymphocyte differentiation. Other studies on the expression of immunoregulatory molecules on normal cells were performed by Borvack et al. (8). They demonstrated that a functional Fc receptor could be isolated from cultured murine endothelial cells by immunoprecipitation and immunofluorescence studies. Agadjanyan et al. (1) have described the use of monoclonal antibodies to precipitate an 80- to 85-kDa antigen from lysates of human T-cell leukemia virus (HTLV) target cells. Likewise, Vogetseder et al. (88) used monoclonal antibodies to isolate a recombinant human endogenous retrovirus-K envelope protein. In many of these studies, immunoprecipitation is combined with protein blotting techniques or Western blotting to fully characterize the isolated analyte. Doolittle and Ben-Zeev (18) have recently reviewed these procedures for the immunological characterization of lipoprotein lipase.

Immunoprecipitation usually employs a specific antibody to catalyze analyte isolation by forming an immune complex consisting of the antibody plus the analyte. This can be achieved by several different approaches. The antibody can be directly attached to the analyte in situ, or the membrane-attached analyte (receptor) can be first reacted with a suitable receptor-binding ligand and then isolated by immunological recognition of the ligand. This reaction can take place either on a cell surface (23) or following cell disruption (Figure 1) (29). Once formed, the immune complex is then precipitated using a number of procedures such as naturally occurring precipitation or precipitation catalyzed by a number of secondary agents including saturated ammonium sulfate, PEG, or another antibody (40). Ayes et al. (4) described the use of a low protein-binding filter for the preparation of cell lysates for immunoprecipitation. This procedure has now been refined and is commercially available from CytoSignal under the trade name, IMMUNOcatcher™. This system involves capturing the immune precipitates on a protein Λ/protein G matrix and then using centrifugation through a filtration system for washing and handling the immune precipitate (Figure 2). The company also issues a similar system, HISTAGcatcher™, for isolating polyhistidine-labeled fusion proteins using iminodiacetic acid resin for capturing the histine-tagged proteins.

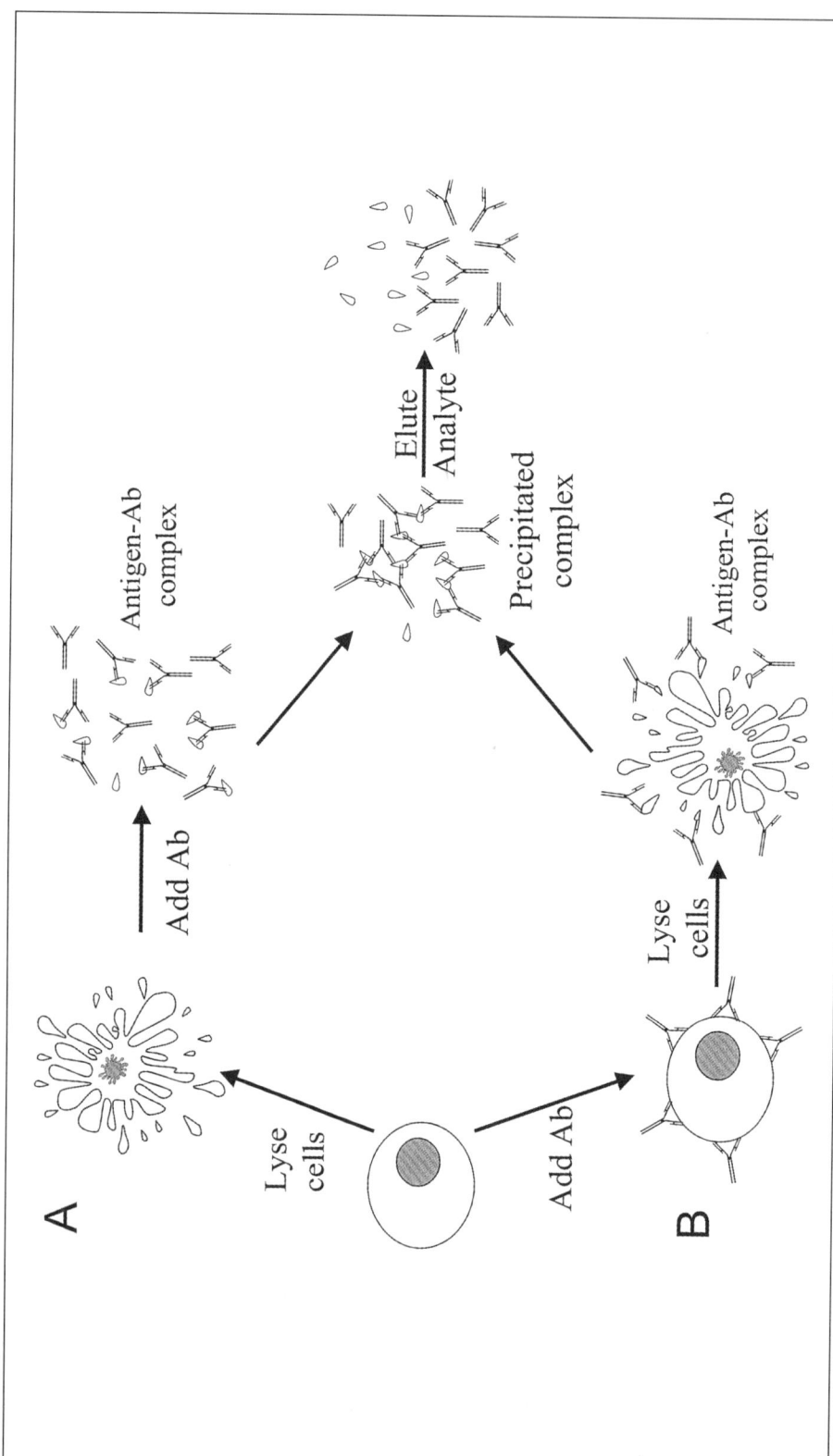

Figure 1. Use of antibodies to recover immunoprecipitated cell surface antigens. (A) Addition of antibodies to the cell lysate. (B) Localization of antibodies on the cell surface prior to cell lysis.

PROTOCOLS

Preparation of Tissue or Cell Extracts

1. Place 1 g of tissue (rinsed in cold 0.01 M sodium phosphate, pH 7.4) or 1×10^8 washed cells in a glass hand homogenizer (i.e., Dounce homogenizer).
2. Add 500 μL of hypotonic buffer (20 mM HEPES, 5 mM KCl, 1.5 mM MgCl$_2$, 1 mM DTT, 1 mM phenylmethylsulfonyl fluoride [PMSF], buffered to pH 7.5 with KOH).
3. Incubate at 4°C for 15 min.
4. Gently disrupt the tissue or cells by applying 30–50 strokes with a tight-fitting pestle.
5. Centrifuge at 1000× g for 5–10 min at 4°C to sediment the cell debris.
6. Collect both the supernatant (membrane fractions) and the pellet (cytosol).
7. Use immediately or store frozen at -80°C.

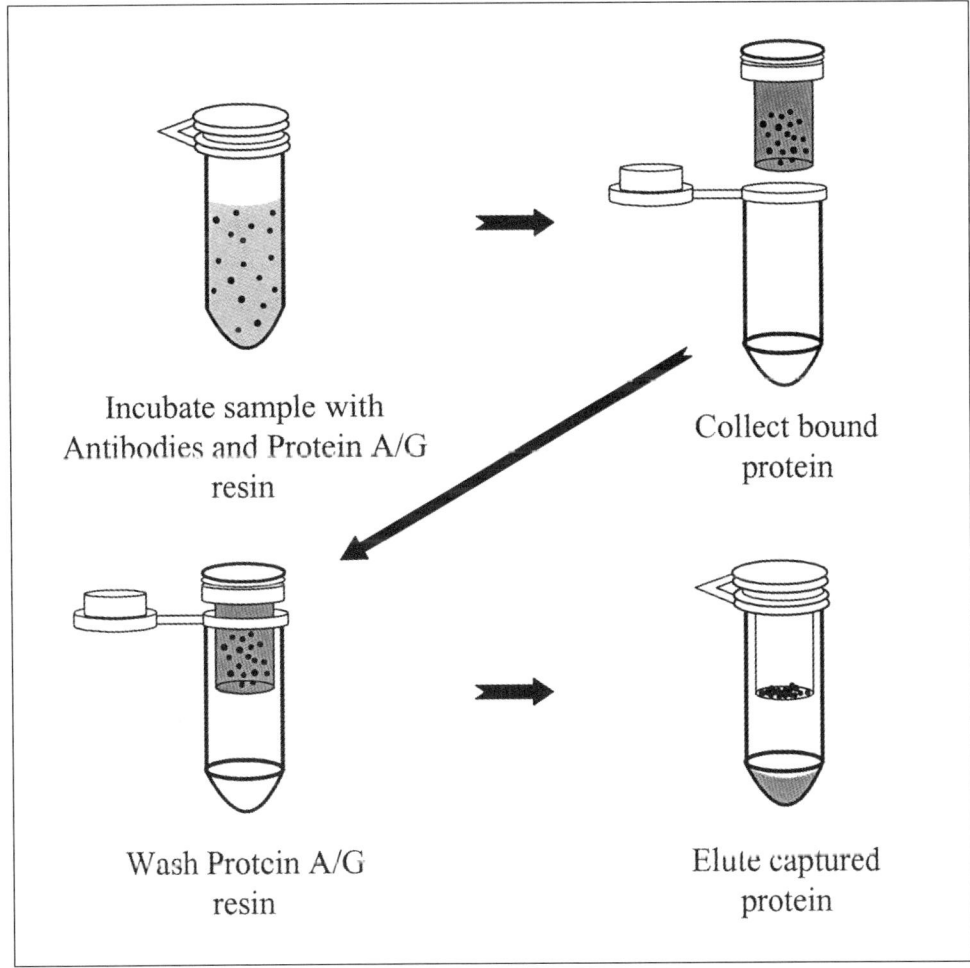

Incubate sample with Antibodies and Protein A/G resin

Collect bound protein

Wash Protein A/G resin

Elute captured protein

Figure 2. IMMUNOcatcher filtration system for the capture of immunoprecipitates and the subsequent elution of bound antigen.

Affinity Precipitation of a Membrane-Bound α-ᴅ-Mannose Glycoprotein

1. Place a 500-μL sample of tissue or cell extract supernatant in a microcentrifuge tube.
2. Add 200 μL of a 1:500 dilution of daffodil (*Narcissus pseudonarcissus*) lectin in hypotonic buffer (see protocol for preparation of tissue or cell extracts, above; other lectins can be used depending on the sugar moiety of the glycoprotein analyte).
3. Incubate at 4°C for 10–16 h on an overhead mixer.
4. Add a further 100 μL of the lectin dilution.
5. Incubate at 4°C for 3 h on an overhead mixer.
6. Centrifuge at 12 000× *g* for 15 min at 4°C.
7. Recover the pellet.
8. Wash three times by centrifugation (12 000× *g* for 15 min at 4°C) in cold 0.01 M phosphate-buffered saline (PBS)/0.01% 0.01 M NP-40.
9. Resolubilize the pellet in 500 μL of cold 0.01 M PBS/ 0.01% (vol/vol) NP-40.

Calculation of Antibody Activity for Immunoprecipitation Studies

1. Add 50 μL of labeled (fluorochrome or radiolabeled) antigen (1 μg/mL) to a series of glass tubes.
2. Carefully label each tube as an antibody dilution.
3. Make dilutions of the test antibody to cover a suitable working range as given below:

    1:10 dilution – add 10 μL antibody in 90 μL of buffer

    1:20 dilution – add 10 μL antibody in 190 μL of buffer

    1:50 dilution – add 10 μL antibody in 490 μL of buffer

    1:100 dilution – add 50 μL of the 1:50 dilution in 50 μL of buffer

    1:200 dilution – add 50 μL of the 1:100 dilution in 50 μL of buffer

    (Repeat this "doubling dilution" until the desired dilution range is achieved.)
4. Add 50 μL of each antibody dilution to the appropriate tube.
5. Mix and incubate at 37°C for 1 h.
6. Cap the tubes tightly and leave for 7 days at 4°C.
7. Centrifuge the tubes at 10 000× *g* for 20 min to recover the precipitates.
8. Determine the amount of labeled antigen present in each precipitate.
9. Construct a precipitin curve, plotting the bound antigen against the antibody dilution.

Dilutions forming the top of the curve are the working dilutions for the antibody.

Immunoprecipitation of a Cell-Surface Molecule (CD4) from Intact Lymphocytes

1. Suspend 1×10^6 lymphocytes in 750 μL of cold 0.01 M PBS, pH 7.4.
2. Wash the cells three times in 0.01 M PBS.
3. Resuspend the cell pellet from the last wash in 100 μL of anti-CD4 antibody solution (the concentration of antibody should be calculated prior to use – see the protocol for the calculation of antibody activity for immunoprecipitation studies, above).
4. Incubate overnight on an overhead mixer at 4°C.
5. Wash the cells in 0.01 M PBS containing 1 mM PMSF.
6. Resuspend the cells in 250 μL of 1% (vol/vol) NP-40 or Triton X-100.
7. Mix rapidly.
8. Incubate for 30 min at 4°C.
9. Centrifuge at 1000× g for 10 min to sediment the nuclei.
10. Dialyze the supernatant against cold 0.01 M PBS/1 mM PMSF for 3 h.
11. Centrifuge at 12 000× g for 15 min.
12. Recover the pellet.
13. Wash three times by centrifugation (12 000× g for 15 min at 4°C) in 0.01 M PBS/0.01% (vol/vol) NP-40.
14. Re-solubilize the pellet in 0.01 M PBS.

Immunoprecipitation of an Intracellular Analyte from Lysed Lymphocytes

1. Prepare a cell extract as described (see the protocol for the preparation of tissue or cell extracts, above) and place 500 μL of the extract in a microcentrifuge tube.
2. Add 100 μL of antibody solution (the concentration of antibody should be calculated prior to use; see the protocol for the calculation of antibody activity for immunoprecipitation studies, above).
3. Incubate overnight on an overhead mixer at 4°C.
4. Centrifuge at 12 000× g for 15 min.
5. Recover the pellet.
6. Wash three times by centrifugation (12 000× g for 15 min) in 0.01 M PBS/0.01% (vol/vol) NP-40.
7. Re-solubilize the pellet in 0.01 M PBS.

Precipitation of Secreted Molecules from Cultured Cells

1. Centrifuge at 800× g for 5–10 min to sediment the cells.
2. Recover the supernatant and add 200 μL of affinity ligand or antibody per mL of supernatant.
3. Place in a capped tube and incubate on an overhead mixer for 24–48 h at 4°C.
4. Centrifuge at 12 000× g for 25 min to sediment the precipitate.
5. Recover the pellet.

6. Wash three times by centrifugation ($12\,000 \times g$ for 15 min) in 0.01 M PBS.

7. Re-solubilize in cold PBS.

Immunoaffinity precipitates possess unique physiochemical properties that allow them to be isolated by a number of other procedures such as those given below.

PEG Recovery of the Immunoprecipitation Product

1. Prepare a 20% (wt/vol) solution of PEG 6000 in 0.05 M sodium phosphate, pH 7.4, by heating the solution to 37°C.

2. Add 2 vol of the PEG solution to 1 vol of reaction solution containing the immunoprecipitate.

3. Gently vortex-mix for 20–30 s.

4. Allow the mixture to stand at room temperature for 15 min.

5. Centrifuge at $2000 \times g$ for 15 min.

6. Discard the supernatant.

7. Gently resuspend the pellet in 100 µL of 0.33 M citric acid, pH 1.5.

8. Run through a protein A column to remove the free antibody.

9. Dialyze the recovered analyte overnight against 0.01 M sodium phosphate, pH 7.4.

Second Antibody Recovery of the Immunoprecipitation Product

1. Allow the immunoprecipitation to take place in a suitable tube.

2. Add 200 µL of precipitating secondary antibody to the reaction tube and mix well.

3. Incubate for 6–12 h at 4°C.

4. Centrifuge at $10\,000 \times g$ for 20 min.

5. Discard the supernatant.

6. Gently resuspend the pellet in 100 µL of 0.33 M citric acid, pH 1.5.

7. Run through a protein A column to remove both the primary and secondary antibody.

8. Collect the first major peak by monitoring a spectrophotometer at 280 nm.

9. Dialyze the recovered analyte overnight against 0.01 M sodium phosphate, pH 7.4.

Protein A Recovery of Immunoprecipitation Products

1. Allow the immunoprecipitation reaction to take place in a suitable tube.

2. Add 100 µL of a 10% (vol/vol) fixed *S. aureus* suspension to the tube and mix well.

3. Incubate for 45 min on a rotating mixer at room temperature.

4. Centrifuge at $10\,000 \times g$ for 10 min.

5. Discard the supernatant.

6. Gently resuspend the bacterial pellet in 500 μL of 0.33 M citric acid, pH 1.5.

7. Incubate for 30 min at room temperature.

8. Centrifuge at 10 000× g and collect the supernatant (containing the free analyte).

9. Dialyze the recovered analyte overnight against 0.01 M sodium phosphate, pH 7.4.

Recovery of Immunoprecipitated Proteins Using the ImmunoCatcher

This protocol was modified from the manufacturer's instructions. We find this modification, which increases incubation times, more suitable for the isolation of trace amounts of analyte or for use with weakly reactive antibodies.

1. Prepare a cell extract as described in the protocol for the preparation of tissue or cell extracts, above.

2. Add 200 μL of antibody solution to a 500-μL sample of tissue or cell extract supernatant and mix well.

3. Incubate for 1 h on an overhead mixer at 4°C.

4. Quickly vortex-mix and add 200 μL of the protein A/protein G resin supplied in the IMMUNOcatcher kit.

5. Incubate for 15 min on an overhead mixer at room temperature.

6. Place the mixture in the IMMUNOcatcher filter unit and place the filter in the microcentrifuge tube.

7. Centrifuge in a microcentrifuge (ca. 16 000× g) for 1.5 min.

8. Wash the resin twice in 200 μL of 0.01 M sodium phosphate, pH 7.4.

9. Add 200 μL of 0.1 M glycine, pH 1.5.

10. Incubate for 10 min with occasional shaking.

11. Centrifuge in a microcentrifuge at 16 000× g for 1 min.

12. Collect the filtrate containing the desired analyte.

13. Dialyze for 2–3 h against 0.01 M sodium phosphate, pH 7.4, to neutralize the acid.

AFFINITY PARTITIONING

Biological molecules will often preferentially partition into one or the other of two nonmiscible phases of a binary liquid mixture. A well-known example of such partitioning is the preferential separation of lipids in the organic phase of an aqueous-organic binary liquid system. In rare cases, the physiochemical properties of a target molecule can lead to its preferential partitioning into one phase of a binary aqueous system. One recent example is the use of a binary detergent system composed of a Triton X-114–rich and a Triton X-114–poor phase to isolate plant cytochrome P450 from green pigments (15). However, the natural properties of the desired target molecule often either do not preferentially partition it into one phase, or if it does, too many of the other molecules present in the heterogeneous mixture have similar partition properties to allow useful separation of the target. Affinity partitioning uses the power of affinity binding to form a ligand-target complex that possesses appropriate physical properties to preferentially partition into one phase of the binary aqueous system. In the most basic form of affinity partitioning (Figure 3), the ligand acts not only to select the target, but also acts as a carrier molecule to preferentially extract the target into one of the two phases of the binary system. However, like all separation techniques, the efficacy of the procedure is dependent on a number of external factors such as pH, temperature, and the properties of the phase system itself (39,47). Affinity partitioning can be achieved using simple aqueous systems composed of dextran and derivatized PEG. The system is usually buffered by salts such as sodium phosphate, sodium citrate, and sodium or magnesium sulfate (all of these salt species being incompatible with PEG, thus allowing the formation of two-phase systems). Dextran (Pharmacia) and PEG (Sigma Chemical) are reasonably pure and can easily be used to make two-phase separation systems. When using these reagents, it is usual

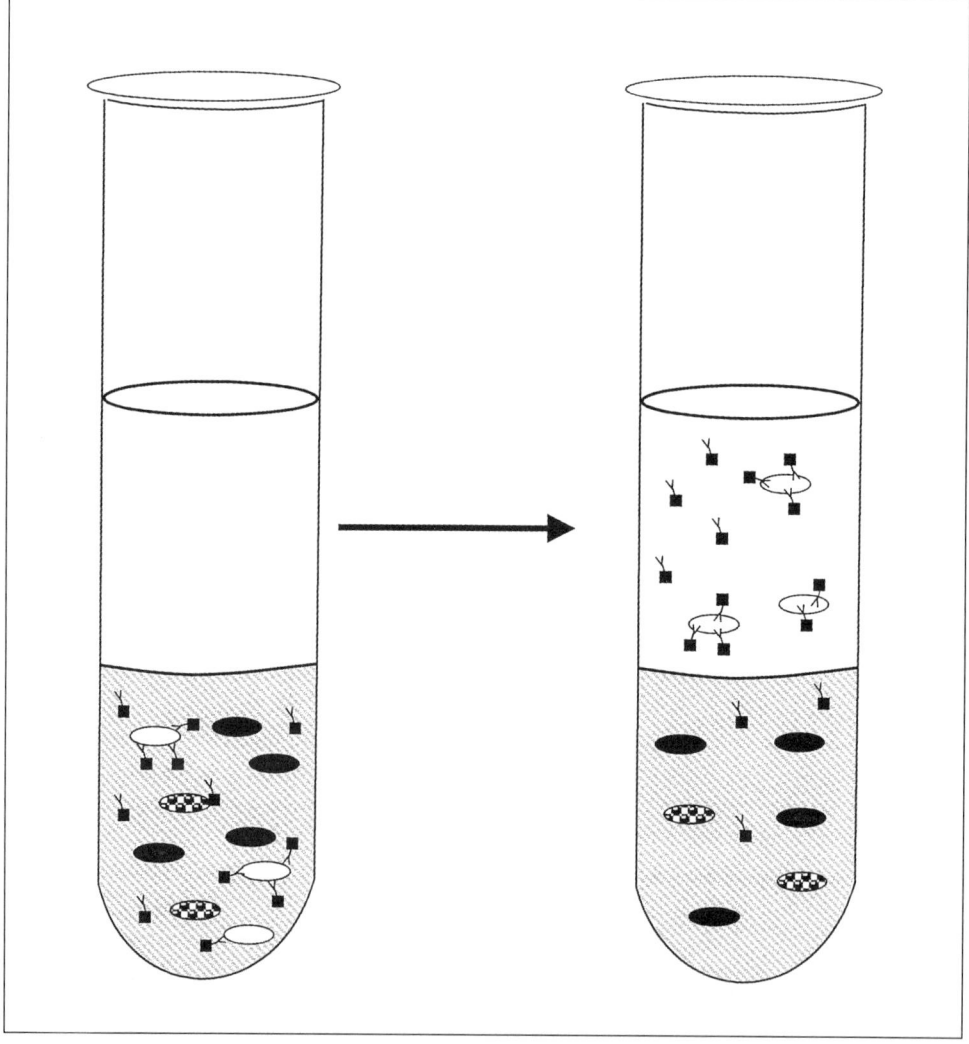

Figure 3. A two-phase aqueous immunoaffinity separation system. The antibodies are bound to moieties that preferentially partition into the upper phase, carrying the bound analytes with it.

to prepare 20%–25% stock solutions of dextran and 30%–40% stock solutions of PEG (both in distilled water). While the two major components of the system (i.e., dextran and PEG) are usually made up as stock solutions and diluted to make the desired phases, a potential difficulty often encountered in preparing the phases is an inaccuracy in transferring stock liquids. This can be overcome by using weight as the indicator of the amount of stock to add to form the right mixture (90). For a more in-depth review, the reader is directed to the book on affinity techniques edited by Mohr and Pommerening (62). The reader is also directed to Volume 228 of *Methods in Enzymology* for a more detailed account of the theory and application of aqueous two-phase systems (89).

Affinity Partitioning

In early work using affinity partitioning, Flanagan and Barondes (26) used affinity partitioning to isolate S-23 myeloma protein. For this study, they prepared a binary mixture composed of dextran-rich and poly(ethylene oxide)-rich phases. Dinitro-phenyl-poly(ethylene oxide), a specific synthetic ligand for S-23 myeloma protein, was added to the binary system containing the target protein. The ligand acted as a carrier molecule to cause the extraction of the target protein into the poly(ethylene oxide) phase. Just as with affinity chromatography, textile dyes have become popular ligands for use in affinity partitioning. Giuliano (28) used several different textile dyes in an aqueous binary system composed of maltodextrin and PVP to obtain a 50-fold enrichment in lysosome preparation from egg whites. Guoqiang and colleagues (33) used Cibacron Blue 3GA as a ligand in the affinity separation of lactate dehydroge-nase from porcine muscle extracts in a PEG-dextran binary system. To assist in the isolation, the dye ligand had EUPERGIT S100 attached to it as a carrier to aid in the affinity separation once the protein had bound the dye-ligand. The use of EUPERGIT S100 as the ligand carrier had two advantages in this isolation. Not only did it help partition the ligand (with the attached lactate dehydrogenase) into the upper phase, its properties were also used to help in extraction of the enzyme after the upper phase (containing the target-ligand-carrier) had been recovered. This was accomplished by taking advantage of a physical characteristic of EUPERGIT; it becomes insoluble at pH 5.1, so that the EUPERGIT-dye-enzyme complex was precipitated by simply low-ering the pH. Another textile dye example was the use of Cibacron Blue F3GA in affinity batch purification of phosphofructokinase from yeast at three different tem-peratures: 20°, 0°, and -18°C (32). To withstand the subzero temperature, the binary aqueous phase system was composed of water, dextran, PEG, ethylene glycol, and buffer. Cibaron Blue, acting as the ligand for the enzyme, increased the partitioning of the phosphofructokinase into the upper phase as the temperature decreased.

Kopperschläger and Birkenmeier (48) reviewed the field of affinity partitioning and made reference to the application of metal chelates as suitable ligands for parti-tioning proteins. Several other investigators have described this approach, especially for the isolation of specific types of enzymes. Kohler et al. (45) employed PEG-potassium phosphate two-phase aqueous systems to isolate four different β-galac-tosidase fusion proteins. Otto and Birkenmeier (70) reported that a copper-iminodi-acetate-PEG complex was successful in isolating lactate dehydrogenase isoenzymes from different species. Dehydrogenase enzymes have also been isolated from skele-tal and heart muscle by a two-phase system containing copper, nickel, zinc, or cad-mium chelated by iminodiacetate-PEG (74). Another application of affinity partitioning is in cell separation. Metal chelates immobilized into PEG complexes have been used extensively for the isolation of cells such as erythrocytes, lymphocytes, and fibroblasts by metal affinity partitioning (9,50,51,65).

A two-phase system using PEG has been applied to batch isolation of murine monoclonal IgG (2). Using a 15% (wt/vol) PEG-rich phase as the first forward extraction, Andrews et al. reported that virtually all of the IgG partitioned to this phase, leaving the contaminants in the lower phase. Back extraction was achieved by the addition of 14% (wt/vol) sodium phosphate in the absence of NaCl. The authors claim that the process was capable of isolating 5.9-fold purification (maximum possible = 7.3) at 80% purity.

A modification of liquid-liquid–based affinity partitioning termed affinity-based reverse micellar extraction and separation (ARMES) has been described by Paradkar and Dordick (72) and used to isolate peroxidase from soybean hulls (73). Instead of a binary aqueous system, these investigators used a binary system composed of an aqueous phase and an organic micelle phase. For this study, con A was used as the ligand and carrier molecule, and as such, required no chemical modification. The extraction system was composed of an aqueous acetate buffer, to which an isooctane solution containing bis(2-ethylhexyl) sulfosuccinate was added. Con A bound the sugar moiety of the soybean peroxidase, and the complex was extracted into the organic micelles. A fresh ionic phase with an altered pH can be used to cause reverse extraction of the con A back into the aqueous phase, which allows soybean lipase purification of nearly 30-fold.

Ekblad et al. (21) have reported that biotinylated, small unilamellar liposomes can be successfully affinity-partitioned in an aqueous PEG-dextran two-phase system using dextran-avidin as the affinity ligand. These investigators reported that more than 95% of the liposomes partitioned into the dextran-rich phase in the presence of 10 mM lithium sulfate, and that when a 6-carbon spacer arm was used, less than two biotin molecules per liposome were required for such partitioning.

Immunoaffinity Partitioning

In addition to affinity ligands, immunological ligands can also be used in a modification of the procedure called immunoaffinity partitioning. This procedure has been used for the isolation of horseradish peroxidase (22,92) and an isoenzyme of porcine lactate dehydrogenase (22). While technically feasible, immunoaffinity partitioning has a number of limitations that generally reduce its suitability for use as a batch separation technique. One limitation is the high cost of antibodies, especially when compared to inexpensive ligands such as textile dyes for use with large volumes of liquids. Another limitation is that the antibody molecule alone seldom has a selective physiochemical property to act as a carrier to aid in the partitioning of the complex into one or the other of the binary phases. This lack of a carrier property requires the addition of a carrier moiety to the antibody to aid in this separation, which often affects the binding of the antibody to its specific target molecule. When sufficient antibodies exist to immuno-extract one of the entire aqueous phases, the need for creating a two-phase system is avoided. The need to modify the antibody by adding a carrier molecule can be negated by simply using immunoprecipitation techniques discussed in *Precipitation Reactions* above, or the other readily available immunoaffinity batch separation techniques to be discussed in the following sections. While the practical usefulness of immunoaffinity partitioning for macromolecule separation is limited, it can be useful in cell isolation techniques (see Chapter 8) and, to a lesser extent, in cell organelle isolation (see Chapter 9).

PROTOCOLS

Formation of a Simple Two-Phase Partitioning System in a Centrifuge Tube

The following protocol will make 10 g of a phase system containing 5% (vol/vol) dextran, 4% (vol/vol) PEG, 90 mM sodium phosphate, and 30 mM sodium chloride.

1. Prepare the following stock solutions:
   20% (wt/wt) solution of dextran (mol. wt. = 500 000) in distilled H_2O
   40% (wt/wt) PEG (mol. wt. = 8000) in distilled H_2O
   0.44 M NaH_2PO_4, pH 6.8
   0.6 M NaCl
2. Mix 2.5 g of the dextran stock, 1 g of the PEG stock, 2.05 g of the NaH_2PO_4 stock, 0.5 g of the NaCl stock, and 3.95 g of distilled H_2O in a capped centrifuge tube.
3. Mix thoroughly by repeated inversion (10 times).
4. Allow the contents of the tube to equilibrate to the temperature at which the separation will take place (either room temperature or 4°C).
5. Mix again, and allow the phases to settle (centrifugation at 1000× g for 10 min will accelerate the phase formation).

When a clear interface is visible, the phase is ready to use.

Preparation of Dye-PEG Compounds

This protocol is modified from Reference 47.

1. Dissolve 5 g of PEG 8000, 0.25 g of anhydrous $NaSO_4$, and 0.5 g of NaOH in 10 mL of distilled H_2O.
2. Slowly add 0.2 g of Cibacron Blue F3GA.
3. Stir the mixture on an overhead mixer for 7 h at room temperature.
4. Adjust the pH to 6.5 with concentrated phosphoric acid.
5. Dialyze the mixture overnight against distilled H_2O.
6. Add 5–8 g of pre-activated DEAE cellulose.
7. Mix by continuous stirring until all of the color is adsorbed.
8. Recover the cellulose on a sintered glass filter by applying a vacuum.
9. Using a vacuum, wash the cellulose exchanger with 100 mL of distilled H_2O.
10. Recover the dye-PEG compound by washing the cellulose with 50 mL of 2 M KCl.
11. Extract the compound by adding 50 mL of chloroform.
12. Dry the compound in anhydrous $NaSO_4$ to evaporate the chloroform.

Preparation of Activated PEG

1. Dissolve 10 g of monomethyl ether PEG in 50 mL of toluene and place in a strong, toluene-resistant tube.
2. Prepare 40 mL of activated dichloromethane by heating at 330°C for 12 h.
3. Add 25 mL of the activated dichloromethane plus 1.2 mL of triethylamine to the PEG solution.
4. Mix carefully and plunge the reaction tube into crushed ice.
5. Under nitrogen, add 0.6 mL of tresyl chloride and stir the mixture overnight at room temperature.
6. Place the mixture in a sintered glass filter and filter the mixture to remove the residue.
7. Add the filtrate to 400 mL of anhydrous ether and place in an ice bath.
8. Recover the precipitate by placing it in a sintered glass filter and dry by applying a vacuum.
9. Store under vacuum in a dessicator.

Immobilization of Ligands to Activated PEG

1. Dissolve 5 mg of activated PEG in 500 µL of 0.01 M PBS, pH 7.4, and place in a small capped glass tube.
2. Dissolve 1 mg of ligand in 500 µL of 0.01 M PBS, pH 7.4.
3. Add the ligand solution slowly to the PEG solution (swirl the solution to ensure adequate mixing during addition).
4. Tightly cap the tube and incubate for 1–3 h at room temperature on an overhead mixer (if the ligand is heat labile [i.e., an antibody], then this step can be performed at 4°C and the incubation time doubled).
5. Store the solution at 4°C until required (leave the ligand in the solution during storage to ensure maximum binding to the PEG).

Affinity Partitioning of α-Mannose–Containing Glycoproteins

1. Form a two-phase system as described in the protocol for the formation of a simple two-phase partitioning system in a centrifuge tube, above, but substitute con A-labeled PEG for regular PEG.
2. Allow the phases to become completely separated and equilibrated to the temperature at which the separation is to take place.
3. Centrifuge at $1000 \times g$ for 10 min to ensure complete phase separation and recover the top phase with a pipet.
4. Centrifuge at $30\,000 \times g$ for 20 min to sediment the PEG.
5. Carefully remove the supernatant and resuspend the pellet in 0.25 M α-mannose (1:1 vol:vol).
6. Incubate at room temperature for 20–30 min to elute the bound glycoprotein.
7. Centrifuge at $30\,000 \times g$ for 20 min and recover the supernatant with a pipet.
8. Dialyze the supernatant against 0.01 M sodium phosphate, pH 7.4, overnight at

4°C (Use a regular dialysis membrane [mol. wt. cut-off 12 000–14 000] and change the phosphate buffer at least 4 times).

Immunoaffinity Partitioning of Proteins

1. Form a two-phase system (see the protocol for the formation of a simple two-phase partitioning system in a centrifuge tube, above), but substitute antibody-labeled PEG for the regular PEG.
2. Allow the phases to become completely separated and equilibrated to 4°C.
3. Centrifuge at 1000× g for 10 min at 4°C to ensure complete phase separation and recover the top phase with a pipet.
4. Add 1 mL of 1 M glycine and incubate at 4°C for 30–60 min to elute the bound protein.
5. Centrifuge at 30 000× g for 20 min at 4°C and recover the supernatant with a pipet.
6. Dialyze the supernatant against 0.01 M sodium phosphate, pH 7.4, overnight at 4°C (use a dialysis membrane that will retain the isolated protein, and change the dialysis buffer at least 4 times).

AFFINITY MEMBRANE SEPARATIONS

The basic principle of affinity membrane chromatography is easy to understand, because it involves only a permeable membrane with affinity ligand properties through which a solution containing an analyte of interest is passed, followed by elution with a dissociation buffer. Membrane chromatography commonly occurs using a rigid disk membrane encased in a holding device that attaches to a syringe (Figure 4). These same devices are also used for ion-exchange membranes, while large-scale units or contactors are used for biotechnology processing. Although membrane chromatography appears to have great advantages, there are downsides to the technique. One major drawback when separating protein solutions is membrane fouling (60). Another disadvantage, especially when using any disk membrane for batch separations, is the limited binding capacity that is obtained in such a small chromatographic thickness. Additionally, when extending membrane technology to affinity or immunoaffinity separations, the method in which ligands are bound to the membrane can have significant effects on the binding capacity of the membrane. A number of companies offer systems designed for use in affinity membrane chromatography. For instance, Millipore and Amicon sell cartridge systems specifically designed for affinity membrane chromatography. These systems are suitable for the direct addition by the investigator of any ligand, especially monoclonal or polyclonal antibodies (Table 1).

Basic Principles

Regular membrane separations are often based upon ultrafiltration using size-exclusion principles, such that molecules too large to pass through pores in the membrane are retained on the inlet side while smaller molecules percolate through the pores. Affinity membrane separations, however, are more akin to column chro-

Table 1. A Selection of Commercially Available Activated Membranes

Company	Product	Use
Amicon	Amino	Protein immobilization
	Carboxyl	Protein immobilization
	Protein A	Antibody immobilization
	Protein G	Antibody immobilization
Millipore	Amino	Protein immobilization
	Carboxyl	Protein immobilization
	Protein A	Antibody immobilization
	Protein G	Antibody immobilization
Whatman	Amino	Protein immobilization
	Carboxyl	Protein immobilization
	Tresyl	Protein immobilization

matography than they are to ultrafiltration techniques. In affinity membrane separations, the membrane possesses affinity-binding properties toward the target of interest. Most often, this is achieved through modification of the surfaces of the membranes by ligand addition to form a matrix with properties necessary for protein (and other analyte) purification. The membranes are most often modified by exactly the same ligands used in normal affinity chromatography, including textile dyes, protein A, protein G, or specific antibodies (61). The target of interest is bound by these added ligands upon the surface of the membrane (including in the pores), and then

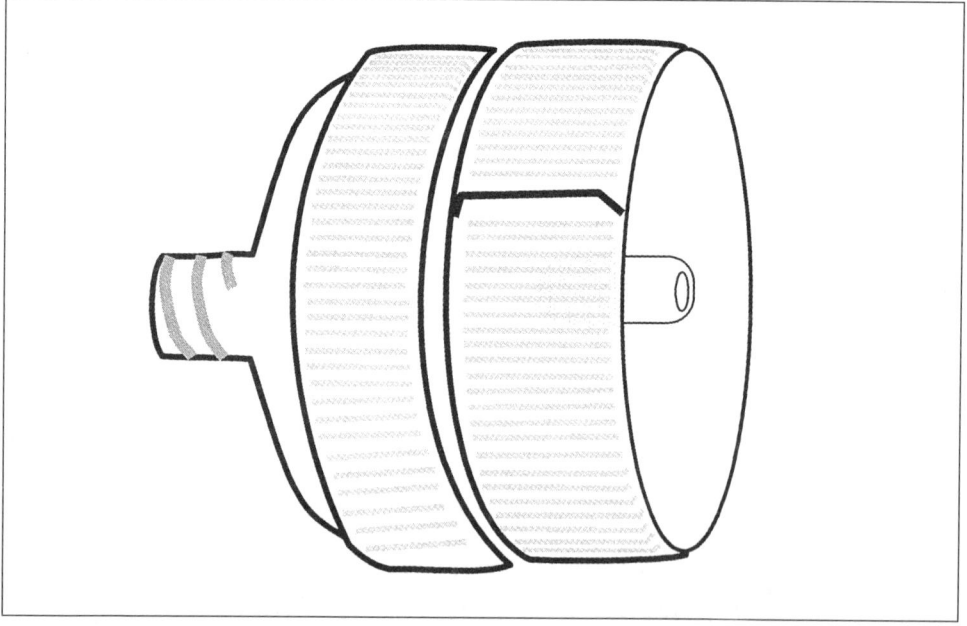

Figure 4. Swinnex syringe membrane holder (Millipore) used to hold affinity and immunoaffinity membranes.

removed with an elution buffer that dissociates the ligand-analyte bond. Thus, affinity membrane techniques are often referred to as membrane chromatography, where the membrane is considered to be a very thin, uniformly packed chromatographic support. Viewed in this way, membranes offer several remarkable advantages for "chromatographic" separations. They allow a very rapid flow rate, allowing for large volumes to be quickly processed. They produce very little back pressure, and consequentially, the attached ligand generally holds up better during batch procedures than in conventional chromatography. Since the ligands line the membrane pores (and, in this location, are freely accessible to the analyte as it filters through), there is often no need for the addition of a spacer arm, simplifying construction of the matrix. Another advantage of membrane chromatography is the very simple and low cost of the "chromatographic" apparatus (a simple filtration device) compared to other chromatographic systems such as HPLC (42). A novel membrane-based affinity system has been described by Najarian and Bellhouse wherein a tubular membrane was used to produce a tangential flow to the membrane surface. Using an immobilized textile dye, they demonstrated that the oscillatory flow produced by this apparatus increased recovery of BSA by a factor of 2 (63). Despite these advantages, there are also disadvantages to membrane chromatography. The major drawback is a limited binding capacity as compared to a conventional affinity chromatography. Another problem that needs to be considered when using macromolecular ligands is that the proper membrane pore size needs to be carefully selected to avoid restricted diffusion that can occur as the size of both target and ligand increases (75).

Affinity Separations

Membrane chromatography is often used in the single-step recovery of biological products from cell culture media. For example, Langlotz and Kroner (52) compared membrane chromatographic techniques using protein A and specific antibodies for the isolation of monoclonal antibodies from tissue culture. Another example of affinity membrane chromatography for the isolation of molecules of biological interest from culture media was the isolation of antithrombin III (56) using heparin (which has a strong affinity for antithrombin III) as the affinity ligand. This study compared the effect of attaching heparin to the membrane by amine modification vs. epoxy-activated membranes. The results showed that amine modification produced a membrane matrix with a three-times higher binding capacity than for antithrombin III when epoxy-activated membranes were used (91 vs. 32 mg/cm^2).

The antithrombin III study showed that while affinity chromatography is an effective method for the recovery of proteins from samples, it is important to take into account not only properties of the membrane but also those of the ligand used. Kasper et al. (44) compared the effect of several attachment methods for both low- and high-molecular-weight ligands. They showed that low-molecular-weight ligands, such as textile dyes, worked well on commercially available affinity membrane supports. These low-molecular-weight ligands allowed for high binding capacity, rapid flow rates, and very low back pressure. Unlike low-molecular-weight ligands, the use of antibodies and con A was much more problematic and could result in low ligand binding and poor antigen binding when compared to other techniques. Like heparin in the antithrombin III study (56), Kasper and colleagues demonstrated that epoxy-activated membranes gave relatively poor results (44). However, this study did demonstrate that excellent results could be obtained with the higher weight ligands if these proteins were immobilized using tresyl- or tosyl-activated membrane surfaces. The importance of selecting the proper immobilization technique was further illustrated by another study by Kasper et al. (43). In this study, human IgG antibodies directed against *E. coli* protein G were immobilized by a one-step reaction between native epoxy groups on a macroporous affinity disk composed of a glycidyl methacrylate-co-ethylene dimethacrylate polymer. In this study, the epoxy-activated membrane provided an excellent single-step isolation of highly purified protein G from bacterial lysates. Finger et al. (24) used thiophilic membranes for the purification of monoclonal antibodies from hybridoma culture media. Using membrane stacks in a cross-flow module with a spiral filtration channel, these investigators demonstrated fivefold purification factors and up to eightfold concentration factors. L-Histidine immobilized on poly(ethylenevinyl alcohol) hollow fiber membranes has been reported as a one-step isolation procedure for IgG from untreated human serum (11). The maximum binding capacity of this system was between 70 and 80 mg of IgG per gram of support. The authors calculate that a cartridge with a surface area of 1 m^2 could potentially isolate up to 1.5 g of IgG per hour from untreated human serum. Discontinuous pH or salt gradient elution could achieve recovery of the bound IgG and, although IgM was also adsorbed, it could be partially separated from the IgG by pH gradient elution.

Metal chelate chromatography can also be performed on membranes. Certain transitional metals, notably zinc, copper, iron, cadmium, and nickel, can form complexes with metal-binding proteins (77). These complexes can be used to isolate a number of key enzymes that bind specifically to these metals. For instance, a recent study combined textile dye with and without various levels of chelated Fe^{3+} ions on single membranes (3). This study showed that adding metal chelation (iron) properties to the membrane improved the maximum binding capacity of the membrane towards iron-binding enzymes. The use of metal chelation chromatography is being rapidly expanded through the use of hexa-histidine affinity-tagged sequences in recombinant-engineered proteins (41). However, a major disadvantage to the use of membrane-based metal-ion chelation is the difficulty in obtaining enough exposure of the target of interest to the metal ions in the small volume of matrix provided. The combination of metal chelation properties with other affinity ligands, as discussed above, is one solution to this drawback. Despite the problem of limited opportunities for metals to interact with larger protein molecules in a membrane system, metal chelation membrane chromatography has been useful, even on a microassay scale. A recent report described a technique for measurement of posttransitional hexa-histidine–tagged proteins using a membrane-based dot blot filtration device (86).

Immunoaffinity Separations

Immobilizing antibodies onto a membrane surface can often improve the specificity and capacity of affinity membranes. The first method to overcome the limited binding capacity of an affinity membrane system involves recycling the sample multiple times through the same column (79). Using such recycling for immunodepletion, the sample can be passed through the membrane, the bound target eluted, and the original sample applied again (recycled) through the membrane to further deplete the target molecule. This recycling can be repeated until the sample is sufficiently free of the extracted target material. This recycling filtration approach can be accelerated by placing a series of essentially identical membranes in sequence to obtain sufficient membrane-binding capacity to remove all of the target of interest in one passage through the assembled membranes. In some cases, these membranes can even be stacked within the same holding device. This approach increases the binding surfaces without dramatically increasing the thickness of the chromatographic bed.

A recent advance in immunoaffinity membrane chromatography (17) offers another way to dramatically improve membrane-binding capacity. This involves the construction of affinity particle-loaded membranes. These membranes are built by entrapping immunoaffinity-labeled beads within a thin (0.5-mm) disk membrane web composed of polytetrafluoroethylene fibers. The addition of beads to the polytetrafluoroethylene fiber membrane dramatically increased the affinity surface area within the membrane over that which could be obtained without adding such particles. In describing this method, the authors entrapped beads coated with polyclonal antibodies directed against 2,4-dichlorophenyoxyacetic acid within the flexible membrane. These immunoaffinity particle-loaded membranes were then compared with anion-exchange and solid-phase extraction for the isolation of the target molecule from test water samples. The immunoaffinity particle-loaded membrane showed a marked increase in binding specificity and relative capacity when compared to anion-exchange particle-loaded membranes and solid-phase extraction techniques. The trapping of affinity-based beads within the structure of a membrane further blurs the distinction between routine affinity chromatography and affinity membrane chromatography.

Just as batch immunological extraction can be improved by the stacking of identical affinity membranes in sequence, batch immunological purification methods can take advantage of a sequence of membrane-based adsorbants. Champluvier and Kula (13) used a sequence of membranes with carefully selected ligands and buffers such that, as a fraction was eluted from one membrane, it was directly applied to the next membrane in sequence. Using this approach, they demonstrated that passing a sample from *Candida boidinii* sequentially through cation exchange, dye-ligand, and anion-exchange membranes allowed for the rapid purification of formate dehydrogenase. This sequential use of an ion-exchange, dye-ligand membrane system was scalable from "micro" to "macro" levels. Another alternative method is to choose a membrane format other than the typical membrane disk. One such option is to use

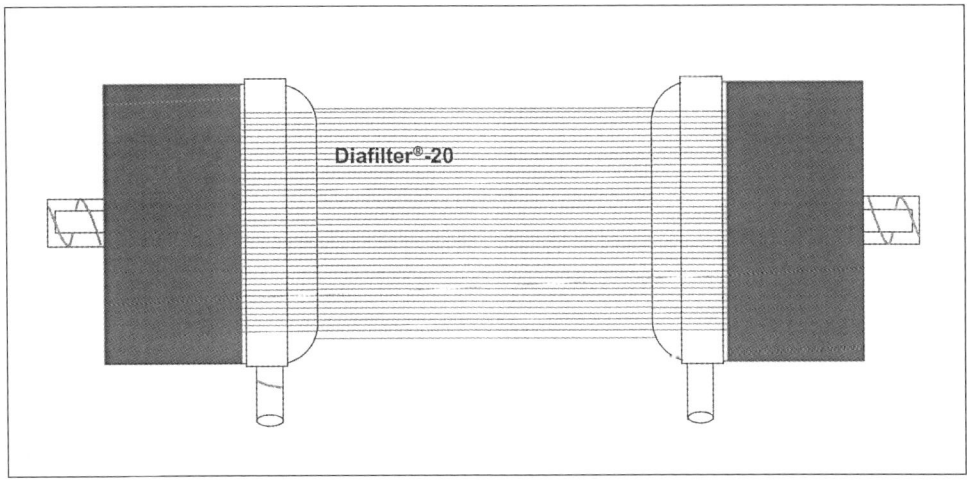

Figure 5. Amicon Diafilter for batch affinity extractions.

pleated sheets of membranes contained within filtration devices such as Amicon's Diafilter®-20 hemofilter (Figure 5), which can be modified with affinity ligands. To increase the size of the membrane surfaces, it is also possible to use filtration devices such as hollow-fiber membranes (36) or spiral-wound membrane cartridges. A recent publication used a combination of larger sheet membranes and material recycling for a clinical application of immunoaffinity membrane chromatography. Using this combined approach, Soltys and Eufemio (85) compared the in vitro immunological depletion of damaging LDL from patient plasma using membrane chromatographic separation, to that obtained with packed affinity columns. They demonstrated that flat sheets of membranes containing covalently immobilized monoclonal antibodies (attached to the membrane through a PEG spacer) provided for the rapid depletion of patient plasma LDL. The bound LDL could be rapidly eluted from the membrane, the affinity surfaces regenerated, and the plasma rapidly cycled through the membrane again. This depletion of LDL by membrane chromatography was equivalent to that achieved by conventional packed columns.

Immunoaffinity membrane chromatography has also been adapted for use in direct immunoassay procedures, unrelated to batch separations. For instance, affinity membranes play a key role in the design and construction of immunosensors (which are discussed in Chapter 10). They have also been used in other novel ways. One example used a membrane with defined regions coated with a monoclonal antibody directed against the C-reactive protein (67). The captured analyte was then measured by a sandwich immunoassay using an intensely blue-colored latex bead with a second monoclonal antibody directed against a different specific site on the C-reactive protein.

PROTOCOLS

Isolation of IgG Using a Protein A Syringe-Membrane Unit

1. Place a membrane coated with immobilized protein A on the retaining support in the membrane holder.
2. Carefully place the top of the holder over the membrane and hand-tighten.
3. Dialyze the sample against 0.01 M sodium phosphate, pH 7.4.
4. Fill a disposable syringe with the dialyzed sample.
5. Attach the membrane holder to the syringe by the Luer-Lok® fitting at the inlet of the holder.
6. Gently push the sample through the membrane holder unit and collect the solution passing through the membrane holder.
7. Wash the membrane by passing 5–10 mL of 0.01 M sodium phosphate, pH 7.4, through the membrane holder. (Discard the wash solution passing through the membrane holder.)
8. Place 2 mL of 0.1 M glycine in the syringe and pass it slowly through the membrane holder, collecting the effluent.
9. Dialyze the effluent against 0.01 M sodium phosphate, pH 7.4, and analyze the recovered sample.

Isolation of a Specific Glycoprotein from a Bioreactor Feedstock Using a Lectin Membrane Unit

1. Prepare a series of membranes (10–25) with immobilized mistletoe lectin (see *Membranes* in Chapter 2 and *Lectins* and *Ligand Immobilization* in Chapter 4).
2. Place the membranes carefully in the main body of the holder (contactor).
3. Gently compress the membranes to ensure that no air bubbles are trapped between the membrane layers.
4. Screw the top on the holder (secure tightly).
5. Connect the inlet of the holder to the outlet port of the bioreactor.
6. Connect the outlet of the holder to either a suitable collection vessel or recycle the fluid back to the bioreactor.
7. Open the bioreactor outlet port and allow the fluid to flow through the membrane holder for 1–2 h in a recycling system or until the reactor is empty.
8. Disconnect the membrane holder.
9. Connect the inlet of the holder to a pump.
10. Pump 100 mL of 0.01 M sodium phosphate, pH 7.4, through the membrane holder.
11. Pump 20–50 mL of 0.5 M β-D-galactosamine through the membrane holder and collect the effluent.
12. Dialyze the effluent overnight against 0.01 M sodium phosphate, pH 7.4, and analyze the recovered glycoprotein.

Isolation of an Antigen Using an Immunoaffinity Membrane

1. Place an antibody-coated membrane on the retaining support in the membrane holder.
2. Carefully place the top of the holder over the membrane and hand-tighten.
3. Fill a disposable syringe with the sample.
4. Attach the membrane holder to the syringe by the Luer-Lok fitting at the inlet of the holder.
5. Slowly push the sample through the membrane holder unit and discard the effluent.
6. Wash the membrane by passing 10 mL of 0.01 M sodium phosphate, pH 7.4, through the membrane holder. Discard the effluent from the washing.
7. Place 2 mL of 0.33 M citric acid or 2.5 M sodium thiocyanate in the syringe and slowly pass through the membrane holder, collecting the effluent.
8. Dialyze the effluent against 0.01 M sodium phosphate, pH 7.4.

MAGNETIC SEPARATION

An affinity magnetic bead is a particle with a magnetic core surrounded by a surface that can be modified by the addition of a ligand. These particles can be made from metal shot (such as BB shot) or plastic beads, although the most popular commercially available particles are based on agarose or polystyrene. The most common use of affinity magnetic beads has been in the immunoaffinity magnetic separation of cells (covered in detail in Chapter 8). Recently, however, the use of magnetic beads for the isolation of macromolecules has gained considerable favor, particularly in the area of molecular biology. In magnetic separation, the affinity ligand is bound to the surface of a magnetic particle (usually a bead with a magnetic core). The affinity magnetic particle is then mixed with a sample containing the analyte of interest. After allowing suitable time for the ligand-analyte reaction to occur, the magnetic beads with their bound analytes are removed from the solution using a magnetic field. Several companies make simple magnetic separators, the most common of which is the apparatus sold by Dynal for use with a variety of different tubes ranging from glass test tubes to microcentrifuge or Beckman tubes (Figure 6).

General Techniques

The popularity of magnetic separation has grown tremendously with the increased commercial availability of agarose-coated magnetic beads. They are available from

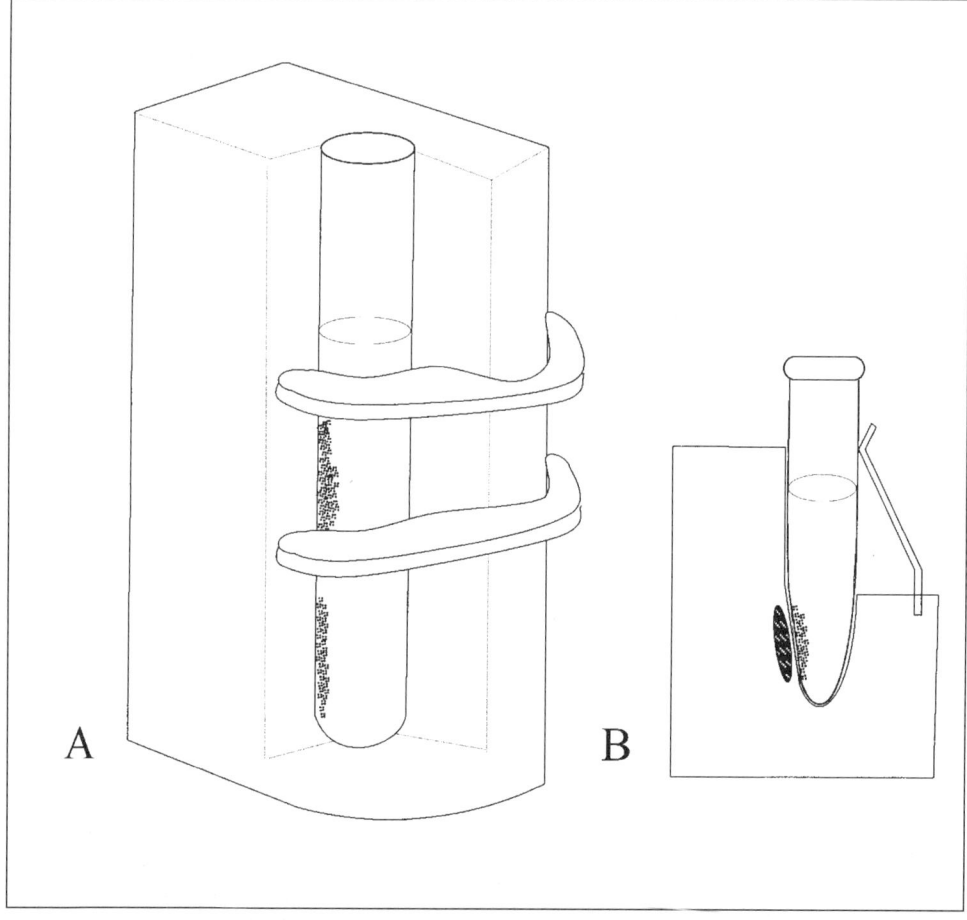

Figure 6. Two types of magnetic particle concentrators available from Dynal. (A) Dynal MPC-1 for tubes from 5 to 50 mL. (B) MPC-E-1 for microcentrifuge tubes. Similar commercial devices are available from other suppliers.

numerous vendors, including Dynal, Chemicon, Roche, and Seradyn (see Appendix III). Many of the commercially available beads have already had their surfaces activated by epoxide, with bound streptavidin, or by other modifications that make them ready for direct immobilization of biological or chemical ligands. Furthermore, several vendors provide magnetic beads that already contain bound ligands and are ready to use in immunomagnetic separation. For example, some offer magnetic beads with antibodies directed against specific cell types of biomedical importance (see Chapter 8). Dynal and other companies offer magnetic beads pre-coated with secondary antibodies directed against primary antibodies from animals like rabbits and mice (see Table 1 in Chapter 8). To use these magnetic beads for magnetic separation, primary antibodies are utilized that are produced in the appropriate species against the target molecule of interest. These beads can be used in either direct or indirect techniques for batch separations. The target can then be removed from the solution by removing the magnetic bead–secondary antibody primary antibody–target complex.

Magnetic beads can be fixed with essentially any ligand and used for the rapid separation of target molecules. When monoclonal antibodies are immobilized on their surface, magnetic beads can be selectively targeted toward highly specific analytes. Using such immunological magnetic separation techniques has proven to be a very rapid technique for the recovery of analytes from solution. Quitadamo and Schelling (78) compared immunologic magnetic separation to conventional immunoaffinity purification of murine IgM. Immunoaffinity supports were prepared with rat anti-mouse IgM monoclonal antibodies for both methods. Under the conditions studied, magnetic bead purification of IgM from crude ascites was more effective than that by the traditional immunoaffinity chromatography. Another example of magnetic bead technology for the immunological extraction of biomolecules is the removal of tumor necrosis factor (TNF) from crevicular fluid of patients, using magnetic beads coated with anti-TNF antibodies (80).

Specialized Techniques

Magnetic separation is not limited to the immobilization of ligands onto the surface of latex particles. Other examples of magnetic particles include bacteria with magnetic properties (64) and ultrafine magnetite particles (46). Ultrafine magnetite latex particles, composed of poly-(styrene/N-isopropylacrylamide/methacrylic acid), undergo temperature-sensitive flocculation. Using such beads on which BSA was immobilized by the carbodiimide method, Kondo et al. (46) isolated antibodies directed against BSA from immune serum. After the dispersed particles bound the antibodies in the serum, the temperature was lowered, which resulted in thermally induced flocculation of the magnetite particles. The flocculated particles were then quickly removed from the serum by applying a magnetic field.

There has been a recent advance that uses uniformly orange-to-red fluorescent polystyrene beads (Luminex) to mimic many of the principles of immunological magnetic bead separation. Beads coated with antibodies directed against targets of interest can be mixed with heterogeneous mixtures of molecules, and the bead-antibody-target complex removed by the FlowMetrix™ (Luminex) modification to a Fluorescence Activated Cell Sorter (68) rather than by magnetic force. The main disadvantage of this approach is the expensive instrumentation required for its application. However, this method has two remarkable advantages over immunological magnetic separations. First, different antibodies can be added to different colored batches of beads and mixed within the same heterogeneous sample. Using this approach, Oliver and colleagues (68) were able to differentially separate macrophage granulocyte colony-stimulating factor, TNF, and interleukins 2 and 4 from a single mixture of cytokines. The second advantage is that by using a series of second reporter antibodies tagged with a green fluorescent moiety, the amount of bound target molecule can be simultaneously quantified as the FlowMetrix-modified FACS separates the beads.

Molecular Biological Applications

While not generally what might be considered a batch technique, magnetic particles have made rapid advances possible in molecular biology studies. One area in which these advances have been made is in the rapid isolation of hexa-histidine–tagged recombinant proteins using agarose magnetic beads coated with nickel-loaded ligands for the hexa-histidine moieties. One such ligand is nitrilotriacetic acid, which is available from Qiagen in their QIA*express*® System. After the recombinant proteins bind the specific nickel ligands, they are isolated by the application of a magnetic field. To aid in the micro-extraction, Qiagen also offers a magnetic separation device for use with 96-well microplates. This 96-well magnetic plate has 24 magnetic "fingers," each of which is used to extract magnetic beads from 4 wells of a 96-well plate. After the magnetic beads are recovered, the recombinant protein can be safely recovered by the addition of a nickel chelator.

Another molecular biological use of magnetic beads is in the isolation and manipulation of DNA and RNA. Deggerdal and Larsen (16) used magnetic beads for the rapid, small-scale isolation of polymerase chain reaction (PCR)-ready DNA from blood, bone marrow, and cultured cells. They isolated DNA from all three sources within 10 min using the Dynabeads® DNA DIRECT kit (Dynal). Another study compared the capture of human cytomegalovirus (CMV) DNA by DNA capture probes covalently bound to the surface of magnetic beads with probes bound to the surface of microplates (94). Both surfaces were aminated to allow the capture of the DNA probes. The magnetic beads proved to provide faster hybridization than the microplates.

Isolation of DNA and subsequent PCR amplification of the genetic material has grown increasingly popular as a means for detection of the presence of pathogenic organisms. The use of magnetic bead technology to concentrate trace amounts of such DNA within a larger sample volume has been growing in popularity (69). It provides considerable savings in time and increased specificity compared to traditional DNA isolations. Niederhauser and colleagues (66) used magnetic beads coated with a specific oligonucleotide sequence directed against a gene found in *Listeria monocytogenes* to examine food and clinical samples for DNA evidence of the presence of this organism. The DNA, isolated by binding to these magnetic beads, was then amplified by a semi-nested PCR procedure. This procedure was able to detect as little as 1 to 10 colony-forming units (cfu) in 50 mL of buffer or in 10 g of food suspended in 40 mL of buffer.

The isolation of PCR-ready DNA is not, however, limited to pathogenic organisms. A recent study examined the usefulness of magnetic bead isolation for PCR-ready DNA and has been tested for a wide variety of organisms, including bacteria, fungi, algae, plants, and vertebrates (including fish and mammals) (81). While the procedure necessary to lyse the various cell types varied somewhat, the magnetic bead technology was shown to be universally applicable for the rapid (within 30 min) isolation of PCR-ready DNA. In fact, magnetic bead technology has been shown to be useful in the detection of circulating prostate cancer cells through the use of reverse transcription-PCR (58). In vitro selection of DNA ligands to 4-chloroaniline (4-CA), 2,4,6-trichloroaniline (TCA), and pentachlorophenol (PCP) was performed by a novel method utilizing magnetic beads that include a linker arm for immobilization. The beads with surface-captured template DNA could be directly added to PCR mixtures. In addition, a simplified PCR scheme requiring only one type of primer and a rapid fluorescence microscopic method for assessing nucleic acid binding after each round of selection was demonstrated (10).

A number of other ligands have been applied to the isolation of genetic materials. Metal affinity interactions have also been described as a tool for isolating molecular biological reagents. Large chromatin fragments released by treating isolated nuclei with restriction endonucleases can be fractionated on mercury-activated agarose magnetic beads. This separation is achieved by interaction of the mercury with accessible histone H3 thiols (present in nucleosomes of active genes but inaccessible in compactly beaded nucleosomes of inactive genes). This technique has successfully been applied to the separation of chromatin restriction fragments of c-*myc* and the growth arrest gene *gas1* (14). Likewise, streptavidin-coated magnetic beads are not only useful in isolating DNA for diagnostic purposes; it has also found use in aiding in the Human Genome Project. Liang (54) described an application of this technique that allowed for the isolation of extended large fragments (>10–20 kb) of cDNA. For these studies, "long" PCR occurred with Alu primers to generate a biotinylated template for cDNA hybridization, and the product was recovered using streptavidin-coated magnetic beads. Dynal produces an excellent monograph on magnetic bead technology and its application to molecular biology (20).

PROTOCOLS

Production of Magnetic Beads from BB Shot

1. Place 500 commercially available BBs into a small low energy ("jewelry-type") sonicator and cover with methylene chloride.
2. Sonicate at full strength for 15 min.
3. Remove the BBs with a magnet and air-dry on a sheet of filter paper.
4. Dissolve 5 g of polycarbonate resin in 50 mL of methylene chloride and add the BBs.
5. Ensure that adequate coating of the BBs with the plastic takes place by vigorously swirling the BBs in the plastic solution.
6. Incubate for 10 min at room temperature, then drain the fluid from the BBs.
7. Spread the BBs on aluminium foil (shiny side uppermost) to air-dry at room temperature.
8. Dissolve 50 µg of antigen or antibody in 50 mL of 0.01 M sodium phosphate, pH 7.2.
9. Add the plastic-coated BBs to the solution and incubate for 30 min at room temperature.
10. Decant the fluid, then wash the BBs five times in 0.1% (wt/vol) BSA in 0.01 M sodium phosphate, pH 7.2.

The coated beads can be stored in 0.1% BSA/0.01 M sodium phosphate, pH 7.2, at 4°C.

Preparation of Magnetic Agarose Beads

1. Dissolve 4 g of agarose in 100 mL of double-distilled H_2O by gently boiling in a round-bottom flask.
2. Prepare a solvent solution of 445 mL of toluene/15 mL CCl_4.
3. Add 0.45 g of sorbitan sesquioleate to the solvent solution and preheat to 50°C.
4. Add the solvent solution to the agarose and stir on an overhead stirrer set at 1700 rpm.
5. After 1 min, cool the solution by placing the flask in cold H_2O.
6. After 5 min, place the beads in a sintered glass filter.
7. Wash the beads with 500 mL of ether, then remove the beads and resuspend in 1000 mL of distilled H_2O.
8. Allow the remaining ether to form a layer, then remove the layer and wash the beads extensively in distilled H_2O.

Coating Commercial Magnetic Beads

1. Place 10 mL of a uniform bead suspension into a 15-mL capped tube.
2. Place a magnet on the side of the tube to retain the beads and decant the fluid.
3. Remove the magnet and resuspend the beads in 10 mL of 7:3 (vol/vol) water:acetone.
4. Mix well and repeat step 2.
5. Remove the magnet and resuspend in 10 mL of 5:5 (vol/vol) water:acetone.
6. Mix well and repeat step 2.
7. Remove the magnet and resuspend in 10 mL of 2:8 (vol/vol) water:acetone.
8. Mix well and repeat step 2.
9. Remove the magnet and resuspend in 10 mL of 100% acetone.
10. Incubate for 15 min at room temperature, then repeat step 2.
11. Add 600 µL of pyridine and 600 mg of p-toluene sulfonyl chloride to 10 mL of acetone, and resuspend the beads in this mixture.
12. Place on an overhead mixer and incubate overnight at room temperature.
13. Using the magnet as in step 2, wash the beads three times in acetone.
14. Hydrate by successively resuspending the beads in 2:8; 5:5, and 8:2 (vol/vol) water:acetone.
15. Resuspend the beads in 10 mL of distilled H_2O.
16. Using the magnet as in step 2, decant the H_2O and resuspend the beads in 10 mL of 1 mM HCl.
17. Dissolve 200 µg of ligand protein in 1 mL of 0.2 M borate, pH 9.5.
18. Resuspend the ligand solution in a 1-mL suspension of activated magnetic beads.
19. Place on an overhead mixer and incubate for 24 h at 4°C.

20. Using the magnet, wash the beads three times in 10 mL of 0.01 M sodium phosphate, pH 7.2.
21. Using the magnet as in step 2, wash the beads in 1 M ethanolamine, pH 9.0, to block nonreacted groups on the bead surface.
22. Wash the beads twice in 0.01 M sodium phosphate, pH 7.2, containing 0.1% (wt/vol) BSA.
23. Store at 4°C in the washing solution.

Recovery of Specific Antigens from Biological Fluids Using Antibody-Coated Magnetic Beads

1. Prepare the coated beads as described in the protocol for the preparation of magnetic agarose beads.
2. Immediately before adding the coated beads to the sample, use a magnet to retain the beads while removing the supernatant.
3. Add 100 µL of the sample to the bead pellet.
4. Remove the magnet and resuspend the beads in the sample.
5. Mix at room temperature for 30–60 min.
6. Replace the magnet to the side of the tube and wait for all the beads to migrate to the magnet.
7. Pour off the fluid, and add 100 µL of 0.01 M sodium phosphate, pH 7.2.
8. Remove the magnet and resuspend the beads.
9. Wash the beads by repeating steps 6 and 7 three times.
10. Resuspend the final bead pellet in 100 µL of 100 mM glycine, pH 2.0.
11. Incubate at room temperature for 30 min.
12. Place the magnet of the side of the tube and collect the beads, then recover the fluid containing the isolated analyte.

Isolation of mRNA from Cell Cultures Using (dT)$_{25}$-Coated Magnetic Beads

Throughout this procedure, ensure that all tubes, reaction vessels, and reagents are RNAse-free.

1. Pellet the cells by centrifugation at 1000× g for 10 min.
2. Wash the cell pellet five times in 200 µL of 0.01 M PBS, pH 7.4.
3. Resuspend the final cell pellet in 1 mL of lysis buffer (10 mM Tris-HCl, pH 7.4, 0.14 M NaCl, and 5 mM KCl, to which 1% (vol/vol) NP-40 is added).

   Note: A viscous extraction fluid should occur and indicates full cell lysis and release of DNA as well as mRNA.

4. Reduce the viscosity by a short sonication in a low-energy sonicator at one-half power.
5. Pre-condition the magnetic beads by suspending a washed aliquot of beads in 200 µL of coupling buffer (20 mM Tris-HCl, pH 7.4, 1 M LiCl, 2 mM EDTA).
6. Immediately before adding the beads to the cell lysate, use a magnet to retain the beads while removing the supernatant.

165

Table 2. A Selection of Magnetic Particles Used in Molecular Biological Techniques

Company	Reagent Name	Use
BioSeparations	GenoMag	DNA separation
	RiboMag	Total RNA separation
	Oligo(dT) microspheres	mRNA separation
	PCR cleanup	Removal of unwanted reagents following PCR amplification
	STREP microspheres	Streptavidin-coated particles suitable for attachment of biotinylated probes
CPG, Inc.	MPG® mRNA Purification Kit	mRNA separation
	MPG DNA TempPrep™ Kit	Preparation of both immobilized and free single-stranded DNA templates
Dynal	DNA DIRECT Kit	DNA separation

7. Add the cell lysate to the beads.

8. Remove the magnet and resuspend the beads.

9. Mix at room temperature for 5 min.

10. Replace the magnet to the side of the vial and wait for all the beads to migrate to the magnet, then pour off the fluid.

11. Remove the magnet and resuspend the beads in 200 µL of 10 mM Tris-HCl, pH 8.0, 0.15 M LiCl, 1 mM EDTA.

12. Wash the beads by repeating steps 10 and 11 three times.

13. Resuspend the final bead pellet in 20 µL of 2 mM EDTA, pH 8.0.

14. Incubate at 65°C for 2 min.

15. Place the magnet on the side of the tube to collect the beads, then recover the fluid containing the mRNA.

Several companies offer magnetic particles suitable for the isolation of both DNA and RNA, as well as products for the cleanup of PCRs prior to further analysis. Table 2 lists a selection of these reagents, most of which are sold as kits.

REFERENCES

1. Agadjanyan, M.G., M.A. Chattergoon, I. Petrushina, M. Bennett, J. Kim, K.E. Ugen, T. Kieber-Emmons and D.B. Weiner. 1998. Monoclonal antibodies define a cellular antigen involved in HTLV-I infection. Hybridoma 17:9-19.

2. Andrews, B.A., S. Nielsen and J.A. Asenjo. 1996. Partitioning and purification of monoclonal antibodies in aqueous two-phase systems. Bioseparation 6:303-313.

3. Arica, M.Y., H.N. Testereci and A. Denizli. 1998. Dye-ligand and metal chelate poly(2-hydroxyethyl-methacrylate) membranes for affinity separation of proteins. J. Chromatogr. A 799:83-91.

4. Ayes, M., H.B. Marshall, J. Ishikawa and R. Takahashi. 1991. Rapid preparation of lysate for immuno-precipitation using a low protein-binding micro-filter. BioTechniques 10:578-580.

5. **Babashak, J.V. and T.M. Phillips.** 1988. Use of avidin-coated glass beads as a support for high-performance immunoaffinity chromatography. J. Chromatogr. *444*:21-28.

6. **Bertocci, B., G. Garotta, M. Da Prada, H.W. Lahm, G. Zurcher, G. Virgallita and V. Miggiano.** 1991. Immunoaffinity purification and partial amino acid sequence analysis of catechol-O-methyltransferase from pig liver. Biochim. Biophys. Acta *1080*:103-109.

7. **Born, T.L., E. Thomassen, T.A. Bird and J.E. Sims.** 1998. Cloning of a novel receptor subunit, AcPL, required for interleukin-18 signaling. J. Biol. Chem. *273*:29445-29450.

8. **Borvak, J., J. Richardson, C. Medesan, F. Antohe, C. Radu, M. Simionescu, V. Ghetie and E.S. Ward.** 1998. Functional expression of the MHC class I-related receptor, FcRn, in endothelial cells of mice. Int. Immunol. *10*:1289-1298.

9. **Botros, H.G., G. Birkenmeier, A. Otto, G. Kopperschläger and M.A. Vijayalakshmi.** 1991. Immobilized metal ion affinity partitioning of cells in aqueous two-phase systems: erythrocytes as a model. Biochim. Biophys. Acta *1074*:69-73.

10. **Bruno, J.G.** 1997. In vitro selection of DNA to chloroaromatics using magnetic microbead-based affinity separation and fluorescence detection. Biochem. Biophys. Res. Commun. *234*:117-120.

11. **Bueno, S.M., K. Haupt and M.A. Vijayalakshmi.** 1995. Separation of immunoglobulin G from human serum by pseudobioaffinity chromatography using immobilized L-histidine in hollow fibre membranes. J. Chromatogr. B Biomed. Appl. *667*:57-67.

12. **Cella, N., B. Groner and N.E. Hynes.** 1998. Characterization of Stat5a and Stat5b homodimers and heterodimers and their association with the glucocortiocoid receptor in mammary cells. Mol. Cell. Biol. *18*:1783-1792.

13. **Champluvier, B. and M.R. Kula.** 1992. Sequential membrane-based purification of proteins, applying the concept of multidimensional liquid chromatography (MDLC). Bioseparation *2*:342-351.

14. **Chen-Cleland, T.A., L.C. Boffa, F.M. Carpaneto, M.R. Mariani, F. Valentin, E. Mendez and V.G. Allfrey.** 1993. Recovery of transcriptionally active chromatin restriction fragments by binding to organomercurial-agarose magnetic beads. A rapid and sensitive method for monitoring changes in higher order chromatin structure during gene activation and repression. J. Biol. Chem. *268*:23409-23416.

15. **Dahl Andersen, M. and B.L. Moller.** 1998. Double Triton X-114 phase partitioning for the purification of plant cytochromes P450 and removal of green pigments. Protein Expr. Purif. *13*:366-372.

16. **Deggerdal, A. and F. Larsen.** 1997. Rapid isolation of PCR-ready DNA from blood, bone marrow and cultured cells, based on paramagnetic beads. BioTechniques *22*:554-557.

17. **Dombrowski, T.R., G.S. Wilson and E.M. Thurman.** 1998. Investigation of anion-exchange and immunoaffinity particle-loaded membranes for the isolation of charged organic analytes from water. Anal. Chem. *70*:1969-1978.

18. **Doolittle, M.H. and O. Ben-Zeev.** 1999. Immunodetection of lipoprotein lipase: antibody production, immunoprecipitation, and western blotting techniques. Methods Mol. Biol. *109*:215-237.

19. **Dunah, A.W., J. Luo, Y.H. Wang, R.P. Yasuda and B.B. Wolfe.** 1998. Subunit composition of N-methyl-D-aspartate receptors in the central nervous system that contain the NR2D subunit. Mol. Pharmacol. *53*:429-437.

20. **Dynal Technical Manual.** 1995. Biomagnetic Techniques in Molecular Biology. Oslo.

21. **Ekblad, L., J. Kernbichler and B. Jergil.** 1998. Aqueous two-phase affinity partitioning of biotinylated liposomes using neutral avidin as affinity ligand. J. Chromatogr. A *815*:189-195.

22. **Elling, L. and M.R. Kula.** 1991. Immunoaffinity partitioning: synthesis and use of polyethylene glycol-oxirane for coupling to bovine serum albumin and monoclonal antibodies. Biotechnol. Appl. Biochem. *13*:354-362.

23. **El-Samalouti, V.T., J. Schletter, H. Brade, L. Brade, S. Kusumoto, E.T. Rietschel, H.D. Flad and A.J. Ulmer.** 1997. Detection of lipopolysaccharide (LPS)-binding membrane proteins by immuno-coprecipitation with LPS and anti-LPS antibodies. Eur. J. Biochem. *250*:418-424.

24. **Finger, U.B., J. Thommes, D. Kinzelt and M.R. Kula.** 1995. Application of thiophilic membranes for the purification of monoclonal antibodies from cell culture media. J. Chromatogr. B Biomed. Appl. *664*:69-78.

25. **Firestone, G.L. and S.D. Winguth.** 1990. Immunoprecipitation of proteins. Methods Enzymol. *182*:688-700.

26. **Flanagan, S.D. and S.H. Barondes.** 1975. Affinity partitioning. A method for purification of proteins using specific polymer-ligands in aqueous polymer two-phase systems. J. Biol. Chem. *250*:1484-1489.

27. **Gayraud, B., B. Hopfner, A. Jassim, M. Aumailley and L. Bruckner-Tuderman.** 1997. Characterization of a 50-kDa component of epithelial basement membranes using GDA-J/F3 monoclonal antibody. J. Biol. Chem. *272*:9531-9538.

28. **Giuliano, K.A.** 1991. Aqueous two-phase protein partitioning using textile dyes as affinity ligands. Anal. Biochem. *197*:333-339.

29. **Gomez-Escobar, N., C.F. Chou, W.W. Lin, S.L. Hsieh and R.D. Campbell.** 1998. The G11 gene located in the major histocompatibility complex encodes a novel nuclear serine/threonine protein kinase. J. Biol. Chem. *273*:30954-30960.

30. **Goretzki, L. and B.M. Mueller.** 1998. Low-density-lipoprotein-receptor-related protein (LRP) interacts

with a GTP-binding protein. Biochem. J. *336*:381-386.

31.**Gouras, G.K., H. Xu, J.N. Jovanovic, J.D. Buxbaum, R. Wang, P. Greengard, N.R. Relkin and S. Gandy.** 1998. Generation and regulation of beta-amyloid peptide variants by neurons. J. Neurochem. *71*:1920-1925.

32.**Grimonprez, B. and G. Johansson.** 1996. Liquid-liquid partitioning of some enzymes, especially phosphofructokinase, from *Saccharomyces cerevisiae* at sub-zero temperature. J. Chromatogr. B Biomed. Appl. *680*:55-63.

33.**Guoqiang, D., R. Kaul and B. Mattiasson.** 1998. Integration of aqueous two-phase extraction and affinity precipitation for the purification of lactate dehydrogenase. J. Chromatogr. A *668*:145-152.

34.**Hua, C.T., J.R. Gamble, M.A. Vadas and D.E. Jackson.** 1998. Recruitment and activation of SHP-1 protein-tyrosine phosphatase by human platelet endothelial cell adhesion molecule-1 (PECAM-1). Identification of immunoreceptor tyrosine-based inhibitory motif-like binding motifs and substrates. J. Biol. Chem. *273*:28332-28340.

35.**Irwin, J.A. and K.F. Tipton.** 1995. Affinity precipitation: a novel approach to protein purification. Essays Biochem. *29*:137-156.

36.**Iwata, H., K. Saito, S. Furusaki, T. Sugo and J. Okamoto.** 1991. Adsorption characteristics of an immobilized metal affinity membrane. Biotechnol. Prog. *7*:412-418.

37.**Jackson, D.E., K.R. Kupcho and P.J. Newman.** 1997. Characterization of phosphotyrosine binding motifs in the cytoplasmic domain of platelet/endothelial cell adhesion molecule-1 (PECAM-1) that are required for the cellular association and activation of the protein-tyrosine phosphatase, SHP-2. J. Biol. Chem. *272*:24868-24875.

38.**Jethmalani, S.M. and K.J. Henle.** 1998. Interaction of heat stress glycoprotein GP50 with classical heat-shock proteins. Exp. Cell. Res. *239*:23-30.

39.**Johansson, G.** 1994. Partitioning procedures and techniques: small molecules and macromolecules. Methods Enzymol. *228*:28-42.

40.**Johnstone, A.P. and R. Thorpe.** 1996. Affinity chromatography and immunoprecipitation, p. 268-276. Immunochemistry in Practice. Blackwell Science, Oxford.

41.**Jones, C., A. Patel, S. Griffin, J. Martin, P. Young, K. O'Donnell, C. Silverman, T. Porter and I. Chaiken.** 1995. Current trends in molecular recognition and bioseparation. J. Chromatogr. A *707*:3-22.

42.**Josic, D., J. Reusch, K. Loster, O. Baum and W. Reutter.** 1992. High-performance membrane chromatography of serum and plasma membrane proteins. J. Chromatogr. *590*:59-76.

43.**Kasper, C., L. Meringova, R. Freitag and T. Tennikova.** 1998. Fast isolation of protein receptors from streptococci G by means of macroporous affinity discs. J. Chromatogr. *798*:65-72.

44.**Kasper, C., O.W. Reif and R. Freitag.** 1998. Evaluation of affinity filters for protein isolation. Bioseparation *6*:373-382.

45.**Kohler, K., A. Veide and S.O. Enfors.** 1991. Partitioning of beta-galactosidase fusion proteins in PEG/potassium phosphate aqueous two-phase systems. Enzyme Microb. Technol. *13*:204-209.

46.**Kondo, A., H. Kamura and K. Higashitani.** 1994. Development and application of thermo-sensitive magnetic immunomicrospheres for antibody purification. Appl. Microbiol. Biotechnol. *41*:99-105.

47.**Kopperschläger, G.** 1994. Affinity extraction with dye ligands. Methods Enzymol. *228*:121-136.

48.**Kopperschläger, G. and G. Birkenmeier.** 1990. Affinity partitioning and extraction of proteins. Bioseparation *1*:235-254.

49.**Kuo, T.H., H.R. Kim, L. Zhu, Y. Yu, H.M. Lin and W. Tsang.** 1998. Modulation of endoplasmic reticulum calcium pump by Bcl-2. Oncogene *17*:1903-1910.

50.**Laboureau, E., J.C. Capiod, C. Dessaint, L. Prin and M.A. Vijayalakshmi.** 1996. Study of human cord blood lymphocytes by immobilized metal ion affinity partitioning. J. Chromatogr. B Biomed. Appl. *680*:189-195.

51.**Laboureau, E. and M.A. Vijayalakshmi.** 1997. Concerning the separation of mammalian cells in immobilized metal ion affinity partitioning systems: a matter of selectivity. J. Mol. Recognit. *10*:262-268.

52.**Langlotz, P. and K.H. Kroner.** 1998. Surface-modified membranes as a matrix for protein purification. J. Chromatogr. *591*:107-113.

53.**Li, C., E. Ioffe, N. Fidahusein, E. Connolly and J.M. Friedman.** 1998. Absence of soluble leptin receptor in plasma from dbPas/dbPas and other db/db mice. J. Biol. Chem. *273*:10078-10082.

54.**Liang, B.C.** 1995. Use of 'long' polymerase chain reaction and magnetic beads for extraction of chromosome-specific cDNAs. Genet. Anal. *12*:105-108.

55.**Lilius, G., M. Persson, L. Bulow and K. Mosbach.** 1991. Metal affinity precipitation of proteins carrying genetically attached polyhistidine affinity tails. Eur. J. Biochem. *198*:499-504.

56.**Lutkemeyer, D., M. Bretschneider, H. Buntemeyer and J. Lehmann.** 1993. Membrane chromatography for rapid purification of recombinant antithrombin III and monoclonal antibodies from cell culture supernatant. J. Chromatogr. *639*:57-66.

57.**MacMillan-Crow, L.A. and J.A. Thompson.** 1999. Immunoprecipitation of nitrotyrosine-containing proteins. Methods Enzymol. *301*:135-145.

58.**Makarovskiy, A.N., W. Ackerley, L. Wojcik, G.K. Halpert, B.S. Stein, M.P. Carreiro and D.C. Hixson.**

1997. Application of immunomagnetic beads in combination with RT-PCR for the detection of circulating prostate cancer cells. J. Clin. Lab. Anal. *11*:346-350.

59. **Mandal, M., X.R. Chen, M.L. Alegre, N.M. Chiu, Y.H. Chen, A.R. Castano and C.R. Wang.** 1998. Tissue distribution, regulation and intracellular localization of murine CD1 molecules. Mol. Immunol. *35*:525-536.

60. **Millesime, L., J. Dulieu and B. Chaufer.** 1996. Fractionation of proteins with modified membranes. Bioseparation *6*:135-145.

61. **Mohan, S.B. and A. Lyddiatt.** 1997. Recent developments in affinity separation technologies, p. 1-38. *In* P Matejtschuk (Ed.), Affinity Separations. A Practical Approach. IRL Press, Oxford.

62. **Mohr, P. and K. Pommerening.** 1985. Affinity partition, p. 253-257. *In* Affinity Chromatography. Practical and Theoretical Aspects. Marcel Dekker, New York.

63. **Najarian, S. and B.J. Bellhouse.** 1997. Effect of oscillatory flow on the performance of a novel cross-flow affinity membrane device. Biotechnol. Prog. *13*:113-116.

64. **Nakamura, N., J.G. Burgess, K. Yagiuda, S. Kudo, T. Sakaguchi and T. Matsunaga.** 1993. Detection and removal of Escherichia coli using fluorescein isothiocyanate conjugated monoclonal antibody immobilized on bacterial magnetic particles. Anal. Chem. *65*:2036-2039.

65. **Nanak, E., M.A. Vijayalakshmi and K.C. Chadha.** 1995. Segregation of normal and pathological human red blood cells, lymphocytes and fibroblasts by immobilized metal-ion affinity partitioning. J. Mol. Recognit. *8*:77-84.

66. **Niederhauser, C., J. Luthy and U. Candrian.** 1994. Direct detection of Listeria monocytogenes using paramagnetic bead DNA extraction and enzymatic DNA amplification. Mol. Cell. Probes *8*:223-228.

67. **Nilsson, S., C. Lager, T. Laurell and S. Birnbaum.** 1995. Thin-layer immunoaffinity chromatography with bar code quantitation of C-reactive protein. Anal. Chem. *67*:3051-3056.

68. **Oliver, K.G., J.R. Kettman and R.J. Fulton.** 1998. Multiplexed analysis of human cytokines by use of the FlowMetrix system. Clin. Chem. *44*:2057-2060.

69. **Olsvik, O., T. Popovic, E. Skjerve, K.S. Cudjoe, E. Hornes, J. Ugelstad and M. Uhlen.** 1994. Magnetic separation techniques in diagnostic microbiology. Clin. Microbiol. Rev. *7*:43-54.

70. **Otto, A. and G. Birkenmeier.** 1993. Recognition and separation of isoenzymes by metal chelates. Immobilized metal ion affinity partitioning of lactate dehydrogenase isoenzymes. J. Chromatogr. *644*:25-33.

71. **Paul, L.C., J. Muralidharan, S.A. Muzaffar, E.H. Manting, J.F. Valentin, E. de Heer and M. Kashgarian.** 1998. Antibodies against mesangial cells and their secretory products in chronic renal allograft rejection in the rat. Am. J. Pathol. *152*:1209-1223.

72. **Paradkar, V.M. and J.S. Dordick.** 1991. Purification of glycoproteins by selective transport using concanavalin-mediated reverse micellar extraction. Biotechnol. Prog. *7*:330-334.

73. **Paradkar, V.M. and J.S. Dordick.** 1993. Affinity-based reverse micellar extraction and separation (ARMES): a facile technique for the purification of peroxidase from soybean hulls. Biotechnol. Prog. *9*:199-203.

74. **Pesliakas, H., V. Zutautas and B. Baskeviclute.** 1994. Immobilized metal-ion affinity partitioning of NAD(+)-dependent dehydrogenases in poly(ethylene glycol)-dextran two-phase systems. J. Chromatogr. *678*:25-34.

75. **Petropoulos, J.H., A.I. Liapis, N.P. Kolliopoulos, J.K. Petrou and N.K. Kanellopoulos.** 1990. Restricted diffusion of molecules in porous affinity chromatography adsorbents. Bioseparation *1*:69-88.

76. **Phillips, T.M., N.S. More, W.D. Queen, T.V. Holohan, N.C. Kramer and A.M. Thompson.** 1984. High-performance affinity chromatography: a rapid technique for the isolation and quantitation of IgG from cerebral spinal fluid. J. Chromatogr. *317*:173-179.

77. **Porath, J. and B. Olin.** 1983. Immobilized metal affinity adsorption and immobilized metal ion affinity chromatography of biomaterials. Serum protein affinities for gel-immobilized iron and nickel ions. Biochemistry *22*:1621-1630.

78. **Quitadamo, I.J. and M.E. Schelling.** 1998. Efficient purification of mouse anti-FGF receptor IgM monoclonal antibody by magnetic beads. Hybridoma *17*:207.

79. **Riese, U., D. Lutkemeyer, R. Heidemann, H. Buntemeyer and J. Lehmann.** 1994. Re-use of spent cell culture medium in pilot scale and rapid preparative purification with membrane chromatography. J. Biotechnol. *34*:247-257.

80. **Rossomando, E.F. and L. White.** 1998. A novel method for the detection of TNF-alpha in gingival crevicular fluid. J. Periodontol. *64*:445-449.

81. **Rudi, K., M. Kroken, O.J. Dahlberg, A. Deggerdal, K.S. Jakobsen and F. Larsen.** 1997. Rapid, universal method to isolate PCR ready DNA using magnetic beads. BioTechniques *22*:506-511.

82. **Sabri, A., G. Govindarajan, T.M. Griffin, K.L. Byron, A.M. Samarel and P.A. Lucchesi.** 1998. Calcium- and protein kinase C-dependent activation of the tyrosine kinase PYK2 by angiotensin II in vascular smooth muscle. Circ. Res. *83*:841-851.

83. **Salih, B.A. and R.F. Rosenbusch.** 1998. Identification and localization of a 94 kDa membrane protein found in Mycoplasma bovoculi strains. Comp. Immunol. Microbiol. Infect. Dis. *21*:281-290.

84. **Sii, D. and A. Sadana.** 1991. Bioseparation using affinity techniques. J. Biotechnol. *19*:83-98.

85. **Soltys, P.J. and M. Eufemio.** 1998. In vitro characterization of a membrane-based low-density lipoprotein affinity adsorption device. Blood Purif. *16*:123-134.

86. **Tao, Y., H.M. Skrenta and K.Y. Chen.** 1995. Deoxyhypusine synthase assay based on the use of polyhistidine-tagged substrate and metal chelate-affinity chromatography. Anal. Biochem. *221*:103-108.

87. **Umeno, D., M. Kawasaki and M. Maeda.** 1998. Water-soluble conjugate of double-stranded DNA and poly(*N*-isopropylacrylamide) for one-pot affinity precipitation separation of DNA-binding proteins. Bioconjug. Chem. *9*:719-724.

88. **Vogetseder, W., J. Feng and M.P. Dierich.** 1995. Reactivity of monoclonal antibodies established against a recombinant human endogenous retrovirus-K (HERV-K) envelope protein. Immunol. Lett. *46*:129-134.

89. **Walter, H. and G. Johansson (Eds).** 1994. Aqueous two-phase systems. Methods Enzymol. Volume 228.

90. **Walter, H. and C. Larsson.** 1994. Partitioning procedures and techniques: cells, organelles, and membranes. Methods Enzymol. *228*:42-63.

91. **Weiss-Messer, E., R. Ber, T. Amit and R.J. Barkey.** 1998. Characterization and regulation of prolactin receptors in MA-10 Leydig cells. Mol. Cell. Endocrinol. *143*:53-64.

92. **Winzor, D.L.** 1991. Evaluation of antigen-antibody affinity constants by partition equilibrium studies with a two-phase aqueous polymer system: a more rigorous analysis. Biochim. Biophys. Acta *1115*:141-144.

93. **Wu, C.W., R. Lovrien and D. Matulis.** 1998. Lectin coprecipitative isolation from crudes by Little Rock Orange ligand. Anal. Biochem. *257*:33-39.

94. **Zammatteo, N., I. Alexandre, I. Ernest, I. Le, F. Brancart and J. Remacle.** 1997. Comparison between microwell and bead supports for the detection of human cytomegalovirus amplicons by sandwich hybridization. Anal. Biochem. *253*:180-190.

7 Chromatographic Techniques

olumn chromatographic techniques are perhaps the most popular approach for affinity and immunoaffinity separations (28,30,45,50,62,68,70,80,93,94,99, 115,117,121,125). This popularity is due to the relative ease with which such separations can be performed. The separation is performed under flow conditions, thereby reducing the incidence of nonspecific binding, with the immobilized ligand isolating and retaining the analyte of interest while nonreactive materials are washed through the column. Recovery of the bound materials is achieved by the introduction of an elution agent, which breaks the analyte-ligand interaction, thus freeing the bound analyte to be washed through the column and recovered. Figure 1 illustrates a typical affinity chromatogram. Further factors adding to the popularity of column chromatographic techniques are the controllability of the separation process and the availability of a variety of commercially available column packing materials, apparatus, and chromatographic systems. Column chromatographic separations can be performed in a variety of ways, ranging from simple low-pressure (i.e., gravimetric) separations (12,58) to highly controlled high-performance procedures with a high degree of sophistication employed in the detection systems (35,57,75,101,126). However, the advantages of this technique can be offset by a number of factors capable of impeding the success of the separation process. These factors are discussed in Chapter 3, but perhaps the most serious of these problems as related to column chromatography is ligand leakage (46,113). This is usually caused by breakdown of the packing matrix, inadequate coupling chemistry, or shearing forces involved during the separation process. Today, this problem has been diminished by the development of new packing matrices and better immobilization chemistries. Despite these improvements, ligand leakage can still pose a serious problem when increased flow rate and pressure are employed, such as conditions used in HPLC (28,43). In such cases, attention must be paid to the selection of suitable chromatographic equipment to overcome or reduce the development of excessive shear forces during the separation process. An overview of the equipment required to perform column chromatography will be discussed in the following sections.

CHROMATOGRAPHIC EQUIPMENT

Column chromatographic separations require a minimum of equipment: a column, a column reservoir, and a rack of collection tubes. If more sophisticated separations

Affinity and Immunoaffinity Purification Techniques
Terry M. Phillips and Benjamin F. Dickens
© 2000 Eaton Publishing, Natick, MA

are required, then additional equipment such as a pump, gradient controller, detection system, and a more sophisticated collection device must be added. A simple guide to the different chromatographic systems available appears below, along with information on their assembly and use.

Columns

The column is the most important component of any system and there are many different commercial sources offering a variety of different columns (Table 1). A column is a tube equipped with a retaining device at one or both ends for retaining the packing matrix. The upper opening of the column is usually called the inlet and the lower opening (which is usually fitted with some type of valve or restriction) is called the outlet. Simple laboratory columns suitable for micro-affinity separations can be constructed from glass Pasteur pipets (90), empty syringes, or open glass tubes (82). In such columns, the lower end is equipped with a sieve capable of retaining the packing material while allowing passage of the running buffer and separated analytes. Although these columns can be simply made, commercial versions are available and many are pre-packed with affinity matrices. An advance over these simple columns is the basic column offered commercially. This apparatus has a frit already attached to the outlet of the column and can be further modified by the addition of a flow adaptor to the inlet or top of the column. The flow adaptor allows the column to be run in an ascending fashion, which makes for more efficient separation (82). Many companies also offer columns with flow adaptors fitted to both the inlet and outlet.

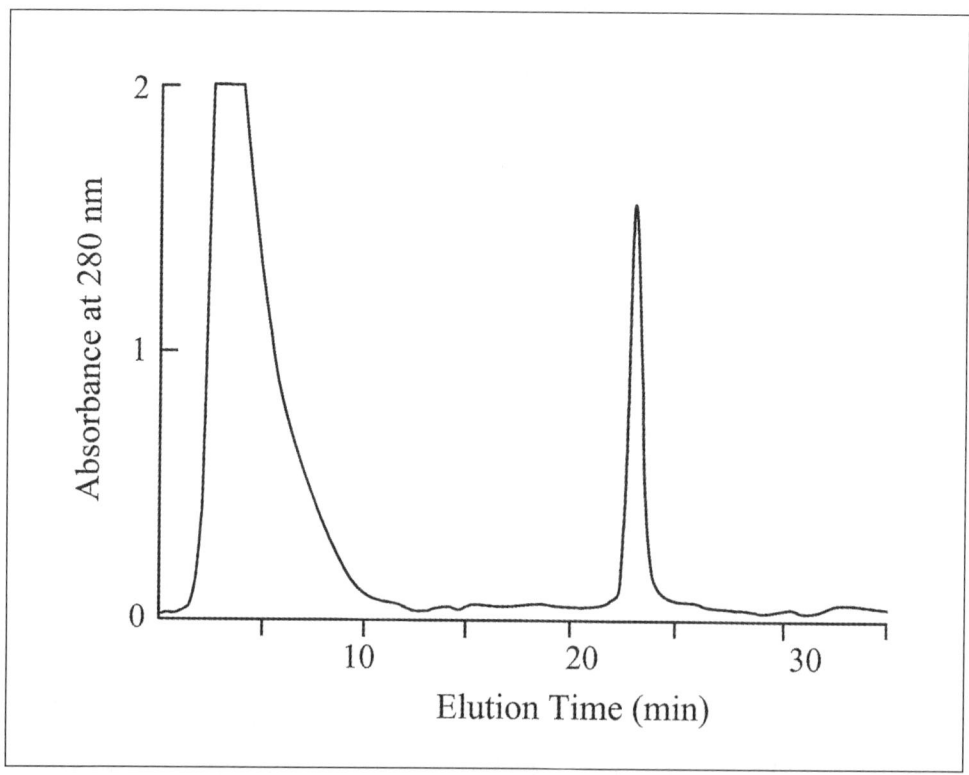

Figure 1. Typical chromatograph obtained using affinity or immunoaffinity chromatography.

Table 1. Selected Commercial Sources of Chromatography Columns Suitable for Packing Affinity and Immunoaffinity Matrices

Alltech Associates	Low-, medium-, and high-pressure columns
Bio-Rad Laboratories	Low- and medium-pressure columns
Kimble/Kontes	Low- and medium-pressure columns
PE Biosystems	Medium- and high-pressure columns
Pharmacia	Low- and medium-pressure columns
Pierce Chemical	Low- and medium-pressure columns
Supelco	Low-, medium-, and high-pressure columns
Upchurch Scientific	Medium- and high-pressure columns

This arrangement greatly improves the versatility of the column, allowing the investigator to adjust the height of the column bed and to run the column in both ascending and descending modes. Another advantage of this type of column is that it is available in a variety of diameters and lengths, thus allowing scale-up of both affinity and immunoaffinity separations. Many of these columns are also equipped with an external water jacket, which is useful when controlled temperature (i.e., 4°C) is required (temperature control is essential for most immunoaffinity separations).

Several companies also offer specialized columns designed for use with chromatography systems available from the same company. Such columns are available from Pharmacia, Bio-Rad, and PE-Biosystems (formerly PerSeptive Biosystems), for use in the Fast Protein Liquid Chromatography (FPLC), ÄKTA™, and BioCAD® chromatographic systems, respectively. Many of these columns are constructed of stainless steel, high-strength plastic, or other polymers and alloys. Such materials capable of withstanding high-pressure are used exclusively for high-performance separations. Both immunoaffinity and affinity separations have been performed in either standard or short HPLC columns, i.e., stainless steel tubes equipped with screw-on end-fittings (83,123). However, these columns are not truly biocompatible and can cause denaturation of sensitive biological analytes. To overcome the problem, the introduction of high-strength plastic polymers such as polyether-polyether-ketone (PEEK) (79) columns has greatly enhanced the field of high-performance affinity and immunoaffinity separations. PEEK columns are available in different lengths and diameters, all suitable for affinity separation of biological materials. To withstand the pressures generated by HPLC systems, the PEEK columns are usually encased in an alloy outer casing. This casing serves as an attachment point for the end-fittings and placement of suitable retaining frits. When the stainless steel tubing of the chromatography system is replaced with PEEK tubing, the system becomes truly biocompatible. In fact, the whole HPLC system can now be constructed of biocompatible tubing, valves, and columns. This is not only a great improvement in biological separation science, but essential for most affinity and all immunoaffinity separations. There has also been a trend towards designing capillary columns for affinity or immunoaffinity extractions. These columns range from 100–250 μm internal diameter (i.d.) silica capillaries capable of handling 30–50-μL samples (29,116). Another adaptation has been described by Zhou et al. (133), who built a short column (4-mm i.d. × 40-mm long) packed with stacks of either protein A- or antibody-coated membranes. This column was used to isolate IgG or measure antibody reactivity in serum.

Although not in common usage in affinity or immunoaffinity procedures, there is a new trend towards the design of chromatography systems "on a chip." He and colleagues (39) have described systems for designing small columns in situ. They have shown that stationary phases can be bound to $5 \times 5 \times 10$-μm collocated monolith structures separated by 1.5×10-μm deep rectangular channels. This design can produce columns of 150 μm × 4.5 cm long with an internal volume of 18 μL. Although designed for capillary electrochromatography, such a system would hold great promise for other separation technology such as affinity or immunoaffinity procedures. Figure 2 illustrates the basic designs of laboratory-built, simple, and complex columns. The reader is also directed to Table 1, which lists selected commercial sources of suitable columns and fittings.

Pumps

Allowing the sample to percolate through the column packing can easily perform low-pressure separations. This is sufficient for small laboratory-packed columns and simple experiments, although gravity does not allow the investigator any control of the separation process. To interject some control over the speed and, to a certain extent, efficiency of the separation, it is necessary to introduce a pump into the system. Peristaltic pumps are usually the choice for low-pressure systems. These instruments must have a variable speed control and a positive pumping action capable of handling back pressures up to 18–20 psi. Peristaltic pumps operate on a series of rollers that exert a squeezing force on a soft plastic tube, thus providing a peristaltic or positive pumping action on the tube with no contact between the liquid being pumped and the moving parts of the pump. This action forces the liquid through the tube and into the column. When coupled with columns equipped with flow adaptors, liquid can be forced upward through the column. This ascending chromatography creates a more consistent front to the separation and more efficient results.

When using medium- or high-performance column separations, a more sophisticated pumping system is required to maintain an accurately calibrated constant flow rate or pressure. Additionally, the instrument components should be compatible with the solvent systems used and the flow rate must be adjustable over a relatively wide range (91). These pumps can range from dual-syringe pumps to piston or diaphragm reciprocating pumps (91). The former are usually the heart of medium-performance systems such as the FPLC and ÄKTA systems, while the latter represents the most common pumps used in high-performance systems. Syringe-based pumps can usually operate at pressures up to 600 psi and deliver high flow rates, but these pumps require a pulse dampener to allow interactions with high-sensitivity detectors. The high-performance pumps use quartz or other crystalline pistons to produce an almost pulse-less flow. These pumps can develop working pressures of 100–1000 psi with corresponding high back pressures. However, the true value of the HPLC pump lies in its versatility and control (91). Although the syringe pumps can also be computer-controlled, the HPLC pumps have a wider operational field. Table 2 lists commercial sources for these different types of chromatography pumps.

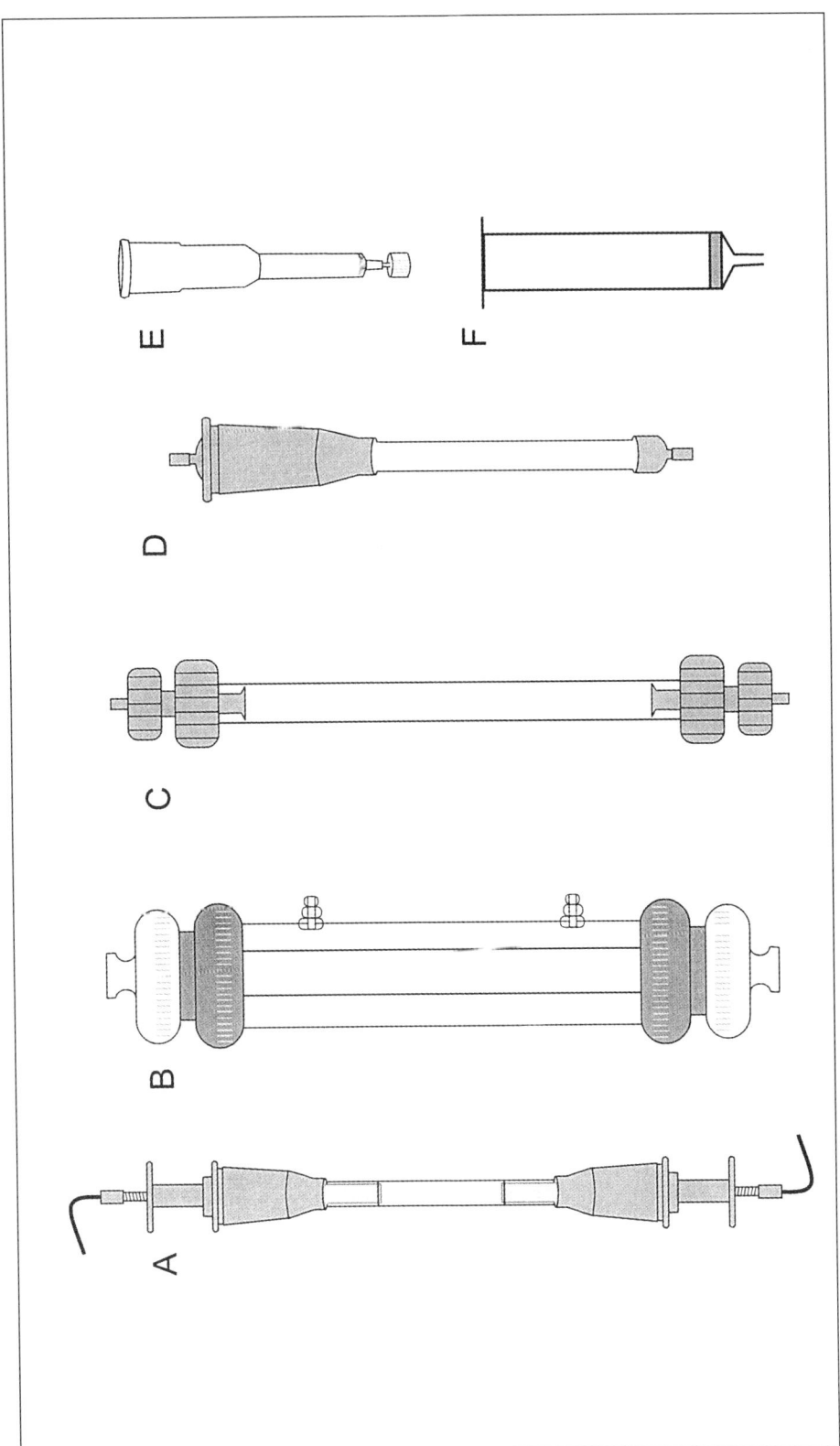

Figure 2. Commercially available chromatography columns. (A) Bio-Rad Econo-Column® with flow adaptors. (B) Kontes CHROMAFLEX column. (C) Amersham Pharmacia Biotech. (D) Basic Bio-Rad Econo-Column without flow adaptors. (E) Disposable Bio-Rad Poly-Prep® column. (E) Basic disposable syringe column.

Gradient Makers and Controllers

To recover bound analytes from affinity and immunoaffinity columns, it is necessary to add a gradient-making device to the chromatography system. Changing the starting buffer to the final (elution) buffer over a set volume or time creates a gradient. Although the simple stepwise addition of increasing-strength elution buffer can produce an effective recovery (see *Recovery of Isolated Analytes*, in this chapter), this procedure is harsh and often causes loss of biological activity. Additionally, this approach produces a chromatogram (Figure 3) in which the recovered peak is often broad, resulting in recovery of a diluted analyte that often requires further preparation (i.e., concentration) before it can be subsequently used. Sharper peak resolution (Figure 4) requires a smooth transition in the addition of the elution agent. This form of elution is called linear gradient elution. Linear gradients can be made with a variety of different devices ranging from a simple laboratory-built apparatus to complicated dual pump systems equipped with mixing chambers and computer control mechanisms. The simplest gradient maker is laboratory-built and can be constructed from two identical beakers (Figure 5) (82). For the investigator who does not wish to construct such a device, there are commercially available devices based on the same principle. These devices incorporate the two chambers plus a regulating valve placed between the chambers (Figure 6). Gradients can also be created using a single pump and a specialized gradient controller (86). This instrument is comprised of a micro-

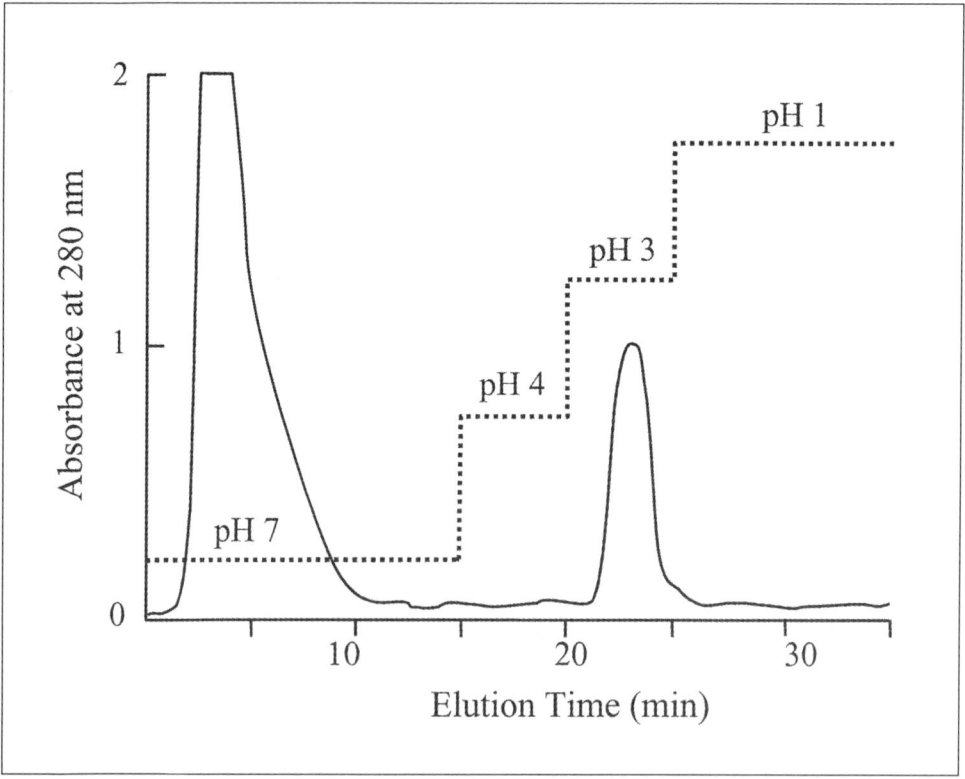

Figure 3. Affinity chromatography showing the use of a multiple step elution gradient to elute bound analyte.
Compare the broad analyte peak (one on right) with the one in Figure 4.

Table 2. Commercially Available Chromatography Pumps

Alltech Associates—Peristaltic and HPLC pumps

Bio-Rad Laboratories—Peristaltic, diaphragm, and HPLC pumps

Pharmacia—Peristaltic and diaphragm pumps

Rainin—Peristaltic and HPLC pumps

Supelco—Peristaltic and HPLC pumps

processor-controlled valve and dual input lines (Figure 7). Line A is placed in the starting buffer while line B is placed in the final buffer. The valve regulates the inflow of the two buffers, thus changing the composition of the column running buffer from A to B. The advantage of these simple gradient controllers is that only a single chromatography pump is required, thus significantly lowering the cost of the system. True linear gradients are produced by a dedicated system employing two pumps and a mixing chamber (Figure 8) (11). This is the traditional gradient maker and there are many versions commercially available. In fact, most chromatography system manufacturers can supply a gradient form of any chromatography system. Both affinity and immunoaffinity chromatography require the use of some form of gradient elution

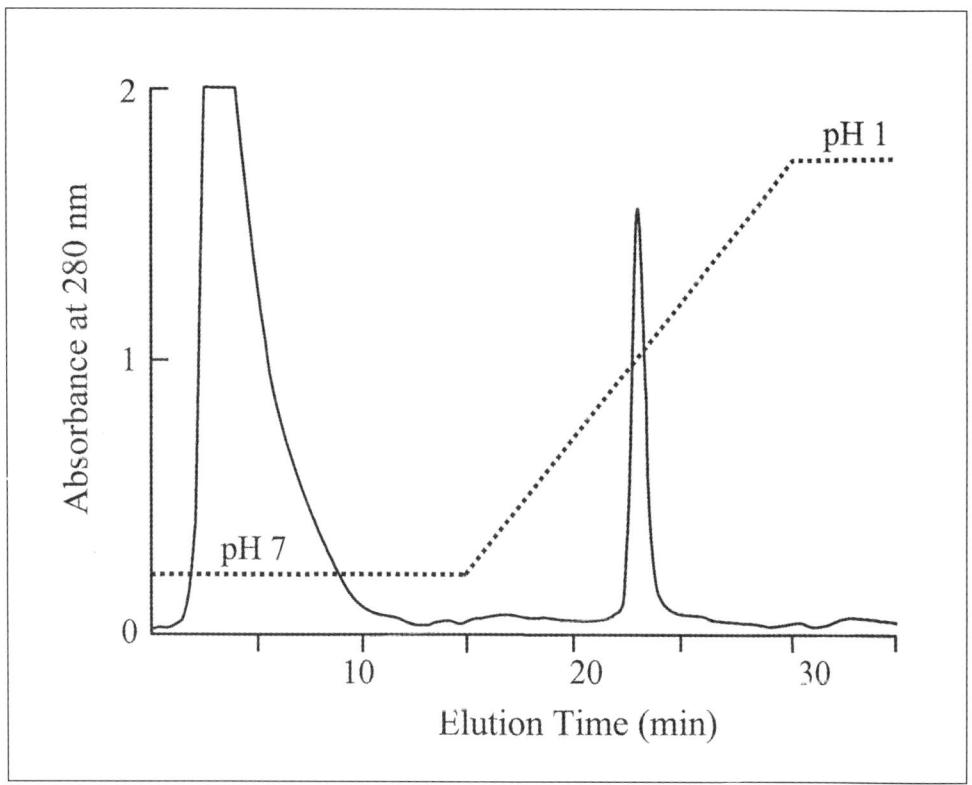

Figure 4. Affinity chromatography showing the use of a linear elution gradient to elute bound analyte. Compare the narrow analyte peak (one on right) with the one in Figure 3.

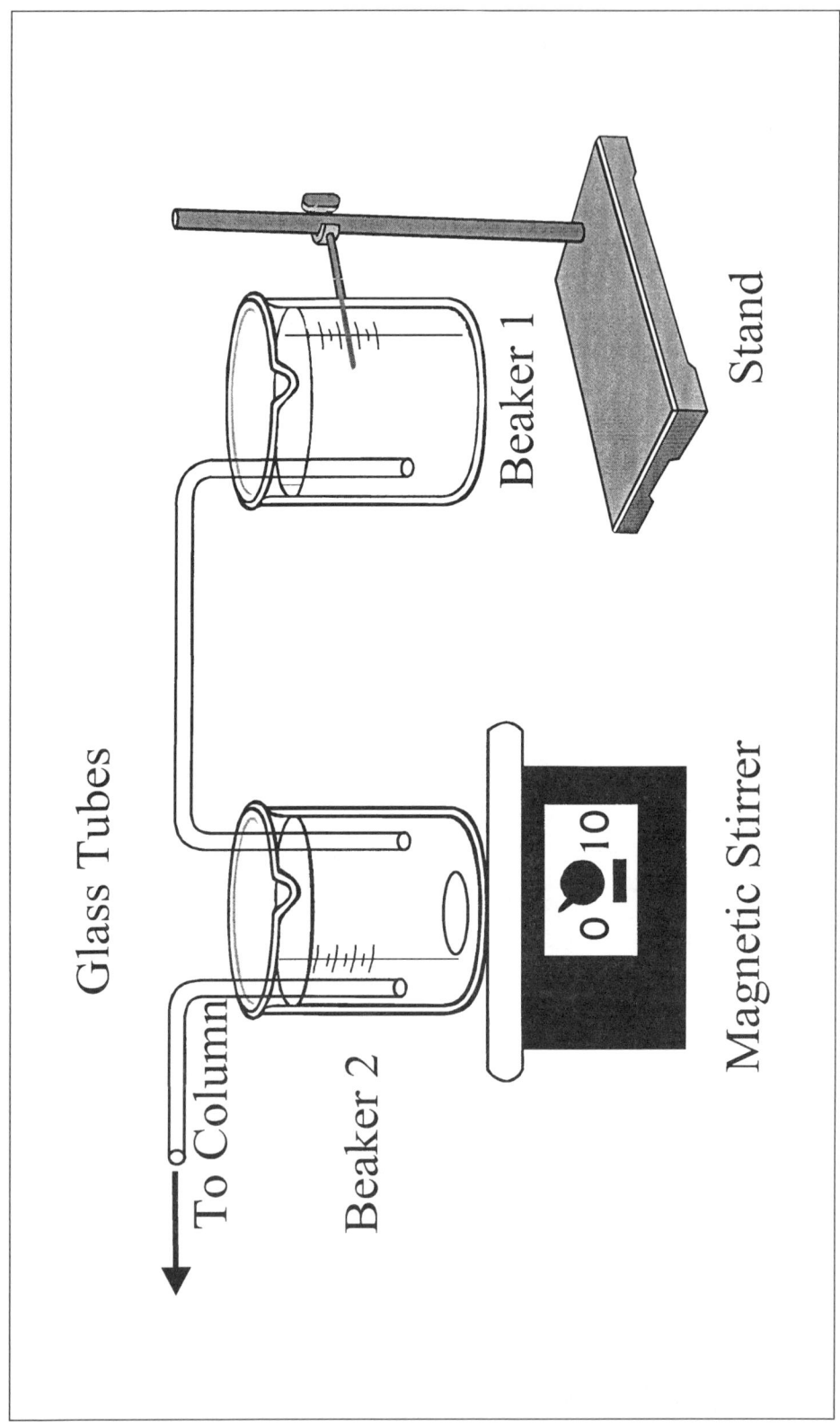

Figure 5. Simple laboratory-constructed linear gradient maker.

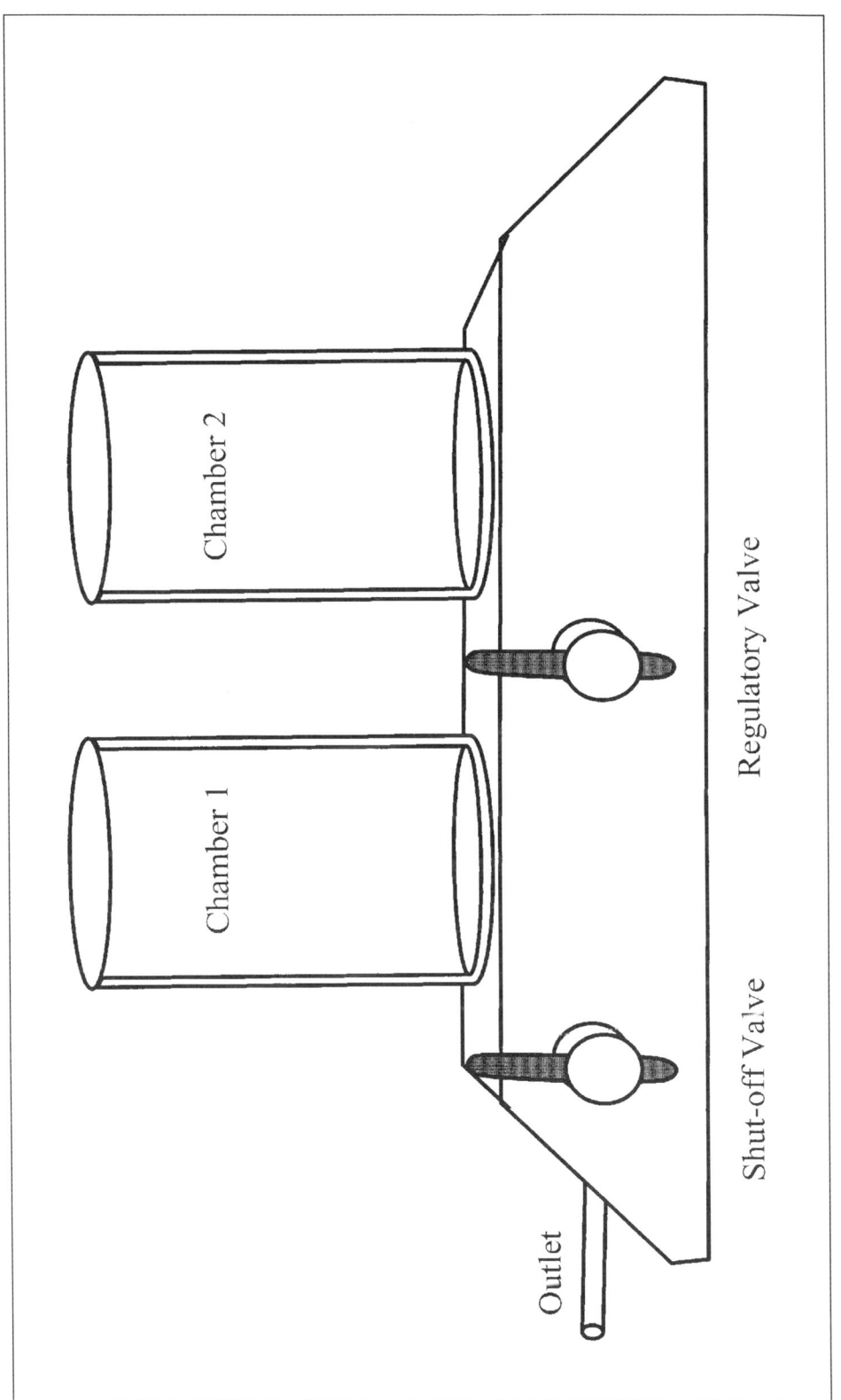

Figure 6. A typical commercial linear gradient maker.

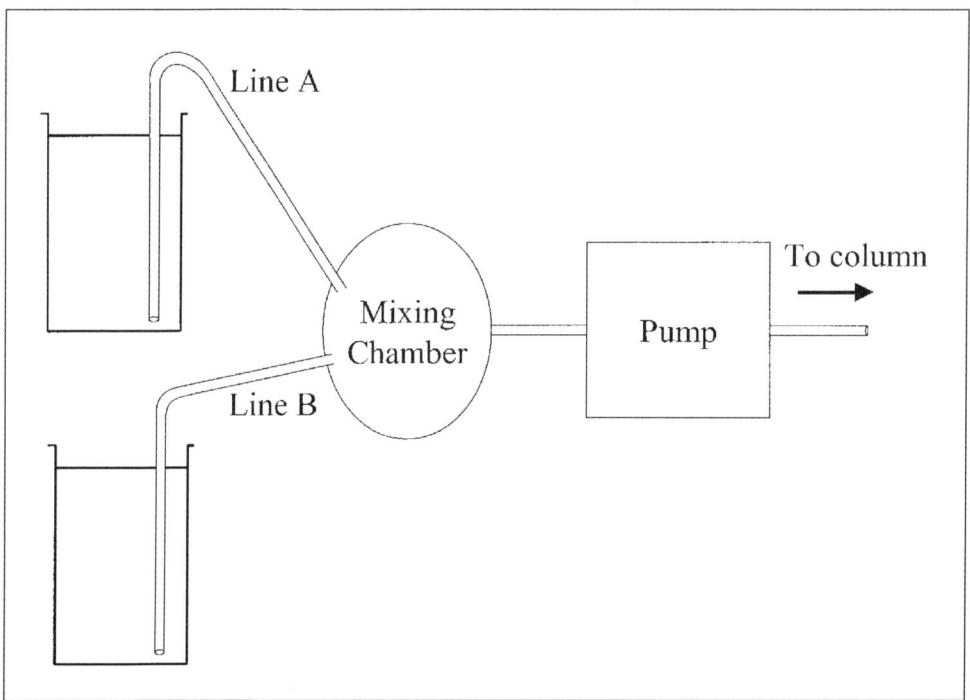

Figure 7. A single-pump gradient system.

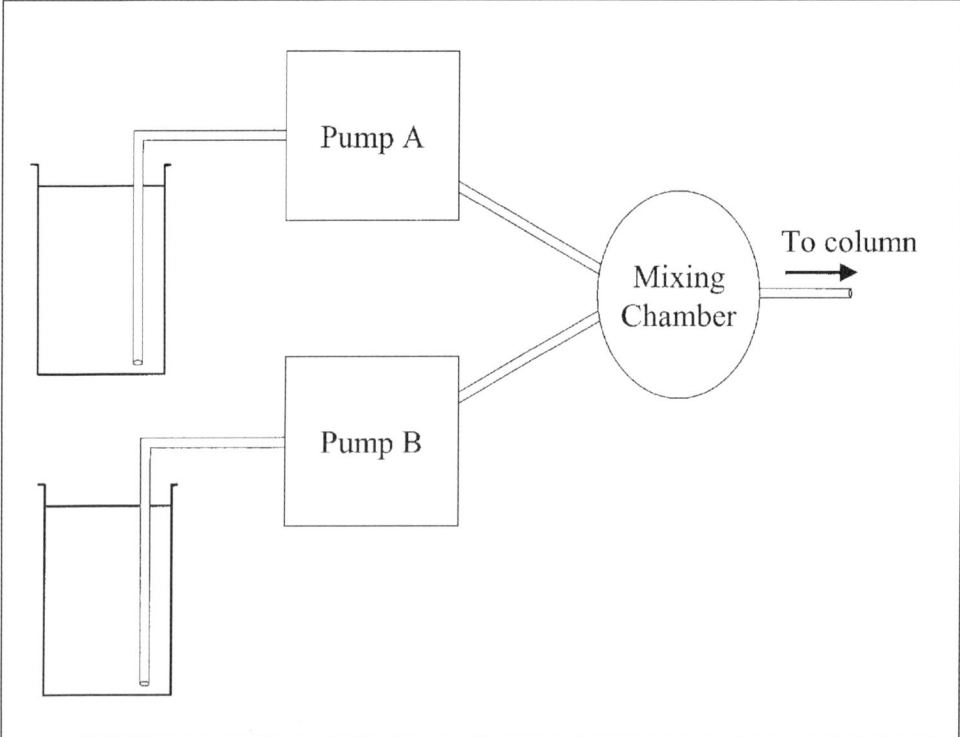

Figure 8. A dual-pump gradient system.

to recover the bound analyte, and investigators acquiring new systems are well-advised to consider the purchase of a dedicated gradient chromatography system, as many systems are difficult to upgrade.

Detectors

Although one can simply collect fractions of the column effluent and determine the concentration of analyte present in each tube by any of the techniques described in Appendix I, direct on-line analyte measurement is preferable. To achieve this, a detector must be placed in-line, immediately following the column outlet. The advantage of adding this instrument is twofold; it enables direct measurement of the eluting analyte on-line and also provides a useful monitor of the entire chromatogram. To monitor the column effluent, the flow must pass through a specialized reservoir called a flow cell in which detection takes place. It is important to match the flow cell volume with the size of the chromatography column: a 10–20-μL flow cell will not register adequate readings for a 2-mm × 2-cm immunoaffinity column. Usually the smaller the flow cell volume, the better the peak resolution. The outlet of the column is connected directly to the flow cell, from which the buffer can flow to either a collection vessel or to waste.

There are a number of different detection instruments available, each using a different physical property of the eluted analyte to detect its presence. The most common is an ultraviolet detector that detects the presence of molecules by their ability to absorb light at a defined wavelength (9,10,91). For protein analysis, wavelengths of 200, 219, 254, and 280 nm are commonly used. A refractive index monitor can be used to detect carbohydrate analytes. This instrument measures the refraction of light in the liquid present in the reference cells and compares it to the liquid flowing through the analytical flow cell (71). Unfortunately, both ultraviolet and refractive index detectors are unable to detect extremely low concentrations of analytes, which can prove to be a serious problem when small volumes of biological samples are being analyzed or in situations where the analyte is present in extremely low concentrations.

In such cases, more sophisticated detectors are required such as those using fluorescence, infrared detection, electrochemical detection, and chemiluminescence as their means of detection. All of these detectors require some form of analyte modification either pre- or post-separation. This modification can entail labeling with light-emitting molecules such as fluorescent probes, or post-columns reactions such as oxidative or reductive reactions (53,95). The use of post-column reactions has also led to the development of procedures that incorporate immunoassays for the detection and measurement of analytes post-separation (60,108). The advantage of all these detectors is that they can achieve a high degree of sensitivity, making them capable of detecting extremely low concentrations of analytes.

When choosing a suitable detector, one has to determine the concentration of the analyte to be measured and the complexity of the matrix from which it has to be extracted. No matter what material is being separated, the major deciding factor will be the limit of detection (or sensitivity) of the instrument. This problem has been studied by several investigators, who have compared the limits of detection demonstrated by different detection systems commonly applied to chromatographic techniques. Murphy et al. (72) compared the efficiency of ultraviolet, light-scattering, and fluorescence detection for the detection of neutral lipids. They found that all of the detectors produced satisfactory standard curves with linearity over a wide concentration range. However, Singh et al. (104) found that a diode-array detector provided simultaneous screening with a limit of detection at 100–500 ng/mL for the detection of several acidic drugs in horse plasma and urine. The limit of detection was slightly improved when a fixed-wavelength ultraviolet detector was used (50–150 ng/mL) but that fluorescence detection was the most sensitive (limit of detection ca. 10 ng/mL for naproxen). Unfortunately, in this study, naproxen was the only drug detected at that level of sensitivity.

Recently, Kehr (51) compared fluorescence and electrochemical detection for glutamate and aspartate in terms of reproducibility, sensitivity, interference, maintenance, and troubleshooting. He found that the correlation coefficients for electrochemical vs. fluorescence detectors were 0.918 for aspartate and 0.988 for glutamate for over 60 samples, and that the coefficients of variation between calibrations were below 3% for both detectors. The limits of detection for both amino acids were about 0.4 picomolar for electrochemical detection and 50 femtomolar for fluorescence detection. However, using microbore columns (such as those often applied to affinity separations), Kehr demonstrated that the limit of detection for fluorescence detection was approximately 20–30 fM.

Davies and Hounsell (21) examined instruments suitable for the detection of carbohydrate analytes. They found that the nonchromogenic nature of most carbohydrates greatly decreases the sensitivity of ultraviolet detection and that refractive index is, in most cases, inadequate for sensitive analysis. Electrochemical and fluorescence (following suitable labeling) detection was found to exhibit a limit of detection in the range of 200–300 fM for oligosaccharides. Finally, these authors suggest that the use of miniaturized separation techniques (such as CE or capillary chromatography) coupled to a laser-induced fluorescence detection unit could vastly improve sensitivity.

Chemiluminescence detection shows promise, with the suggestion that this form of detector may be able to achieve greater levels of sensitivity, although some reports have indicated that this may not be the case. Fluorescence and chemiluminescence detection were compared for HPLC analysis of a fluorescamine derivative of histamine (122). This study concluded that fluorescence detection was superior to chemiluminescence. The investigators report that fluorescence detection exhibited a linear range of 166–1666 pg with a limit of detection of 13 pg, while the chemiluminescence response was linear over a more restricted range of 1.66–16.6 ng with a limit of detection of 1.0 ng. Other investigators have reported more promising results with chemiluminescence detection. Limits of detection of 2–50 fM have been reported for the detection of nitropolycyclic aromatic hydrocarbons in benzene:ethanol extracts from airborne particulates (38) and 2 pM with recovery of 84% for tea catechin (74).

Advances in detector design continue to be driven by new analytical fields such as CE and capillary electrochromatography. These techniques have driven detector design towards miniaturized instruments for use with microcolumns and capillaries (65,107). The advent of micromachining or micro-engineering has revolutionized the design of detectors (110). This technology has advanced to produce microchip detectors capable of quantitatively measuring molecules labeled with luminescent or radioactive probes (25). Another advance in miniaturized detectors has come about because of the introduction of fiber optic-based spectrometers such as the Model S200 spectrometer available from Ocean Optics. This instrument uses a series of charge-coupled device (CCD) cameras to capture signals delivered by a fiber optic (88) (see the protocol for building a basic fluorescence fiber-optic immunosensor in

Chapter 10, or the protocol for building a basic fluorescence fiber-optic immunosensor in Chapter 10, or the protocol for building a fiber-optic fluorescence detector for increased sensitivity, in Appendix II). An advantage of this type of instrument is that it can easily be adapted for fluorescence or chemiluminescence detection. Table 3 lists commercial sources of some commonly used detectors.

Fraction Collectors

The effluent from all columns can be collected in even aliquots or fractions for either chemical analysis or further study. Although these fractions can be collected manually, this usually becomes a tedious task, often producing less than optimal results. Fraction collection can be simplified and automated by the addition of a fraction collector to the chromatography system. These instruments are designed to automatically collect a series of identical samples from the column effluent based on time or drop count. Timed sample collection is simply based on changing the tube at a preset time, such as every 5 min. The time is set by experimentation with column flow and kinetics. At each time point, a new tube is placed under the column outlet, thereby collecting a new fraction. Drop counting instruments employ a simple light detector to record the drops formed at the column outlet. These instruments are calibrated to count a given number of drops before activating the moving mechanism to place a new tube under the drop dispenser. In this way, the chromatography run is divided into a number of time points and compared with the detector analysis of the effluent. In cases where no detector is in-line, then each fraction is simply analyzed for the presence and concentration of the analyte. Fraction collectors are made in several different designs, ranging from circular collection racks with a drop-dispensing arm traveling in a concentric path to a square with the collection tubes passing in a serpentine manner under a fixed dispenser. In both cases, collection can usually be automated by computer control.

SAMPLE PREPARATION

In both affinity and immunoaffinity procedures, attention to sample preparation prior to injection into the analytical system is often required. For instance, it is important to adequately remove detergents such as SDS, deoxycholate, and Triton X-100, which will interfere with the binding of the analyte to the immobilized ligand. The presence of such agents will also reduce the efficiency of the column by unfolding complex protein and glycoprotein ligands such as antibodies. Simple dialysis can easily overcome the problem of removing these and other contaminating agents. Preparing small volumes of samples has become easier because of the introduction of new, easily used devices designed for this purpose. For instance, the sample may be passed through a small, commercially available desalting column and equilibrated to the running buffer used in the analytical system. Millipore now offers a new tool that could become extremely useful to investigators performing ultramicroanalyses. This tool is called the ZipTip^{TM}$_{C18}$, is designed around a micropipet tip, and can be used to desalt femtomolar quantities of peptides present in a just a few µL of sample (31).

RECOVERY OF ISOLATED ANALYTES

Recovery of the isolated analyte is one of the most important aspects of either affinity or immunoaffinity chromatography. Several approaches have been taken to achieve this end. These range from changes in the pH of the running buffer to the introduction of disruption agents (chaotropic ions) and competing ligands into the running buffer.

The simplest way to recover materials from affinity and immunoaffinity columns is to change the pH by introducing a gradient. Acid elution is the most common approach to this form of elution, although there is some debate on the actual degree of acidity required for a good recovery. Parekh et al. (80)

Table 3. Selected Commercial Sources of Chromatography Detectors

Alltech Associates	Ultraviolet
	Refractive Index
	Fluorescence
	Electrochemical
Bio-Rad	Ultraviolet
Pharmacia	Ultraviolet
	Refractive Index
Rainin	Ultraviolet
	Refractive Index
	Fluorescence
	Electrochemical
Supelco	Ultraviolet
	Refractive Index
	Fluorescence
	Electrochemical

used a pH of 3.0 to recover antibodies that were reactive against immobilized peptides derived from either lymphocytic choriomeningitis virus or the human acetylcholine receptor. Likewise, acid elution with mild biological acids such as citric acid has been used to recover a number of analytes from immunoaffinity columns (84,124). Although acid elution is highly effective, there are reports that alkaline elution can also be equally effective. Logeat et al. (63) used alkaline conditions to recover progesterone receptor from an immobilized monoclonal antibody column. They demonstrated that changing the pH of the running buffer to pH 10.5 resulted in a complete reversal of the binding between the receptor and the immobilized antibody. The choice of the form of elution was governed by both the binding characteristics of the immobilized antibody and the stability of the isolated receptor at this pH. Under such conditions, the authors demonstrated recovery of a biologically active receptor.

Although pH manipulation is highly successful, many investigators prefer the addition of chaotropic ions as the elution agent. These chemicals disturb the structural integrity of the ligand-analyte complex, thus ensuring a gentle and complete breakdown of the immobilized complex. Chaotropic elution results in a high degree of recovery of the bound analyte, coupled with minimal damage of the immobilized ligand. Although many chemicals exhibit chaotropic activity, sodium thiocyanate is the most popular. We have demonstrated that thiocyanate gradients are proficient in recovering biologically active receptors from lymphocytes (85). This type of elution can also be applied to most immunoaffinity techniques including recovery of recombinant hepatitis B surface antigen from yeast cultures (1), and recovery of specific antibodies from immobilized antigen columns (48). In addition to thiocyanate, it has been reported that 200 mM sodium chloride is also an effective affinity elution agent (123).

Elution can also be achieved by apply a gradient of denaturation agents such as 8

M urea (1), and 6 M guanidine hydrochloride (61). Kim et al. (52) report a unique system for recovering hepatitis B surface antigen from an immobilized antibody column. They used an anti-idiotypic antibody (one which competes with the antigen for binding to the immobilized antibody) as an effective elution agent. Another unique approach is described by Yu and colleagues (130) who used 100% methanol to release cyclopiazonic acid from an immobilized antibody column. Although effective, it should be noted that such elution conditions reduced the active life of the column to 10 cycles. Elution can also be achieved by employing elution systems containing the analyte of interest or a similar molecule. Katoh et al. (49) used the immunizing antigen to elute recombinant chimeric α-amylase from an immobilized antibody column. Table 4 summarizes the most common chemicals used as elution agents.

PROTOCOLS

Dialysis Cleanup of a Sample

1. Cut a length of dialysis tubing sufficient to hold the volume of the sample (ensure that adequate tubing is available for sealing both ends with either a clip or a knot).
2. Wet the cut tubing in distilled water (wear gloves; do not contaminate the tubing surface with proteins from your hands).
3. Clip or knot the lower end of the tubing and fill with distilled H_2O.
4. Carefully squeeze the tubing to ensure that there are no punctures.
5. Drain the water and insert the sample.
6. Seal the top end.
7. Place the sealed tubing in a container that holds at least 10–50 vol of buffer.
8. Leave for 6–12 h at 4°C with constant stirring.
9. Retrieve the sample by puncturing the tubing and squeezing the sample into a clean container.

Desalting a Sample Using a Commercially Available Column

1. Clamp the column in a vertical position and place a 100-mL beaker under the column.
2. Remove the top cap and the bottom plug of the column.
3. Fill the top of the column (above the fitted top frit) with distilled H_2O.
4. Pass 10 mL of 0.01 M sodium phosphate, pH 7.4, through the column.
5. Allow all the fluid to pass through the top frit and add the sample. Allow the sample to pass into the column (the top frit becomes barcly moist).
6. Fill the top portion of the column with 0.01 M sodium phosphate, pH 7.4, and run the column by gravity, keeping the top portion filled with buffer.
7. Collect 0.25-mL fractions of the column effluent and analyze the fractions at 280 nm in a spectrophotometer.

Removal of Detergent from Samples Using a Pierce Extracti-Gel® Column

1. Clamp the column in a vertical position and remove the top and bottom caps or plugs.
2. Regenerate the column bed by the following steps:
   a. Add 1 mL of distilled H_2O
   b. Add 1 mL of 100% ETOH
   c. Add 2 mL of 100% butanol
   d. Add 1 mL of 100% ETOH
   e. Add 4 mL of distilled H_2O.
   Run the column after each addition until the flow stops.
3. Add 2 mL of 0.01 M sodium phosphate, pH 7.4.
4. Add the sample and run into the column bed.
5. Add 1.25 mL of 0.01 M sodium phosphate, pH 7.4.
6. Collect the column effluent, which contains the detergent-free sample.

Construction of a Simple Gradient Maker

1. Place a 200-mL glass beaker on a magnetic stirrer (beaker 1).
2. Clamp another 200-mL glass beaker (beaker 2) to a laboratory stand (make sure that the two beakers are at the same height [Figure 5]).
3. Using a propane torch, heat and bend a piece of thin glass tubing to form a "U" (make one arm the length of the beaker side and the other 1/2 inch shorter).
4. Position the tube with the long arm reaching to the bottom of beaker 2 and the short arm in beaker 1.
5. Place the column inlet tube in beaker 1.

Producing a Linear Gradient with a Laboratory-Built Gradient Maker

1. Fill beaker 1 with 0.01 M sodium phosphate, pH 7.4.
2. Fill beaker 2 with 0.01 M sodium phosphate, pH 1.5.
3. Start the column and add the sample.
4. Connect the tube from beaker 1 to the column.
5. Run the column for 10–20 min using the buffer in beaker 1.
6. Set a magnetic stirrer under beaker 1 and set the stirrer at approximately ¼ maximum speed.
7. Connect beaker 1 to beaker 2 by the bridging tube.
8. Collect 0.25–0.5 mL fractions of the column effluent.
9. Continue running until all the fluid in beaker 2 has passed into beaker 1.
10. Stop the column and analyze the fractions in a spectrophotometer.

Table 4. Common Elution Agents for Affinity and Immunoaffinity Chromatography

Acid Buffers	
0.1 M Tris/HCl	pH 2.0–1.0
0.1 M Tris/glycine	pH 1.5–1.0
0.1 M Glycine	pH 1.5–1.0
0.33 M Citric acid	pH 2.0–1.5
20% Formic acid	pH 1.8–2.0
1 M Acetic acid	pH 2.5–1.5
Alkaline Buffers	
0.1 M Sodium hydroxide	pH 10.0–12.0
0.1 M Potassium hydroxide	pH 10.0–12.0
Chaotropic Ion Buffers	
Sodium thiocyanate	2.0–3.0 M
Sodium iodide	2.5–3.0 M
Sodium chloride	2.0–8.0 M
Denaturing Agents	
Urea	6.0–8.0 M
Guanidine hydrochloride	4.0–6.0 M
Polarity-Reducing Buffers	
Dioxane	5%–10%
Ethylene glycol	up to 50%
Competition	
Various monosaccharides	0.25–2.0 M
Peptide ligands	0.1–2.5 M

SIMPLE OR LOW-PRESSURE TECHNIQUES

Affinity and immunoaffinity separations can easily be performed in short columns, under gravitational flow. Such columns are commercially available from a number of commercial sources including Bio-Rad, Pierce Chemical, and Kimble/Kontes, and are easy to assemble and use. Additionally, simple columns can be constructed from glass tubes or syringe barrels. In these latter columns, the outlet has to be equipped with a frit or mesh designed to retain the packing material. Recovery of bound analytes from such columns is simply achieved by changing the pH by the stepwise addition of more acidic buffer.

Low-pressure systems have been used for several different types of affinity chromatography and are usually performed on agarose, dextran, or other soft gel beads (82). In fact, this form of affinity chromatography was popular until the medium- and high-pressure systems became available. Although performed less frequently these days, low-pressure systems can still effectively be used as cleanup or pre-analysis procedures. However, it must not be forgotten that, no matter what system is used, affinity and immunoaffinity separations can be used as single-step separation procedures, resulting in the recovery of reasonably pure materials. Dourges et al. (24) employed heparin-Sepharose beads to isolate heparin-binding growth factors from bovine brain homogenates. Glyoxalase I has been isolated and purified from human red blood cells using S-hexylglutathione as the affinity ligand (3). The investigators of this study used low-pressure chromatography with a concentration gradient of S-hexylglutathione as the elution agent. This procedure produced a modest but highly purified protein suitable for biochemical characterization.

A new apparatus has recently been introduced that could revive the interest in low-pressure column separations. Spin-columns are more commonly used for separating materials by molecular weight, but they can also be used to isolate nucleic acids by their affinity to immobilized ligands (40). In our laboratory, we have adapted the spin column system produced by Sialomed for immunoaffinity separations of low-volume samples. The columns are packed with antibody-coated glass beads and immunoaffinity isolations are performed in the usual way. The advantage of these spin columns is that all of the washing is performed by centrifugation, therefore ensuring maximal recovery during the elution step.

PROTOCOLS

Construction of a Simple Syringe Column

1. Cut a disk from a sheet of Whatman 3MM filter paper with a #8 cork borer.
2. Remove the plunger from a 5-mL syringe and push the filter-paper disk to the bottom of the syringe (ensure that the disk lies flat).
3. Attach a 23-gauge needle to the syringe outlet.
4. Clamp the syringe in a vertical position and add 2 mL of 0.01 M sodium phosphate, pH 7.4.
5. Add 2 mL of affinity ligand or antibody-coated matrix (see Chapters 2–4) and allow the beads to pack by gravity.
6. Add additional beads until the column is packed to the 2.5-mL mark.
7. Stop the flow by pushing the needle tip into a small rubber stopper. The column is now ready to use or it can be stored at 4°C.

Gravimetric Separations in a Syringe Column

1. Remove the fluid from the top of the column and layer the sample onto the bead bed.
2. Remove the stopper and allow the sample to run completely into the bead bed.
3. Add 2 mL of 0.01 M sodium phosphate, pH 7.4, and allow the buffer to pass into the column bed.
4. Add an additional 2 mL of 0.01 M sodium phosphate, pH 7.4, and stop the flow when the bead bed appears just moist.
5. Prepare five 1-mL tubes of 0.01 M sodium phosphate, pH 7.4.

6. Adjust the pH of the tubes with 0.1 M HCl in the following manner:
   a. tube 1: pH 6.0
   b. tube 2: pH 5.0
   c. tube 3: pH 4.0
   d. tube 4: pH 3.0
   e. tube 5: pH 1.5.
7. Add the contents of tube 1 to the syringe.
8. When the top of the column bed appears barely moist, add the contents of tube 2. Repeat for all tubes.
9. Collect 0.25-mL fractions in clean tubes and dialyze each fraction overnight against 0.01 M sodium phosphate, pH 7.4.
10. Analyze each fraction for the presence of the analyte.

Simple Immunoaffinity Separation in a Commercial Gravimetric Column

1. Remove the cap from the column (Figure 9) and clamp the column in a vertical position.
2. Prepare antibody-coated beads (see Chapters 2–4).
3. Place 2 mL of the bead matrix in the column and allow the beads to settle.
4. Layer the sample onto the bead bed.
5. Remove the snap-on tip and allow the sample to run completely into the bead bed.
6. Add 2 mL of 0.01 M sodium phosphate, pH 7.4, and allow the buffer to pass into the column bed.
7. Place 10 mL of 0.01 M sodium phosphate, pH 7.4, into the reservoir and run the buffer through the column.
8. Add 2 mL of 0.33 M citric acid, pH 1.5, to the reservoir and allow the acid to run through the column.
9. Collect 0.1-mL fractions in clean tubes.
10. Dialyze each fraction for 2–3 h against 0.01 M sodium phosphate, pH 7.4, and analyze each fraction for the presence of the analyte.

Packing a 1-cm × 30-cm Commercial Column with a Column Reservoir

1. Attach a 1-cm × 30-cm column to a laboratory stand, ensuring that the column is vertical.
2. Attach a short length of silicon tubing to the outlet.
3. Close the outlet by attaching a clamp to the silicon tubing.
4. Fill the column with 0.01 M sodium phosphate, pH 7.4.
5. Attach a packing reservoir to the top of the column.
6. Prepare a suitable packing matrix (see Chapters 2–4).
7. Fill the reservoir with a bead suspension (usually a 3:1 bead:buffer slurry).
8. Open the column outlet and allow the beads to pack the column by gravity. Pack until the bead bed reaches the top of the column.

9. Stop the column flow, and empty and remove the reservoir.

10. Adjust the top of the bead bed until it is level with the top of the column by carefully adding beads.

11. Place a top on the column.

12. Attach the inlet of the column to the packing reservoir by a short length of tubing.

13. Fill the reservoir with 0.01 M sodium phosphate, pH 7.4, and open the column outlet to start the column running.

14. Pass 200–300 mL of 0.01 M sodium phosphate, pH 7.4, through the column, then stop the flow by closing the outlet port.

Isolating an Analyte by Immunoaffinity Spin Chromatography

1. Prepare antibody-coated glass beads (see Chapters 2–4) and place 250 µL of the beads into a spin column (Figure 10).

2. Allow the matrix to settle, then add 100–200 µL of sample.

3. Place the spin column in a collection tube and incubate at room temperature for 10–20 min.

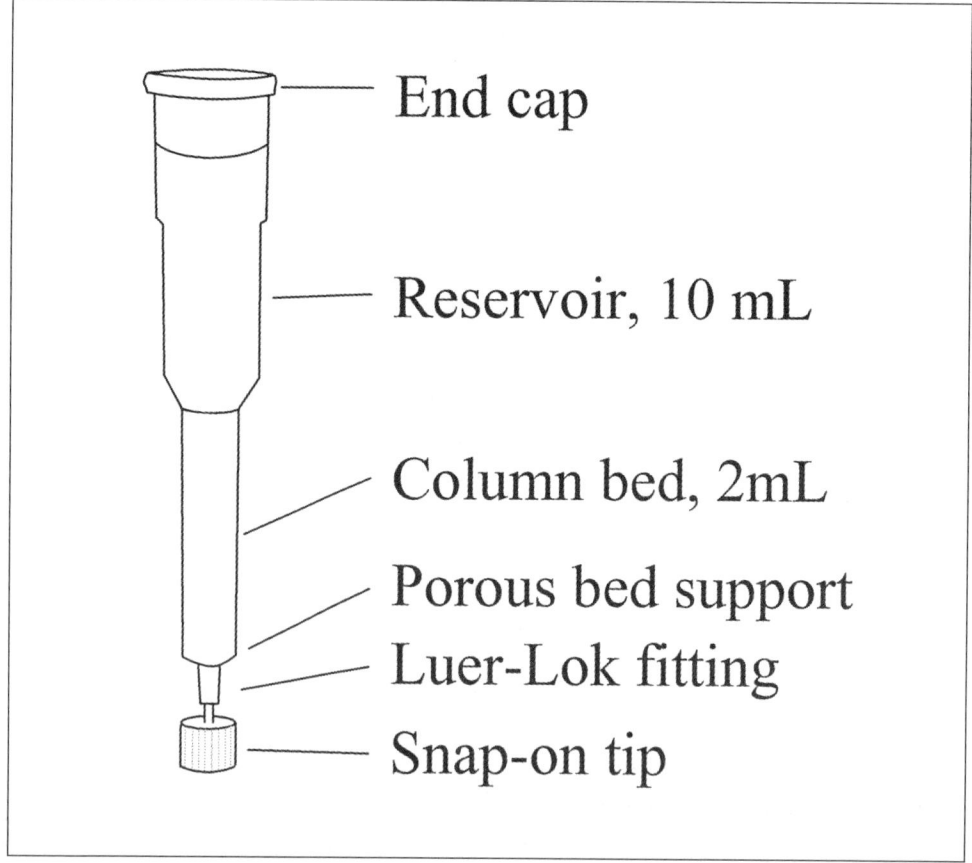

Figure 9. Components of a simple, disposable column.

4. Place the spin column and collection tube assembly in a tube holder (supplied with the empty spin columns).

5. Place the tube holder in a microcentrifuge and spin for 2 min.

6. Add 200 µL of 0.01 M sodium phosphate, pH 7.4.

7. Place in a microcentrifuge and spin for 2 min.

8. Repeat steps 4–7 three times (these are wash steps; discard the effluent following step 7 in each procedure).

9. Add 200 µL of 0.33 M citric acid, pH 1.5, (as an elution buffer) and incubate for 5 min at room temperature.

10. Place in a microcentrifuge and spin for 2 min.

11. Collect the effluent in the collection tube and dialyze against 0.01 M sodium phosphate, pH 7.4, overnight.

MEDIUM-PERFORMANCE TECHNIQUES

The addition of a pump and, therefore, control of the running conditions can easily turn a simple chromatography system into a medium-performance one. Although it is a simple matter to construct such an instrument from the required laboratory equipment, most investigators tend to purchase commercially available systems. The advantage of this approach is that the units are integrated and often under computer control. Additionally, these units come equipped with accessory units such as switching valves and injection ports (the addition of the latter unit ensuring controlled and reliable injection of known volumes of sample). This now introduces a factor of reproducibility and precision, which are both essential for the validation of a procedure.

Medium-performance techniques (such as FPLC) are generally used for protein purification by conventional techniques. However, this procedure has occasionally been used for affinity or immunoaffinity chromatography procedures, although usually as a secondary procedure to purify materials initially isolated by immunoaffinity (15,18). Xue et al. (128) used FPLC to prepare purified recombinant hemoglobin by absorbing it to the ion exchanger DEAE and then selectively eluting the hemoglobin with pyrophosphate. This approach, although not truly affinity chromatography, produced a highly pure product from mixtures of *E. coli* extract. Affinity isolations have been performed with medium-pressure or FPLC systems. Proost and colleagues (92) used a combination of techniques including heparin affinity columns to purify a bovine chemotactic protein from bovine monocytes. Trost et al. (111) used a Blue Sepharose 6B column to purify the plant cytosolic protein NAD(P)H:(quinone-acceptor) oxidoreductase from cultured sugarbeet cells. The enzyme was recovered using a NADPH gradient and further characterized by electrophoresis. Zhuang et al. (134) used immobilized bovine submaxillary mucin to isolate a lectin from the mushroom *Amantia pantherina*.

Sepharose columns containing protein A or protein G have been used in FPLC systems for the isolation of a number of immunoglobulin classes from different species. Mazza et al. (67) isolated four fractions of IgG from normal dog serum using a combination of gel filtration and protein A and protein G affinity chromatography using an FPLC system. In a similar manner, Sheoran and Holmes (102) used protein A and protein G FPLC columns to isolate and study the binding characteristics of horse immunoglobulins. Protein G columns attached to medium-pressure systems

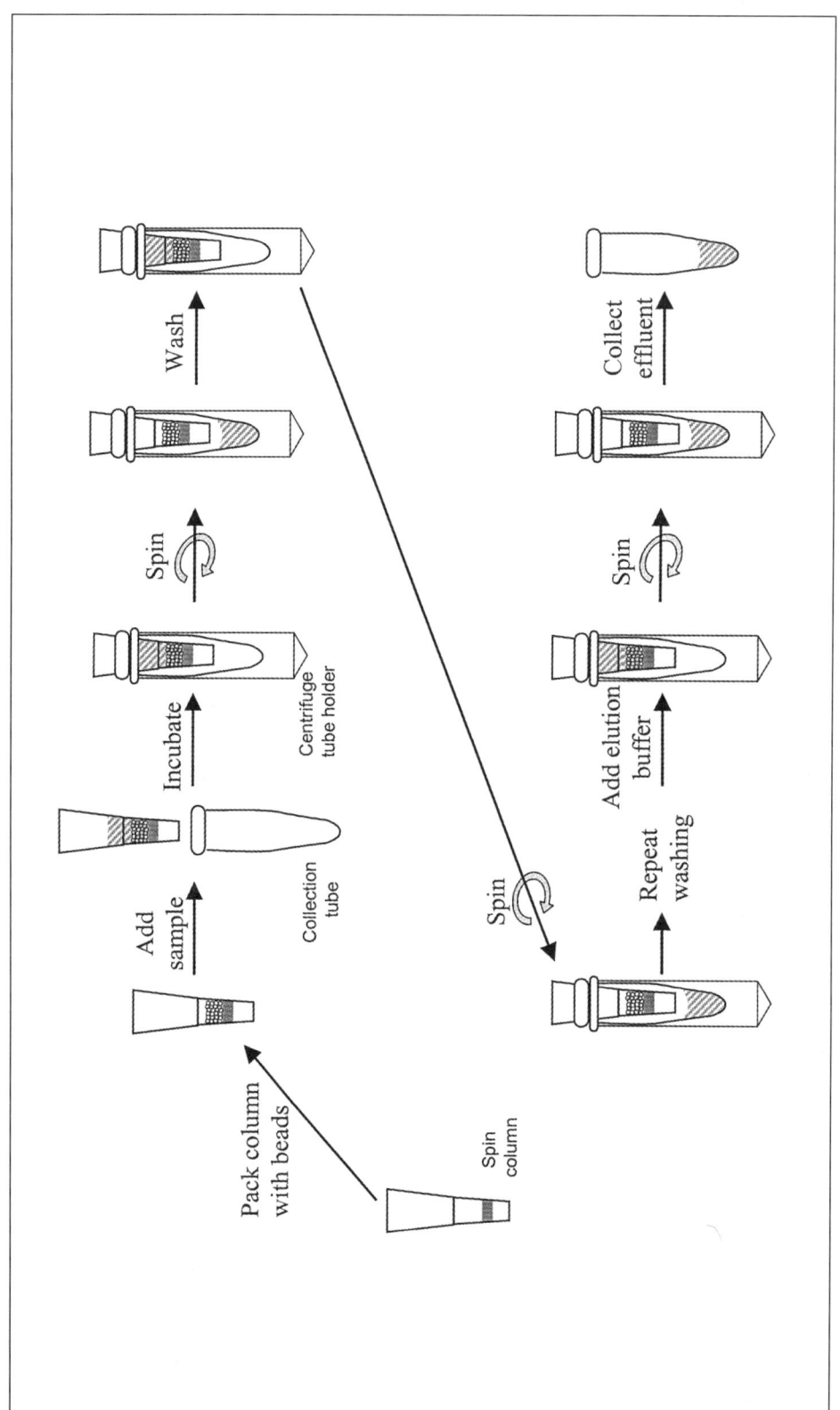

Figure 10. General procedure for immunoaffinity chromatography using commercially available spin columns.

have been used to isolate monoclonal antibodies (6) and to isolate IgG fractions from dogs (67). These authors, like Sheoran and Holmes (102), also employed the differences in protein A and protein G binding to further purify the different IgG fractions.

PROTOCOLS

Isolation of IgG on a Protein A Affinity Column Connected to a Commercial Medium-Pressure System

1. Add 10 mL of saturated NH_2SO_4 to 10 mL of sample.
2. Immediately spin at $12\,000\times g$ for 5 min and discard the supernatant.
3. Redissolve the pellet in 1 mL of distilled H_2O.
4. Filter through a 0.22-μm filter.
5. Connect either a laboratory-built or commercially available protein A column to the FPLC instrument.
6. Equilibrate the instrument by pumping 50 mL of 0.01 M sodium phosphate, pH 8.0, through the system.
7. Set the instrument to isocratic mode at a flow rate of 1.5 mL/min.
8. Set the fraction collector to collect 1-min timed fractions.
9. Program the initial phase to 10 min, followed by a linear gradient that will change the pH of the running buffer from 8.0 to 1.5 in 10 min.
10. Set the detector to 280 nm, 1.5 absorbance unit full scale (auf).
11. Inject 500 μL of the redissolved precipitate and allow the instrument to complete its program.
12. Recover the collected fractions, select the fractions containing the second peak, and dialyze them against 0.01 M sodium phosphate, pH 7.4 (the fractions are simply found by studying the chromatogram produced by the pen recorder or computer).

Isolation of a Specific Analyte by Immunoaffinity in a Medium-Pressure System

1. Connect a laboratory-built immunoaffinity column to the instrument and equilibrate the instrument by pumping 50 mL of 0.01 M sodium phosphate, pH 7.4, through the system.
2. Program the instrument to run in isocratic mode at a flow rate of 0.5 mL/min for 10 min, followed by a linear gradient mode for a further 10 min.
3. Set the fraction collector to collect 1-min timed fractions.
4. Fill buffer reservoir A with 0.01 M sodium phosphate, pH 7.4.
5. Fill buffer reservoir B with 0.01 M sodium phosphate, pH 1.5.
6. Set the detector to 280 nm, 1.5 auf.
7. Inject 100 μL of the sample and allow the instrument to complete its program.
8. Recover the collected fractions, select the fractions containing the second peak, and dialyze them against 0.01 M sodium phosphate, pH 7.4 (the fractions are simply found by studying the chromatogram produced by the pen recorder or computer).

HIGH-PERFORMANCE LIQUID CHROMATOGRAPHY

The control and precision of HPLC is ideally suited to affinity and immunoaffinity procedures. However, until suitable packing materials were available, this marriage could not take place (22,66,69,82,127). The development of plastic, glass, and other materials capable of withstanding the pressures of high-performance separations quickly led to their application as affinity and immunoaffinity supports (see Chapter 2). Once this had been achieved, affinity and immunoaffinity separations took on a new and still expanding role.

High-Performance Affinity Chromatography

The new family of techniques commonly termed high-performance affinity chromatography (HPLAC) are now popular techniques being applied to the separation of molecules of biological interest in a wide range of different fields (22,66,69). HPLAC has been used as a pretreatment or cleanup technique prior to subjecting the eluted material to further analysis by reverse-phase HPLC (RP-HPLC) or mass spectroscopy. Apffel et al. (5) used a combination of HPLAC, RP-HPLC, and electrospray ionization mass spectroscopy to study peptide maps of both glycosylated and nonglycosylated peptides derived from an endopeptidase (LysC) digest of a recombinant DNA-derived glycoprotein. Although these studies were only partially successful, they demonstrated the potential use of such multidimensional approaches to structure chemistry. HPLAC has also been employed as a cleanup technique for trace drug analysis in animal tissues and foods. Stubbings et al. (109) used an immobilized Copper (II) metal chelate column to extract tetracycline, oxytetracycline, chlortetracycline, and demeclocycline from sheep liver and bovine kidney extracts. Following EDTA elution, the analytes retained on the metal affinity column were subsequently analyzed by conventional RP-HPLC. This approach was also used by Croubels et al. (20); however, these authors obtained greater sensitivity with fluorescence detection.

HPLAC has also been employed as the sole analytical process for drug or drug metabolite monitoring. Immobilized HSA is a useful binding ligand for a number of clinically important drugs, and several investigators have used this parameter to study protein binding and other drug-binding characteristics. Hage and Sengupta (36) employed immobilized HSA as an affinity ligand to examine the binding characteristics of clomiphene. Using zonal elution, they were able to determine the binding of racemic forms of the drug to the immobilized ligand. They were also able to use competition-binding studies to map the binding sites involved. Further studies by Chattopadhyay et al. (14) demonstrated that chemical modification of the tryptophan residue (Trp-214) of HSA helped to resolve the steroselective binding of the protein for racemic forms of the drug warfarin. These studies made the important point that chemically modified proteins could be used as chiral separators. Similar studies (132) have also demonstrated that HSA could be used to study the steroselective binding of ketoprofen. Kaliszan (47) used both immobilized HSA and a series of other proteins [α_1-acid glycoprotein, keratin, collagen, melanin, and amylose tris(3,5-dimethylphenylcarbamate)] to study the quantitative structure-retention relationships (QSRR) of a series of drugs.

Other ligands used in HPLAC range from anticoagulants to lectins and protein A or protein G. Sluzky et al. (105) employed immobilized heparin to study the stability and aggregation of recombinant human basic growth factor. Clairbois et al. (17)

used immobilized heparin as the affinity ligand capable of isolating heparin-binding proteins from detergent-solubilized vascular smooth muscle cell membranes. Chemical modification of the packing surface has also proven useful in the development of affinity ligands. Such modifications include grafting porous zirconia with polyphosphates of varying length. These packings were capable of retaining single-stranded RNA and DNA and separating them from double-stranded DNA (64). Coating of packing medium with small chemical ligands has also been used by Lakhiari et al. (56). They employed silica beads coated with dextran prior to attachment of *N*-acetylneuraminic acid to isolate insulin. Lin et al. (59) approached the isolation of bioactive human transthyretin to a high degree of purity.

Lectins are popular ligands for affinity separations and many such materials have been applied to HPLAC. The affinity of certain lectins for sugar groups can be used as a selective ligand to isolate a number of different molecules. Con A bound to nonporous, microparticulate silica has been used to isolate horseradish peroxidase (4), and wheat germ lectin bonded to diol-silica has been used to isolate bone and liver alkaline phosphatases in human tissues (33). Jacalin is a lectin that has high specificity and affinity for the core disaccharide, 1-β-galactopyranosyl-3-(α-2-acetamido-2-deoxygalactopyranoside), in *O*-linked oligosaccharides. This property has been shown to be useful for the isolation of IgA from human serum (16). Ohlson et al. (76) have used wheat germ agglutinin as a weak affinity ligand for the analysis of *N*-acetyl derivatives of different sugar-containing saccharides.

The immunoglobulin-binding proteins of bacterial origin, namely protein A and protein G, have also been used widely in HPLAC techniques (32,112). Nadler et al. (73) used protein A in the second phase of a two-dimensional assay for determining the concentration of IgG aggregates in antibody- or immunoglobulin-containing pharmaceutical preparations. They demonstrated that the system described in their report could be automated, performing reliable assays within 1 h. Likewise, Stahl and colleagues (106) used a Pharmacia HiTrap® Protein G column to isolate and identify IgG alloantigens. They concluded that the isolation of such IgG antibodies easily facilitated their detection, even in the presence of other reactive antibodies (i.e., IgM). Protein G columns have also been used in a semi-automated system for the on-line monitoring and harvesting of monoclonal antibodies from a hollow fiber fermenter (131).

High-Performance Immunoaffinity Chromatography

High-performance immunoaffinity chromatography (HPIAC) has gained considerable popularity in recent years. The procedure is now being used to study analytes in many different fields ranging from analytical chemistry to molecular biology. The field of drug residue analysis is an area where immunoaffinity chromatography is becoming popular (30). Eller at al. (26) have used immunoaffinity chromatography for the isolation of propranolol metabolites from human plasma, while Ikegawa et al. (42) have successfully employed immunoaffinity separations to the isolation of the racemic forms of bufuralol. Urine drug testing using immunoaffinity isolation has also gained some interest. Kussak et al. (55) employed immunoaffinity columns as a pre-analytical cleanup for the detection of aflatoxins B_1, B_2, G_1, G_2, M_1, and Q_1 in human urine. Fluorescence detection, following RP-HPLC analysis of the immunoaffinity isolated materials, demonstrated detection limits of aflatoxins to be 6.8 pg/mL for B_1, B_2, G_1, and G_2, and 18 pg/mL for M_1 and Q_1. Other workers have approached drug monitoring by a combination of immunoaffinity RP-HPLC and mass spectroscopy. This combination has been successfully used to measure propanolol and lysergic acid (LSD) (97). This system is reported to be capable of detecting 2.5 ng/mL of propranolol and 500 pg/mL of LSD in phosphate-buffered urine samples. Creaser and colleagues (19) examined horse urine for the presence of corticosteroids using a similar multiple analytical approach. They report that their system was able to detect dexamethasone and flumethasone with limits of detection in the 3–4 ng/mL range.

Other examples of HPIAC applications include immuno-extraction of contaminants from immunoglobulin preparations (29), isolation of hormones and growth factors (7,27), and the preparation of viral proteins (98,120). HPIAC has also been used to isolate specific receptors and membrane-associated proteins (13,81,89), and enzymes like glutamine synthetase (2). HPIAC has also been applied to the recovery and measurement of urinary proteins such as urine protein 1 (78) and urinary albumin (96). Recently, Ohlson and colleagues (77) demonstrated that high affinity antibodies were not necessary for successful immunoaffinity separations. This group showed the potential of using weak affinity monoclonal antibodies as immunoaffinity ligands and the fact that the system allowed carbohydrate analytes with similar structures to easily be separated by isocratic elution. One of the emerging fields of importance in which immunoaffinity separation techniques hold great potential is in the environmental sciences (93,103,118). Initially, immunoaffinity chromatography was applied to the detection of nortestosterone in beef muscle (119). The results of these studies indicated that immunoaffinity separations could be used to detect the presence of selected analytes at 100 ng/kg concentrations. Bianchini et al. (8) used immunoaffinity techniques as a pre-analytical cleanup for the measurement of 7-methylguanine as a marker of exposure to environmental methylating agents. Likewise, Yin and colleagues (129) applied a combination of immunoaffinity chromatography and immunoassay to the detection of 8-hydroxydeoxyguanosine in human placental tissue. Halmes et al. (37) used immunoaffinity to isolate trichloroethylene-protein adducts in experimentally exposed rats. However, these studies did not provide conclusive evidence of a link between exposure and adduct formation. Recently, Ueno et al. (114) studied the presence of ochratoxin A in a healthy population of Japanese workers. These investigators employed a combination of immunoassay and immunoaffinity-linked HPLC to determine the presence of the analyte in plasma samples.

Immunoaffinity columns have also been applied for pre-analytical sample cleanup by several investigators. Scudamore and MacDonald reported that such columns were used by a cooperative of laboratories in the UK to measure ochratoxin A contamination in samples of wheat (100). They concluded that the incorporation of the immunoaffinity step was reliable and compared favorably with other published procedures. Other workers have also reported that immunoaffinity pretreatment can negate the need for extraction of many important molecules. Deinl and colleagues (23) reported that this approach was successful in the analysis of flunitrazepam and its metabolites. Kobayashi et al. (54) employed immunoaffinity pretreatment as a means to improve a radio-receptor assay for human plasma 1-α, 25-dihydroxyvitamin. Janis and colleagues (44,45) described a further variation of the use of immunoaffinity cleanup procedures. They used column-switching techniques to successfully perform on-line immunoassays for a number of different analytes including measurement of rabbit anti-human transferrin antibody, human transferrin, and variants of egg white lysozyme. Hage (34) recently published an excellent review of the advances in analytical immunoaffinity procedures.

Column Packing

The packing of high-performance affinity or immunoaffinity columns is an area where serious problems can arise. Once the ligand has been attached to the packing medium by any of the procedures described in Chapters 4 and 5, it can be packed into suitable columns using conventional column-packing systems (Figure 11). Usually, pumped slurry packers are gentler than gas-activated ones as long as care is taken to adequately control the pressure in the slurry packer. Most affinity- and all immunoaffinity-packing materials require aqueous solvents, which cause the buildup of friction. This can be overcome to some extent by using low salt buffers that reduce both the friction caused by flow and the denaturation of the immobilized ligand. This is important when delicate antibody or antibody fragments are immobilized on the packing medium. The short columns used in immunoaffinity techniques can easily be slurry-packed by hand, thus overcoming the problems associated with commercial slurry-packing instruments. Additionally, many affinity and immunoaffinity packing materials can be "dry-packed," which further ensures minimal damage to the immobilized ligand.

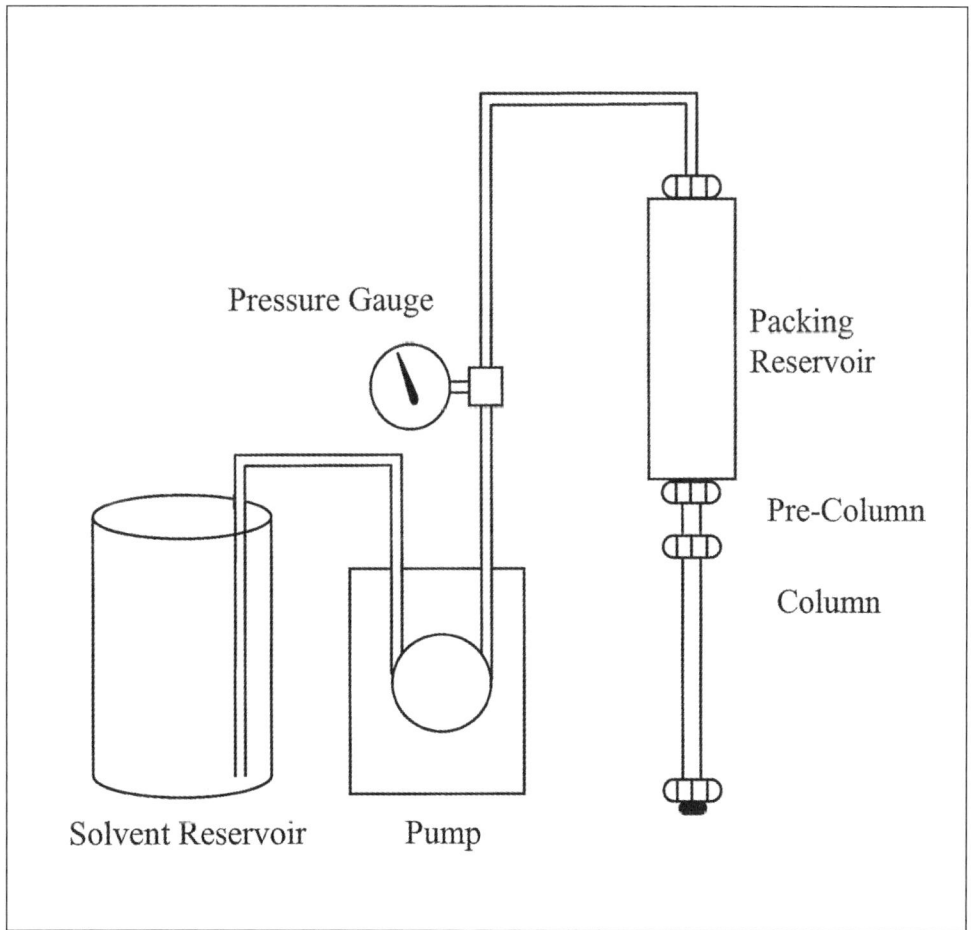

Figure 11. A conventional slurry packing system.

Operational Conditions

Generally, the operational conditions required for successful high-performance affinity or immunoaffinity procedures are mild compared to conventional techniques such as RP-HPLC. The operational temperatures are usually maintained at room temperature or below. Temperatures greater than 30°C can eventually denature delicate proteins, which in turn shortens the active life of the ligand. Immobilized antibodies usually maintain a longer active life when kept at 4°C (66,82,83). This can easily be achieved by surrounding the column with a water jacket and a circulating ice bath. Alternatively, the HPLC system can be operated within a controlled environmental cabinet.

The flow rate used in affinity or immunoaffinity techniques should be kept to a rate between 0.5 and 1.5 mL/min (82,83), because the faster the flow rate, the less time the analyte has to interact with the immobilized ligand. Thus, short retention times often result in less separation efficiency. However, some antibodies can react quickly (34) and prior knowledge of the affinity characteristics of the antibody can help to determine the flow rate and efficiency of the separation process.

Another important operational parameter when considering affinity and immunoaffinity HPIAC separations is column pressure. This factor quickly increases, especially when high concentrations of analyte form dense complexes with the immobilized ligand. Care should be taken to watch the pressure during a chromatographic run. Pressures greater than 400–500 psi can cause serious structural damage to immobilized ligands and, in some cases, cause ligand leakage (82,83). This is often caused by shearing forces generated by a combination of flow rate and pressure, resulting in shearing of the ligand from its support.

Active Column Life

There have been a number of studies on the active life expectancy of high-performance affinity and immunoaffinity columns. From these studies, it appears that the number of cycles is not constant and that different ligands express different life spans. Much of this variability will depend on the conditions used to elute the bound analyte and the running or storage conditions of the columns. Huschka and colleagues (41) reported that anti-human IgM antibodies immobilized on controlled glass beads were stable for a minimum of 50 adsorption/elution cycles. A similar finding for anti–interleukin-2 antibodies immobilized on glass beads was made in our laboratory (87). In studies on the isolation of total interleukin 2 from human samples, we found that our columns were stable for 200 cycles. Maintenance of the columns at 4°C was found to greatly extend their active lives. Janis and Regnier (45) reported that immunoaffinity columns used as preselection devices prior to conventional HPLC analysis were stable over 77 cycles. They immobilized their ligands on porous plastic beads and did not appear to use any specialized conditions for the maintenance of their bioactivity. Other workers (23) have also reported that immobilized antibodies used in pre-analytical columns have reasonable life expectancy. These authors report that greater than 88 cycles can be expected from their immunoaffinity columns without the loss of activity or selectivity. We have continued to keep our immunoaffinity columns cold during both working and storage periods. Under this condition, we have found that our columns exhibit working lives of up to 200 cycles.

PROTOCOLS

Cleaning Conventional Stainless Steel Columns

1. Remove the column end-fittings and soak the column tube and end-fittings in 10% (vol/vol) Decon™ detergent (Fisher Scientific) for 24 h.
2. Scrub the interior of the tube and the interior of the end-fittings with a test-tube brush.
3. Thoroughly rinse in distilled H_2O.
4. Soak for 1 h in absolute ethanol, then dry in an oven at 100°C and reassemble the column.

Assembly of a Commercial PEEK Column

The following protocol describes assembly of an Omegachrom™ PEEK column available from Upchurch Scientific.

1. Insert a "fingertight" ferrule connector into the column end-fitting and tighten by finger pressure.
2. Insert a "fingertight" connector into the ferrule connector and tighten by finger pressure.
3. Place a retaining frit in the assembled column end-fitting (make sure that the frit lies flat in the fitting).
4. Carefully attach the assembled end-fitting to one end of the main column tube and tighten by finger pressure.
5. Use a star wrench and a suitable adjustable wrench to securely tighten the fittings. The column is now ready to pack.
6. Following packing, the top end-fitting is assembled as described in steps 1–5 above.

Dry-Packing an Affinity or Immunoaffinity Column

1. Clamp the column in a vertical position and remove the top end-fitting.
2. Attach a small-bore funnel to the top of the column.
3. Slowly add the dried packing material to the funnel. Occasionally, tap the sides of the column with a spatula handle to loosen material from the side of the column.
4. Continue to pack the column until the level of packing material stops moving.
5. Remove the column from the clamp and slowly bounce the lower end of the column on a flat surface.
6. Repeat steps 4 and 5 until the packing level stays constant.
7. Remove the funnel and reattach the top end-fitting to the column.

Slurry-Packing a Small Affinity or Immunoaffinity Column by Hand

1. Assemble the lower half of the PEEK column by attaching the end fitting onto

the column (see the protocol for the assembly of a commercial PEEK column, above).

2. Clamp the column to a laboratory stand (a small workshop vise is excellent for this job), ensuring that the column is vertical.
3. Place 250 μL of 0.01 M sodium phosphate, pH 7.4, in the column tube.
4. Slowly add the antibody-coated beads.
5. Occasionally, cover the top of the column tube with Parafilm and bounce the assembly on the bench top to eliminate air bubbles.
6. Continue adding beads until the column tube is completely packed.
7. Using a spatula blade, smooth the packed bead surface so that it is level with the top of the column tube.
8. Carefully attach the top end-fitting and tighten.

Slurry-Packing an Affinity or Immunoaffinity Column with a Slurry Packer

1. Remove the top end-fitting of the column and fill the column with 10% (vol/vol) glycerol.
2. Attach a "pre-column" to the main body of the column to be packed.
3. Fill the packing reservoir of the packing instrument with a 1:3 matrix:water slurry.
4. Attach the pre-column to the outlet of the packing instrument and pump at 0.2–0.5 mL/min (the pressure must not exceed 500 psi).
5. Pump for 5 min, then disconnect the pre-column/column assembly from the packing instrument.
6. Gently disconnect the pre-column from the column (take care not to disturb the top surface of the column bed).
7. With a flat spatula blade, gently smooth the top surface level of the column matrix (add matrix if required).
8. Reattach the top end-fitting to the column and tighten.

Isolation of a Glycoprotein on a Con A Affinity Column

1. Prime the HPLC system with 0.01 M sodium phosphate, pH 7.4.
2. Connect the packed HPLAC (lectin) column inlet to the outlet of the injection port.
3. Connect the column outlet to the inlet of the detector flow-cell.
4. Encase the column in a water jacket and cool to 4°C.
5. Set the pump speed to 0.5 mL/min and run 10–20 mL of 0.01 M sodium phosphate, pH 7.4, through the system.
6. Inject 100 μL of sample into the injection port and run for 10 min.
7. Program the gradient controller to produce a linear gradient from 0%–100% of the elution buffer (α-D-mannose) over 10 min. Monitor the entire chromatogram using a suitable detector and collect the second peak.

Isolation of a Specific Analyte by Immobilized Antibody Immunoaffinity Chromatography

1. Prime the HPLC system with 0.01 M sodium phosphate, pH 7.4.

2. Connect a short (5-cm × 2-mm i.d.), packed HPIAC column in-line between the injection port and the detector flow-cell.

3. Encase the column in a water jacket and cool to 4°C.

4. Set the pump speed to 0.5 mL/min and run 10 mL of 0.01 M sodium phosphate, pH 7.4, through the system.

5. Inject 25 µL of sample and run the system in an isocratic mode for 5 min.

6. Program the gradient controller to produce a linear gradient from 0–2.5 M sodium thiocyanate over 5 min. Monitor the entire chromatogram with an ultraviolet detector and collect the second peak.

7. Dialyze against 0.01 M sodium phosphate, pH 7.4, overnight.

REFERENCES

1. **Agraz, A., C.A. Duarte, L. Costa, L. Perez, R. Paez, V. Pujol and G. Fontirrochi.** 1994. Immunoaffinity purification of recombinant hepatitis B surface antigen from yeast using a monoclonal antibody. J. Chromatogr. *672*:25-33.

2. **Alhama, J., J. Lopez-Barea, F. Toribio and J.M. Roldan.** 1992. Purification and determination of glutamine synthetase by high-performance immunoaffinity chromatography. J. Chromatogr. *589*:121-126.

3. **Allen, R.E., T.W. Lo and P.J. Thornalley.** 1993. A simplified method for the purification of human red blood cell glyoxalase. I. Characteristics, immunoblotting, and inhibitor studies. J. Protein Chem. *12*:111-119.

4. **Anspach, F.B., H.J. Wirth, K.K. Unger, P. Stanton, J.R. Davies and M.T. Hearn.** 1989. High-performance liquid affinity chromatography with phenylboronic acid, benzamidine, tri-L-alanine, and concanavalin A immobilized on 3-isothiocyanatopropyltriethoxysilane-activated nonporous monodisperse silicas. Anal. Biochem. *179*:171-181.

5. **Apffel, A., J.A. Chakel, W.S. Hancock, C. Souders, T. M' Timkulu and E. Pungor.** 1996. Application of multidimensional affinity high-performance liquid chromatography and electrospray ionization liquid chromatography-mass spectrometry to the characterization of glycosylation in single-chain plasminogen activator. Initial results. J. Chromatogr. *750*:35-42.

6. **Basta, M., V.D. Miletic, C.H. Hammer and M.M. Frank.** 1996. Production and characterization of the monoclonal antibody against SGP120, a novel serum protein. Hybridoma *15*:69-75.

7. **Berghman, L.R., O. Lescroart, I. Roelants, F. Ollevier, E.R. Kuhn, P.D. Verhaert, A. De Loof, F. van Leuven and F. Vandesande.** 1996. One-step immunoaffinity purification and partial characterization of hypophyseal growth hormone from the African catfish, Clarias gariepinus (Burchell). Comp. Biochem. Physiol. B Biochem. Mol. Biol. *113*:773-780.

8. **Bianchini, F., R. Montesano, D.E. Shuker, J. Cuzick and C.P. Wild.** 1993. Quantification of 7-methyldeoxyguanosine using immunoaffinity purification and HPLC with electrochemical detection. Carcinogenesis *14*:1677-1682.

9. **Black, J.A. and R.S. Hodges.** 1991. Peak identification in HPLC of peptides using spectral analysis and second order spectroscopy of aromatic amino acid residues, p. 563-570. In C. Mant and R.S. Hodges (Eds.), HPLC of Peptides and Proteins: Separation, Analysis and Conformation. CRC Press, Boca Raton.

10. **Blain, R.** 1993. Absorbance detection, p. 39-66. In D. Parriott (Ed.), A Practical Guide to HPLC Detection. Academic Press, New York.

11. **Bruckner, C.A., M.D. Foster, L.R. Lima and R.E. Synovec.** 1994. Column liquid chromatography: equipment and instrumentation. Anal. Chem. *66*:1R-16R.

12. **Burden, D.W. and D.B. Whitney.** 1995. Protein purification by column chromatography, p. 108-123. Biotechnology: Proteins To PCR. Birkhäuser, Boston.

13. **Burrows, G.G., K. Ariail, B. Celnik, J.E. Gambee, H. Offner and A.A. Vandenbark.** 1997. Multiple class I motifs revealed by sequencing naturally processed peptides eluted from rat T cell MHC molecules. J. Neurosci. Res. *49*:107-116.

14. **Chattopadhyay, A., T. Tian, L. Kortum and D.S. Hage.** 1998. Development of tryptophan-modified human serum albumin columns for site-specific studies of drug-protein interactions by high-performance

affinity chromatography. J. Chromatogr. B Biomed. Sci. Appl. *715*:183-190.

15. **Chen, R.H. and Y.C. Chen.** 1989. Isolation of an acidic phospholipase A2 from the venom of Agkistrodon acutus (five pace snake) and its effect on platelet aggregation. Toxicon *27*:675-682.

16. **Chui, S.H., C.W. Lam, W.H. Lewis and K.N. Lai.** 1990. High-performance liquid affinity chromatography for the purification of immunoglobulin A from human serum using jacalin. J. Chromatogr. *514*:219-225.

17. **Clairbois, A.S., D. Letourneur, D. Muller and J. Jozefonvicz.** 1998. High-performance affinity chromatography for the purification of heparin-binding proteins from detergent-solubilized smooth muscle cell membranes. J. Chromatogr. B Biomed. Sci. Appl. *706*:55-62.

18. **Corti, A., M.L. Nolli, A. Soffientini and G. Cassani.** 1986. Purification and characterization of single-chain urokinase-type plasminogen activator (pro-urokinase) from human A431 cells. Thromb. Haemost. *56*:219-224.

19. **Creaser, C.S., S.J. Feely, E. Houghton and M. Seymour.** 1998. Immunoaffinity chromatography combined on-line with high-performance liquid chromatography-mass spectrometry for the determination of corticosteroids. J. Chromatogr. *794*:37-43.

20. **Croubels, S.M., K.E. Vanoosthuyze and C.H. van Peteghem.** 1997. Use of metal chelate affinity chromatography and membrane-based ion-exchange as clean-up procedure for trace residue analysis of tetracyclines in animal tissues and egg. J. Chromatogr. B Biomed. Sci. Appl. *690*:173-179.

21. **Davies, M.J. and E.F. Hounsell.** 1996. Carbohydrate chromatography: towards yoctomole sensitivity. Biomed. Chromatogr. *10*:285-289.

22. **Dean, P.D.G., W.S. Johnson and F.A. Middle (Eds.)** 1985. Affinity Chromatography. A Practical Approach. IRL Press, Oxford.

23. **Deinl, I., L. Angermaier, C. Franzelius and G. Machbert.** 1997. Simple high-performance liquid chromatographic column-switching technique for the on-line immunoaffinity extraction and analysis of flunitrazepam and its main metabolites in urine. J. Chromatogr. B Biomed. Sci. Appl. *704*:251-258.

24. **Dourges, M.A., D. Gulino, J. Courty, J. Badet, D. Barritault and J. Jozefonvicz.** 1990. Affinity chromatography of fibroblast growth factors on substituted polystyrene. J. Chromatogr. *526*:35-45.

25. **Eggers, M., M. Hogan, R.K. Reich, J. Lamture, D. Ehrlich, M. Hollis et al.** 1994. A microchip for quantitative detection of molecules utilizing luminescent and radioisotope reporter groups. BioTechniques *17*:516-525.

26. **Eller, T.D., U.K. Walle and T. Walle.** 1993. Immunoaffinity isolation of the sulfate conjugate of 4'-hydroxypropranolol from plasma. J. Chromatogr. *612*:320-325.

27. **Feng, W. and X. Geng.** 1993. Studies on silica-bonded monoclonal antibody packing material for separation of recombinant interferon by high performance immunoaffinity chromatography. Biomed. Chromatogr. *7*:317-320.

28. **Fleminger, G., T. Wolf, E. Hadas and B. Solomon.** 1990. Eupergit C as a carrier for high-performance liquid chromatographic-based immunopurification of antigens and antibodies. J. Chromatogr. *510*:311-319.

29. **Flurer, C.L. and M. Novotny.** 1993. Dual microcolumn immunoaffinity liquid chromatography: an analytical application to human plasma proteins. Anal. Chem. *65*:817-821.

30. **Giuliano, K.A.** 1992. Chromatography of proteins on columns of polyvinylpolypyrrolidone using adsorbed textile dyes as affinity ligands. Anal. Biochem. *200*:370-375.

31. **Gobom, J., E. Nordhoff, E. Mirgorodskaya, R. Ekman and P. Roepstorff.** 1999. Sample purification and preparation technique based on nano-scale reversed-phase columns for the sensitive analysis of complex peptide mixtures by matrix-assisted laser desorption/ionization mass spectrometry. J. Mass Spectrom. *34*:105-116.

32. **Godfrey, M.A.** 1997. Immunoaffinity and IgG receptor technologies, p. 141-195. *In* P. Matejtschuk (Ed.), Affinity Separations. A Practical Approach. IRL Press, Oxford.

33. **Goncharoff, D.G., E.L. Branum, S.L. Cedel, B.L. Riggs and J.F. O'Brien.** 1991. Clinical evaluation of high-performance affinity chromatography for the separation of bone and liver alkaline phosphatase isoenzymes. Clin. Chim. Acta *199*:43-50.

34. **Hage, D.** 1998. Survey of recent advances in analytical applications of immunoaffinity chromatography. J. Chromatogr. B Biomed. Sci. Appl. *715*:3-28.

35. **Hage, D.S.** 1999. Affinity Chromatography: A Review of Clinical Applications. Clin. Chem. *45*:593-615.

36. **Hage, D.S. and A. Sengupta.** 1998. Studies of protein binding to nonpolar solutes by using zonal elution and high-performance affinity chromatography: interactions of cis- and trans-clomiphene with human serum albumin in the presence of beta-cyclodextrin. Anal. Chem. *70*:4602-4609.

37. **Halmes, N.C., E.J. Perkins, D.C. McMillan and N.R. Pumford.** 1997. Detection of trichloroethylene-protein adducts in rat liver and plasma. Toxicol. Lett. *92*:187-194.

38. **Hayakawa, K., N. Terai, P.G. Dinning, K. Akutsu, Y. Iwamoto, R. Etoh and T. Murahashi.** 1996. An on-line reduction HPLC/chemiluminescence detection system for nitropolycyclic aromatic hydrocarbons and metabolites. Biomed. Chromatogr. *10*:346-350.

39. **He, B., N. Tait and F.E. Regnier.** 1998. Fabrication of nanocolumns for liquid chromatography. Anal.

Chem. *70*:3790-3797.

40. **Hirsch, H.H. and W. Bossart.** 1999. Two-centre study comparing DNA preparation and PCR amplification protocols for herpes simplex virus detection in cerebrospinal fluids of patients with suspected herpes simplex encephalitis. J. Med. Virol. *57*:31-35.

41. **Huschka, U., J. Maier, E.W. Rauterberg and H.W. Doerr.** 1982. Rapid separation of immunoglobulin M by immunoaffinity chromatography for detection of specific antibodies to rubella and Treponema pallidum. Eur. J. Clin. Microbiol. *1*:118-121.

42. **Ikegawa, S., K. Matsuura, T. Sato, N.M. Isriyanthi, T. Niwa, S. Miyairi, H. Takashina, Y. Kawashima and J. Goto.** 1998. Enantioselective immunoaffinity extraction for simultaneous determination of optically active bufuralol and its metabolites in human plasma by HPLC. J. Pharm. Biomed. Anal. *17*:1-9.

43. **Jack, G.W.** 1994. Immunoaffinity chromatography. Mol. Biotechnol. *1*:59-86.

44. **Janis, L.J., A. Grott, F.E. Regnier and S.J. Smith-Gill.** 1989. Immunological-chromatographic analysis of lysozyme variants. J. Chromatogr. *476*:235-244.

45. **Janis, L.J. and F.E. Regnier.** 1988. Immunological-chromatographic analysis. J. Chromatogr. *444*:1-11.

46. **Johansson, B.L., U. Hellberg and O. Wennberg.** 1987. Determination of the leakage from Phenyl-Sepharose Cl-4B, Phenyl-Sepharose FF and Phenyl-Superose in bulk and column experiments. J. Chromatogr. *403*:85-98.

47. **Kaliszan, R.** 1998. Retention data from affinity high-performance liquid chromatography in view of chemometrics. J. Chromatogr. B Biomed. Sci. Appl. *715*:229-244.

48. **Kannan, K., P. Lalitha, K.V. Rao, R.B. Narayanan and P. Kaliraj.** 1997. Optimisation of immunoaffinity purification of Wuchereria bancrofti specific antibodies from human sera. Indian J. Exp. Biol. *35*:1076-1079.

49. **Katoh, S., M. Terashima and K. Miyaoku.** 1997. Purification of alpha-amylase by specific elution from anti-peptide antibodies. Appl. Microbiol. Biotechnol. *475*:521-524.

50. **Katz, S.E. and M. Siewierski.** 1992. Drug residue analysis using immunoaffinity chromatography. J. Chromatogr. *624*:403-409.

51. **Kehr, J.** 1998. Determination of glutamate and aspartate in microdialysis samples by reversed-phase column liquid chromatography with fluorescence and electrochemical detection. J. Chromatogr. B Biomed. Sci. Appl. *708*:27-38.

52. **Kim, W.B., B.M. Kim, E.C. Choi and B.K. Kim.** 1990. Immunoaffinity purification of hepatitis B surface antigen using anti-idiotype as a specific eluent. J. Immunol. Methods *126*:119-124.

53. **Kissinger, P.T.** 1996. Electrochemical detection in bioanalysis. J. Pharm. Biomed. Anal. *14*:871-880.

54. **Kobayashi, N., H. Mano, T. Imazu and K. Shimada.** 1995. Tandem immunoaffinity chromatography for plasma 1 alpha,25-dihydroxyvitamin D3 utilizing two antibodies having different specificities: a novel and powerful pretreatment tool for 1 alpha,25-dihydroxyvitamin D3 radioreceptor assays. J. Steroid Biochem. Mol. Biol. *54*:217-226.

55. **Kussak, A., B. Andersson and K. Andersson.** 1995. Immunoaffinity column clean-up for the high-performance liquid chromatographic determination of aflatoxins B1, B2, G1, G2, M1 and Q1 in urine. J. Chromatogr. B Biomed. Appl. *672*:253-259.

56. **Lakhiari, H., E. Legendre, D. Muller and J. Jozefonvicz.** 1995. High-performance affinity chromatography of insulin on coated silica grafted with sialic acid. J. Chromatogr. B Biomed. Appl. *664*:163-173.

57. **Lambert, W.E., J.F. Van Bocxlaer and A.P. De Leenheer.** 1997. Potential of high-performance liquid chromatography with photodiode array detection in forensic toxicology. J. Chromatogr. B Biomed. Sci. Appl. *689*:45-53.

58. **Lillehoj, E.P. and V.S. Malik.** 1989. Protein purification. Adv. Biochem. Eng. Biotechnol. *40*:19-71.

59. **Lin, Q., T. Su, G. Liu and J. Gu.** 1996. Purification of transthyretin by high performance affinity chromatography from human plasma. Prep. Biochem. Biotechnol. *26*:245-257.

60. **Lindgren, A., J. Emneus, G. Marko-Varga, H. Irth, A. Oosterkamp and S. Eremin.** 1998. Optimisation of a heterogeneous non-competitive flow immunoassay comparing fluorescein, peroxidase and alkaline phosphatase as labels. J. Immunol. Methods *211*:33-42.

61. **Liu, C.L. and L.D. Bowers.** 1996. Immunoaffinity trapping of urinary human chorionic gonadotropin and its high-performance liquid chromatographic-mass spectrometric confirmation. J. Chromatogr. B Biomed. Appl. *687*:213-220.

62. **Liu, J., N.W. Shworak, L.M.S. Fritze, J.M. Edelberg and R.D. Rosenberg.** 1996. Purification of heparan sulfate D-glucosaminyl 3-O-sulfotransferase. J. Biol. Chem. *271*:27072-27082.

63. **Logeat, F., R. Pamphile, H. Loosfelt, A. Jolivet, A. Fournier and E. Milgrom.** 1985. One-step immunoaffinity purification of active progesterone receptor. Further evidence in favor of the existence of a single steroid binding subunit. Biochemistry *24*:1029-1035.

64. **Lorenz, B., S. Marme, W.E. Muller, K. Unger and H.C. Schroder.** 1994. Preparation and use of polyphosphate-modified zirconia for purification of nucleic acids and proteins. Anal. Biochem. *216*:118-126.

65. **MacTaylor, C.E. and A.G. Ewing.** 1997. Critical review of recent developments in fluorescence detec-

tion for capillary electrophoresis. Electrophoresis *18*:2279-2290.

66. **Matejtschuk, P. (Ed.)** 1997. Affinity Separations. A Practical Approach. IRL Press, Oxford.

67. **Mazza, G., W.P. Duffus, C.J. Elson, C.R. Stokes, A.D. Wilson and A.H. Whiting.** 1993. The separation and identification by monoclonal antibodies of dog IgG fractions. J. Immunol. Methods *161*:193-203.

68. **Mislovicova, D., M. Chudinova, P. Gemeiner and P. Docolomansky.** 1995. Affinity chromatography of invertase on concanavalin A-bead cellulose matrix: the case of an extraordinary strong binding glycoenzyme. J. Chromatogr. B Biomed. Sci. Appl. *664*:145-153.

69. **Mohr, P. and K. Pommerening.** 1985. Affinity Chromatography. Practical and Theoretical Aspects. Chromatogr. Sci. Series, Vol 33. Marcel Dekker, New York.

70. **Moreira, R.A., C.C. Castelo-Branco, A.C. Monteiro, R.O. Tavares and L.M. Beltramini.** 1998. Isolation and partial characterization of a lectin from Artocarpus incisa L. seeds. Phytochemistry *47*:1183-1188.

71. **Munk, M.** Refractive index detection, p. 5–38. *In* D. Parriott (Ed.), A Practical Guide to HPLC Detection. Academic Press, New York.

72. **Murphy, E.J., T.A. Rosenberger and L.A. Horrocks.** 1996. Separation of neutral lipids by high-performance liquid chromatography: quantification by ultraviolet, light scattering and fluorescence detection. J. Chromatogr. B Biomed. Appl. *685*:9-14.

73. **Nadler, T.K., S.K. Paliwal and F.E. Regnier.** 1994. Rapid, automated, two-dimensional high-performance liquid chromatographic analysis of immunoglobulin G and its multimers. J. Chromatogr. *676*:331-335.

74. **Nakagawa, K. and T. Miyazawa.** 1997. Chemiluminescence-high-performance liquid chromatographic determination of tea catechin, (-)-epigallocatechin 3-gallate, at picomole levels in rat and human plasma. Anal. Biochem. *248*:41-49.

75. **Nice, E.C. and R.J. Simpson.** 1989. Micropreparative high-performance liquid chromatography of proteins and peptides. J. Pharm. Biomed. Anal. *7*:1039-1053.

76. **Ohlson, S., M. Bergstrom, L. Leickt and D. Zopf.** 1998. Weak affinity chromatography of small saccharides with immobilized wheat germ agglutinin and its application to monitoring of carbohydrate transferase activity. Bioseparation *7*:101-105.

77. **Ohlson, S., M. Bergstrom, P. Pahlsson and A. Lundblad.** 1997. Use of monoclonal antibodies for weak affinity chromatography. J. Chromatogr. A *758*:199-208.

78. **Okutani, R., Y. Itoh, H. Hirata, T. Kasahara, N. Mukaida and T. Kawai.** 1992. Simple and high-yield purification of urine protein 1 using immunoaffinity chromatography: evidence for the identity of urine protein 1 and human Clara cell 10-kilodalton protein. J. Chromatogr. *577*:25-35.

79. **Oliver, R.W.A. and B.W. King.** 1998. Microbore and packed capillary HPLC, p. 1–14. *In* R.W.A. Oliver (Ed.), HPLC of Macromolecules: A Practical Approach. IRL Press, Oxford.

80. **Parekh, B.S., P.W. Schwimmbeck and M.J. Buchmeier.** 1989. High efficiency immunoaffinity purification of anti-peptide antibodies on thiopropyl sepharose immunoadsorbants. Pept. Res. *2*:249-252.

81. **Phillips, T.M.** 1989. The isolation and recovery of biologically active proteins by high-performance immunoaffinity chromatography, p. 129-154. *In* A. Kerlavage (Ed.), The Use of HPLC in Receptor Purification and Characterization. Alan R. Liss, New York.

82. **Phillips, T.M.** 1992. Analytical Techniques in Immunochemistry. Marcel Dekker, New York.

83. **Phillips, T.M.** 1991. High-performance immunoaffinity chromatography: theory and practical aspects, p. 507–515. *In* C. Mant and R.S. Hodges (Eds.), HPLC of Peptides and Proteins: Separation, Analysis and Conformation. CRC Press, Boca Raton.

84. **Phillips, T.M.** 1991. Isolation and measurement of membrane proteins high-performance immunoaffinity chromatography, p. 517-525. *In* C. Mant and R.S. Hodges (Eds.), HPLC of Peptides and Proteins: Separation, Analysis and Conformation. CRC Press, Boca Raton.

85. **Phillips, T.M.** 1991. Isolation of an interleukin 2-binding receptor from activated lymphocytes by high-performance immunoaffinity chromatography. J. Chromatogr. *550*:742-749.

86. **Phillips, T.M.** 1992. Measurement of recombinant interferon levels by high-performance immunoaffinity chromatography in body fluids of cancer patients on interferon therapy. Biomed. Chromatogr. *6*:287-290.

87. **Phillips, T.M.** 1997. Measurement of total and bioactive interleukin-2 in tissue samples by immunoaffinity-receptor affinity chromatography. Biomed. Chromatogr. *11*:200-204.

88. **Phillips, T.M. and B.F. Dickens.** 1998. Analysis of recombinant cytokines in human body fluids by immunoaffinity capillary electrophoresis. Electrophoresis *19*:2991-2996.

89. **Phillips, T.M. and S.C. Frantz.** 1988. Isolation of specific lymphocyte receptors by high-performance immunoaffinity chromatography. J. Chromatogr. *444*:13-20.

90. **Phillips, T.M. and M.G. Lewis.** 1971. A system for the elution of immunoglobulins from the surface of living cells. Rev. Europ. Etud. Clin. Biol. *16*:1052-1055.

91. **Poppe, H.** 1992. Column liquid chromatography, p. A151–A225. *In* E. Heftmann (Ed.), Chromatography. J. Chromatography Library, Vol. 51A. Elsevier, Amsterdam.

92. **Proost, P., A. Wuyts, J.P. Lenaerts and J. Van Damme.** 1994. Purification, sequence analysis, and bio-

logical characterization of a second bovine monocyte chemotactic protein-1 (Bo MCP-1B). Biochemistry *33*:13406-13412.

93. **Przybycien, T.M.** 1998. Protein-protein interactions as a means of purification. Curr. Opin. Biotechnol. *9*:164-170.

94. **Renauer, D., F. Oesch, J. Kinkel, K.K. Unger and R.J. Wieser.** 1985. Fractionation of membrane proteins on immobilized lectins by high-performance liquid affinity chromatography. Anal. Biochem. *151*:424-427.

95. **Rossi, T.M.** 1990. Laser-based detectors in chromatographic analysis. J. Pharm. Biomed. Anal. *8*:469-476.

96. **Ruhn, P.F., J.D. Taylor and D.S. Hage.** 1994. Determination of urinary albumin using high-performance immunoaffinity chromatography and flow injection analysis. Anal. Chem. *66*:4265-4271.

97. **Rule, G.S. and J.D. Henion.** 1992. Determination of drugs from urine by on-line immunoaffinity chromatography-high-performance liquid chromatography-mass spectrometry. J. Chromatogr. *582*:103-112.

98. **Santucci, A., M. Rustici, L. Bracci, L. Lozzi, P. Soldani and P. Neri.** 1990. HPLC immunoaffinity purification of rabies virus glycoprotein using immobilized antipeptide antibodies. J. Immunol. Methods *127*:131-138.

99. **Santucci, A., P. Soldani, L. Lozzi, M. Rustici, L. Bracci, S. Petreni and P. Neri.** 1988. High performance liquid chromatography immunoaffinity purification of antibodies and antibody fragments. J. Immunol. Methods *114*:181-185.

100. **Scudamore, K.A. and S.J. MacDonald.** 1998. A collaborative study of an HPLC method for determination of ochratoxin A in wheat using immunoaffinity column clean-up. Food Addit. Contam. *15*:401-410.

101. **Shaw, C.** 1994. Peptide purification by reverse-phase HPLC. Methods Mol. Biol. *32*:275-287.

102. **Sheoran, A.S. and M.A. Holmes.** 1996. Separation of equine IgG subclasses (IgGa, IgGb and IgG(T)) using their differential binding characteristics for staphylococcal protein A and streptococcal protein G. Vet. Immunol. Immunopathol. *55*:33-43.

103. **Sherry, J.** 1997. Environmental immunoassays and other bioanalytical methods: overview and update. Chemosphere *34*:1011-1025.

104. **Singh, A.K., Y. Jang, U. Mishra and K. Granley.** 1991. Simultaneous analysis of flunixin, naproxen, ethacrynic acid, indomethacin, phenylbutazone, mefenamic acid and thiosalicylic acid in plasma and urine by high-performance liquid chromatography and gas chromatography-mass spectrometry. J. Chromatogr. *568*:351-361.

105. **Sluzky, V., Z. Shahrokh, P. Stratton, G. Eberlein and Y.J. Wang.** 1994. Chromatographic methods for quantitative analysis of native, denatured, and aggregated basic fibroblast growth factor in solution formulations. Pharm. Res. *11*:485-490.

106. **Stahl, D., H. Kreft, H. Hack, B. Schraven and D. Roelcke.** 1998. Serum affinity chromatography for the detection of IgG alloantibodies in a patient with high-titer IgM cold agglutinins. Vox Sang. *74*:253-255.

107. **Staller, T.D. and M.J. Sepaniak.** 1997. Chemiluminescence detection in capillary electrophoresis. Electrophoresis *18*:2291-2296.

108. **Strong, R.A., B.-Y. Cho, D.H. Fisher, J. Nappier and I.S. Krull.** 1996. Immunodetection approaches and high-performance immunoaffinity chromatography for an analog of bovine growth hormone releasing factor at trace levels. Biomed. Chromatogr. *10*:337 345.

109. **Stubbings, G., J.A. Tarbin and G. Shearer.** 1996. On-line metal chelate affinity chromatography clean-up for the high-performance liquid chromatographic determination of tetracycline antibiotics in animal tissues. J. Chromatogr. B Biomed. Appl. *679*:137-145.

110. **Suda, M., T. Sakuhara and I. Karube.** 1993. Miniaturized detectors for a chemical analysis system. Appl. Biochem. Biotechnol. *41*:3-10.

111. **Trost, P., P. Bonora, S. Scagliarini and P. Pupillo.** 1995. Purification and properties of NAD(P)H: (quinone-acceptor) oxidoreductase of sugarbeet cells. Eur. J. Biochem. *234*:452-458.

112. **Turkova, J.** 1999. Oriented immobilization of biologically active proteins as a tool for revealing protein interactions and function. J. Chromatogr. B Biomed. Sci. Appl. *722*:11-31.

113. **Ubrich, N., P. Herbert, V. Regault, E. Dellacherie and C. Rivat.** 1992. Compared stability of Sepharose-based immunoadsorbents prepared by various activation methods. J. Chromatogr. *584*:17-22.

114. **Ueno, Y., S. Maki, J. Lin, M. Furuya, Y. Sugiura and O. Kawamura.** 1998. A 4-year study of plasma ochratoxin A in a selected population in Tokyo by immunoassay and immunoaffinity column-linked HPLC. Food Chem. Toxicol. *36*:445-449.

115. **Usha, R. and M. Singh.** 1999. Purification of a multicatalytic protease complex from developing winged bean seeds by indirect immunoaffinity chromatography. Protein Expr. Purif. *15*:48-56.

116. **van der Heeft, E., G.J. ten Hove, C.A. Herberts, H.D. Meiring, C.A. van Els and A.P. de Jong.** 1998. A microcapillary column switching HPLC-electrospray ionization MS system for the direct identification of peptides presented by major histocompatibility complex class I molecules. Anal. Chem. *70*:3742-3751.

117. **Van Emon, J.M., C.L. Gerlach and K. Bowman.** 1998. Bioseparation and bioanalytical techniques in environmental monitoring. J. Chromatogr. B Biomed. Sci. Appl. *715*:211-228.

118. **van Ginkel, L.A.** 1991. Immunoaffinity chromatography, its applicability and limitations in multi-residue

analysis of anabolizing and doping agents. J. Chromatogr. *564*:363-384.

119. **van Ginkel, L.A., R.W. Stephany, H.J. van Rossum, H.M. Steinbuch, G. Zomer, E. Van de Heeft and A.P. De Jong.** 1989. Multi-immunoaffinity chromatography: a simple and highly selective clean-up method for multi-anabolic residue analysis of meat. J. Chromatogr. *489*:111-120.

120. **van Sommeren, A.P., P.A. Machielsen, W.J. Schielen, H.P. Bloemers and T.C. Gribnau.** 1997. Purification of rubella virus E1-E2 protein complexes by immunoaffinity chromatography. J. Virol. Methods *63*:37-46.

121. **Vockley, J. and H. Harris.** 1984. Purification of human adult and foetal intestinal alkaline phosphatases by monoclonal antibody immunoaffinity chromatography. Biochem. J. *217*:535-541.

122. **Walters, D.L., J.E. James, F.B. Vest and H.T. Karnes.** 1994. A comparison of fluorescence versus chemiluminescence detection for analysis of the fluorescamine derivative of histamine by HPLC. Biomed. Chromatogr. *8*:207-211.

123. **Wheatley, J.B., M.K. Kelley, J.A. Montali, C.O. Berry and D.E. Schmidt.** 1994. Examination of glutathione S-transferase isoenzyme profiles in human liver using high-performance affinity chromatography. J. Chromatogr. *663*:53-63.

124. **Wieczorek, E., J.M. Parkitna, J. Szkudlarek, A. Ozyhar and M. Kochman.** 1996. Immunoaffinity purification of juvenile hormone-binding protein from Galleria mellonella hemolymph. Acta Biochim. Pol. *43*:603-610.

125. **Williams, S.C. and R.B. Sim.** 1993. Dye-ligand affinity purification of human complement factor B and beta 2 glycoprotein I. J. Immunol. Methods *157*:25-30.

126. **Wisniewski, R.** 1992. Principles of the design and operational considerations of large scale high performance liquid chromatography (HPLC) systems for proteins and peptides purification. Bioseparation *3*:77-143.

127. **Wu, D. and R.R. Walters.** 1988. Protein immobilization on silica supports: a ligand density study. J. Chromatogr. *458*:169-174.

128. **Xue, H., H. Zheng and X.F. Wu.** 1998. Preparation of highly purified hemoglobin by affinity elution. Artif. Cells Blood Substit. Immobil. Biotechnol. *26*:317-327.

129. **Yin, B., R.M. Whyatt, F.P. Perera, M.C. Randall, T.B. Cooper and R.M. Santella.** 1995. Determination of 8-hydroxydeoxyguanosine by an immunoaffinity chromatography-monoclonal antibody-based ELISA. Free Radic. Biol. Med. *18*:1023-1032.

130. **Yu, W., J.W. Dorner and F.S. Chu.** 1998. Immunoaffinity column as cleanup tool for a direct competitive enzyme-linked immunosorbent assay of cyclopiazonic acid in corn, peanuts, and mixed feed. J. AOAC Int. *81*:1169-1175.

131. **Zhang, J., H. Zhou, Z. Ji and F.E. Regnier.** 1998. Monoclonal antibody production with on-line harvesting and process monitoring. J. Chromatogr. B Biomed. Sci. Appl. *707*:257-265.

132. **Zhivkova, Z.D. and V.N. Russeva.** 1998. Stereoselective binding of ketoprofen enantiomers to human serum albumin studied by high-performance liquid affinity chromatography. J. Chromatogr. B Biomed. Sci. Appl. *714*:277-283.

133. **Zhou, D., H. Zou, J. Ni, L. Yang, L. Jia, Q. Zhang and Y. Zhang.** 1999. Membrane supports as the stationary phase in high-performance immunoaffinity chromatography. Anal. Chem. *71*:115-118.

134. **Zhuang, C., T. Murata, T. Usui, H. Kawagishi and K. Kobayashi.** 1996. Purification and characterization of a lectin from the toxic mushroom Amanita pantherina. Biochim. Biophys. Acta *1291*:40-44.

8 Affinity Cell Separations

Affinity cell separation techniques are used to quickly and efficiently isolate specific mammalian and other cell types from heterogeneous cellular suspensions, based on ligand-specific binding involving cell surface molecules. These affinity techniques are variations on the basic affinity methods used in previous chapters for molecular separations, except that one of the participants in the affinity binding interaction is present on the cell's surface. The interaction involves either a cell surface ligand and its immobilized complementary binding molecule, or specific ligand-targeting molecules on the cell surface and immobilized ligands. Affinity cell isolation will be presented based on the way the cell surface molecular participants in the affinity interaction are bound during cell separation. In *Cell Panning Techniques* (in this chapter), the affinity interaction involves cell adherence to a flat surface (e.g., a culture dish). *Rosetting Techniques* involve cell surface affinity binding to free floating microparticles. In *Cell Chromatography Techniques*, the affinity interaction occurs between cell surface molecules and complementary agents immobilized on a support matrix within a chromatographic column. Finally, in *Cell Affinity Partitioning Techniques*, the affinity binding at the cell surface occurs within a biphasic aqueous liquid suspension such that the partitioning properties of the cell are altered.

There are an almost endless number of uses for cells isolated by affinity techniques. Cell biologists use them to establish pure cell culture lines as a research tool. Clinicians use affinity cell separation techniques as a diagnostic tool for cancer or bacterial diseases by quickly screening patient specimens for the presence of low-abundance cancer cells or bacteria. Affinity cell isolation has become increasingly important in therapeutic treatments, an example being the isolation of stem cells from a transplant patient for ex vivo expansion and autologous transplantation.

The uses for affinity cell separations are based on the remarkable variability in cell surface components that can be utilized in affinity separations. Virtually any cell surface macromolecule is a possible candidate for use as a ligand in affinity cell separation techniques, including surface proteins, glycoproteins, carbohydrates, and other constituents. In addition, cell surface molecules used for natural cell binding, such as surface antibodies on immune cells, lectins, and receptor proteins, can be used for affinity binding to immobilized ligands. Since cellular functions associated with specific cell types are directly related to the expression of unique surface macromolecules, affinity cell separations based on biological functions are possible by targeting these molecules. For example, cells expressing the adhesive surface molecule LFA-1 function through their adherence to cells expressing the LFA-1–specific ligand,

Affinity and Immunoaffinity Purification Techniques
Terry M. Phillips and Benjamin F. Dickens
© 2000 Eaton Publishing, Natick, MA

ICAM-1. Researchers interested in studying cell-cell adhesion involving the LFA-1–ICAM-1 interaction can use affinity techniques to isolate cells expressing either of these molecules. This chapter provides a general description of the wide variety of affinity cell separation techniques that specifically target such cell surface components for cell selection and separation.

AFFINITY CELL SEPARATION VS. CELL CLONING

There has been a steady increase in the biomedical use of cell lines prepared by affinity separation techniques. Many reasons exist for this increase, including the needs of: *(i)* immunologists, to better understand the cells of the immune system and to develop monoclonal antibodies; *(ii)* geneticists, to study cell-cell fusion and various aspects of molecular biology; *(iii)* virologists, to develop antiviral vaccines; *(iv)* oncologists, to study the molecular mechanism of cell transformation; *(v)* microbiologists, to isolate and study new strains of bacteria; and *(vi)* for all biomedical researchers, to prevent the unnecessary use of experimental research animals.

Affinity cell separation techniques are the tools of choice for the isolation of cells to establish pure cell lines. However, before turning to affinity cell separations, it is useful to determine if there are alternative sources of the cell types to be studied. One common source of established, well-characterized cell lines is the national cell banks including ATCC, BioWhittaker, or Cell Systems. Jay (49) recently reviewed the commercially available sources for a wide variety of cell lines. Researchers frequently find that one or more of these cell banks offer established cell lines that meet their research needs. Clearly however, cell banks cannot offer all the possible cell lines that researchers would want to grow in culture, nor are such banks useful when freshly isolated primary cell lines are required. In these cases, cells will have to be isolated using appropriate techniques that provide the desired level of isolated cell purity. When a high degree of purity is required, the application of affinity separation techniques will provide the desired specificity.

Perhaps the ideal method to establish pure cell lines is by cell cloning from a single, original parental cell of the desired lineage. This simple sounding approach, however, has many technical barriers that greatly reduce its applicability. First, the techniques to obtain the single cell for cloning are often labor-intensive, and require expensive and complicated equipment. For instance, laser-assisted devices exist to aid in the recovery of a single cell, but these are very expensive and require significant expertise to operate. An alternative method that overcomes the high cost of equipment is the use of limiting dilution techniques. The drawback to this approach is that if the desired cell is a minor component in the original mixed-cell sample, finding a desired clone following dilution can be extremely labor-intensive and can be compared to looking for "a needle in a haystack." A recent publication described a simplified device, combining microdissection and aspiration to rapidly obtain single cells from clinical blood smears (3), that solves many of the problems of cost and effort associated with isolating a single cell. However, even when using one of these techniques to obtain a single mammalian cell, technical problems may still prevent its cloning to establish a cell line. Cloning from a single mammalian cell frequently fails because these cells typically survive for a very limited number of cell passages and have poor cloning efficiencies.

When single-cell cloning is not feasible, affinity cell separation techniques provide an easy, inexpensive alternative for the isolation of sufficiently purified cell populations. The initial isolation and in vitro cultivating of cells results in the establishment of a primary culture. Primary cell cultures were originally obtained by tissue explant techniques, in which pieces of tissue were placed on a culture dish and cells were allowed to migrate from the tissue to the surface. This was soon followed by procedures in which the tissue was first disrupted by enzymatic or mechanical means, and the resulting cell suspension was plated onto the surface of a dish. Primary cultures prepared from both explant techniques and simple cell plating from disrupted tissues, while still useful (78,91,99), resulted in mixed-cell populations. Further improvements in cell separation techniques were designed to improve the recovery of a population of cells composed of a single cell type (such as endothelial cells or lymphocytes) that is essentially free of other cells. Today, even such "pure" cell populations are divided into subpopulations based on specific cell surface markers associated with their differential cellular functions. An example is T lymphocytes, which once isolated, are routinely subjected to affinity cell separation techniques to further fractionate them into CD4+ and CD8+ subsets, and these populations can be even further subdivided. Cell immunotyping allows the fractionation of "pure" cell populations into subpopulations based on surface structural differences. Immunotyping cells has proven to be useful not only in the understanding of basic cell functioning, but also in both the diagnosis and prognosis of various types of cancer (82).

One of the main uses of the affinity cell separation techniques described in the following sections is for the isolation of a pure population of cells from a heterogeneous cell population. While the procedures described are useful in the establishment of cell lines, important topics such as how to harvest and dissociate tissue, optimal cell culture conditions, cell viability assays, and aseptic culture protocols are beyond the scope of this chapter. Readers unfamiliar with such basic topics are encouraged to refer to general texts on these subjects (26,44,116).

CELL PANNING TECHNIQUES

The various cell types comprising a heterogeneous cell mixture will often adhere to an unmodified glass, plastic, or nylon surface at differential rates. Researchers have exploited such differential adhesion to "naked" (or nonligand-coated) surfaces as a crude method to selectively isolate certain cell types. While we refer to such surfaces as being naked or uncoated and to the cells as directly attaching to the plastic or glass, the surfaces are often first coated with nonspecific proteins, most commonly by incubation with serum. Because such differential attachment represents the relative affinity of the various cell types to bind the surface, this is the simplest affinity cell separation technique. One such approach is cell panning, which is the process by which cells can be selected based on their differential rate of adherence to a flat surface, typically that of a culture flask or dish. Cells with the highest binding affinity for the unmodified surface will rapidly adhere to it, while cells with lower binding affinity will take longer. Panning got its name because of its similarity to the process of panning for gold.

Positive and Negative Cell Panning

To pan for cells, the cell suspension is allowed to remain in contact with the glass or plastic for a short period of time, followed by removal of the liquid containing the unbound cells. This results in two cell populations, one bound to the surface of the plate that is enriched in cells with the highest affinity for the support used, and a second population of cells remaining in the recovered fluid, which is enriched in cell types with lower affinity for that surface. Alternatively, the attached cells can be viewed as being partially depleted of cells with low affinity for the plate surface, while the unbound cells can be viewed as being partially depleted of cells with high affinity for the plate surface.

The two cell populations obtained from the simple panning example in Figure 1 illustrate an important concept in affinity cell isolation; positive vs. negative cell selection. In positive selection, cells are isolated through their sequestration by direct binding to surface ligands. In this example, cells that adhered to the surface of the dish have been positively selected. Negative selection removes unwanted cells through affinity binding on the surface of the unwanted cells, thus depleting them and enriching the solution for cells whose surface has not been involved in any affinity binding interactions. In Figure 1, unbound cells in the retrieved fluid from the dish have been negatively selected by removing a subpopulation of cells that bound to the surface. Negative cell selection is often critically important to experimental design because the binding of cell surface macromolecules frequently causes undesired changes in the properties of the cell being isolated. However, while negative affinity selection is often necessary to avoid "activating" the cell being isolated for immediate use, it is just as useful to use positive selection when establishing cell culture lines.

Naked cell panning is a binary fractionation of heterogeneous cell populations into rapidly adhering and nonrapidly adhering subpopulations. Macrophages rapidly and tenaciously adhere to plastic or glass surfaces and can be positively selected, using panning techniques. Mosier (81) described a positive selection panning technique to obtain a macrophage-enriched population from two murine sources, peritoneum and spleen. While the isolation of a cell population by panning can result in fairly remarkable cell enrichments, insufficient specificity exists to obtain pure cultures. For instance, Mosier (81) reported that approximately 15% of the cells in the macrophage-enriched cell population were contaminating T lymphocytes. To further enrich the macrophage content, he recommended the use of anti-Thy–1.2 antibodies and complement to selectively lyse the contaminating T lymphocytes. Antibodies and complement (117) or other antibody-directed cytotoxic methods [such as anti-Thy–1 antibody-ricin conjugate (92)] are used to selectively target and kill contaminating cells, and as such, represent an additional negative immunoaffinity cell selection process. The protocol for the cell panning of macrophages with uncoated plastic petri dishes (below) describes panning for the positive selection of macrophages.

Cell panning is also useful for negative cell selection, especially for the isolation of immunocompetent cells. For example, while the macrophage cell panning protocol with uncoated plastic petri dishes was described as a positive selection method to isolate macrophages, it can also be used to deplete macrophages from mixed-cell populations. While negative selection is often required to avoid activation of specific cell types, it is also useful as a simple, quick, and inexpensive means of isolating cells that just happen to adhere slowly to surfaces. The following two examples exploited differential adhesion rates to remove major contaminants that adhere more rapidly to the surface of uncoated plates than the desired cell types. Suzumura et al. (128) took advantage of the low affinity of oligodendrocytes present in mouse brain cellular extracts to negatively select for these cells from astrocytes, fibroblasts, and microglia, all of which more readily bind to an uncoated culture disk. After allowing the cell suspension to react with the surface of the plastic, the oligodendrocyte-enriched non-adherent cell fraction was removed by pipeting and plated on poly-L-lysine–coated coverslips. Another example of negative selection panning using uncoated dishes was the isolation of endothelial cells from cell suspensions prepared from neonatal mice hearts (67). In this study, the slowly attaching neonatal heart vascular endothelial cells were separated from fibroblast and smooth muscle cells, two more rapidly attaching contaminates. The heterogeneous cell mixture was initially incubated in a culture flask for 1 h at 37°C. The non-adherent cells, primarily composed of endothelial cells, were then transferred to a second culture flask that had been precoated with 1% gelatin and incubated overnight at 37°C. This resulted in a highly endothelial cell-enriched population being bound to the gelatin-coated flask.

Cell Panning Using Nonligand-Coated Surfaces

Frequently, the cells in a heterogeneous population will lack sufficient differential adherence rates to naked surfaces to allow cell panning on uncoated plates. However, the rate of nonspecific cell adhesion to culture dishes may be modified by carefully choosing the source of protein used to coat the plates. For example, cell types with high surface expression of β_1 integrins will rapidly adhere to nonspecific extracellular matrix proteins such as fibronectin and type IV collagen. Therefore, coating plastic surfaces with such nonspecific extracellular proteins has been useful in increasing the specificity of panning for a variety of cells, including human (52) and murine (5) stem cells and rat lung endothelial cells (58).

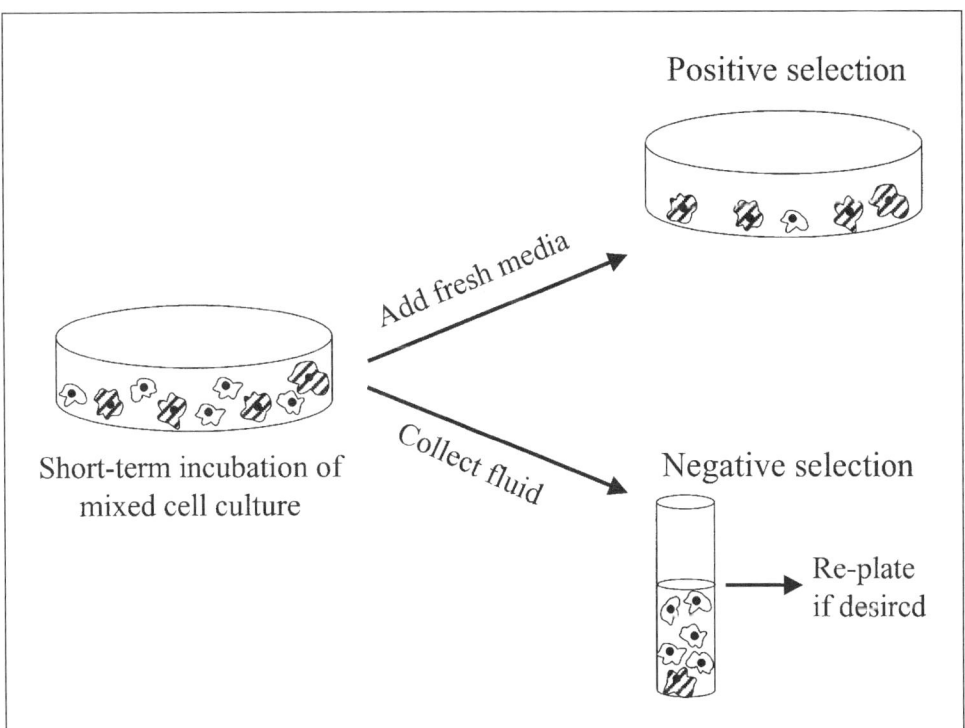

Figure 1. The basic principles of positive and negative selection by cell panning.

Affinity Ligand-Assisted Cell Panning

While differential binding to a naked plate, even when coated with nonspecific extracellular matrix proteins, can give rise to an enriched cell population, more specific cell selection techniques are often needed. Panning can be improved by coating the surface with molecules that convey affinity specificity for the desired cell type. Such molecules include lectins, protein A, receptor molecules, and cell-specific monoclonal antibodies. The adherence of cells to plates coated with molecules used in affinity binding (e.g., specific antibodies) occurs much more rapidly than the adherence to naked plates. Thus, typically short incubation periods are used with affinity molecule-specific–coated plates to limit nonspecific cell adhesion to the surface.

One selective modification of cell panning exploits specific cell-cell interactions in which cells are isolated based on their affinity for specific target cells. In this case, the panning surface is a monolayer either grown on top of the dish surface or chemically attached to the surface. Mage (68) provided a review of panning for a variety of cell types with attached cells, either grown in monolayers (fibroblasts and macrophages) or adhered in a monolayer using poly-L-lysine (for red blood cells [RBCs] and spleen cells), anti-immunoglobulins (for B cells), or lectins (for lymphocyte monolayers). When using negative selection, cell-cell adhesion merely removes contaminant cells that bind the specific target cells, and the recovered unbound cells can be processed as if a naked surface was used. However, if positive selection was used, the elution method must be carefully selected to limit contamination of the selected cells by the spontaneous release of the monolayer cells from the plate.

Lectins are multivalent sugar-binding proteins or glycoproteins that combine noncovalently with monosaccharides or oligosaccharides. Cell separation exploits lectin binding of specific carbohydrate moieties (112,143) located on cellular surfaces. While lectins lack the very high specificity of monoclonal antibodies, they have several distinct advantages over antibodies for use in affinity cell isolations. First, they are typically less expensive. Second, unlike antibodies that bind cells so strongly that it is difficult to recover the bound cell, cells bound to lectins can be gently recovered by simply adding the free competing sugar in excess. Finally, lectins can be used to select and isolate a cell population that lacks specific antibodies against it. Lectins have been used to isolate duck lymphocytes (45), chick embryonic muscle cells (126), and mesenchyme cells from sea urchin embryos (24). In addition, cell panning with lectins has been used to isolate immunocompetent cells, often by negative selection. Examples include negative selection of peripheral blood mononuclear cells with the lectin from *Erythrina cristagalli* to obtain a population of cells highly enriched for natural killer cells (38), and panning with wheat germ agglutinin for bovine T lymphocytes (19).

Immunoaffinity Cell Panning

The specificity of the antibody-ligand interaction is particularly well suited for use in all forms of affinity cell separations. Panning with monoclonal antibody-coated plates allows the rapid, positive selection of highly purified cell populations. One example of this selection was the panning of fresh blood mononuclear cells using flasks that were covalently coated with monoclonal antibodies directed against either CD4+ or CD8+ to fractionate T lymphocytes (80). The resulting cell populations were highly purified, positively selected adherent CD4+ and CD8+ cell subsets. Virtually any cell type to which antibodies can be raised may be a candidate for positive selection using panning with antibody-coated plates. Examples include the isolation of endothelial cells using EN4 [an anti-endothelial cell monoclonal antibody (37)], human melanoma cells from surgically removed lesions using a monoclonal antibody specific for a high-molecular-weight membrane-associated antigen (88), and human bone marrow CD34+ hematopoietic cells using monoclonal anti-CD34 antibodies (11).

This last example (11) was performed using an antibody-coated polystyrene plate (the CELLector®) developed by Applied Immune Sciences (AIS) (90). These plates were designed to isolate CD8+, CD4+, or CD34+ cells. AIS has been purchased by Rhône-Poulenc Rorer, and the technology is being prepared for clinical use. While it is unclear whether or not such plates will be available in the future, it is easy to prepare similar antibody-coated plates for cell panning. Like other proteins, antibodies will coat the surface of plastic tissue dishes when incubated for an hour at room temperature or overnight in the refrigerator. The protocol for the cell panning of mouse natural killer T cells using target cell monolayers (below) is a simple procedure for preparing an antibody-coated plate for use in antibody-directed cell panning. As was the case with the attachment of antibodies to other solid surfaces (see Chapter 4), such incubation does not result in the ideal orientation of all the antibody molecules. Frequently, this will not be a problem in panning, since antibodies on the plate and antigens on the surface of the cell are usually present in excess. However, for the antibody to have the highest binding affinity for the cell surface ligand, it is necessary to optimize the antibody orientation when it is bonded to the surface, and it may be necessary to use spacer molecules to avoid steric hindrance as discussed in Chapter 2. Thus, the same techniques can be used to prepare the surface for direct chemical linking of the antibody to the surface of the plate.

Antibody-coated plates can also be used for negative selection. Mage (69) described the use of antibody-coated plates for the binary fractionation of mouse splenic lymphocytes into two populations based on the expression of surface immunoglobulins. The plates were coated with goat anti-mouse antibodies, and spleen lymphocytes were then panned. Cells that expressed immunoglobulins were captured, and the non-immunoglobulin–expressing cells were negatively selected for, by recovering the non-adherent cells. A modification of this panning method (Figure 2) uses anti-immunoglobulin–coated plates and the addition of anti-cell antibodies to the cell mixture. In this modification, "primary antibodies" directed against surface antigens on either the cells to be isolated (positive selection) or depleted (negative selection) are added to the cell suspension. After these antibodies have bound to the cell surface antigens, the cells are panned using plates with "secondary antibodies" directed specifically against the primary antibodies used to coat the cells. Monoclonal antibody MAb H28 (as a primary antibody) directed against the mouse neuronal cell adhesion molecule N-CAM was used for positive selection of murine C2 myogenic cells, and the cells were isolated by panning, using a plate coated with anti-IgG antibodies (83). Tai and Spry (129) described a further modification related to the indirect use of antibody-coated plates in primary cell separation. In their study, petri dishes were coated with human IgG, and then rabbit anti-human IgG was added to bind the human IgG, thereby exposing the Fc portion of the rabbit molecules. Neutrophils and monocytes possess Fc-binding sites, which interacted with the exposed Fc portion of the secondary rabbit antibodies. The affinity of these two cell types for the Fc portion of the bound antibodies allowed the depletion of neutrophils and monocytes in a negative selection procedure for the isolation of eosinophils from human peripheral blood.

Instead of using anti-immunoglobulin–coated plates to recover cells that had been coated in solution with primary antibodies, immunoglobulin-binding protein A or protein G can be used instead. Plates can be coated with protein A directly (87), or with cell monolayers of *S. aureus* or RBCs coated with protein A (31).

Bousso et al. (8) described a panning method for isolating T lymphocytes that exploited the T cell-specific MHC-peptide interaction. In their study, the MHC-peptide complex was immobilized, which allowed the isolation of an enriched population of antigen-specific T lymphocytes from a heterogenous CD8+ population based on adherence to this complex. Despite such recent advances and the relative ease of differential adhesion methodology, most laboratories are turning to techniques based on principles of rosette formation, which are discussed later in this chapter.

Cell Recovery in Affinity Cell Panning

The ease of detaching cells involved in affinity binding to solid supports depends on the nature and strength of the cell adhesion. When cells are attached by lectins, they can be gently released by the addition of high concentrations of the free specific carbohydrate that the lectin binds (112). However, after cells are securely attached to a surface through high affinity interactions, drastic treatment is often required to elute the cells. Such rough handling can have detrimental effects on the isolated cell. Mechanical shearing (by the use of a "rubber policeman" or a cell-scraping device) can be used on cells attached to a culture flask. Even cells captured on affinity columns (see *Cell Chromatography Techniques*, below) are sometimes released by recovery of the support material and mechanical methods used to release the cells. Another common approach for cell recovery is to use hydrolytic enzymes with metal chelators. Such enzymatic digestion causes cell injury, as shown by the rounding up of cells prior to their release and evidence that pieces of the cell membrane remained attached to the immobilization substrate after such procedures (47).

The recovery of viable, uninjured (or minimally injured) cells is even more of a problem when cells are positively selected using antibodies, due to the strong antibody-antigen binding. The fewer antibody attachments to each cell, the more likely the bond can be broken with minimal cell damage. When antigen is present in high concentrations on the cell surface, panning with antibody-coated plates has the potential to form many such antibody-antigen bonds. Therefore when positively selecting cells with a high abundance of surface antigen, one approach is to coat the surface of the plate with an antibody solution diluted with nonspecific antibodies. By using diluted antibodies, the active binding sites can be spaced sufficiently far apart on the plate to allow only a few of the antigen sites on each cell to be bound. On the other hand, when the concentration of surface antigens per cell is low, coating the dish with dilute antibody solution may result in insufficient antibody-ligand bonds to firmly attach the cell to the plate. For such cells, the antibody solution should be concentrated and the orientation of the antibody on the plate optimized to maximize the formation of such bonds. Antibody-bound cells can be released by proteolytic enzymes, high salt treatment (1 M sodium chloride), acidic elution (pH 1.0), or with high concentration of haptens (Reference 109). Because these treatments frequently cause cell injury, methods have been designed that bind the immobilizing antigens or antibodies to solid supports using easily cleavable bonds. One such approach is to link the immobilized molecule to the support using disulfide bridges that can be broken by thiol treatment. Cambier and Neale (9) captured BALB/c lymphocytes that bound

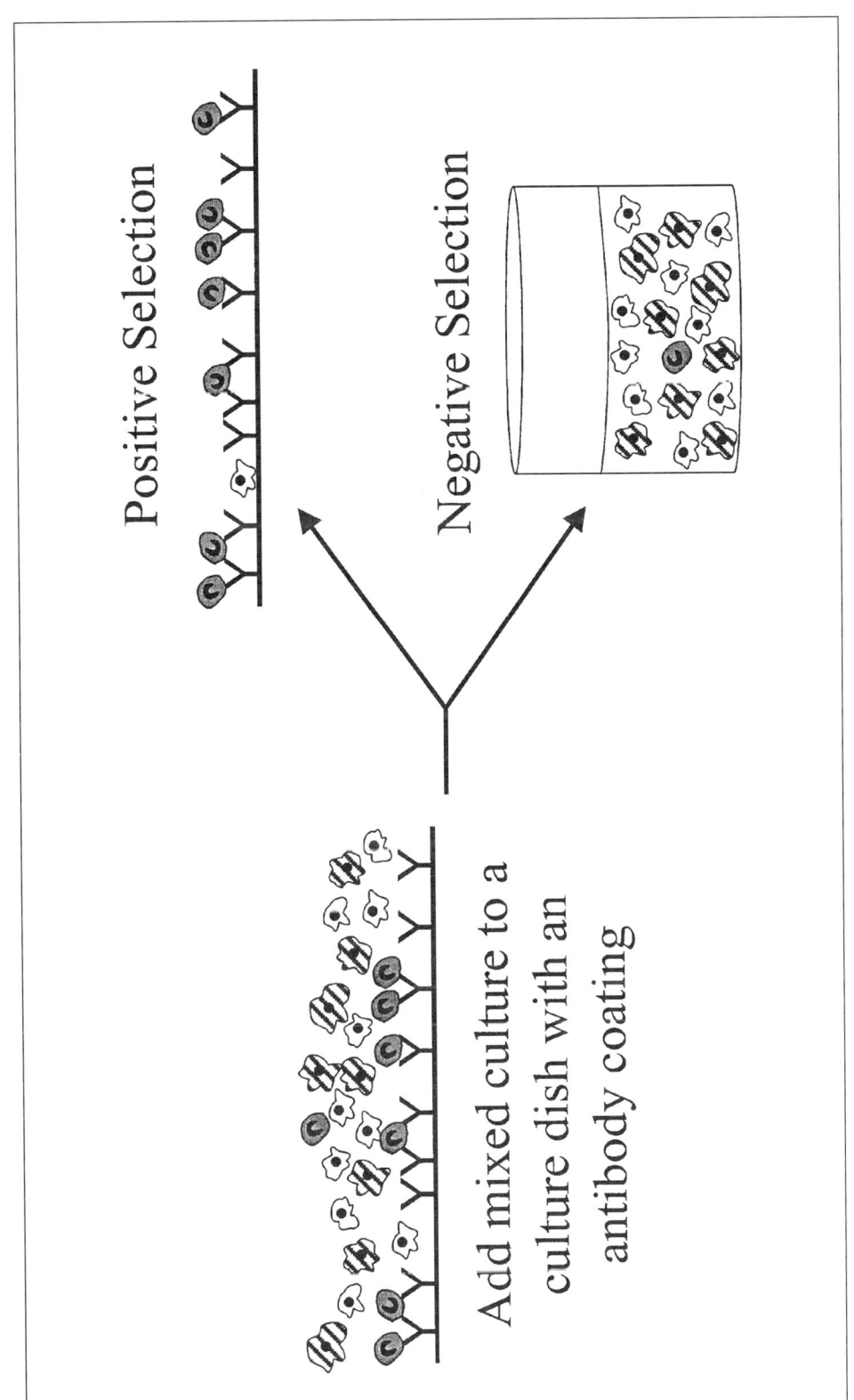

Figure 2. Positive and negative selection using immunoaffinity cell panning.

specific phosphorylcholine by using *N*-succinimidyl 3-(2-pyridyldiothio)propionate to link the phosphorylcholine by an easily cleaved bond to gelatin-coated dishes. Bonnafous et al. (7) used thiolated ligands immobilized on solid matrices to form easily cleavable bonds for affinity cell separations. They linked a lectin, con A, using the same reagent as Cambier and Neale, and anti-dinitrophenyl antibodies using S-acetylmercaptosuccinic anhydride. The lectin was used to capture mouse thymocytes, and the antibody was used to capture modified sheep RBCs targeted by the antibody. Thiols were then used to release the captured cells. Grimsly et al. (35) described a method of linking murine monoclonal antibodies to surfaces in a three-layer immune complex. The cell-capturing antibody was attached to a F(Ab$'$)$_2$ fragment directed against it, and this complex was bound to the immobilized matrix by a specific myeloma protein. In this orientation, the disulfide bond in the hinge region of the F(Ab$'$)$_2$ could be reduced with DTT, resulting in the release of the captured cell. Thomas et al. (131) used a similar method of linking the cell-recognizing antibody to a glass bead using F(Ab$'$)$_2$ fragments that were susceptible to DTT.

Ljungquist et al. (66) described an affinity separation method for positive selection of human lymphocytes and recombinant proteins, in which the specific antibody-antigen bond was very gently broken. In this study, they attached the antibodies to the solid matrix by a DNA link, which could then be broken by treatment with DNase. Using this method, they showed the gentle and efficient recovery of human CD37+ B lymphocytes from peripheral blood. Commercial immunoaffinity separation products are now available that take advantage of easily cleaved bonds to gently recover cells isolated by the use of a monoclonal antibody (see *Rosetting Techniques*, below).

PROTOCOLS

Cell Panning of Macrophages with Uncoated Plastic Petri Dishes

This protocol was abbreviated from the method described by Mosier (81).

1. Prepare sterile macrophage culture media containing 15%–20% (vol/vol) heat-inactivated serum, 50 µg/mL gentamycin, and 10 µg/mL streptomycin.
2. Incubate sterile glass petri dishes or plastic culture flasks with the sterile culture media for 30 min at 37°C in a 5% CO_2–95% air atmosphere.
3. Immediately before plating, suspend the cell mixture in warm (37°C) culture media at 5×10^6 to 3×10^7 cells/mL.
4. Remove the sterile culture media from the culture dishes and immediately add the warmed cell mixture in a minimum volume to just cover the surface.
5. Swirl the dishes to assure the liquid comes in contact with the entire surface.
6. Incubate the dishes at 37°C for 1 h in a 5% CO_2 environment.
7. Remove the dishes from the incubator and rapidly swirl to loosen non-adherent cells.
8. Aspirate the media.

   Note: The aspirated media contains a population of negatively selected cells that are macrophage depleted.
9. Wash the surface of the plate several times with cold, sterile 0.01 M PBS (or 0.01 M Hank's balanced salt solution [HBSS]) to remove non-adherent and

loosely attached cells.

Note: For immediate use, skip to step 12 to harvest the freshly isolated macrophages.

10. Add sterile culture media to the dish and incubate at 37°C in a 5% CO_2–95% air atmosphere to culture the cells.

11. Add warm 0.01 M PBS containing 10 µM EDTA to the dishes and incubate for 15 min at 37°C.

12. Vigorously pipet the PBS-EDTA solution over the surface of the cells. A physical scrapping device, such as a rubber policeman, can be used if necessary to aid in detaching the cells from the surface.

13. Collect the cells by aspirating the PBS-EDTA solution into a sterile centrifuge tube containing an equal volume of the culture media.

14. Repeat the wash of the dish (steps 11–13), if necessary, to recover more cells.

All the procedures should be performed under aseptic conditions. When panning for cells, recovery yield can be increased by increasing the contact time with the surface of the plate, but this also increases contamination by allowing for the attachment of other, slower-attaching cells. During this protocol, the surface of the plate was coated with nonspecific protein (step 2). The surface could have been coated with commercially available matrix proteins instead (as discussed in *Cell Panning Using Nonligand-Coated Surfaces*, above), if desired.

Cell Panning of Mouse Natural Killer T Cells Using Target Cell Monolayers

1. Sensitize a mouse with an intraperitoneal (i.p.) injection of 1×10^7 target tumor cells.

2. After 14 days, anesthetize the mouse and aseptically prepare a single-cell suspension by harvesting peritoneal lymphocytes (steps 3–6) and/or spleen (steps 7–15).

3. To harvest peritoneal lymphocytes, inject 5 mL of heparinized (5 U/mL) 0.01 M PBS into the peritoneal cavity using a syringe with a 22-gauge needle.

4. Vigorously massage the abdomen and withdraw the fluid from the peritoneal cavity using the syringe.

5. Wash the cells twice with isolation media (RPMI culture media supplemented with 50 µg/mL gentamycin, 10 µg/mL streptomycin, 10% [vol/vol] heat-inactivated serum, 10 mM HEPES, 2 mM L-glutamine, and 25 mM sodium bicarbonate).

6. Resuspend the peritoneal lymphocytes in isolation media at 5×10^6 cells/mL and store at 4°C until ready for use.

Note: Skip to step 16 if spleen cells are not to be used.

7. To harvest spleen lymphocytes, place the aseptically excised spleen into a sterile petri dish containing isolation media (see step 5).

8. Using aseptic technique, surgically de-encapsulate the spleen.

9. Carefully filter the spleen cells through a fine nylon mesh with the aid of a sterile spatula, collecting the cell filtrate in a sterile centrifuge tube.

10. Wash the cell filtrate 3 times in the isolation media.

11. Resuspend the cells in sterile 0.83% (vol/vol) Tris-buffered ammonium chloride, pH 7.4, in a centrifuge tube.

12. Gently swirl the cell suspension for 5 min at room temperature to lyse contaminating RBCs.

13. Centrifuge at 100× g for 10 min to separate the cells from the red cell ghost.

14. Wash the cells in isolation media and remove macrophages and monocytes by the protocol for macrophage cell panning with uncoated petri dishes, described above, and collect the lymphocytes by aspirating the macrophage-depleted media.

15. Pellet the lymphocytes (800× g for 10 min), resuspend them in isolation media at 5×10^6 cells/mL, and store at 4°C until ready for use.

16. Add a 0.01 M PBS solution containing 5 µg/mL of poly-L-lysine to a 35-mm polystyrene culture dish.

17. Incubate at room temperature for 1 h.

18. Aspirate the solution, and wash the surface of the culture dish 3 times with excess PBS.

19. Add 2 mL of a 0.01 M PBS solution containing chicken RBCs (5% vol/vol).

20. Incubate for 1 h at room temperature.

21. Aspirate the solution and gently wash with 0.01 M PBS.

22. Add a 2 mL of 0.01 M PBS solution containing 0.025% (vol/vol) glutaraldehyde.

23. Incubate at room temperature for 10 min.

24. Aspirate the solution and wash 3 times with 2 mL of 0.01 M PBS. If desired, the dishes can be stored under PBS at 4°C for a few days.

25. Add 2 mL of a 0.01 M PBS solution containing 20 µg/mL of phytohemagglutinin (PHA).

26. Incubate at 37°C for 1 h in a humid 5% CO_2–95% air atmosphere.

27. Aspirate the solution and wash 5 times with excess 0.01 M PBS.

28. Add 1 mL of the desired target tumor cell-line suspension suspended at 3×10^7 cells/mL in the isolation media. If possible, centrifuge the dish for 5 min at 100× g at room temperature.

29. Incubate for 1 h at 37°C in a humid 5% CO_2–95% air atmosphere.

30. Gently swirl the flask and aspirate the liquid.

31. Gently wash the flask 4 times with excess 0.01 M PBS to remove unattached cells.

32. Add 2 mL of a 0.01 M PBS solution containing 2% (vol/vol) mouse RBC. If possible, centrifuge the dish for 5 min at 100× g.

33. Incubate at 37°C in a humid 5% CO_2–95% air atmosphere for 1 h.

34. Swirl the dish, aspirate the solution, and then wash 5 times with excess 0.01 M PBS.

35. Add 1 mL of the peritoneal or spleen lymphocyte preparation from step 6 or 15 to the dish. Centrifuge at 800× g for 10 min at 37°C if possible.

36. Incubate the dish with the culture media for 2 h at 37°C in a humid 5% CO_2–95% air atmosphere.

37. Vigorously swirl the dish, aspirate the unattached cells, and then collect the aspirate.

38. Wash 3 times in 2 mL of 0.01 M PBS, and collect the aspirate.

39. Pool the aspirates in steps 37 and 38, and calculate the percentage of cell attachment by the following equation:

$$\text{\% of attached cells} = \frac{\text{No. of cells recovered in the aspirate}}{\text{No. of cells added to the dish (which was } 5 \times 10^6 \text{ cells)}} \times 100 \qquad \text{[Eq. 1]}$$

The captured cells on the surface of the target monolayer can be recovered by gently swirling them during incubation at 37°C for 10 min with 5 mL of isolation media containing 5 mM EDTA.

Direct Cell Panning Using Antibody-Coated Plates

1. Prepare an aseptic solution of antibody (1–15 µg/mL) against the cell of interest in sterile 50 mM sodium carbonate, pH 9.5.

2. Add sufficient volume of the antibody solution (ca. 1 mL/25-mm surface) to 2 (or more) polystyrene culture dishes to fully coat the surface.

3. Incubate the dish overnight in the refrigerator. Alternatively, the dish may be incubated at room temperature for 1 h immediately prior to use.

4. Aspirate the liquid.
   Note: The antibody solution can be saved and used to coat additional plates until the antibody content is depleted.

5. Wash the plate 3 times with excess sterile 0.01 M PBS and once with 0.01 M PBS containing 5% (vol/vol) heat-inactivated serum. The plate can be stored in 0.01 M PBS at 4°C for up to a month.

6. Suspend the cell mixture in sterile culture media containing 10%–15% (vol/vol) heat-inactivated serum, 50 µg/mL gentamycin, and 10 µg/mL streptomycin at 5×10^6 to 3×10^7 cells/mL.

7. Remove the PBS and add the mixed-cell suspension in a minimal volume necessary to coat the surface (typically 1 mL/25-mm surface).

8. Swirl the dishes to assure that the liquid comes in contact with the entire surface.

9. Incubate at room temperature for no more than 40 min.
   Note: Timing should be determined empirically, but the longer the incubation, the more nonspecific binding of contaminating cells will occur.

10. Swirl and rapidly remove the cell suspension and add it to the second dish coated with the same antibodies, then repeat steps 4 and 5.

11. Wash the plates 3 times with 0.01 M PBS.

12. Wash the plate with 0.01 M PBS supplemented with 5% (vol/vol) heat-inactivated serum.

13. Release the immuno-selected cells by adding 10 mL of culture media or 0.01 M PBS containing 10% heat-inactivated serum, and then use forceful pipeting to dislodge the cells, followed by scraping with a rubber policeman.

14. Collect the cells by aspiration.

Indirect Cell Panning Using Anti-Mouse IgG-Coated Plates

1. Prepare rabbit anti-mouse IgG antibody solution-coated plates as described in steps 1–5 of the protocol for direct cell panning using antibody-coated plates.

2. Suspend a mixed-cell suspension at 1–3×10^7 cells/mL in culture media or 0.01 M PBS.

3. To the mixed-cell suspension, add 1 mL of mouse antibodies directed against the cell type of interest.

4. Incubate at room temperature for 20 min with gentle swirling.

5. Pellet the cells by gentle centrifugation ($800 \times g$ for 10 min).

6. Wash the cells 3 times by resuspending them in 0.01 M PBS containing 5% (vol/vol) heat-inactivated serum and recentrifuging them ($800 \times g$ for 10 min).

7. Resuspend the cells in 0.01 M PBS containing 5% (vol/vol) heat-inactivated serum at a concentration of approximately 1×10^7 cells/mL.

8. Add the mixed-cell suspension to the antibody-coated plates at 0.5–1.0×10^7 cells/25 mm of plate surface.

9. Swirl the dishes to assure that the liquid comes in contact with the entire surface.

10. Incubate at room temperature for no more than 40 min.
    Note: Timing should be determined empirically, but the longer the incubation, the more nonspecific binding of contaminating cells will occur.

11. Swirl and rapidly remove the cell suspension, and add it to the second dish coated with the same antibodies, and then repeat steps 4–6.

12. Immediately wash the plate 3 times with 0.01 M PBS.

13. Wash the plate with 0.01 M PBS supplemented with 5% (vol/vol) heat-inactivated serum.

14. Release the immuno-selected cells by adding 10 mL of culture media or 0.01 M PBS containing 5% heat-inactivated serum, and then use forceful pipeting to dislodge the cells, followed by scraping with a rubber policeman.

15. Collect the cells by aspiration.

ROSETTING TECHNIQUES

The second general method of affinity cell separation involves the addition of an indicator cell or immuno-adsorbent bead to which a specific cell subpopulation will bind to form "rosettes." A rosette is merely the indicator cell (or immuno-adsorbent bead) with a cluster of target cells attached to it (Figure 3). The rosetting technique has several advantages for cell separation, not the least of which is that rosettes have a higher density than free, monodispersed cells in suspension, making it relatively easy to isolate rosettes by sedimentation velocity or density gradient centrifugation. A similar affinity technique is the agglutination of cells by lectins (100). Once again, the aggregated cells can be isolated from the suspension by the same relatively conventional techniques.

Red Blood Cell Rosette Formation

Elliott and Pross (23) reviewed the use of erythrocytes (RBCs) in rosette formation assays to capture lymphocytes. Native sheep RBCs will form rosettes with most human T lymphocytes, while mouse RBCs will form rosettes with human IgM-bearing B lymphocytes. In addition, RBCs are often chemically modified prior to use to enhance rosette formation. Rosette formation with either native or modified RBCs is often an initial negative selection technique to enrich for the desired cell types before applying more specific isolation techniques. An example includes the initial rosette formation step using papain-treated rabbit erythrocytes during the isolation of rabbit B spleen cells (12). The resulting T-lymphocyte rosettes were easily removed, enriching the B-lymphocyte content in the remaining cell suspension.

The surface of the RBC can be modified by the addition of antibodies or protein A to allow rosette formation with cells that otherwise would fail to bind to RBCs. Protein A-coated RBCs are used to separate cells that are either naturally expressing immunoglobulins, or to which antibodies against specific cell surface markers had previously been added. After formation, the rosettes can be isolated by density gradient centrifugation, and the erythrocytes can be lysed by osmotic shock to recover the isolated cell with the cell surface immunoglobulin-protein A complex attached (32). An example would be the use of RBCs coated with monoclonal antibodies for the negative selection of hematopoietic stem cells from human donors (98). In this study, RBCs were treated with chromium chloride, and then coated with murine monoclonal antibodies against human myeloid cells (CD11b) and T lymphocytes (CD6). These antibody-coated RBCs were then used for rosette depletion of unwanted cells from a suspension derived from leukapheresis. The resulting rosettes were then removed by Ficoll®-Hypaque (Amersham Pharmacia Biotech) density gradient centrifugation. The negatively selected stem cells were then further purified by immunomagnetic techniques.

Magnetic Particle-Based Rosettes

While rosetting separation using RBCs remains a useful preparative tool, a versatile modern refinement of this technique uses affinity-based magnetic particles for cell separation. Affinity magnetic separation combines the principle of an affinity absorbent particle that targets and specifically binds cells in solution with an isolation technique based on magnetic force (135). Thus, after affinity binding has occurred, the magnetic particles and attached cells are then recovered by the application of a magnetic field. The magnetic beads can be coated with virtually any ligand or ligand-specific binding molecules, such as monoclonal antibodies, lectins, antigens, or protein A, to afford selectivity in cell binding. This method requires no expensive equipment, can convey remarkable selectivity for the desired cell type, and frequently results in highly purified cell populations requiring only one or two separation steps. Just as was the case with normal rosetting, affinity magnetic separation can be used to either positively or negatively select for the desired cell types (Figure 4).

Antibody-coated magnetic particles are widely used for isolation techniques in a process often referred to as immunomagnetic cell separation. Widder et al. (139,140) described rosette formation using 1-μm magnetic albumin microspheres (composed of iron oxide plus protein A) and antibodies directed against chicken or rat immunoglobulins to separate immunoglobulin-bearing cells from mixed-cell cul-

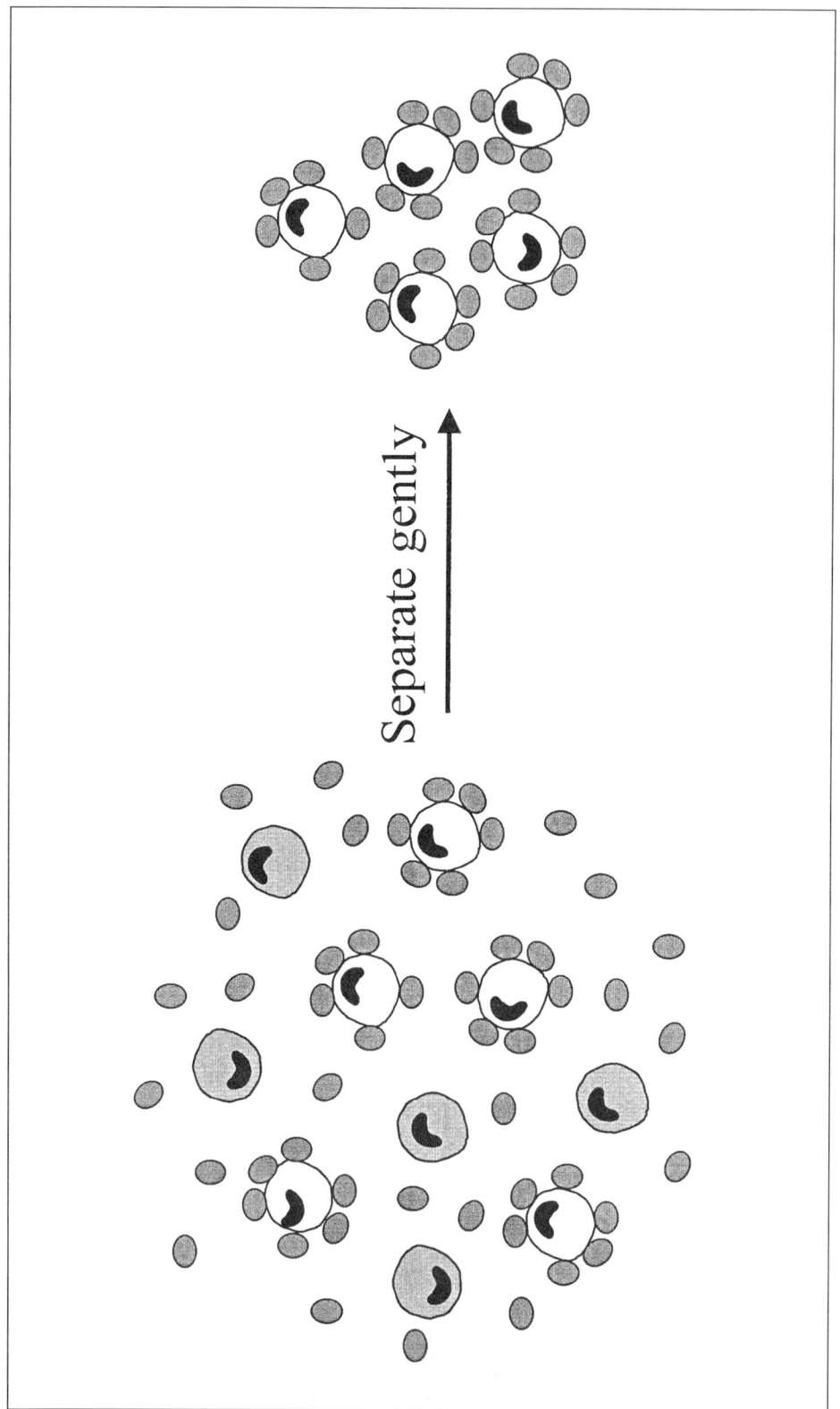

Figure 3. Positive cell selection by RBC rosette formation.

tures. Two advantages exist for preparing beads with protein A included within the bead (140). First, protein A will bind the Fc portion of most IgG antibodies, providing an easy means of obtaining immunospecific magnetic beads. Second, by including the protein at the time of bead formation, there is no need to chemically bond the protein to the bead later, since it is already incorporated in the structure. Positive selection of cells using monoclonal antibody-coated magnetic beads is frequently the method of choice for rapid one- or two-step cell isolations. This procedure was used to isolate human CD34+ mononuclear cells from liver cell suspensions using monoclonal anti-CD34–coated magnetic beads (64), and mast cells from a number of human sources using beads coated with monoclonal antibodies specific for *c-kit* receptor tyrosine kinase (110). When sufficient cells are present in the starting cell suspension, it is possible to use immunomagnetic separation to obtain pure populations of cells without cultivation in vitro (28).

Magnetic particles coated with monoclonal antibodies against T and B lymphocytes were used to isolate unstimulated monocytes from human peripheral blood by negative selection (27). Such negative selection should be used for the fractionation of other nucleated cell types when it is important to avoid cell surface changes that might occur during the binding or elution of the cells from the matrix surface. However, the major disadvantage to such negative selection is that the bound cell population left behind is most often more pure than the nonbound cell population obtained.

A useful adaptation of immunomagnetic separation is the use of magnetic beads with high binding affinity for antibody-coated cells. This approach uses magnetic beads that are coated with antibody-recognizing molecules (such as anti-immunoglobulins, protein A, or protein G) to bind cells that either have naturally occurring surface immunoglobulins or that previously have been labeled with primary antibodies (Figure 5). Wright et al. (142) used secondary antibody-coated magnetic beads to rapidly separate glial cells from rat central nervous system (CNS) tissue. In this study, cells enzymatically dissociated from the CNS were incubated with primary antibodies directed specifically against cell surface antigens on two target cell types within the cell suspension. Magnetic beads coated with secondary antibodies directed against the primary antibodies rapidly isolated a nearly pure population of glial cells (99%). A similar immunomagnetic separation using primary antibody-labeled cells and immunomagnetic beads coated with anti-immunoglobulin antibodies was used to study macrophages from atherosclerotic tissues (75). In this paper, Mattson et al. used these techniques for both negative and positive macrophage selection. A cell suspension obtained from humans was sequentially fractionated using immunomagnetic separation and monoclonal antibodies against T lymphocytes, HLA-DR–expressing cells, and macrophages. Macrophages were first enriched by negative selection by incubating the cell suspension with antibodies against T lymphocyte and HLA-DR–expressing cells. The T lymphocyte and HLA-DR–expressing cells were removed from the solution by incubating the cell suspension with magnetic particles coated with secondary antibodies (goat anti-mouse IgG). Macrophages were then recovered by positive selection, by first adding anti-macrophage antibodies to the mixture, followed by magnetic particles coated with anti-immunoglobulin–labeled magnetic beads. Protein A– or anti-immunoglobulin–coated beads can also be used to fractionate cells with naturally occurring surface immunoglobulins. Some strains of *Plasmodium falciparum*-infected RBCs have non-immune human immunoglobulins bound to their surfaces. This observation allows these parasitized RBCs to be easily separated from nonparasitized cells by using

magnetic particles coated with either antibodies raised against human immunoglobulins or with immunoglobulin-binding protein A or protein G (40).

There has been a rapid expansion in the clinical use of immunomagnetic and other
affinity cell separation techniques in recent years. One area that has received a lot of
attention is hematologic engraftment. In the study by Rambaldi et al. (98) mentioned
above, in which most of the T lymphocytes and myeloid cells were removed by RBC
rosette formation, immunomagnetic cell separation was used to further purify the
CD34+ stem cells. Negative selection of stem cells by rosette formation was subsequently followed by a further round of negative selection to remove contaminating T
lymphocytes with magnetic beads coated with anti-CD2 and anti-CD7 antibodies.
Contaminating B lymphocytes were also removed with anti-CD19–coated beads. The
resulting cell suspension, obtained by the two-step negative selection process, was
virtually depleted of T and B lymphocytes (on the order of 99.8%). After confirming
that the circulating progenitor cell (CPC) isolates were disease-free by molecular
biology techniques, nine patients were transplanted with either autologous or
allosteric CPCs, and all nine patients showed prompt and sustained hematologic
engraftment. In a similar study, immunomagnetic separation was used for autologous
bone marrow transplantation in children with acute lymphoblastic leukemia who
lacked an HLA-matched sibling donor (10). In this study, T and B lymphocytes were
depleted using a large panel of mouse antibodies directed against human B and T
lymphocyte antigens followed by immunomagnetic depletion using magnetic beads
coated with sheep anti-mouse antibodies. Twenty-seven children were then trans-

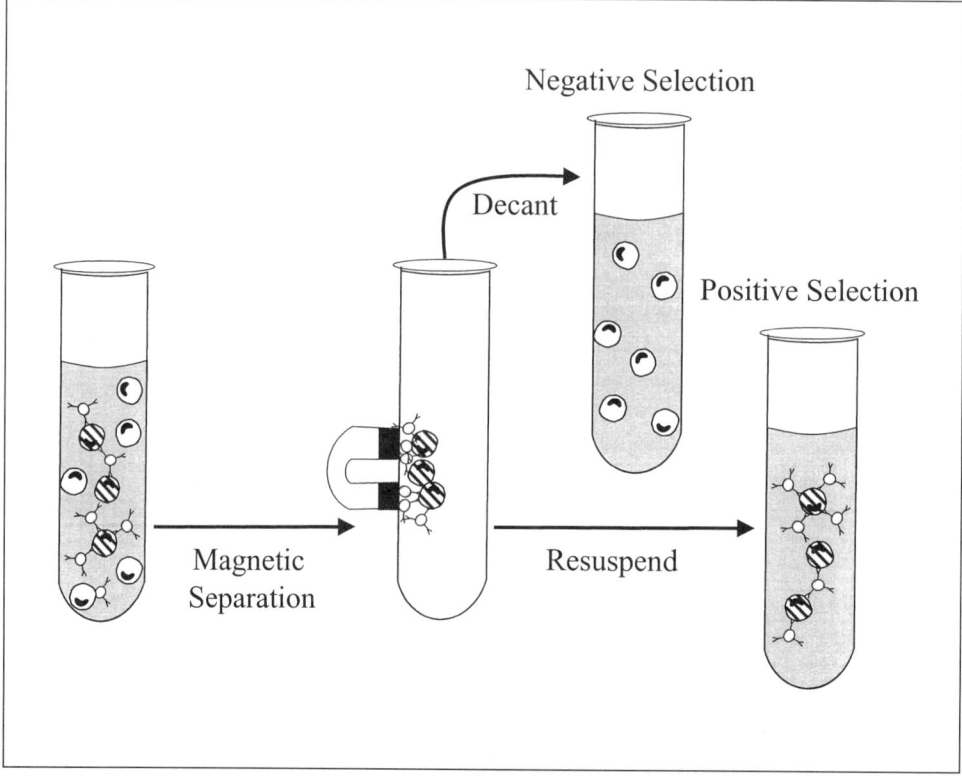

Figure 4. Positive and negative cell selection using affinity-labeleed magnetic particles.

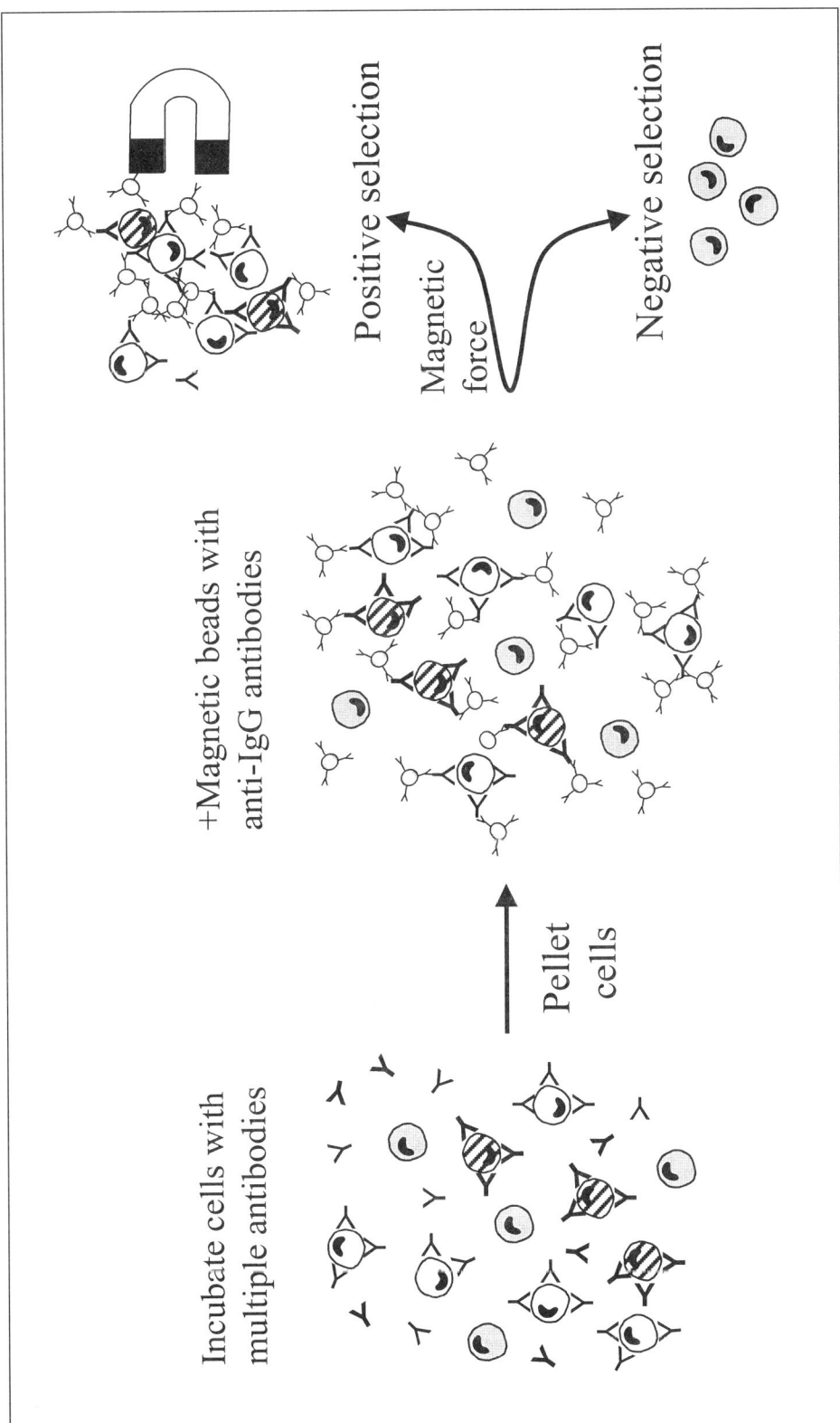

Figure 5. Magnetic immunoaffinity selection of antibody-coated cell populations.

planted with the negatively selected cells, all with satisfactory clinical results.

The success of these and other stem cell recovery and transplantation experiments has led to the development of automated clinical instruments for stem cell recovery. One such device is the Ceprate® SC stem cell concentrator developed by CellPro for the treatment of cancer (2) and transplant patients (33). Another automatic cell concentrator, the CliniMACS™, is produced by AmCell for the therapeutic-scale cell selection of hematopoietic progenitor cells. The CliniMACS uses immunomagnetic separation with superparamagnetic iron–dextran particles covalently conjugated to mouse anti-human CD34+ antibodies. The system automatically retains the magnetically labeled cells and allows the unlabeled cells to flow through, after which the magnetic force is removed and the retained CD34+ cells are recovered.

Immunomagnetic cell separations are also being used in clinical and environmental microbiology. Using magnetic beads coated with antibodies against specific microbes, it is possible to concentrate and detect even trace pathogenic organisms in clinical and environmental samples. Immunomagnetic separation of bacteria, with colorimetric PCR detection, is becoming a very useful diagnostic tool. Stark et al. (123) used immunomagnetic separation and solid-phase detection to diagnose the presence of *Bordetella pertussis*. In this study, magnetic beads precoated with polyclonal antibodies directed against surface antigens on the bacteria were added directly to nasopharyngeal aspirates to capture the bacteria. The beads were recovered and then directly transferred to a PCR tube for a colorimetric determination of the presence of *B. pertussis*. Mazurek et al. (76) used this approach to detect the slow growing *Mycobacterium tuberculosis* in cerebrospinal fluid, greatly shortening the time for confirmation of the organism in patient specimens, and Hedrum et al. (43) used the approach to detect *Chlamydia trachomatis* in patient urine samples. Instead of PCR, Yu (144) demonstrated a fluorogenic and electrochemiluminescent solid-phase detection system for the rapid measurement of *E. coli* O157 and *Bacillus anthrax* endospores captured on the surface of magnetic particles coated with specific antibodies. Magnetic beads coated with specific antibodies have also been used to investigate *E. coli* O157 in bovine fecal samples (14) and infectious *Cryptosporidium* oocyts from seeded environmental water supplies (103). The advantage of magnetic beads over normal rosette formation is shown in both of these cases, as bacteria would not form rosettes of sufficient added density to be separated by traditional differential sedimentation methods. The fact that the bacteria are attached to a magnetic particle, however, allows them to be readily recovered by the application of magnetic force. Also, as the isolation of pathogenic *E. coli* from fecal sample shows, magnetic particle separation has advantages over cell affinity chromatography in that it can be used with samples that would clog conventional chromatography columns.

Commercial suppliers offer a growing list of microbial detection kits containing magnetic particles precoated with antibacterial monoclonal antibodies. Dynal offers microbial detection kits that contain magnetic beads precoated with antibodies to such pathogens as *Salmonella*, *E. coli* O157, *Listeria*, and *Cryptosporidium*. However, uncoated magnetic particles can be obtained from Dynal or other suppliers and coated with other antibodies or lectins that are useful in the isolation of microbes. For example, Porter et al. (96) recently examined the feasibility of coating magnetic particles with 16 different lectins as an alternative approach to the isolation of enteric bacteria. They report that magnetic particles coated with con A were effective in recovering enteric bacteria from both laboratory cultures and the natural enteric bacterial population present in rivers.

Bacteria have themselves been used as immunomagnetic particles for the separation and quantification of molecules including IgG (85) and other bacteria. A monoclonal antibody immobilized on a bacterial magnetic particle (BMP) was also applied to the specific removal of *E. coli* from a bacterial suspension (84). This study used these bacterial particles and a fluoroimmunoassay method to develop a highly sensitive bacterial detection system. Fluorescein isothiocyanate (FITC)-conjugated monoclonal anti-*E. coli* antibody was immobilized onto BMPs. A suspension of *E. coli* was allowed to react with the FITC-antibody–BMP conjugates for 15 min in a magnetic field that enhanced aggregation. As the cell-BMP complex sedimented, the solution's fluorescence intensity decreased, allowing the quantification of the concentration of *E. coli* in solution.

While the normal method of separating immunomagnetic particles is through the application of magnetic force, the large size of the magnetic rosette allows other separation methods to be applied. Rye et al. (106) used antibody-coated magnetic beads to isolate tumor cells from a variety of sources, including blood and bone marrow. However, instead of using magnetic force to separate these beads, they filtered the suspension through 20-μm nylon membrane filters. Unattached cells and immunomagnetic beads without attached cells passed through the filters, while the tumor cell–bead rosettes were trapped and concentrated within the filter. The filter-trapped cells could be cultured and studied directly attached to the filter, or could be harvested by digestion with trypsin/EDTA.

Immunomagnetic cell separation is less sensitive to the effect of low antigen density on the cell surface than immunoaffinity cell panning with antibody-coated plates, immunoaffinity cell chromatography, or complement-mediated cell lysis. Gee et al. (29) demonstrated that the efficiency of immunomagnetic depletion can be adversely affected by high expression density of the cellular target antigen through either crowding-induced steric hindrance or poorly accessible antibody active sites. Such observations point out the need to optimize the experimental conditions to assure adequate antibody-cell surface antigen interactions during immunomagnetic cell separations.

As was the case with cell panning, once cells adhere to the magnetic particle through antibody-antigen reactions, it is often difficult to recover the cell without injuring it. Commercial suppliers of magnetic beads have devised kits to gently release cells from immunomagnetic beads. Dynal offers two useful means to do this. The first uses a DNA linker to attach recombinant streptavidin to the surface of the magnetic particle. The DNA-streptavidin is then used to attach a biotinylated antibody to the bead. The antibodies can be attached either prior to adding the bead to the cell suspension, or soluble biotinylated antibodies can be mixed with the cells and the biotinylated antibody-cell complex can then be captured by exploiting the affinity of streptavidin for biotin. After the streptavidin-antibody-cell surface antigen complex is formed and the bead with its attached cell is recovered by the application of magnetic force, the DNA linker can be easily cleaved with a DNase releasing buffer. Dynal offers the magnetic bead with DNA-linked streptavidin, along with the DNase releasing buffer, as their CELLection™ Biotin Binder Kit. They also offer a bead with this same type of DNA link that comes with a pre-attached antibody directed against mouse IgG, as the CELLection Pan Mouse IgG Kit. Using the Pan Mouse IgG kit, cells in suspension are selected using any monoclonal or polyclonal mouse IgG antibodies (nonbiotinylated). The cells labeled with the mouse antibodies are then captured with the anti–mouse-IgG–labeled magnetic bead, and the bound cells are eventually released from the beads with DNase releasing buffer. Dynal's second method of gently releasing cells attached to their magnetic particles is a polyclonal anti-FAb antibody called DETACHaBEAD®, which is specific for several of the antibodies they supply on antibody precoated magnetic particles. When this anti-FAb is added to magnetic beads with cells firmly attached by the appropriate antibody, the anti-FAb competes with the antibody-antigen binding and releases the antibody from the bead.

Protein A- or protein G-coated magnetic particles are also used to separate cells with either investigator-added or natively expressed surface immunoglobulins. The effectiveness of commercially available protein A- or protein G-coated magnetic beads was compared to beads coated with anti-immunoglobulin antibodies (141), and protein A-coated beads provided superior recovery of *Salmonella*. In this study for the recovery of bacteria from heterogeneous suspensions, secondary antibody-coated beads were found to inadvertently bind to bacteria that possessed surface proteins with Fc-binding activity. While such an interaction can be an undesired side effect, researchers can exploit the Fc-binding proteins on the surface of some strains of bacteria in their experimental protocols. For instance, magnetic beads coated with sheep IgG were used in a rapid and simple agglutination test for detecting surface protein A in staphylococci (50).

Another use of protein A-coated magnetic beads is as a binding platform for properly oriented capture antibodies. Protein A on the surface of the beads will bind the Fc region of antibodies such that the two antigen-binding sites are correctly exposed for cell capture. A similar orientation is obtained by using avidin- or streptavidin-coated beads and antibodies that have been biotinylated on their Fc region. An example of antibodies attached by protein A to magnetic particles involves the use of microspheres prepared with a mixture of albumin, protein A, and iron oxide, allowing the protein A to strongly bind the Fc region of the antibody (54,139,140). Using antibodies against human peripheral blood lymphocytes, it was possible to deplete nearly 99% of the target cells with this technique (54).

Richard et al. (102) used an immunomagnetic separation technique based on cytokine-induced expression of surface antigens that has exciting implications for cell biology. In this paper, a subconfluent primary mixed-cell culture composed of endothelial cells, fibroblasts, and keratinocytes was obtained from neonatal foreskins. These cultures were then treated for 6 h with TNF-α to induce endothelial expression of E-selectin. Magnetic particles coated with anti–E-selectin monoclonal antibodies were then used to selectively bind and recover a homogenous population of microvascular endothelial cells. The use of biological agents to cause selective expression of surface molecules on cells of interest, followed by the positive selection of responding cells by immunomagnetic separation, may find many useful future applications.

Another modification of the magnetic rosetting technique is the use of lectin-coated beads to provide cell recognition specificity based on surface-expressed carbohydrates. Grupp et al. (36) used lectin-coated magnetic beads to isolate cells from rat inner medulla. They used the lectin *Dolichos biflorus* agglutinin, which selectively binds inner medullary collecting duct cells. By adding magnetic beads coated with this lectin to a cell suspension, they obtained a cell fraction containing 98% pure inner medullary collecting duct cells in a single step. By repeating the isolation step, the purity could be improved. Lectin-coated magnetic particles have become a method of choice for the isolation of endothelial cells, particularly using the lectin *Ulex europaeus* agglutinin-1, which binds α-linked fucose residues on the cell surface. Magnetic beads coated with this lectin have been used successfully to isolate microvascular endothelial cells from human umbilical vein (16), human fetal skin

(13), human ventricle (77), and primate corpus luteum (15). Choroidal endothelial cells were easily and quickly isolated using a similar fucose-binding lectin prepared from *Lycopersicon esculentum*, which overcomes problems normally associated with endothelial cell isolation from this source (46). In addition, antibody-coated magnetic particles are also used for endothelial cell isolation. One example used magnetic beads coated with rat anti-mouse PECAM-1 monoclonal antibodies to isolate murine lung and murine subcutaneous sponge implants (20). The authors of this study used two cycles of immunomagnetic separation followed by limiting dilution to be assured of establishing endothelial cell-derived cell lines.

A wide variety of other ligand and ligand-binding molecules can be used to coat beads for affinity magnetic separations. Molecules recognized by cell surface receptors can be immobilized on the magnetic particle and used as a ligand for cells with surface receptors targeting these molecules. One such example targeted the surface molecule LFA-1 using its specific ligand, ICAM-1. In this study, Hedman et al. (42) isolated an ICAM-1 fusion protein with binding characteristics similar to a native molecule from a Chinese hamster ovary (CHO) cell culture supernatant that contained ICAM-1. Magnetic beads were coated with this peptide and used to isolate cells with LFA-1 on their surface, including B (41) and T lymphocytes (25).

As noted above, lectins are frequently used to coat magnetic beads to selectively bind sugar moieties on the surface of cells. However, useful magnetic separations have also been performed by coating the surface of the magnetic bead with a sugar as a ligand for isolating cells that bind specific sugars. Rye and Bovin (105) attached biotinylated pseudo-polysaccharides to streptavidin-coated magnetic beads to test the rosette formation of seven human cell lines with known anti-carbohydrate reactivity (four cancer cell lines, two lymphoblastic leukemia lines, and a hybridoma cell line). Another study used carbohydrate-coated magnetic beads for positive selection of transfected mammalian cells. In this study, kidney fibroblast-like cells from the African green monkey were lectin-transfected with a plasmid containing the cDNA encoding for the rat Kupffer cell surface lectin (94), a lectin found only on resident liver macrophages. Glycosphingolipids were adsorbed to carboxylated magnetic polystyrene microbeads, then mixed with the lectin-transfected monkey cells to positively select for transfected cells. These studies suggest that carbohydrate-coated magnetic beads may prove to be a useful tool in future cell biology applications.

A range of methods have been used to produce magnetic particles (135) that can vary tremendously in size. The protocol for the preparation of protein A-coated magnetic particles (below) provides an easy procedure to produce magnetic particles. However, carefully controlled magnetic particles with superior magnetic properties and uniform particle size are available from many companies. A recent article listed 11 companies offering such magnetic particles (119). Among these suppliers, by far the most popular is Dynal, the maker of Dynabeads®. Dynabeads are uniform, paramagnetic polystyrene beads that Dynal offers with a wide variety of affinity coatings. The basic Dynabead is uncoated to allow it to be derivatized by researchers using any method that fits their research needs. They also offer both streptavidin-coated and tosyl-activated beads that simplify the process of adding antibodies, lectins, or other molecules to the surface. In addition, they provide a wide variety of beads precoated with antibodies against an impressive array of human and mouse surface markers. For instance, at the time of preparation of this manuscript, Dynal sold beads precoated with antibodies directed against human cell subsets CD2+, CD3+, CD4+, CD8+, CD19+, CD14+, CD15+, and CD71+. Dynal also offers Dynabeads coated with highly specific antibodies against murine IgG. Using these beads, any cell to which murine monoclonal or polyclonal antibodies exist can be immunomagnetically separated. The mouse antibodies are added to a heterogeneous cell suspension, and cells tagged with mouse antibodies are recovered using Dynabeads coated with anti-mouse IgG.

While Dynal may be arguably the most popular supplier of magnetic particles for cell separations, other companies offer comparable products (Table 1). This field is growing rapidly, and soon, these companies will be offering more products and additional companies may introduce competing products. Therefore, this list is not meant to be exhaustive, but rather to simply provide evidence of the rapid growth in this area of technology and to illustrate some of the commercial options available. Seradyn offers Sera-Mag™, a 1-μm magnetic particle with either a carboxylate-modified or streptavidin-modified surface. ProZyme offers avidin-, streptavidin-, protein A-, protein G-, and mixed protein A/protein G-coated magnetic beads. Advanced Biotechnologies Ltd. offers 3.5- and 1.0-μm uncoated magnetic beads, as well as a range of magnetic beads precoated with streptavidin, protein A, or antibodies directed against mouse, human, sheep, or rabbit IgG. (The US distributor of Advanced Biotechnologies' products is Marsh Biomedical Products.) Miltenyi Biotec offers a wide range of antibody-coated magnetic particles for the isolation of hematopoietic progenitor cells, T and B lymphocytes, monocytes, granulocytes, fibroblast, and more. Spherotech offers uniformed sized polystyrene magnetic particles that are useful for cell separations that come in six different size ranges of diameter from 1–1.4 μm up to 6.0–8.0 μm. Like the other suppliers, they also offer these paramagnetic particles with a range of precoatings, including avidin, streptavidin, anti-mouse IgG, anti-rabbit IgG, and biotin. Vector Laboratories offers SCIGEN magnetic beads composed of magnetite and either cellulose or agarose. Pierce Chemical offers MagnaBind™ magnetic beads coated with either streptavidin or biotin. Roche offers magnetic polystyrene particles whose surface has been coated with anti-dioxigenin IgG antibodies or streptavidin. Immunicon offers magnetic ferrofluid particles that are much smaller than Dynabeads, but just as effective in isolating cells. Immunicon offers a variety of antibody-labeled ferrofluids for the isolation of human CD4+, CD8+, CD34+, CD25+, and CD19+ cells. In addition, a large variety of companies offer magnetic particles primarily designed for the isolation of macromolecules, such as DNA. A typical example would be the streptavidin MagneSphere® Paramagnetic Particles Plus M13 Oligo from Promega. Table 1 provides a summary of the types of beads currently available from these suppliers.

Most of the magnetic particles used in cell isolation studies require fairly strong magnetic fields to capture them from solution. Many of the same companies that provide magnetic particles also provide magnetic separators designed specifically for cell separation techniques. These separators come in a wide variety of designs, from ones designed to separate cells in multiwell plates (including 96-well plates), to single or multiple centrifuge tubes. Investigators should be careful to choose a magnetic separator that fits their experimental needs. In some cases, more complicated magnetic separation devices are used. Sun et al. (127) built a dipole magnetic flow sorter to separate T lymphocyte subpopulations into multiple fractions according to the abundance of cell surface markers bound by immunomagnetic colloids. Two additional devices for magnetic cell sorting were reviewed by Sharpe (115).

Another type of affinity separation involves the treatment of cells with FITC-labeled cell-specific antibodies. These cells are then individually monitored for their fluorescence with a laser and subsequently separated from nonfluorescent cells by

Table 1. Suppliers of Magnetic Particles for Immunomagnetic Separations and the Coatings Provided

Supplier	Non–Antibody-Coated Beads	Antibody-Coated Beads
Advanced Biotechnologies	Uncoated Streptavidin Protein A	Sheep anti-FITC Sheep anti-mouse IgG Rabbit anti-mouse IgG Donkey anti-sheep IgG Sheep anti-human IgG Sheep anti-human IgM Sheep anti-rabbit IgG
Bangs Laboratories	Uncoated Streptavidin Protein A	Goat anti-mouse IgG
BioSource International	Streptavidin	Mouse anti-human CD4 Mouse anti-human CD8a Mouse anti-human CD14 Mouse anti-human CD19 Mouse anti-human CD34 Rat anti-mouse CD4 Goat anti-rat IgG
Cortex Biochem	Uncoated Streptavidin Protein A Protein G	Donkey anti-goat IgG Donkey anti-sheep IgG Donkey anti-sheep IgG (Fc specific) Donkey anti-chicken IgG Sheep anti-guinea pig IgG Sheep anti-human IgG Sheep anti-human IgG (Fc specific) Sheep anti-mouse IgG Sheep anti-mouse IgG (Fc specific) Sheep anti-rabbit IgG Sheep anti-rabbit IgG (Fc specific) Sheep anti-rat IgG
CPG, Inc.	Uncoated Glyceryl-coated Long chain alkylamine Hydrazide-coated	None
Dynal	Uncoated Streptavidin Tosyl-activated	Anti-mouse IgG Anti-CD45 Anti-CD10

Table 1. (Continued)

Supplier	Non–Antibody-Coated Beads	Antibody-Coated Beads
Dynal (continued)	Anti-Salmonella Anti-E. coli O157 Anti-Listeria Anti-Cryptosporidium Anti-Cryptosporidium +anti-Giardia	Anti-human epithelial cell (anti-BerEP4) Anti-human CD19 Anti-mouse CD4 Sheep anti-mouse IgG Rat anti-mouse IgG (Fc specific) Goat anti-mouse IgG Sheep anti-rabbit IgG Rat anti-mouse IgM Mouse Pan T (Thy 1.2) Mouse CD4 (L3T4) Mouse CD8 (Lyt-2) Mouse Pan B (B220) Human—CD4, CD8, CD19, CD14 (monocytes), CD15 (myeloid cells), CD71, HLA Class II, CD34, CD2, CD3
Immunicon	Streptavidin	Goat anti-mouse IgG Mouse anti-human EPCAM (epithelial cells)
Miltenyi Biotech (50-nm micro- beads)	Streptavidin Biotin Uncoated Annexin V Glycophorin A	Anti-FITC Human CD34 Human AC133 CD2, CD3, CD45RA, CD45RO, TCR Pan T cell (CD4, CD8, CD19, CD22, CD128) B-cell isolation kit Mouse anti-human IgG CD11b, CD14, CD33, CD15, CD16, CD33 Monocyte isolation kit Basophil isolation kit CD56 CD71 HLA-DR CD45 Anti-fibroblast Anti-human epithelial antigen CD138

Table 1. (Continued)

Supplier	Non–Antibody-Coated Beads	Antibody-Coated Beads
Miltenyi Biotech (50-nm microbeads) (continued)		CD90, CD4, CD5, CD8a, CD62L
		CD19, CD43, CD45R, CD11b, DX5 (for NK cells), CD11c
		Rat anti-mouse IgG$_1$
		Rat anti-mouse IG2$_{a+b}$
		Rat anti-mouse IgM
		Mouse anti-rat Kappa
		Goat anti-mouse IgG (H+L)
		Goat anti-rabbit IgG (H+L)
		Mouse anti-human IgG
		NK cell isolation kit
		Blood dendritic cell isolation kit
PE Biosystems	Protein A	Goat anti-human IgG
	Protein G	Goat anti-human IgG (Fc specific)
	Biotin	Goat anti-human IgM
	Streptavidin	Goat anti-rat IgG
	Carboxylterminated	Goat anti-rat IgG (Fc specific)
	Amine-terminated	Goat anti-rat IgM
		Goat anti-mouse IgG
		Goat anti-mouse IgG (Fc specific)
		Goat anti-mouse IgM
		Goat anti-rabbit IgG
		Goat anti-hamster IgG
		Donkey anti-sheep IgG
		Sheep anti-Fluorescein
		Rabbit anti-goat IgG
		Mouse anti-human CD3
		Mouse anti-human CD4
		Mouse anti-human CD5
		Mouse anti-human CD8
		Mouse anti-human CD14
		Mouse anti-human CD16
		Mouse anti-human CD19
		Mouse anti-human C34
		Mouse anti-human CD45
		Mouse anti-human CD56
		Mouse anti-human CD71
		Rat anti-mouse CD4
		Rat anti-mouse CD8a

Table 1. (Continued)

Supplier	Non–Antibody-Coated Beads	Antibody-Coated Beads
PE Biosystems (continued)		Rat anti-mouse CD45R
Pierce Chemical	Streptavidin Biotin Amine-derivatized	Goat anti-mouse IgG Goat anti-rabbit IgG Goat anti-rat IgG
Roche	Streptavidin	Anti-dioxigenin IgG
Promega	Streptavidin	None
Prozyme	Streptavidin Avidin Protein A Protein G Protein A + Protein G	None
Seradyn	Carboxylate-modified Streptavidin	None
Spherotech	Uncoated (polystyrene, carboxyl, amino) Avidin Streptavidin Biotin Protein A	Goat anti-mouse IgG Goat anti-rabbit IgG Goat anti-human IgG Gamma globulin
Vector Laboratories	Uncoated	

differential deflection in an electric field using the fluorescence-activated cell sorting (FACS) technique (115). When the density of surface antigens is too low to allow adequate fluorescence discrimination, fluorescently labeled or magnetically labeled immunomicrospheres can be used to label the cell surface to enhance cell fractionation using FACS or free-flow electrophoresis, respectively (101). Cell isolation by immunofluorescence-enhanced FACS or immunomagnetic electrophoresis are both useful tools, but require expensive equipment and considerable user training, and are thus beyond the scope of this chapter.

PROTOCOLS

Isolation of Human T Lymphocytes by RBC Rosette Formation

1. Collect peripheral human blood by venous puncture with a syringe containing 10 U/mL of heparin.
2. Add 10 mg of carbonyl iron/mL of heparinized blood.
3. Incubate at 37°C for 30 min with gentle continuous swirling (or intermittent shaking at 5-min intervals).
4. Capture the iron-loaded monocytes in a strong magnetic field using a magnetic particle concentrator (i.e., Dynal MPC®-1 or similar product).
5. Decant the blood into a fresh tube and discard monocytes.
6. Repeat the magnetic capture (steps 4 and 5), again collecting the decanted liquid and discarding the magnetically trapped monocytes.
7. Dilute the heparinized blood 1:1 with sterile 0.01 M PBS, pH 7.4.
8. Layer 5 mL of cell suspension over 3 mL of LysoPrep™ (Ficoll-Isopaque from Nycomed) in a 12-mL centrifuge tube (for lymphocyte isolations from animal species, adjust the specific gravity to 1.09).
9. Centrifuge at $600\times g$ for 30 min at room temperature.
10. Carefully withdraw the cells at the Ficoll-plasma interface with a plastic pipet and dilute 1:1 into sterile isolation media (RPMI culture media supplemented with 50 µg/mL gentamycin, 10 µg/mL streptomycin, 5% [vol/vol] heat-inactivated serum, 10 mM HEPES, 2 mM L-glutamine, and 25 mM sodium bicarbonate).
11. Centrifuge the cells at $200\times g$ for 5 min at room temperature.
12. Wash the cells twice in 0.01 M PBS with 5% (vol/vol) heat-inactivated serum by centrifugation at $150\times g$ for 3 min at room temperature.
13. Resupend the cell pellet at a final concentration of 2×10^6 cells/mL in the isolation media.
14. Sediment whole sheep blood by centrifugation at $800\times g$ for 10 min at room temperature. Discard the plasma and resuspend the cell pellet in the original volume of 0.99% NaCl. Repeat the wash five times.
15. Wash the SRBC 3 times in HBSS, pH 6.5, and prepare a 5% (vol/vol) SRBC suspension in HBSS.
16. Dilute neuraminidase to 50 U/mL in HBSS at pH 6.5.
17. Mix 0.2 mL of the neuraminidase solution with 5 mL of the 5% SRBC solution.
18. Incubate for 1 h at room temperature.
19. Centrifuge the RBC at $800\times g$ for 10 min at room temperature.
20. Wash the cell pellet 3 times in isolation media.
21. Resuspend the cells in fresh sterile isolation media at a final concentration of 5% (vol/vol).
22. Add 1 mL of the neuraminidase-treated SRBCs to 5 mL of isolation media.
23. Incubate for 30 min at room temperature.
24. Centrifuge at $800\times g$ for 10 min and discard the pellet.

25. In a 12-mL plastic centrifuge tube, dilute an aliquot of the remaining neuraminidase-treated SRBCs in isolation media to obtain 5 mL of a 1% solution of cells.

26. Add 5 mL of the lymphocyte preparation from step 14 to the centrifuge tube containing the neuraminidase-treated SRBCs.

27. Incubate at 37°C for 15 min.

28. Centrifuge at 200× g for 5 min at room temperature.

29. Place the centrifuge tube in an ice bath, and incubate for an additional 2 h at 0°C.

30. Remove the top 5 mL of the supernatant above the settled rosettes with a pipet.

31. Very gently twirl the centrifuge tube between your fingers to resuspend the rosettes.

32. Place a wide-bore pipet in the centrifuge tube and allow it to fill by capillary action, tilting the pipet and tube as necessary.

33. Gently layer the rosettes (ca. 5 mL) over 3 mL of LysoPrep in a 12-mL centrifuge tube.

34. Centrifuge at 800× g for 20 min at room temperature.

35. Collect the cells in the pellet and at the interface by resuspending in a minimal volume of isolation media and recovering with a pipet.

36. Dilute the cells 1:1 in isolation media, and centrifuge at 300× g for 10 min at room temperature.

37. Resuspend the cells in fresh sterile isolation media, and centrifuge at 150× g for 5 min at room temperature.

38. Resuspend the cell pellet in 0.83% (vol/vol) Tris-buffered ammonium chloride, pH 7.4.

39. Swirl the cell suspension for 5 min at room temperature to lyse the SRBC.

40. Centrifuge at 100× g for 10 min to separate the lymphocytes from the red cell ghost.

Isolation of Lymphocytes with Surface-Expressed IgG Using Protein A-Labeled Erythrocytes

1. Prepare a single-cell suspension of lymphocytes as described in earlier protocols (for example, the protocol for the isolation of human T cells by RBC rosette formation, above).

2. Suspend 0.5 mL of packed RBC (sheep, human, or other) in 1 mL of 0.15 M NaCl containing 5 mg of protein A.

3. Add 0.5 mM chromium chloride in 10 mL of 0.01 M PBS for each 1 mL of packed heparinized blood.

4. Incubate for 1 h at 30°C.

5. Wash twice by centrifugation (800× g for 10 min) with 0.15 M NaCl as the wash buffer.

6. Resuspend to a final concentration of 5% RBC in isolation media (RPMI culture media supplemented with 50 µg/mL gentamycin, 10 µg/mL streptomycin, 5% [vol/vol] heat-inactivated serum, 10 mM HEPES, 2 mM L-glutamine, and

25 mM sodium bicarbonate).

Note: Cells can be stored at 4°C for up to 2 weeks by adding 1 mg/mL sodium azide to the solution. (Please note, however, that sodium azide is highly toxic; inhalation, ingestion, or skin contact with this material can cause severe injury or death.)

7. Add 1 mL of the RBC preparation and a fresh 4 mL of the isolation buffer to a 12-mL tube.

8. Add 5 mL of lymphocyte cell suspension (2×10^6 cells/mL in isolation media) to the same 12-mL tube.

9. Continue from step 27 of the protocol for the isolation of human T lymphocytes by RBC rosette formation.

Rosettes formed by the attachment of lymphocytes expressing surface IgG to protein A-coated erythrocytes will not be as fragile as the ones prepared in the human T lymphocyte isolation protocol. This does not mean, however, that pipeting and other handling should be done any less gently.

Preparation of Protein A-Coated Albumin Magnetic Particles

This protocol is based on the method in Reference 140.

1. Suspend 144 mg of HSA, 36 mg of ferroso-ferric oxide, and 28.5 mg of protein A in 0.5 mL of distilled H_2O.

2. Add 60 mL of cottonseed oil.

3. Sonicate the emulsion for 1 min at 60 W.

4. Add the emulsion to 200 mL of hot (120°C) cottonseed oil and stir constantly for 10 min.

5. Recover the microspheres by magnetic force (be careful of the hot oil).

6. After the beads have cooled, wash them 4 times with excess ether, and separate them by either magnetic force or centrifugation at $2000 \times g$ for 15 min.

7. Dry the microspheres and store at 4°C until use.

This method prepares microspheres in the range of 0.2–2.0 µm. Commercially available protein A-coated microspheres are typically more uniform in size. The protocols for the production of magnetic beads from BB shot and the preparation of magnetic agarose beads in Chapter 6 describe the preparation of magnetic BBs and agarose that can also be used in immunomagnetic separation.

Adding Antibodies to Dynal M-450 Magnetic Particles

The protocol for coating commercial magnetic beads in Chapter 6 described one method for activating magnetic particles and adding antibodies directly to their surface. The following is a similar procedure using uncoated commercial beads. Note that the orientation of the antibodies is not optimized using this procedure. Antibodies can be attached more specifically by first coating the beads with streptavidin or protein A, and then linking the antibodies to these molecules.

1. Wash the beads (Dynal M-450 or similar commercial bead) in wash buffer/ (0.01 M PBS, pH 7.4, containing 0.1% [wt/vol] BSA). Use a commercially available magnetic particle concentrator (i.e., Dynal MPC-1 or similar product) to recover the beads during this and all subsequent washings.
2. Suspend the beads in a minimal volume of the wash buffer.
3. Add 0.1–3.5 μg of antibody/10^7 magnetic particles to the bead suspension. The concentration of the antibody used depends upon the nature of the magnetic particle, the density of surface antigens on the cell of interest, and the nature of the antibody.
4. Incubate the bead-antibody suspension for 30 min at 4°C with constant, gentle swirling.
5. Sequester the magnetic beads by magnetic force, pour or pipet off the incubation buffer, and resuspend the beads in 200 μL of wash buffer.
6. Repeat the wash a total of 3 times by repeating step 5.
7. Resuspend the beads in wash buffer. These antibody-coated magnetic beads can be stored at 4°C for as long as 2 weeks prior to use.

Coating of Magnetic Particles with Streptavidin

1. Suspend the magnetic beads in excess 50 mM sodium carbonate, pH 9.5.
2. Place the tube with the beads in a magnetic cell concentrator for 2–3 min.
3. Discard the supernatant, and resuspend the beads in 50 mM sodium carbonate, pH 9.5 (for Dynabeads and similar sized particles, suspend at 4×10^8 beads/mL).
4. Prepare a solution of streptavidin at 0.2 mg/mL in 50 mM sodium carbonate, pH 9.5.
5. Add equal volumes of streptavidin and beads, and mix for 24 h at room temperature with end-over-end rotation.
6. Place the tube with the beads in a magnetic cell concentrator for 2–3 min.
7. Discard the supernatant, and wash 3 times by resuspension of the beads in 0.01 M PBS containing 0.1% (wt/vol) BSA, and repeat step 6.
8. Resuspend the beads in 0.01 M PBS containing 0.1% (wt/vol) BSA, and store up to 2 weeks at 4°C until ready to use.

Coating of Magnetic Particles with Protein A and/or Protein G

1. Suspend the magnetic beads in excess 50 mM sodium carbonate, pH 9.5.
2. Place the tube with the beads in a magnetic cell concentrator for 2–3 min.
3. Discard the supernatant, and resuspend the beads in 50 mM sodium carbonate, pH 9.5 (for Dynabeads and similar sized particles, suspend at 4×10^8 cells/mL).
4. Prepare a solution of protein A (and/or protein G) at 0.2 mg/mL in 50 mM sodium carbonate, pH 9.5.
5. Add equal volumes of protein A and beads, and mix for 24 h at room temperature with end-over-end rotation.

6. Place the tube with the beads in a magnetic cell concentrator for 2–3 min.

7. Discard the supernatant, and wash 3 times by resuspension of the beads in 0.01 M PBS containing 0.1% (wt/vol) BSA, and repeating from step 6.

8. Resuspend in 0.01 M PBS containing 1% 0.1% (wt/vol) BSA, and store up to 2 weeks at 4°C until ready to use.

Addition of Biotinylated Antibodies to Streptavidin-Coated Magnetic Particles

1. Place 1 mL of 0.01 M sodium phosphate, pH 7.2, containing biotinylated antibody (1 mg polyclonal antibody or 250 µg of monoclonal antibody) into a 12-mL tube.

   Note: Use either commercially available biotinylated antibodies, or biotinylate antibodies or FAbs by the protocols for the hydroxysuccinimide biotinylation of antibodies or for the hydrazine biotinylation of antibodies in Chapter 5 of this volume.

2. Add the streptavidin-coated magnetic particles (4×10^8 beads in 1 mL).

3. Mix for 1 h at 4°C in an overhead mixer.

4. Recover the beads by the application of magnetic force with a commercial magnetic concentrator.

5. Wash 3 times in 0.01 M sodium phosphate, pH 7.2, recovering the magnetic particles each time by magnetic force.

6. Block the nonreactive biotin receptors by incubating the beads for 1 h at 4°C in 5 mL of a solution of 1 mg/mL biotin dissolved in 0.01 M sodium phosphate, pH 7.2.

7. Wash the beads 5 times in 0.01 M sodium phosphate, pH 7.2.

The simplest approach to adding antibodies to commercially available magnetic particles is to purchase particles precoated with streptavidin. Streptavidin-coated magnetic particles can be prepared by starting with uncoated beads and adding streptavidin by following the protocol for the coating of magnetic particles with streptavidin (in this chapter).

Addition of Antibodies to Protein A (or Protein G)-Coated Magnetic Particles

1. Prepare a solution of antibodies directed against cell surface markers at a concentration of 25 µg/mL in cold (4°C) wash buffer (0.01 M PBS containing 0.1% [wt/vol] BSA).

2. Add the antibody solution to a concentrated solution of protein A-coated magnetic particles and incubate at 4°C for 1 h with gentle tilting and swirling.

3. Recover the antibody-coated beads by magnetic force, discarding the supernatant.

4. Wash the magnetic beads 3 times in wash buffer, recovering the beads each time by magnetic force.

5. Resuspend the beads in wash buffer, and store at 4°C until ready to use.

Direct Negative Cell Separation Using Antibody-Coated Magnetic Particles (Cell Depletion)

1. Cool a mixed-cell suspension containing 3×10^7 cells/mL (or less) to 4°C. If isolating blood cells, cell recovery can be improved by washing once and resuspending in wash buffer (0.01 M PBS containing 0.1% [wt/vol] BSA) to remove soluble antigens.

2. Determine the amount of magnetic particles required. When using Dynabeads and similar sized particles, the ratio of beads to target cells in the sample is recommended to be 4:1 or greater. If using smaller or larger beads, check the manufacturer's instructions.

3. Based on the calculations from step 2, add the desired number of antibody-coated beads (commercial products, or purchased prepared as described in previous protocols) to a small vial or test tube.

4. Resuspend the beads in 2–3 mL of wash buffer.

5. Place the container with the beads in a magnetic cell concentrator until the beads settle onto the magnet.

6. Remove the fluid with a pipet and discard the fluid.

7. Remove the tube from the magnetic field, and add sufficient wash buffer to adequately suspend the beads (generally 2–3 mL).

8. Repeat steps 5 and 6.

9. Remove the beads from the magnetic field, and add the cooled cell suspension from step 1. Using Dynabeads, positive cell selection requires 1×10^7 beads/mL of cell suspension, while negative cell selection requires 2×10^7 beads/mL.

10. Incubate for 30 min at 4°C with continuous gentle tilting and rotation.

11. Place the container with the beads in a magnetic cell concentrator for 2–3 min.

12. Transfer the supernatant containing the cells of interest to a fresh tube, and discard the beads and bound depleted cells.

Direct Positive Cell Separation Using Antibody-Coated Magnetic Particles

1. Follow steps 1–10 of the protocol for cell depletion, above.

2. Place the container with the beads in a magnetic cell concentrator for 2–3 min.

3. Discard the supernatant by careful pipeting while the rosetted cells are held to the wall of the tube by the magnetic concentrator.

4. While still within the magnetic field, gently trickle 2–3 mL of wash buffer (0.01 M PBS containing 0.1% [wt/vol] BSA) down the opposite side of the tube from which the beads are held.

5. Remove the tube from the magnetic field, and gently mix by mild swirling.

6. Return the tube to the magnetic concentrator for 1–2 min.

7. While still in the magnetic field, gently remove the supernatant with a pipet.

8. Repeat the washing steps (steps 3–6) 2 more times.

The rosetted cells are ready for further study. They can be lysed for molecular analysis or released from the beads (see the protocol for the recovery of antibody-bound cells from magnetic beads, below) for cell culture.

Indirect Cell Isolation Using Streptavidin-Coated Beads

1. Prepare (or purchase) biotinylated antibodies directed against cell surface markers at a concentration of 25 µg/mL in cold (4°C) wash buffer (0.01 M PBS containing 0.1% [wt/vol] BSA).

2. Prepare a mixed single-cell suspension, and wash twice with wash buffer at 4°C.

3. Resuspend the cells in wash buffer at a final cell concentration of $5–10 \times 10^7$ cells/mL and maintain at 4°C.

4. Add excess biotinylated antibody solution to the washed cell suspension (at 10 µg of antibody/10^7 target cells). Mix and incubate at 4°C for 30 min with gentle tilting and swirling.

   Note: For negative selection, multiple biotinylated antibodies can be added to the solution to remove several types of unwanted cells at once.

5. Dilute with an excess volume of cold (4°C) wash buffer and pellet the cells by mild centrifugation ($300–400\times g$ for 10 min). Discard the supernatant.

6. Resuspend the cells in wash buffer at a concentration of 3×10^7 cells/mL (or less).

7. Add streptavidin-coated magnetic particles at a concentration of 4–5 beads/target cell.

8. Incubate for 30 min at 4°C with continuous gentle tilting and swirling.

9. Use a magnetic force to recover the cells by either positive cell selection (starting at step 2 in the protocol for direct positive cell separation using antibody-coated magnetic particles) or negative selection (starting at step 11 of the cell-depletion protocol).

Because streptavidin will bind any biotinylated antibody, it is not necessary to add the biotinylated antibody to the bead first. Instead, the biotinylated antibody can be incubated with the heterogeneous cell suspension containing the cells that are targeted by the antibody. After the antibody binds the cell surface antigens, streptavidin-coated beads can then be added to capture the antibody-cell complex, and a magnetic field can be used to recover the cells. For positive cell selection (or negative cell selection using a single antibody), this procedure makes little sense as compared to simply attaching the biotinylated antibody to the bead first (see the protocol for the addition of biotinylated antibodies to magnetic particles). However, this procedure is useful when preparing a cocktail of biotinylated antibodies to remove multiple cell types during negative cell selection. Therefore, a different batch of beads need not be prepared for each cell type targeted for depletion.

Indirect Cell Isolation Using Protein A (and/or Protein G)-Coated Beads

1. Prepare a solution of antibodies directed against cell surface markers at a concentration of 25 µg/mL in cold (4°C) wash buffer (0.01 M PBS containing 0.1% [wt/vol] BSA).

2. Prepare a mixed single-cell suspension, and wash twice with wash buffer at 4°C.

3. Resuspend the cells in wash buffer at a final cell concentration of $5–10 \times 10^7$

cells/mL and maintain at 4°C.

4. Add excess antibody solution (10 µg of antibody/10^7 target cells) to the washed cell suspension. Mix and incubate at 4°C for 30 min with gentle tilting and swirling.

   Note: For negative selection, multiple antibodies can be added to the solution to remove several types of unwanted cells at once.

5. Dilute with an excess volume of cold (4°C) wash buffer, and pellet the cells by mild centrifugation (300 or 400× g for 10 min at 4°C). Discard the supernatant.

6. Resuspend the cells in wash buffer at a concentration of 3×10^7 cells/mL (or less).

7. Add protein A (and/or protein G)-coated magnetic particles at a concentration of 4–5 beads/total target cell.

8. Incubate for 30 min at 4°C with continuous gentle tilting and swirling.

9. Use a magnetic force to recover the cells by either positive cell selection (starting at step 2 in the direct positive cell separation protocol) or negative selection (starting at step 11 of the cell-depletion protocol).

This procedure is best used for negative selection of desired cell types using a cocktail of antibodies directed against a variety of major contaminant cell types in the solution. Compare this method with the similar procedure using biotinylated antibodies and streptavidin-coated particles (see the protocol for indirect cell isolation using streptavidin-coated beads, above).

Indirect Cell Isolation Using Secondary Antibody-Coated Beads

1. Prepare a solution of primary antibodies directed against cell surface markers at a concentration of 25 µg/mL in cold (4°C) wash buffer (0.01 M PBS containing 0.1% [wt/vol] BSA).

2. Prepare a mixed single-cell suspension, and wash twice with wash buffer at 4°C.

3. Resuspend the cells in wash buffer at a final cell concentration of $5–10 \times 10^7$ cells/mL and maintain at 4°C.

4. Add excess antibody (2–5 µg/mL for polyclonal antibody and 0.5–1 µg/mL for monoclonal antibody/10^7 target cells) to the washed cell suspension. Mix and incubate at room temperature for 30 min or at 4°C overnight, with continuous gentle tilting and swirling.

   Note: For negative selection, multiple antibodies can be added to the solution to remove several types of unwanted cells at once.

5. Dilute with an excess volume of cold (4°C) wash buffer, and pellet the cells by mild centrifugation (300 or 400× g for 10 min). Discard the supernatant.

6. Resuspend the cells in wash buffer at a concentration of 3×10^7 cells/mL (or less).

7. Add magnetic beads coated with a secondary antibody directed against the primary antibody used in step 1, at a concentration of 4–5 beads/total target cell.

8. Incubate for 30 min at 4°C with continuous gentle tilting and swirling.

9. Use a magnetic force to recover the cells by either positive cell selection (start-

ing at step 2 in the direct positive cell separation protocol) or negative selection (starting at step 11 of the cell-depletion protocol).

Recovery of Antibody-Bound Cells from Magnetic Beads

1. Recover (by a magnetic separator) positively selected rosetted cells that are attached to the magnetic bead.
2. Resuspend the rosetted cells in 0.01 M PBS containing an excess of the free antibody that was used to capture the cell.
3. Incubate at room temperature for 20 min with constant rotation.
4. Place the container with the beads and cells in a magnetic cell concentrator and let stand for 2–3 min.
5. With the beads held to the side of the tube, remove (and save) the supernatant with a pipet.
6. Add 5 mL of 0.33 M citric acid with the excess antibody to the tube, and repeat steps 3–5, pooling the supernatants removed.
7. Centrifuge the pooled supernatants ($800\times g$ for 10 min) and resuspend in desired cell culture media or cell maintenance media.

Mechanical force (vortex mixing), trypsin digestion, or an EDTA solution can also be used to release the cells from the magnetic particles. The antibody elution listed above is the gentlest of these approaches, but gives only modest cell recoveries. In addition, the cells isolated are coated with the antibodies used in the elution step. The other methods lead directly to cell injury and should be used only as a method of last resort. The best approach to isolating cells by antibody-based magnetic bead technology is through negative selection so that the risk of activating the cell through the interaction of the antibody with the cell surface, or the possibility of inducing cell injury during harsh conditions required to break the antibody-antigen bond, can be avoided. Lacking that, novel approaches such as the DNase method or the DETACH-aBEAD method (both described below) to gently release the cells can be utilized.

DNase Method of Cell Release Using the CELLection Pan Mouse IgG Kit

Dynal's CELLection Pan Mouse IgG Kit contains magnetic beads coated with human anti-mouse IgG antibodies that will bind any mouse IgG. Check the detailed instructions that are provided in each CELLection kit for any specific details that are related to changes in the product or specific for your application.

1. Follow steps 1–6 of the protocol for indirect cell isolation using secondary antibody-coated beads (above), using mouse IgG to positively select for the cell of interest.
2. Wash the human anti-mouse IgG from the CELLection kit in cold wash buffer (0.01 M PBS containing 1% 0.1% [wt/vol] BSA), and resuspend at 4×10^8 beads/mL.
3. Add the magnetic beads at 5 beads/target cell, and and at a final concentration of at least 1×10^7 beads/mL.
4. Gently rotate the bead–cell suspension at 4°C for 15–30 min.
5. Place the container with the beads in a magnetic cell concentrator for 2–3 min.

6. Remove the supernatant with a pipet, and gently add fresh RPMI media with 1% (vol/vol) heat-inactivated serum down the side of the tube opposite the beads.

7. Remove from the magnetic field, gently mix, and transfer to a new tube.

8. Repeat steps 5–7 twice more.

9. Resuspend in 200 μL of RPMI media containing 1% (vol/vol) heat-inactivated serum. If the total number of beads is higher than 5×10^7, increase this volume proportionally.

10. Follow the instructions in the CELLection kit to prepare the DNase solution. Add 4 μL of DNase solution for every 10^7–10^8 magnetic beads used. (Avoid rigorous pipetting of the DNase solution.)

11. Seal the tube, and incubate at 37°C for 15 min at room temperature with gentle rotation. Take care to keep the very small volume used at the bottom of the tube during rotation.

12. Repeatedly draw the volume in-and-out of a pipet (10–20 times).

13. Place the container with the beads in a magnetic cell concentrator for 2–3 min.

14. Collect the released cells in the supernatant with a pipet.

15. Add 200 μL of RPMI media with 1% (vol/vol) heat-inactivated serum to the tube containing the beads, and repeatedly draw the solution in-and-out of a narrow-tipped pipet.

16. Collect the released cells in the supernatant with a pipet, and pool with the cells from step 14.

Dynal also offers streptavidin attached to magnetic beads by a DNA linker. Using this product, the biotinylated antibody of choice can be attached. When using these beads in procedures such as the direct positive cell separation protocol, the cell release instructions in that protocol should be followed, starting at step 9. Dynal also offers DETACHaBEAD, which is a polyclonal anti-FAb antibody that is specific for several of the primary antibodies on Dynabeads. When DETACHaBEAD is added to the bead-bound cells, it competes with antibody-antigen binding at the cell surface and releases the antibody and bead from the cells.

DETACHaBEAD Method of Cell Release from Antibody-Coated Magnetic Beads

1. Use specific antibody-coated beads supplied by Dynal to form magnetic rosettes as described in the direct positive cell separation protocol.

2. Resuspend the positively selected, magnetic-based cell rosettes in isolation media (100 μL of RPMI culture media supplemented with 50 μg/mL gentamycin, 10 μg/mL streptomycin, 5% [vol/vol] heat-inactivated serum, 10 mM HEPES, 2 mM L-glutamine, and 25 mM sodium bicarbonate).

3. Add 1 U (100 μL) of DETACHaBEAD solution.

4. Incubate for 1 h at room temperature with gentle tilting and swirling.

5. Place the container with the beads in a magnetic cell concentrator for 2–3 min.

6. Collect the released cells in the supernatant with a pipet, and remove from the magnetic field.

7. Add 200 µL of RPMI media with 1% (vol/vol) heat-inactivated serum to the tube containing the beads, and repeatedly draw the solution in-and-out of a narrow-tipped pipet.

8. Place the container with the beads in a magnetic cell concentrator for 2–3 min.

9. Collect the released cells in the supernatant and pool with the cells from step 6.

DETACHaBEAD is a polyclonal anti-FAb antibody from Dynal, which has binding specificity towards several of the primary antibodies on commercial antibody-coated Dynabeads. DETACHaBEAD competes with antibody-antigen binding at the cell surface of bound cells, causing the release of the antibody and bead from the cells. The general applicability of the DETACHaBEAD approach is limited because these anti-FAb antibodies only work with Dynabeads coated for human CD4, CD8, CD19, and CD34 cells, plus mouse CD4 cells. However, Dynal's CELLection kits can be adapted to virtually any cell, making it the method of choice for routine positive cell selection by immunomagnetic techniques.

Isolation of Endothelial Cells Using Lectin-Coating of Magnetic Particles

1. Suspend magnetic beads in excess 50 mM sodium carbonate, pH 9.5.

2. Place the tube with the beads in a magnetic cell concentrator for 2–3 min.

3. Discard the supernatant, and resuspend the beads in 50 mM sodium carbonate, pH 9.5 (for Dynabeads and similar sized particles, suspend at 4×10^8 cells/mL).

4. Prepare a solution of *Ulex europaeus*-1 lectin (Sigma Chemical) at 0.2 mg/mL in 50 mM sodium carbonate, pH 9.5.

5. Add equal volumes of lectin and beads, and mix for 24 h at room temperature with end-over-end rotation.

6. Place the tube with the beads in a magnetic cell concentrator for 2–3 min.

7. Discard the supernatant, and resuspend the beads in 0.01 M PBS containing 0.1% (wt/vol) BSA. Store at 4°C until ready to use.

8. Add beads to a mixed single-cell suspension containing endothelial cells (bead:target-cell ratio of 5:1).

9. Incubate at room temperature for 20–30 min with constant tilting and swirling.

10. Place the tube with the beads in a magnetic cell concentrator for 2–3 min.

11. Discard the supernatant, and resuspend the beads in 0.01 M PBS containing 0.1% (wt/vol) BSA.

12. Repeat steps 10 and 11 three times.

13. After the last discard of supernatant, resuspend the rosetted cells in ice-cold 0.01 M PBS containing 0.1% (wt/vol) BSA and 0.1 M fucose, and incubate for 10 min at 4°C.

14. Place the container with the beads in a magnetic cell concentrator for 2–3 min.

15. Collect the released cells in the supernatant with a pipet, and remove from the magnetic field.

16. Add 200 µL of RPMI media supplemented with 50 µg/mL gentamycin, 10 µg/mL streptomycin, 5% (vol/vol) heat-inactivated serum, 10 mM HEPES, 2

mM L-glutamine, and 25 mM sodium bicarbonate to the tube containing the beads and repeatedly draw the solution in-and-out of a narrow-tipped pipet.

17. Place the container with the beads in a magnetic cell concentrator for 2–3 min.

18. Collect the released cells in the supernatant, and pool with cells from step 15.

CELL CHROMATOGRAPHY TECHNIQUES

Cell chromatography uses the same principles of affinity chromatography that were used for the isolation of macromolecules (see Chapter 6). In both, the analyte or the cell is specifically and reversibly adsorbed by an immobilized complementary affinity-binding molecule. The major difference between these similar techniques is that positive selection is used almost exclusively for affinity chromatography when isolating macromolecules, while negative selection is frequently used for cell chromatography. Using negative chromatographic selection, cell ligands on the surface of unwanted (contaminating) cells interact with their complementary epitope immobilized on the column support material. The unwanted cells are retained within the column, while the column effluent is enriched with the desired cells. Another difference is that greater care must be exercised when choosing the chromatographic support for affinity cell isolation. First, the matrix chosen must not allow unwanted, nonspecific cell adhesion. Second, the support matrix must be sufficiently porous to allow the cells to pass through without physically trapping them. Large glass or plastic beads, loosely packed fibers, and cross-linked dextrans or agarose are typically used.

Differential Attachment Chromatography

Immunologists have long used affinity cell chromatography to fractionate lymphocytes into B and T lymphocyte populations. These separations are based on the initial fortuitous observation by Julius et al. (53) that the cell population obtained in the effluent after passing murine spleen cells through a nylon fiber column was B-lymphocyte depleted. It is well established that B lymphocytes and monocytes will rapidly adhere to nylon fibers, but most T lymphocytes will not. Thus, one way to deplete B lymphocytes and monocytes from a mixed lymphocyte population is to pass it through such a nylon-filled column (34). The isolation principle is similar to that of differential cell attachment used for cell panning, except that the process takes place in a column rather than on a flat surface. This negative selection procedure prevents the unwanted activation of the T lymphocytes that is often observed when T lymphocytes are isolated by positive selection. Recovery of the non-adherent T lymphocytes from a nylon affinity column could not be easier; simply collect the effluent. Recovery of the B lymphocytes is more tedious. Trizio and Cudkowicz (133) recovered B lymphocytes by removing the nylon material and then teasing, shaking, and squeezing the cells loose from the matrix.

Other column materials possess similar binding characteristics to that of nylon fibers. Tsuru et al. (134) found that hydroxyapatite columns would effectively deplete immunoglobulin-bearing cells and monocytes from peripheral blood lymphocytes, resulting in the negative selection enrichment of helper and suppressor T lymphocyte populations. Maruyama and coworkers (72) used glass beads coated with poly (2-hydroxyethyl methacrylate)/polyamine copolymers for cell chromatography of rat lymphocyte suspension. These columns also bound the B lymphocytes, allowing for the negative selection of T lymphocytes. In a later study, the same group examined the effect of varying operating conditions on the separation efficiency of these columns (73). The rate of cell infusion into the column had a significant effect on the separation efficiency, which increased with decreasing flow rate. In addition, they showed that the B cells could be recovered from the glass beads mechanically by gentle pipeting.

Affinity Chromatography

While nylon fibers and nonspecific coated matrices have proven useful in cell chromatography, normal affinity cell chromatography requires the adsorption of a specific ligand or ligand-binding molecule onto the chromatographic support. The specificity afforded by these adsorbed moieties is then used for cell isolation. Sharma and Mahendroo (111) provided a review of the early history of affinity cell chromatography, focusing primarily on chromatography using lectin-coated supports.

Schrempf-Decker et al. (108) compared the isolation of T and B lymphocytes by rosetting techniques to affinity cell chromatography using Sepharose beads coated with a *Helix pomatia* lectin. Compared to rosetting, the immunoaffinity chromatographic cell isolation resulted in higher yields, superior viability, and, in the case of B lymphocytes, a significantly higher purity (82%). Tlaskalova-Hogenova et al. (132) also demonstrated the usefulness of affinity cell chromatography for the isolation of T and B lymphocytes from human tonsils and peripheral blood by comparing three affinity cell separation techniques. They compared cells fractionated using nylon wool columns to those obtained using anti-human immunoglobulins attached to Sephron™ (hydroxyethyl methacrylate) or Sepharose column supports, and to cell panning using the same anti-human immunoglobulins immobilized on culture dishes. Kataoka et al. (57) used immunoaffinity chromatography with anti-rat IgG immobilized on a matrix of poly(2-hydroxyethyl methacrylate) to rapidly separate lymphocytes from rat mesenteric lymph node into populations with and without surface IgG molecules.

In a study of the binding properties of numerous plant lectins to feline peripheral blood lymphocytes, Whitehurst et al. (138) found that *Pisum sativum* agglutinin bound feline B lymphocytes much more readily than T lymphocytes. Affinity chromatography using *P. sativum* agglutinin-coated supports was then used to deplete feline peripheral blood of B lymphocytes to obtain pure subpopulations of feline T lymphocytes. Immobilized wheat germ agglutinin on Sepharose supports was used to fractionate mouse bone marrow cells (89). Most of the cells applied were initially retained by the column, although a subpopulation of lymphocytes depleted of cells with a high density of surface immunoglobulins passed directly though the column. Elution of the column using different concentrations of N-acetyl-D-glucosamine gave cell fractions enriched in different cell populations. However, mechanical agitation of the Sepharose packing was required to recover lymphocytes with a high density of surface immunoglobulin. Pereira and Kabat (93) used lectins bound to Sephadex® or Sepharose to retain erythrocytes on cell affinity columns.

Protein A is also an excellent ligand-binding molecule for all types of affinity chromatography (95) including affinity cell chromatography (63). In an early study, Ghetie et al. (30) showed that protein A-coated Sepharose beads could be used for cell separations in which cells were initially incubated with IgG antibodies directed against specific cell surface markers. Immunoglobulin-bearing cells from mouse spleen were treated with rabbit IgG anti-mouse immunoglobulins, and the labeled cells were separated by positive selection cell chromatography. Duffey et al. (21) also used protein A-coated affinity supports to fractionate antibody-coated lymphocytes.

Immobilized antibodies are frequently used in cell chromatography. Kondorosi et al. (60) prepared columns with packing coated with non-immune rat immunoglobulin to select for cells containing immunoglobulin receptors. They used these columns to obtain, by negative selection, a highly enriched T-lymphocyte population that lacked surface immunoglobulin receptor molecules. Manderino et al. (70) immobilized anti-rabbit FAb antibodies on CNBr-activated Sepharose beads to negatively select for T lymphocytes from lymph nodes of rabbits immunized with keyhole limpet hemocyanin. van Overveld et al. (136) used antihuman IgE-Sepharose affinity cell chromatography as the last step in fractionating human mast cells from lung tissue. Scott (109) introduced the use of antifluorescein antibodies to isolate immunocompetent lymphocytes. In this method, lymphocytes are exposed to fluorescein-labeled antigens, and cells that bind these labeled antigens are separated by the use of antifluorescein affinity columns. Irlé et al. (48) used affinity chromatography to isolate murine thymocytes that had been incubated with peanut agglutinin, followed by chromatography through an anti-peanut agglutinin-coated Sepharose column.

Mandrusov et al. (71) used a membrane-based modification of the immunoaffinity chromatographic technique to positively select for murine B lymphocytes. Anti-murine IgG was chemically bonded to a cellophane membrane support, over which murine splenocytes were pumped. The B lymphocytes were captured by the immobilized antibody under low flow conditions, and then the flow rate was increased to remove cells that had adhered nonspecifically to the membrane surface. B lymphocytes could be eluted by the application of hydrochloric acid (pH 1.0) supplied to the side of the membrane opposite the cells. Cells that bind lectins can also be isolated by the use of affinity chromatography techniques using immobilized antibodies directed against the surface-bound lectin.

Tassi et al. (130) used a column with an avidin-coated polyacrylamide support to bind and retain cells marked with biotinylated antibodies. Human bone marrow samples were incubated with monoclonal mouse antibodies to select CD34+ or S313+ cell types, followed by incubation with biotinylated goat anti-mouse immunoglobulins. Because of the affinity of avidin for biotin, the cells were positively selected by the presence of surface biotinylated antibodies. Cottler-Fox et al. (17) used a similar approach but with affinity chromatography using avidin-coated beads and biotinylated anti-CD34+ antibodies for positive selection of stem cells. They used this method to deplete contaminating T lymphocytes from specimens obtained from human peripheral blood and bone marrow harvest obtained from six myeloma and five breast cancer patients. Commercially available immunoaffinity columns are available from a number of suppliers. One such system is Ceprate, an avidin-biotin immunoaffinity column from Cellpro. Johnson et al. (51) used Ceprate and CD34+ selection procedures to harvest peripheral blood stem cells from eight patients with myeloma. Kogler et al. (59) also used the Ceprate system to isolate CD34+ hematopoietic stem cells from full-term umbilical cord blood.

R&D Systems offers a series of cell separation columns for the isolation of T lymphocytes from humans, rats, or mice, using principles of negative selection to remove monocytes, B lymphocytes, NK cells, and granulocytes. Cytovax Biotechnologies offers a number of pre-prepared immunoaffinity columns (cellect™) composed of glass beads coated with specific antibodies for the negative selection of human CD4+ and CD8+ T lymphocytes (97), B lymphocytes, and trophoblasts (145), as well as murine B lymphocytes (107). Hawkins et al. compared the effectiveness of three commercially available laboratory-grade selection columns for the separation of CD34+ (39).

A wide variety of antigens, haptens, and receptor molecules have been used in affinity cell chromatography. Bernd et al. (4) used Sepharose beads coated with thyroglobulin to separate thyroid follicular and parafollicular cells. Soderman et al. (122) immobilized insulin by coupling it to Sepharose to isolate adipocytes by affinity chromatography. Dvorak et al. (22) obtained a 95% pure fraction of chick embryo neuronal cells using affinity chromatography with Sepharose containing immobilizing α-bungarotoxin. Rutishauser et al. (104) immunized BALB/c mice with *Limulus* hemocyanin and used antigen-coated fibers to capture antigen-binding cells.

Another type of affinity cell chromatography is based on a combination of column chromatography and two-phase aqueous affinity partitioning (discussed in *Cell Partitioning Techniques* below, along with other examples of affinity cell partitioning). These separations typically use beads coated in an aqueous phase with an increased hydrophobic nature over the mobile phase (typically PEG or polypropylene glycol [PPG]).

PROTOCOLS

Negative Separation of T Lymphocytes by Nylon Wool Chromatography

1. Prepare nylon fibers by washing them 3 times in 0.25 M HCl, followed by extensive flushing with water to remove all traces of acid.

2. Wrap the fibers in a clean cloth, squeeze out excess water, and air-dry the fibers at 37°C for 2 days.

3. For aseptic cell separations, microwave the fibers for 10 min on high in a conventional microwave oven. Alternatively, the fibers can be wrapped in aluminium foil and sterilized by dry heat at 120°C overnight.

4. Take 200 mg of dry, acid-washed nylon fibers and tease into individual fibers under aseptic conditions in a petri dish containing isolation buffer (RPMI media supplemented with 50 μg/mL gentamycin, 10 μg/mL streptomycin, 5% [vol/vol] heat-inactivated serum, 10 mM HEPES, 2 mM L-glutamine, and 25 mM sodium bicarbonate).

5. Pack the fibers into a disposable plastic column with a frit, and cap the column so that the fluid does not pass through. Keep the fibers wet at all times.

6. Wash the fibers by allowing 30 mL of warm (37°C) isolation buffer to pass through the column.

7. Cap and cover the column, and incubate it at 37°C in a 5% CO_2 atmosphere.

8. Wash the column again with 30 mL of isolation buffer, and then cap the outflow.

9. Add 1 mL of lymphocyte cell suspension (ca. 3×10^7 cells) to the top of the column with the outflow still capped.

10. Add 2 mL of isolation buffer, cover, and incubate for 1 h at 37°C under a 5% CO_2 atmosphere.

11. Add a further 10 mL of isolation buffer and collect the non-adherent cells as a single fraction.

The non-adherent cells will be mostly T lymphocytes, although Fc receptor-positive T lymphocytes will be retained on the column. The adhered cells can be recovered by tapping the column and compressing the fibers with a syringe plunger. The compressed fiber can then be teased apart directly in the column, and the cells eluted with an additional 30 mL of isolation buffer.

Positive Cell Selection Using Lectin-Based Cell Chromatography

1. Prepare lectin-coated Sepharose beads as described in *Agaroses* in Chapter 2 or *Ligand Immobilization* in Chapter 4. For cell isolations, beads such as Sepharose 4B (Amersham Pharmacia Biotech), Bio-Gel A-15m (Bio-Rad), or 4% XC Agarose Beads (XC Corporation) should be chosen.

2. Add the cell mixture onto the top of the column in a volume of approximately 30% of the total column volume and allow the cell suspension to drain into the column.

3. Stop the flow while the cells are still in contact with the affinity absorbent column packing, and allow the column to incubate for 10–30 min.

4. Uncap the outflow and elute the unbound cells by washing with 20 column volumes of 0.01 M PBS containing 0.1% (wt/vol) BSA.

5. Flush the column with 20 volumes of elution buffer (0.01 M PBS containing 0.1% [wt/vol] BSA and 1–10 mg/mL of the specific carbohydrate targeted by the lectin used), collecting the effluent.

6. Centrifuge the effluent (800× *g* for 10 min) to pellet the lectin-positive cells.

Glass and polystyrene beads with diameters of 200–400 nm can be used in affinity columns as well. Affinity and immunoaffinity ligands can be attached to these beads as described in *Ligand Immobilization* in Chapter 4.

Negative Cell Selection Using Protein A-Coated Beads

1. Prepare a packed column using a protein A-labeled affinity column as described in the protocol for slurry-packing a small affinity or immunoaffinity column by hand in Chapter 7, but use a bead such as Sepharose 4B, Bio-Gel A-15m, or 4% XC Agarose.

2. Prepare a solution of primary antibodies directed against cell surface markers at a concentration of 25 μg/mL in cold (4°C) wash buffer (0.01 M PBS containing 0.1% [wt/vol] BSA).

   Note: Multiple antibodies can be added to the solution to remove several types of unwanted cells at once.

3. Prepare a mixed single-cell suspension, and wash twice with wash buffer at 4°C.

4. Resuspend the cells in wash buffer at a final cell concentration of 5–10 × 10^7 cells/mL and maintain at 4°C.

5. Add excess antibody (2–5 μg/mL for polyclonal antibodies and 0.5–1 μg/mL for monoclonal antibodies/10^7 target cells) to the washed cell suspension. Mix and incubate either at room temperature for 30 min or at 4°C overnight, with continuous gentle tilting and swirling.

6. Dilute with an excess volume of cold (4°C) wash buffer and pellet the cells by mild centrifugation (300 or 400× g for 10 min at 4°C). Discard the supernatant.

7. Resuspend the cells in wash buffer in a volume equal to 40% of the column to be used.

8. Add the cell mixture onto the top of the column in a volume of approximately 30% of the total column volume, and allow the cell suspension to drain into the column.

9. Stop the flow while the cells are still in contact with the affinity absorbent column packing, and allow the column to incubate for 10–30 min.

10. Uncap the outflow, and elute and capture the unbound negatively selected cells by washing with 20 column volumes of 0.01 M PBS containing 0.1% 0.01 M BSA.

    Note: The column can be regenerated for reuse by washing with 10 column volumes of 100 mM glycine-HCl buffer (pH 2.5) containing 500 mM NaCl, followed by equilibration back to neutral pH with PBS.

An alternative approach is to immobilize the specific antibodies directly within the column. The disadvantage to this second approach is that specific columns have to be prepared for each cocktail of antibodies to be used.

Positive Cell Selection Using Immunoaffinity Chromatography

1. Prepare a packed column using an antibody-coated support as described in the protocol for slurry-packing a small affinity or immunoaffinity column by hand in Chapter 7. The support should be a product such as Sepharose 4B, Bio-Gel A-15m, or 4% XC Agarose.

2. Add the cell mixture onto the top of the column in a volume of approximately 30% of the total column volume, and allow the cell suspension to drain into the column.

3. Stop the flow while the cells are still in contact with the affinity absorbent column packing, and allow the column to incubate for 20–30 min.

4. Uncap the outflow and elute the unbound cells by washing with 20 column volumes of 0.01 M PBS containing 0.1% (wt/vol) BSA.

5. Transfer the beads from the column to a test tube, and vortex-mix for 10–15 s. Return to the column, and collect the elute from the column wash using 10–20 column volumes of 0.01 M PBS containing 0.1% (wt/vol) BSA.

6. Centrifuge the effluent (800× g for 10 min) to pellet the cells, and resuspend the cells in appropriate media.

An alternative elution method at step 5 is to elute using competitive binding by flushing the column with excess immunoglobulin that was used to coat the support. Neither of these elution methods is ideal. Cell injury occurs because of the vortex-mixing in the first method; the second method can be expensive in terms of the antibodies used in the elution buffer, and the isolated cells are recovered from the column fully coated with elution antibody. For this reason, negative cell selection is the preferred choice when isolating cells by immunoaffinity chromatography.

251

CELL PARTITIONING TECHNIQUES

Two-phase binary aqueous solutions produced using high-molecular-weight polymers (such as PPG, polyvinyl alcohol, methylcellulose, and dextran) can be used to fractionate cells (and macromolecules or cell organelles) based on their differential partitioning between the two phases (for a review, see References 1 and 114). Cells typically distribute at the interface or in one of the two phases based on the surface properties of the cell (137). While not based on affinity binding, native aqueous two-phase partitioning can still be very useful in fractionating cells (65). Skuse et al. (121) demonstrated cell and protein partitioning in a novel aqueous two-phase system composed of hydroxypropyl cellulose and PEG/PPG. *Affinity Partitioning* in Chapter 6 of this volume described how aqueous soluble macromolecules are selectively separated using a modified two-phase aqueous system by the addition of affinity ligands. By judicious use of surface-binding ligands that specifically alter the surface properties of cells of interest, cell-surface properties and, subsequently, their partitioning behavior can be altered. The use of such ligand-binding molecules in the two-phase aqueous system is referred to as cell affinity partitioning.

Antibodies are a useful tool in the affinity partitioning of molecules (Chapter 6), organelles (Chapter 9), and cells. The binding of native antibodies to the surface of bacteria is sometimes sufficient to alter their partitioning properties, as shown for *S. aureus* (74) and *Salmonella typhimurium* (124). More commonly however, the antibody has been modified by the addition of a functional group designed specifically as an aid in altering the partitioning of the cell. A frequently used cell affinity technique involves the attachment of a PEG moiety to a specific antibody (or other ligand-binding agent) that targets the cell of interest. Sharp et al. (113) demonstrated the usefulness of PEG-labeled antibodies for immunologically specific cell partitioning. In this study, the PEG-modified anti-human RBC antibodies caused the partitioning of human (but not rabbit) RBCs to shift from the lower to the upper phase of a two-phase dextran/PEG system. In a similar study, Karr et al. (55) partitioned human (but not sheep) RBCs into a rich upper phase using PEG-labeled anti-human RBC antibodies. This approach has been modified by first reacting the antibodies with tresylated monomethoxy (TMM)-PEG (18). In this study, anti-human RBC antibodies were modified with monomethoxy PEG, and the antibodies were allowed to react with a mixture of cells containing a very small percentage of erythrocytes. The PEG-antibody–labeled erythrocytes were then separated in a two-phase system composed of PEG and dextran. This procedure resulted in a highly enriched cellular fraction of erythrocytes (ca. 95% pure). The authors report that the cell populations could be completely separated by countercurrent distribution. While these examples use positive cell selection, negative selection can also be used with cell affinity partitioning, because either phase can be recovered for further study after the antibody-PEG complex drives the bound cells into the upper phase.

The use of antibodies modified by the addition of PEG requires specific chemical modification of every antibody that is going to be used in the selection process. An alternative approach when multiple antibodies are going to be used is to label a mixture of cells in suspension with underivatized antibodies, and then use PEG-labeled anti-immunoglobulins or PEG-labeled protein A to alter the partitioning properties of the cell-antibody complex. This approach eliminates the need to prepare a PEG-modified antibody for every cell type to be fractionated by aqueous affinity partitioning. In one study, a mouse monoclonal antibody directed against sheep IgG was labeled with PEG (125). Sheep anti-human RBC antibodies were then applied together with the PEG-labeled anti-sheep antibodies to a suspension of human RBCs. This combination increased the partitioning of human (but not rabbit) RBCs into the upper phase of a two-phase dextran-PEG system. Karr et al. (56) showed that PEG-modified protein A would shift the partitioning of cells bound by normal, unmodified antibodies.

A variety of other agents have been used to assist in affinity cell partitioning. One example used transferrin covalently bound to poly-L-lysine to alter the partitioning of rat erythroblasts in an aqueous two-phase system composed of dextran and PEG (79). Zijlstra et al. (146) described an affinity technique using an aqueous two-phase system for the separation of hybridoma cells from their IgG product. Since both the IgG and the hybridoma partition into the lower phase, they separated the cells from the IgG by using a commercially available dye coupled to PEG to specifically bind the IgG antibody and partition it into the upper phase.

PEG-metal chelates have been used for the partitioning extraction of erythrocytes (6). A more recent variation using metal ions is the work of Laboureau et al. (62), who used Sepharose-immobilized metal ions as an affinity ligand for cell partitioning in a dextran-PEG two-phase system. In one study, this system was shown to distinguish normal and sickle cell anemia RBCs, or normal and malaria-infected RBCs (86), depending on the metal ion concentration used. They could also use this system to distinguish the partitioning pattern of B lymphocytes, T lymphocytes, monocytes, and stem cells using different partitioning conditions (61). Their subsequent studies have demonstrated that cell separation in immobilized metal ion affinity partitioning systems is correlated with the affinity between the cell surface and the ligand (62).

An alternative approach for affinity cell partitioning is to perform the separation within a chromatographic column. In this case, one of the two liquid phases is immobilized on the surface of a chromatographic packing. Typically, this packing material is a chromatographic bead to which a phase such as PPG or PEG is chemically bonded. The bonded phase is more hydrophobic in nature than the mobile phase containing the applied cell suspension. Shibusawa (118) used this approach with columns packed with either bonded methoxyethoxymethyl groups or chemically bonded with PEG/PPG to separate human peripheral blood cells. Granulocytes and lymphocytes were more strongly retained than RBCs or platelets in the PEG/PPG columns, allowing fractionation of these cell populations. Methoxyethoxymethyl-Sepharose columns were used to isolate platelets and lymphocytes from human leukocyte-rich plasma using a sucrose elution buffer, and platelets could be selectively recovered from the same column by elution with α-methyl D-mannoside. Skuse and Brooks (120) immobilized the dextran-rich phase of a two-phase system onto chromatographic beads and used the PEG-rich phase as an eluent. This system effectively separated a 1:1 mixture of canine and human RBCs.

PROTOCOLS

Preparation of a Two-Phase System

1. Add 306 g of Dextran T-500 (Amersham Pharmacia Biotech) to 1 L of distilled H_2O. Add a stirrer bar, cover the beaker, and heat on a hot plate equipped with a magnetic stirrer, with constant stirring until the dextran is in solution (this may take several hours).

2. Add 667 g of PEG 6000 to 1 L of distilled H_2O in a beaker and stir with a stirrer bar until the PEG is in solution (this may also take several hours).

3. Separate the dextran and PEG solutions into convenient volumes (50–100 mL each), and store in a freezer until ready to use.

4. Warm aliquots of PEG and Dextran in a water bath to 25°C, along with 0.2 M sterile sodium phosphate (pH 7.2), 1 M sterile sodium chloride, 5% (wt/vol) sterile fetal calf serum (FCS), and sterile distilled H_2O.

5. Add the following weights in grams of each component to a container in the specific order given: 72 g PEG, 153 g Dextran T-500, 52 g sodium phosphate, 30 g sodium chloride, 29 g (solid) sucrose, 256 g of distilled H_2O, and 30 g of 5% (vol/vol) FCS.

   Note: As each component is added, care should be taken to avoid disturbing the lower layers.

6. Cover and warm completely to 25°C in a water bath.

7. Mix the solution in the tube thoroughly and centrifuge at 3000× *g* for 30 min.

8. Carefully collect the top and bottom layers separately, leaving behind a little of each layer near the interface.

   Note: Filter-sterilize the two layers through a 0.22-μm filter (Millipore) if the cells to be isolated are to be maintained in culture, and use standard aseptic techniques to avoid contamination.

9. Store the separate top and bottom layers at 4°C until ready to use.

Separation of Cells Using a Single-Step Procedure

1. Stir the separated top and bottom phases prepared in the protocol for the preparation of a two-phase system, and pre-warm the phases separately to 25°C in a water bath.

2. Transfer 1 mL of the well-mixed bottom phase to each of 4 test tubes.

3. Add 5 mL of the well-mixed top phase to a separate test tube. Add 50 μL of the packed mixed-cell solution containing a known number of cells, and mix gently.

4. Add 1 mL of the cell–top-layer mixture to each of the 4 tubes prepared in step 2.

5. Cap the tubes and return to the 25°C water bath for 10–15 min.

6. Mix the solutions by repeated gentle inversion (30 inversions/min) for 1-min intervals, then return the tubes to the water bath for 1-min intervals. Repeat this inversion/incubation process over a 10-min (total) time period.

7. Allow the phases to separate for 10 min.

8. Collect 10 μL from each of the upper and lower phases for cell counting.

9. Estimate the total number of cells in the top and bottom layer to determine the partitioning coefficient.

   Note: A coefficient of 50% would mean one-half of the added cells were located in the top phase. The number of cells remaining at the interface between the two layers can be estimated by subtracting the estimated total number of cells in each of the two layers from the total number of cells added.

The partitioning of cells between the upper and lower phase is governed by many variables including temperature, ionic strength, polymer concentration and composition in the two layers, and the presence of polymer ligands (such PEG-labeled antibodies or PEG-labeled fatty acids). The goal of aqueous partitioning is to find a mixture of these variables that will allow approximately 50% of the desired cell type to partition in the upper phase. Depending on the properties of the cell to be isolated, the composition may need to be altered to find a two-phase system that gives such a partitioning. The recommended way is to prepare a series of tubes containing different ratios of dextran (4%–5.5%) and PEG (3.5%–4.5%) to determine a polymer mixture that gives the best partitioning. Care should be taken to ensure that the cell population required is differentially partitioned from other cells in the mixture.

Separation of Cells Using a Multiple-Step Procedure

1. Recover and retain the top layer from a two-phase system (see the protocol for the separation of cells using a single-step).
2. Add 1 mL of fresh top layer (without any added cells) to the bottom layer.
3. Repeat steps 5–7 of the protocol for the separation of cells in a single-step.
4. Recover the top layer, and pool with that recovered in step 1.
5. Repeat steps 2–4 of this protocol.

This multistep procedure assumes that the desired cells are preferentially enriched in the upper phase. In the event that the desired cell is enriched relative to the contaminating cells in the lower phase, the lower phase should be collected and pooled, then fresh lower phase should be added to re-extract the upper phase.

Preparation of PEG-Labeled Antibodies

1. Dissolve 100 mg of TMM-PEG in 1 mL of 0.01 M PBS and mix.
2. Prepare 0.5 mL PBS containing 3 mg/mL of polyclonal antibody directed against the cells of interest.
3. Add 0.5 mL of the TMM-PEG solution to the 0.5-mL solution containing the antibodies.
4. Incubate for 2 h at room temperature on a rotary mixer. This solution can be stored at 4°C until ready for use.

Use of PEG-Labeled Antibodies in Aqueous Partitioning

1. Prepare a small volume of a 10% (vol/vol) heat-inactivated serum in 0.01 M PBS (PBS-10%S).
2. Add 1 mL of PBS-10%S to the TMM-PEG–antibody mixture (see the protocol for the preparation of PEG-labeled antibodies).
3. Incubate at room temperature for 2 h on a rotary mixer.
4. Add 2 mL of fresh 0.01 M PBS.
5. Suspend the mixed-cell suspension at 3×10^7 cells/mL in 0.01 M PBS.
6. Combine 1 mL of cell solution with 1 mL of the TMM-PEG solution (from step 4).

7. Incubate at 30°C for 30 min with gentle rotation.

8. Collect the cells by centrifugation at 500× *g* for 10 min at 4°C, and resuspend in 5 mL of top phase (see the protocol for the preparation of a two-phase system) pre-warmed to 25°C.

9. Add 5 mL of bottom phase (see the two-phase system protocol) pre-warmed to 25°C, and mix by gently inverting 60–100 times.

10. Allow the phases to separate at 25°C (in a water bath).

11. Recover and retain the top layer.

12. Add 5 mL of fresh top layer to the remaining bottom layer, mix by gently inverting 60–100 times, and allow the phases to separate at 25°C (in a water bath).

13. Recover the top layer and pool with the previously retained phase.

14. Repeat steps 12 and 13.

Using PEG-labeled ligands, it is useful to build a two-phase system with a low partitioning coefficient for all cells into the upper phase in the absence of added ligand. The addition of the PEG-labeled antibody will coat the targeted cell with PEG molecules and drive its partitioning into the upper, PEG-rich layer. With the introduction of magnetic bead technology, the isolation of cells from small volumes of starting material by aqueous partitioning has gone out of favor. However, the aqueous two-phase partitioning system is still very useful (26,116) if a thin-layer countercurrent distribution apparatus is available.

REFERENCES

1. **Albertsson, P.A., G. Johansson and F. Tjerneld.** 1990. Separation processes in biotechnology. Aqueous two-phase separations. Bioprocess Technol. *9*:287-327.
2. **Bachier, C.R., R.E. Giles, D. Ellerson, E.G. Hanania, F. Garcia-Sanchez, M. Andreeff, F. Cabanillas, R. Champlin et al.** 1999. Hematopoietic retroviral gene marking in patients with follicular non-Hodgkin's lymphoma. Leuk. Lymphoma *32*:279-288.
3. **Beltinger, C.P. and K.M. Debatin.** 1998. A simple combined microdissection and aspiration device for the rapid procurement of single cells from clinical peripheral blood smears. Mol. Pathol. *51*:233-236.
4. **Bernd, P., M.D. Gershon, E.A. Nunez and H. Tamir.** 1981. Separation of dissociated thyroid follicular and parafollicular cells: association of serotonin binding protein with parafollicular cells. J. Cell Biol. *88*:499-508.
5. **Bickenbach, J.R. and E. Chism.** 1998. Selection and extended growth of murine epidermal stem cells in culture. Exp. Cell Res. *244*:184-195.
6. **Birkenmeier, G., H. Walter and K.E. Widen.** 1994. Factors in the affinity extraction of red blood cells using poly(ethylene glycol)-metal chelate. Methods Enzymol. *228*:368-377.
7. **Bonnafous, J.C., J. Dornand, J. Favero, M. Sizes, E. Boschetti and J.C. Mani.** 1983. Cell affinity chromatography with ligands immobilized through cleavable mercury-sulfur bonds. J. Immunol. Methods *58*:93-107.
8. **Bousso, P., F. Michel, N. Pardigon, N. Bercovici, R. Kiblau, P. Kourilsky and J.P. Abastado.** 1997. Enrichment of antigen-specific T lymphocytes by panning on immobilized MHC-peptide complexes. Immunol. Lett. *59*:85-91.
9. **Cambier, J.C. and M.J. Neale.** 1982. Isolated phosphorylcholine binding lymphocytes. I. Use of a cleavable crosslinking reagent for solid-phase adsorbent isolation of functional antigen binding cells. J. Immunol. Methods *51*:209-221.
10. **Canals, C., C. Torrico, M. Picon, B. Amill, J.A. Cancelas, G. Fraga, I. Badell, J. Cubells et al.** 1997. Immunomagnetic bone marrow purging in children with acute lymphoblastic leukemia. J. Hematother. *6*:261-268.
11. **Cardoso, A.A., S.M. Watt, P. Batard, M.L. Li, A. Hatzfeld, H. Genevier and J. Hatzfeld.** 1995. An improved panning technique for the selection of CD34+ human bone marrow hematopoietic cells with high recovery of early progenitors. Exp. Hematol. *23*:405-412.

12. **Cavaillon, J.-M., B. Cinader, C.T. Chou and S. Dubiski.** 1979. Purification of rabbit B spleen cells by removal of adherent and of T cells. J. Immunol. Methods *26*:1-10.

13. **Cha, M.S., D.K. Rah and K.J. Lee.** 1996. Isolation and pure culture of microvascular endothelial cells from the fetal skin. Yonsei Med. J. *37*:186-193.

14. **Chapman, P.A., A.T. Malo, C.A. Siddons and M. Harkin.** 1997. Use of commercial enzyme immunoassays and immunomagnetic separation systems for detecting *Escherichia coli* O157 in bovine fecal samples. Appl. Environ. Microbiol. *63*:2549-2553.

15. **Christenson, L.K. and R.L. Stouffer.** 1996. Isolation and culture of microvascular endothelial cells from the primate corpus luteum. Biol. Reprod. *55*:1397-1404.

16. **Conrad-Lapostolle, V., L. Bordenave and C. Baquey.** 1996. Optimization of use of UEA-1 magnetic beads for endothelial cell isolation. Cell Biol. Toxicol. *12*:189-197.

17. **Cottler-Fox, M., K. Cipolone, M. Yu, R. Berenson, J. O'Shaughnessy and C. Dunbar.** 1995. Positive selection of CD34+ hematopoietic cells using an immunoaffinity column results in T cell-depletion equivalent to elutriation. Exp. Hematol. *23*:320-322.

18. **Delgado, C., R.J. Anderson, G.E. Francis and D. Fisher.** 1991. Separation of cell mixtures by immunoaffinity cell partitioning: strategies for low abundance cells. Anal. Biochem. *192*:322-328.

19. **Djilali, S., A.L. Parodi and D. Levy.** 1987. Wheat germ agglutinin (WGA): a bovine T lymphocyte marker. Vet. Immunol. Immunopathol. *16*:151-154.

20. **Dong, Q.G., S. Bernasconi, S. Lostaglio, R.W. De Calmanovici, I. Martin-Padura, F. Breviario, C. Garlanda, S. Ramponi, A. Mantovani and A. Vecchi.** 1997. A general strategy for isolation of endothelial cells from murine tissues. Characterization of two endothelial cell lines from the murine lung and subcutaneous sponge implants. Arterioscler. Thromb. Vasc. Biol. *8*:1599-1604.

21. **Duffey, P.S., D.L. Drouillard and C.P. Barbe.** 1981. Lymphocyte sorting on albuminated CIBA blue dextran-staphylococcal protein A-conjugated sepharose 6MB affinity columns. J. Immunol. Methods *45*:137-151.

22. **Dvorak, D.J., E. Gipps and C. Kidson.** 1978. Isolation of specific neurones by affinity methods. Nature *271*:564-566.

23. **Elliott, B.E. and H.F. Pross.** 1984. Rosetting techniques to detect cell surface markers on mouse and lymphoreticular cells. Methods Enzymol. *108*:49-64.

24. **Ettensohn, C.A. and D.R. McClay.** 1987. A new method for isolating primary mesenchyme cells of the sea urchin embryo. Panning on wheat germ agglutinin-coated dishes. Exp. Cell Res. *168*:431-438.

25. **Fischer, H., A. Gjorloff, G. Hedlund, H. Hedman, E. Lundgren, T. Kalland, H.O. Sjogren and M. Dohlsten.** 1992. Stimulation of human naive and memory T helper cells with bacterial superantigen. Naive CD4+45RA+ T cells require a costimulatory signal mediated through the LFA-1/ICAM-1 pathway. J. Immunol. *148*:1993-1998.

26. **Fisher, D., G.F. Francis and D. Rickwood.** 1998. Cell Separation: A Practical Approach, p. 1-288. Oxford University Press, Oxford.

27. **Flo, R.W., A. Naess, F. Lund-Johansen, B.O. Maehle, H. Sjursen, V. Lehmann and C.O. Solberg.** 1991. Negative selection of human monocytes using magnetic particles covered by anti-lymphocyte antibodies. J. Immunol. Methods *137*:89-94.

28. **Formanek, M., A. Temmel, B. Knerer, M. Willheim, W. Millesi and J. Kornfehl.** 1998. Magnetic cell separation for purification of human oral keratinocytes: an effective method for functional studies without prior cell subcultivation. Eur. Arch. Otorhinolaryngol. *255*:211-215.

29. **Gee, A.P., V.H. Mansour and M.B. Weiler.** 1991. Effects of target antigen density on the efficacy of immunomagnetic cell separation. J. Immunol. Methods *142*:127-136.

30. **Ghetie, V., G. Mota and J. Sjoquist.** 1978. Separation of cells by affinity chromatography on SpA-sepharose 6MB. J. Immunol. Methods *21*:133-141.

31. **Ghetie, V. and J. Sjoquist.** 1975. Separation of lymphocytes by specific adherence to cellular monolayers containing protein A of *Staphylococcus aureus*. J. Immunol. *115*:659-664.

32. **Ghetie, V., G. Stalenheim and J. Sjoquist.** 1975. Cell separation by staphylococcal protein A-coated erythrocytes. Scand. J. Immunol. *4*:471-477.

33. **Gorczynska, E., J. Toporski, J. Boguslawska-Jaworska and M. Slociak.** 1998. High-dose chemotherapy with autologous hematopoietic progenitor cell transplantation in children with high-risk NHL and ALL-preliminary results. Bone Marrow Transplant. *22* Suppl *4*:S107-S109.

34. **Greaves, M.F. and G. Brown.** 1974. Purification of human T and B lymphocytes. J. Immunol. *112*:420-423.

35. **Grimsley, P.G., T.A. Amos, M.Y. Gordon and M.F. Greaves.** 1993. Rapid positive selection of CD34+ cells using magnetic microspheres coated with monoclonal antibody QBEND/10 linked via a cleavable disulphide bond. Leukemia *7*:898-908.

36. **Grupp, C., I. Troche, J. Steffgen, S. Langhans, D.I. Cohen, L. Brandl and G.A. Muller.** 1998. Highly specific separation of heterogeneous cell populations by lectin-coated beads: application for the isolation of inner medullary collecting duct cells. Exp. Nephrol. *6*:542-550.

37. **Gupta, K., S. Ramakrishnan, P.V. Browne, A. Solovey and R.P. Hebbel.** 1997. A novel technique for

culture of human dermal microvascular endothelial cells under either serum-free or serum-supplemented conditions: isolation by panning and stimulation with vascular endothelial growth factor. Exp. Cell Res. *230*:244-255.

38. **Harris, D.T., J.L. Iglesias, S. Argov, J. Toomey and H.S. Koren.** 1987. Heterogeneity of human natural killer (NK) cells: enrichment of NK by negative-selection with the lectin from *Erythrina cristagalli*. J. Leukoc. Biol. *42*:163-170.

39. **Hawkins, T.E., S.B. Marley, S.G. O'Brien, M.Y. Gordon and J.M. Goldman.** 1997. CD34+ cell selection in chronic phase chronic myeloid leukaemia: a comparison of laboratory grade columns. Bone Marrow Transplant. *20*:409-413.

40. **Heddini, A., C.J. Treutiger and M. Wahlgren.** 1998. Enrichment of immunoglobulin binding Plasmodium falciparum-infected erythrocytes using anti-immunoglobulin-coated magnetic beads. Am. J. Trop. Med. Hyg. *59*:663-666.

41. **Hedman, H. and E. Lundgren.** 1992. Regulation of LFA-1 avidity in human B cells. Requirements for dephosphorylation events for high avidity ICAM-1 binding. J. Immunol. *149*:2295-2299.

42. **Hedman, J., H. Branden and E. Lundgren.** 1992. Physical separation of ICAM-1 binding cells. J. Immunol. Methods *146*:203-211.

43. **Hedrum, A., J. Lundeberg, C. Pahlson and M. Uhlen.** 1992. Immunomagnetic recovery of Chlamydia trachomatis from urine with subsequent colorimetric DNA detection. PCR Methods Appl. *2*:167-171.

44. **Helmrich, A. and D. Barnes.** 1998. Animal cell culture equipment and techniques, p. 1-368. *In* J.P. Mather and D. Barnes (Eds.), Animal Cell Culture Methods. Academic Press, San Diego.

45. **Higgins, D.A. and W.K. Ko.** 1995. Duck lymphocytes. VII. Selection of subpopulations using lectin-coated magnetic beads. Vet. Immunol. Immunopathol. *44*:181-195.

46. **Hoffmann, S., C. Spee, T. Murata, J.Z. Cui, S.J. Ryan and D.R. Hinton.** 1998. Rapid isolation of choriocapillary endothelial cells by Lycopersicon esculentum-coated Dynabeads. Graefes Arch. Clin. Exp. Ophthalmol. *236*:779-784.

47. **Hynes, R.O., A.T. Destree and I. Mauve.** 1976. Spatial organization at the cell surface, p. 189-201. *In* V.T. Marchesi (Ed.), Membranes and Neoplasia: New Approaches and Strategies. Alan R. Liss, New York.

48. **Irlé, C., P.F. Piguet and P. Vassalli.** 1978. In vitro maturation of immature thymocytes into immunocompetent T cells in the absence of direct thymic influence. J. Exp. Med. *148*:32-45.

49. **Jay, R.J.** 1998. Cell line availability: where to get the cell lines you need, p. 31-47. *In* J.P. Mather and D. Barnes (Eds.), Animal Cell Culture Methods. Academic Press, San Diego.

50. **Johne, B. and J. Jarp.** 1988. A rapid assay for protein-A in Staph. aureus strains, using immunomagnetic monosized polymer particles. APMIS *96*:43-49.

51. **Johnson, R.J., R.G. Owen, G.M. Smith, M. Galvin, L.J. Newton, A. Rawstron, K. Major, V. Woodhead et al.** 1996. Peripheral blood stem cell transplantation in myeloma using CD34 selected cells. Bone Marrow Transplant. *17*:723-727.

52. **Jones, P.H. and F.M. Watt.** 1993. Separation of human epidermal stem cells from transit amplifying cells on the basis of differences in integrin function and expression. Cell *73*:713-724.

53. **Julius, M.H., E. Simpson and L.A. Herzenberg.** 1973. A rapid method for the isolation of functional thymus-derived murine lymphocytes. Eur. J. Immunol. *3*:645-649.

54. **Kandzia, J., W. Scholz, M.J. Anderson and W. Muller-Ruchholtz.** 1985. Magnetic albumin/protein A immunomicrospheres. II. Specificity, reproducibility, and resolution of the magnetic cell separation technique. Diagn. Immunol. *3*:83-88.

55. **Karr, L.J., S.G. Shafer, J.M. Harris, J.M. Van Alstine and R.S. Snyder.** 1986. Immuno-affinity partition of cells in aqueous polymer two-phase systems. J. Chromatogr. *354*:269-282.

56. **Karr, L.J., J.M. Van Alstine, R.S. Snyder, S.G. Shafer and J.M. Harris.** 1988. Cell separation by immunoaffinity partitioning with polyethylene glycol-modified protein A in aqueous polymer two-phase systems. J. Chromatogr. *442*:219-227.

57. **Kataoka, K., Y. Sakurai, T. Hanai, A. Maruyama and T. Tsuruta.** 1988. Immunoaffinity chromatography of lymphocyte subpopulations using tert-amine derived matrices with adsorbed antibodies. Biomaterials *9*:218-224.

58. **Kim, N.S. and S.J. Kim.** 1991. Isolation and cultivation of microvascular endothelial cells from rat lungs: effects of gelatin substratum and serum. Yonsei Med. J. *32*:303-314.

59. **Kogler, G., T. Somville and P. Werner.** 1994. Evaluation of the Ceprate LC34 affinity system for the enrichment of CD34+ hematopoietic stem cells from cord blood. Blood Cells *20*:371-375.

60. **Kondorosi, E., J. Nagy and G. Denes.** 1977. Optimal conditions for the separation of rat T lymphocytes on anti-immunoglobulin–immunoglobulin affinity columns. J. Immunol. Methods *16*:1-13.

61. **Laboureau, E., J.C. Capiod, C. Dessaint, L. Prin and M.A. Vijayalakshmi.** 1996. Study of human cord blood lymphocytes by immobilized metal ion affinity partitioning. J. Chromatogr. B Biomed. Appl. *680*:189-195.

62. **Laboureau, E. and M.A. Vijayalakshmi.** 1997. Concerning the separation of mammalian cells in immobilized metal ion affinity partitioning systems: a matter of selectivity. J. Mol. Recognit. *10*:262-268.

63. **Langone, J.J.** 1982. Applications of immobilized protein A in immunochemical techniques. J. Immunol. Methods 55:277-296.

64. **Lemmer, E.R., E.G. Shepard, K. Blakolmer, R.E. Kirsch and S.C. Robson.** 1998. Isolation from human fetal liver of cells co-expressing CD34 haematopoietic stem cell and CAM 5.2 pancytokeratin markers. J. Hepatol. 29:450-454.

65. **Levy, E.M., S. Zanki and H. Walter.** 1981. Countercurrent distribution of human peripheral blood lymphocytes: isolation of a subpopulation enriched with natural killer and K cells. Eur. J. Immunol. 11:952-955.

66. **Ljungquist, C., J. Lundeberg, A.M. Rasmussen, E. Hornes and M. Uhlen.** 1993. Immobilization and recovery of fusion proteins and B-lymphocyte cells using magnetic separation. DNA Cell Biol. 12:191-197.

67. **Lodge, P.A., C.E. Haisch and F.T. Thomas.** 1992. A simple method of vascular endothelial cell isolation. Transplant. Proc. 24:2816-2817.

68. **Mage, M.G.** 1984. Cell separation on cellular monolayers. Methods Enzymol. 108:125-132.

69. **Mage, M.G.** 1984. Separation of lymphocytes on antibody-coated plates. Methods Enzymol. 108:118-124.

70. **Manderino, G.L., G.T. Gooch and A.B. Stavitsky.** 1978. Preparation, characterization, and functions of rabbit lymph node cell populations. I. Preparation of KLH primed T and B memory cells with anti-fab' affinity columns. Cell Immunol. 41:264-275.

71. **Mandrusov, E., A. Houng, E. Klein and E.F. Leonard.** 1995. Membrane-based cell affinity chromatography to retrieve viable cells. Biotechnol. Prog. 11:208-213.

72. **Maruyama, A., T. Tsuruta, K. Kataoka and Y. Sakurai.** 1988. Separation of B- and T-lymphocytes by cellular adsorption chromatography using poly(2-hydroxyethyl methacrylate)/polyamine graft copolymer as column adsorbent. J. Biomed. Mater. Res. 22:555-571.

73. **Maruyama, A., T. Tsuruta, K. Kataoka and Y. Sakurai.** 1989. Separation of B and T lymphocytes by cellular adsorption chromatography with polyamine graft copolymers as column matrices. II. Recovery of adsorbed B cell enriched populations from the column. Biomaterials 10:393-399.

74. **Mattiasson, B., T.G. Ling and M. Ramstorp.** 1981. Application of partition affinity ligand assay (PALA) in quick test for quantitation of *Staphylococcus aureus* bacterial cells. J. Immunol. Methods 41:105-114.

75. **Mattsson, L., G. Bondjers and O. Wiklund.** 1991. Isolation of cell populations from arterial tissue, using monoclonal antibodies and magnetic microspheres. Atherosclerosis 89:25-34.

76. **Mazurek, G.H., V. Reddy, D. Murphy and T. Ansari.** 1996. Detection of Mycobacterium tuberculosis in cerebrospinal fluid following immunomagnetic enrichment. J. Clin. Microbiol. 34:450-453.

77. **McDouall, R.M., M. Yacoub and M.L. Rose.** 1996. Isolation, culture, and characterisation of MHC class II-positive microvascular endothelial cells from the human heart. Microvasc. Res. 51:137-152.

78. **McGuire, P.G. and R.W. Orkin.** 1987. Methods in laboratory investigation. Isolation of rat aortic endothelial cells by primary explant techniques and their phenotypic modulation by define substrata. Lab. Invest. 57:94-105.

79. **Mendieta, J. and G. Johansson.** 1992. Affinity-mediated modification of electrical charge on a cell surface: a new approach to the affinity partitioning of biological particles. Anal. Biochem. 200:280-285.

80. **Morecki, S., S.L. Topalian, W.W. Myers, D. Okrongly, T.B. Okarma and S.A. Rosenberg.** 1990. Separation and growth of human CD4+ and CD8+ tumor-infiltrating lymphocytes and peripheral blood mononuclear cells by direct positive panning on covalently attached monoclonal antibody-coated flasks. J. Biol. Response Mod. 9:463-474.

81. **Mosier, D.E.** 1984. Separation of macrophages on plastic and glass surfaces, p. 294–297. *In* G. Di Sabato, J.J. Langone and H. van Vunakis (Eds.), Immunochemical Techniques. Part G. Separation and Characterization of Lymphoid Cells. Academic Press, Orlando.

82. **Muller, P., D. Hempel, D. Oruzio, S. Ehnle, K. Kolloch and G. Schlimok.** 1998. Sequential immunotyping and genotyping of tumor cells in bone marrow of cancer patients: a model study. Cytometry 33:429-497.

83. **Murphy, S.J., D.J. Watt and G.E. Jones.** 1992. An evaluation of cell separation techniques in a model mixed cell population. J. Cell Sci. 102:789-798.

84. **Nakamura, N., J.G. Burgess, K. Yagiuda, S. Kudo, T. Sakaguchi and T. Matsunaga.** 1993. Detection and removal of Escherichia coli using fluorescein isothiocyanate conjugated monoclonal antibody immobilized on bacterial magnetic particles. Anal. Chem. 65:2036-2039.

85. **Nakamura, N., K. Hashimoto and T. Matsunaga.** 1991. Immunoassay method for the determination of immunoglobulin G using bacterial magnetic particles. Anal. Chem. 63:268-272.

86. **Nanak, E., M.A. Vijayalakshmi and D.E. Chadwick.** 1995. Segregation of normal and pathological human red blood cells, lymphocytes and fibroblasts by immobilized metal-ion affinity partitioning. J. Mol. Recognit. 8:77-84.

87. **Nash, A.A.** 1976. Separation of lymphocyte sub-populations using antibodies attached to staphylococcal protein A-coated surfaces. J. Immunol. Methods 12:149-161.

259

88. **Natali, P.G., O. Segatto, G. Zupi, R. Cavaliere, P. Giacomini and S. Ferrone.** 1983. Isolation of viable melanoma cells from surgically removed lesions using dishes coated with monoclonal antibody to a high molecular weight melanoma associated antigen. J. Immunol. Methods *62*:337-346.

89. **Nicola, N.A., A.W. Burgess, D. Metcalf and F.L. Battye.** 1978. Separation of mouse bone marrow cells using wheat germ agglutinin affinity chromatography. Aust. J. Exp. Biol. Med. Sci. *56*:663-679.

90. **Okarma, T., J. Lebkowski, L. Schain, M. Harvey, G. Tricot, E. Srour, W.G. Meyers, A. Burnett, I. Sniecinski and R.J. O'Reilly.** 1992. The AIS CELLector: a new technology for stem cell purification. Prog. Clin. Biol. Res. *377*:487-504.

91. **Opas, M. and E. Dziak.** 1988. Effects of substrata and method of tissue dissociation on adhesion, cytoskeleton, and growth of chick retinal pigmented epithelium in vitro. In Vitro Cell Dev. Biol. *24*:885-892.

92. **Paraskeva, C., B.G. Buckle and P.E. Thorpe.** 1985. Selective killing of contaminating human fibroblasts in epithelial cultures derived from colorectal tumours using an anti Thy-1 antibody-ricin conjugate. Br. J. Cancer *51*:131-134.

93. **Pereira, M.E. and E.A. Kabat.** 1979. A versatile immunoadsorbent capable of binding lectins of various specificities and its use for the separation of cell populations. J. Cell Biol. *82*:185-194.

94. **Persson, A., B. Johansson, H. Olsson and B. Jergil.** 1991. Purification of rat liver plasma membranes by wheat-germ-agglutinin affinity partitioning. Biochem. J. *273*:173-177.

95. **Phillips, T.M., W.D. Queen, N.S. More and A.M. Thompson.** 1985. Protein A-coated glass beads. Universal support medium for high-performance immunoaffinity chromatography. J. Chromatogr. *327*:213-219.

96. **Porter, J., J. Robinson, R. Pickup and C. Edwards.** 1998. An evaluation of lectin-mediated magnetic bead cell sorting for the targeted separation of enteric bacteria. J. Appl. Microbiol. *84*:722-732.

97. **Rader, R.K., L.E. Kahn, G.D. Anderson, C.L. Martin, K.S. Chinn and S.A. Gregory.** 1996. T cell activation is regulated by voltage-dependent and calcium-activated potassium channels. J. Immunol. *156*:1425-1430.

98. **Rambaldi, A., G. Borleri, G. Dotti, P. Bellavita, R. Amaru, A. Biondi and T. Barbui.** 1998. Innovative two-step negative selection of granulocyte colony-stimulating factor-mobilized circulating progenitor cells: adequacy for autologous and allogeneic transplantation. Blood *91*:2189-2196.

99. **Reisner, A., J.J. Olson, J. Yang, R. Assietti, J.M. Klemm and P.R. Girard.** 1995. Isolation and culture of bovine intracranial arterial endothelial cells. Neurosurgery *36*:806-812.

100. **Reisner, Y. and N. Sharon.** 1984. Fractionation of subpopulations of mouse and human lymphocytes by peanut agglutinin or soybean agglutinin. Methods Enzymol. *108*:168-179.

101. **Rembaum, A. and W.J. Dreyer.** 1980. Immunomicrospheres: reagents for cell labeling and separation. Science *208*:364-368.

102. **Richard, L., P. Velasco and M. Detmar.** 1998. A simple immunomagnetic protocol for the selective isolation and long-term culture of human dermal microvascular endothelial cells. Exp. Cell Res. *240*:1-6.

103. **Rochelle, P.A., R. De Leon, A. Johnson, M.J. Stewart and R.L. Wolfe.** 1999. Evaluation of immunomagnetic separation for recovery of infectious cryptosporidium parvum oocysts from environmental samples. Appl. Environ. Microbiol. *65*:841-845.

104. **Rutishauser, U., P. D'Eustachio and G.M. Edelman.** 1973. Immunological functions of lymphocytes fractionated with antigen-derivatized fibers. Proc. Natl. Acad. Sci. USA *70*:3894-3898.

105. **Rye, P.D. and N.V. Bovin.** 1998. Selection of carbohydrate-binding cell phenotypes using oligosaccharide-coated magnetic particles. Glycobiology *7*:179-182.

106. **Rye, P.D., H.K. Hoifodt, G.E. Overli and O. Fodstad.** 1997. Immunobead filtration: a novel approach for the isolation and propagation of tumor cells. Am. J. Pathol. *150*:99-106.

107. **Sad, S., R. Marcotte and T.R. Mosmann.** 1995. Cytokine-induced differentiation of precursor mouse CD8+ T cells into cytotoxic CD8+ T cells secreting Th1 or Th2 cytokines. Immunity *2*:271-279.

108. **Schrempf-Decker, G.E., D. Baron and P. Wernet.** 1980. Helix pomatia agglutinin (HpA) affinity chromatography: the isolation of pure B and T cell populations and their use for the routine HLA-DR (Ia) serology. J. Immunol. Methods *32*:285-296.

109. **Scott, D.W.** 1976. Antifluorescein affinity columns. Isolation and immunocompetence of lymphocytes that bind fluoresceinated antigens in vivo or in vitro. J. Exp. Med. *144*:69-78.

110. **Shah, P.M., S. Husby, T.E. Damsgaard, H.V. Nielsen and P.O. Schiotz.** 1998. Purification of human colonic and gastric mast cells. J. Immunol. Methods *214*:141-148.

111. **Sharma, S.K. and P.P. Mahendroo.** 1980. Affinity chromatography of cells and cell membranes. J. Chromatogr. *184*:471-499.

112. **Sharon, N.** 1984. Use of lectins for separation of cells, p. 13-52. *In* T.G. Pretlow and T.P. Pretlow (Eds.), Cell Separation Methods and Selected Applications, Vol. 3. Academic Press, New York.

113. **Sharp, K.A., M. Yalpani, S.J. Howard and D.E. Brooks.** 1986. Synthesis and application of a poly(ethylene glycol)-antibody affinity ligand for cell separations in aqueous polymer two-phase systems. Anal. Biochem. *154*:110-117.

114. **Sharpe, P.T.** 1988. Aqueous two-phase partition, p. 107-142. *In* R.H. Burdon and P.H. van Knippenberg

(Eds.), Methods in Cell Separation. Elsevier, Amsterdam.

115. **Sharpe, P.T.** 1988. Flow sorting, p. 208-237. *In* R.H. Burdon and P.H. van Knippenberg (Eds.), Methods in Cell Separation. Elsevier, Amsterdam.

116. **Sharpe, P.T.** 1988. Methods of Cell Separation. Elsevier, Amsterdam.

117. **Shen, F.W.** 1984. Use of antisera and complement for the purification of lymphocyte subpopulations. Methods Enzymol. *108*:249-253.

118. **Shibusawa, Y.** 1999. Surface affinity chromatography of human peripheral blood cells. J. Chromatogr. B Biomed. Sci. Appl. *722*:71-88.

119. **Sinclair, B.** 1998. To bead or not to bead. The Scientist *12*:17.

120. **Skuse, D.R. and D.E. Brooks.** 1988. Column-based separation of erythrocytes using aqueous polymeric two-phase systems. J. Chromatogr. *432*:127-135.

121. **Skuse, D.R., R. Norris-Jones, M. Yalpani and D.E. Brooks.** 1992. Hydroxypropyl cellulose/poly(ethylene glycol)-co-poly(propylene glycol) aqueous two-phase systems: system characterization and partition of cells and proteins. Enzyme Microb. Technol. *14*:785-790.

122. **Soderman, D.D., J. Germershauden and H.M. Katzen.** 1973. Affinity binding of intact fat cells and their ghosts to immobilized insulin. Proc. Natl. Acad. Sci. USA *70*:792-796.

123. **Stark, M., E. Reizenstein, M. Uhlen and J. Lundeberg.** 1996. Immunomagnetic separation and solid-phase detection of Bordetella pertussis. J. Clin. Microbiol. *34*:778-784.

124. **Stendahl, O., C. Tagesson, K.E. Magnusson and L. Edebo.** 1977. Physiochemical consequences of opsonization of Salmonella typhimurium with hyperimmune IgG and complement. Immunology *32*:11-18.

125. **Stocks, S.J. and D.E. Brooks.** 1988. Development of a general ligand for immunoaffinity partitioning in two phase aqueous polymer systems. Anal. Biochem. *173*:86-92.

126. **Stringa, E., J.M. Love, S.C. McBride, E. Suyama and R.S. Tuan.** 1997. In vitro characterization of chondrogenic cells isolated from chick embryonic muscle using peanut agglutinin affinity chromatography. Exp. Cell Res. *232*:287-294.

127. **Sun, L., M. Zborowski, L.R. Moore and J.J. Chalmers.** 1998. Continuous, flow-through immunomagnetic cell sorting in a quadrupole field. Cytometry *33*:469-475.

128. **Suzumura, A., S. Bhat, P.A. Eccleston, R.P. Lisak and D.H. Silberberg.** 1984. The isolation and long-term culture of oligodendrocytes from newborn mouse brain. Brain Res. *324*:379-383.

129. **Tai, P.C. and C.J. Spry.** 1977. Purification of normal human eosinophils using the different binding capacities of blood leucocytes for complexed rabbit IgG. Clin. Exp. Immunol. *28*:256-260.

130. **Tassi, C., A. Fortuna, A. Bontadini, R.M. Lemoli, M. Gobbi and P.L. Tazzari.** 1991. CD34 or S313 positive cells selection by avidin-biotin immunoadsorption. Haematologica 76 Suppl *1*:41-43.

131. **Thomas, T.E., H.J. Sutherland and P.M. Lansdorp.** 1989. Specific binding and release of cells from beads using cleavable tetrameric antibody complexes. J. Immunol. Methods *120*:221-231.

132. **Tlaskalova-Hogenova, H., V. Vetvicka, M. Pospisil, L. Fornusek, L. Prokesova, J. Coupek, A. Frydrychova, J. Kopecek, H. Fiebig and J. Brochier.** 1986. Separation of human lymphoid cells by affinity chromatography and cell surface labelling by hydroxyethyl methacrylate particles using monoclonal antibodies. J. Chromatogr. *376*:401-408.

133. **Trizio, D. and G. Cudkowicz.** 1974. Separation of T and B lymphocytes by nylon wool columns: evaluation of efficacy by functional assays in vivo. J. Immunol. *113*:1093-1097.

134. **Tsuru, S., M. Taniguchi, M. Tsugita, S. Sekiguchi and K. Nomoto.** 1988. A rapid method for the isolation of functional human T lymphocytes using hydroxyapatite column fractionation. J. Immunol. Methods *106*:169-174.

135. **Ugelstad, J., P. Stenstad, L. Kilaas, W.S. Prestvik, R. Herje, A. Berge and E. Hornes.** 1993. Monodisperse magnetic polymer particles. New biochemical and biomedical applications. Blood Purif. *11*:349-369.

136. **van Overveld, F.J., G.K. Terpstra, P.L. Bruijnzeel, J.A. Raaijmakers and J. Kreukniet.** 1988. The isolation of human lung mast cells by affinity chromatography. Scand. J. Immunol. *27*:1-6.

137. **Walter, H.** 1994. Analytical applications of partitioning: detection of differences or changes in surface properties of mammalian cell populations. Methods Enzymol. *228*:299-320.

138. **Whitehurst, C.E., N.K. Day and N. Gengozian.** 1994. A method of purifying feline T lymphocytes from peripheral blood using the plant lectin from Pisum sativum. J. Immunol. Methods *175*:189-199.

139. **Widder, K.J., A.E. Senyei, H. Ovadia and P.Y. Paterson.** 1979. Magnetic protein A microspheres: a rapid method for cell separation. Clin. Immunol. Immunopathol. *14*:395-400.

140. **Widder, K.J., A.E. Senyei, H. Ovadia and P.Y. Paterson.** 1981. Specific cell binding using staphylococcal protein A magnetic microspheres. J. Pharm. Sci. *70*:387-389.

141. **Widjojoatmodjo, M.N., A.C. Fluit, R. Torensma and J. Verhoef.** 1993. Comparison of immunomagnetic beads coated with protein A, protein G, or goat anti-mouse immunoglobulins. Applications in enzyme immunoassays and immunomagnetic separations. J. Immunol. Methods *165*:11-19.

142. **Wright, A.P., J.J. Fitzgerald and R.J. Colello.** 1997. Rapid purification of glial cells using immunomagnetic separation. J. Neurosci. Methods *74*:37-44.

143. **Wu, A.M., S.J. Sugii, A. Herp, A.A. Cardoso, S.M. Watt, P. Batard, M.L. Li, A. Hatzfeld, H. Genevier and J. Hatzfeld.** 1995. A guide for carbohydrate specificities of lectins. An improved panning technique for the selection of CD34+ human bone marrow hematopoietic cells with high recovery of early progenitors. Exp. Hematol. *23*:407-412.

144. **Yu, H.** 1998. Comparative studies of magnetic particle-based solid phase fluorogenic and electrochemiluminescent immunoassay. J. Immunol. Methods *218*:1-8.

145. **Yui, J., M. Garcia-Lloret, T.G. Wegmann and L.J. Guilbert.** 1994. Cytotoxicity of tumour necrosis factor-alpha and gamma-interferon against primary human placental trophoblasts. Placenta *15*:819-835.

146. **Zijlstra, G.M., M.J. Michielsen, C.D. de Gooijer, L.A. van der Pol and J. Tramper.** 1996. Separation of hybridoma cells from their IgG product using aqueous two-phase systems. Bioseparation *6*:201-210.

9 | Affinity Subcellular Separations

Subcellular organelles such as mitochondria, plasma membranes, lysosomes, and nuclei are routinely isolated by physical techniques following cell lysis under conditions that do not disrupt their integrity. Traditionally, these isolation techniques have involved cell homogenization followed by separation techniques based on differences in organelle size, shape, density, surface charge, or a combination of these physical properties. While remarkable enrichments in these cytoplasmic organelles can be achieved through the use of physical separation techniques, advances in affinity and immunoaffinity technology offer an opportunity for gentler and more rapid organelle isolation, which also generally results in higher enrichment for the desired organelle than traditional methods. The general approaches for the isolation of membrane organelles by affinity or immunoaffinity techniques are virtually identical to the techniques discussed in Chapter 8 for the isolation of cells, except that the affinity–ligand reaction takes place on the surface of the organelle of interest. Despite similarities with the well-established affinity procedures used for cell isolation and the increased speed and ease of these techniques (77), cytoplasmic organelle isolation by differential centrifugation still remains the most popular technique. Rather than provide affinity isolation techniques for all of the various subcellular organelles, this chapter will highlight advances in affinity and immunoaffinity isolation of cytoplasmic organelles that may someday help affinity techniques become as popular in membrane organelle isolation as the physical approaches are today.

DIFFERENTIAL CELL FRACTIONATION

The initial steps in subcellular membrane organelle isolation protocols include the recovery of the cell or tissues from which organelles will be prepared, followed by cell or tissue homogenization to expose the individual organelles for isolation. The next step is usually designed to remove any remaining nonlysed cells or tissue debris. This step is often accomplished by very low speed centrifugation, which can be preceded by initially passing the cell homogenate through loosely packed gauze when the disrupted cells are part of intact tissues such as liver or myocardium. At this point, traditional or affinity and immunoaffinity techniques often diverge. While traditional

Affinity and Immunoaffinity Purification Techniques
Terry M. Phillips and Benjamin F. Dickens
© 2000 Eaton Publishing, Natick, MA

isolation protocols continue to use a wide variety of physical techniques (see Reference 39 for a review), affinity separation techniques rely on the specificity of ligand binding to recover specific cytoplasmic organelles. This means that the organelle of interest can sometimes be directly recovered from the cell lysate by appropriate affinity or immunoaffinity protocols. However, it can also be useful to perform differential subcellular fractionation based on physical characteristics of the membrane organelle being isolated prior to the application of affinity or immunoaffinity techniques.

Cell Homogenization

Table 1 provides a list of some of the many different general approaches that can be used to rupture cells or disrupt tissues. Researchers must carefully choose which of these techniques are most appropriate for their needs based on a wide variety of concerns including the degree of intact organelles they wish to recover, the natural resistance of the cells or tissue of interest to homogenization techniques, and the available volume of starting material. A general trade-off deals with how harsh a homogenization procedure to use; the harsher the chosen technique, the greater the cell disruption achieved, while the milder the disruption, the greater the percentage of intact organelles recovered per cell lysed. Thus, while very harsh cellular disruption techniques will maximize the release of intracellular components from a sample, the quality of the organelles isolated may be poor. This is often not a concern when isolating membrane fragments from some organelles (such as the plasma membrane), but may be a major consideration if an intact organelle (such as the nucleus) is desired. It should be noted, however, that even when isolating membrane fragments, the cell disruption technique can still be a concern. For instance, a gentle disruption technique is the only way to obtain large plasma membrane vesicles. In addition, harsh cell disruption techniques can result in isolated plasma membranes that have lost or reduced sensitivity to stimulatory agents. This is a generality, however, because Bauer et al. (4) showed that even gentle osmotic lysis can cause loss of con A responsiveness in membrane fractions isolated from thymocytes. In addition, the harshness of the cell disruption technique chosen depends on the hardiness of the tissue or cell being used. Some cell types are easily disrupted by gentle methods, while cells contained within highly structured tissues sometimes require very vigorous disruption techniques. In general, the gentlest tissue or cell disruption technique that still provides an adequate yield of the cytoplasmic organelle of interest should be selected. Another consideration is the selection of an appropriate cell lysis or homogenization buffer for use during the cell lysis protocol. Fragile cytoplasmic organelles can be easily disrupted if the cytosol is released into a buffer with an inappropriate ionic strength. Thus, osmotic cell lysis that is otherwise gentle might be inappropriate if it also disrupts the organelles of interest. All of these concerns require that the investigator pay specific attention to both the cell type and the specific cytoplasmic organelle to be isolated when choosing a cell disruption technique. This is why the disruption technique used for mitochondrial isolation would be very different for rat myocardial tissue when compared to that for the yeast *Saccharomyces cerevisiae*. Because of the wide variability in disruption techniques, starting biological materials, organelles to be isolated, and the need for intact cytoplasmic organelles vs. organelle membrane fractions, further review of the techniques for cell homogenization is well beyond the scope of this chapter. Graham (36) has provided a recent comprehensive review of considerations in choosing the appropriate cell homogenization technique.

Table 1. Common Cell Disruption Techniques

Osmotic lysis

Enzymatic lysis

Chemical disruption

Gentle liquid shear homogenization (i.e., Dounce homogenizer)

Liquid shear homogenization (i.e., Potter-Elvehjem homogenizer)

Large-volume vigorous mechanical shearing (i.e., Waring blender)

Small-volume vigorous mechanical shearing (i.e., Polytron homogenizer)

Grinding with abrasives (i.e., mortar and pestle with acid-washed sand)

Freeze-thawing

Nitrogen cavitation

Sonication

Bead milling

AFFINITY DENSITY PERTURBATIONS

The principle of affinity isolation suggests that once cells are disrupted, immobilized ligands specific for the desired organelle can be used for its direct capture and recovery from the cell lysates. The most common affinity separation technique used in the isolation of cytoplasmic membrane organelles is a simple modification of the differential centrifugation protocol. In this approach, the specific affinity or immunoaffinity ligand binding to the surface of a cytoplasmic organelle results in a change in the organelle's density. Once the density is specifically altered, the change in density afforded to the organelle by the attached ligand easily allows separation based on isopycnic density centrifugation (82). Wallach coined the term "affinity density perturbations" for this method (89). Westwood et al. (93) altered rat liver cell plasma membrane density by exposing the membrane to a primary rabbit anti-rat antibody that recognized the plasma membrane. The plasma membranes were then attached to an immunoabsorbent consisting of aminocellulose with an attached secondary sheep anti-rabbit IgG antibody. A similar cellulose immunoabsorbent was used with a number of different primary antibodies (40) that recognize hepatocyte antigens for the successful isolation of plasma membranes from the Fao cell line.

Density perturbations in organelle isolation do not necessarily require specific affinity ligand binding to select the organelle of interest. For example, the phagocytosis of latex beads has been used not only to identify phagocytic vesicles that have taken up the beads, but the isolation of these phagosomes exploits the change in their density afforded by the entrapped particles (21,94). Beaumelle et al. (5) exploited a variety of density shifting techniques to isolate endosomes, plasma membrane, and lysosomes from a mouse cell line. Plasma membrane was isolated using a surface-binding lectin–gold complex, endosomes were isolated by a specific anti-mouse transferrin receptor antibody–gold complex, and lysosmes were isolated through a gold–LDL complex (following an 18-h incubation to allow LDL uptake). Along similar lines, the change in endosomal and lysosomal density associated with the uptake of iron during iron overloading has been used to aid in the isolation of these compartments (3,33,41).

As was the case in cell isolation, virtually all the affinity and immunoaffinity isolation techniques can use either negative or positive selection to enrich for the cytoplasmic organelle of interest. The use of differential cell fractionation techniques prior to the application of affinity separation can be particularly useful when using negative affinity selection protocols based on affinity density perturbations. Typical of this general approach is Luzio and Stanley's (62) modification of the protocol of Westwood (93). Instead of positive selection of plasma membranes, they used an immunoabsorbent to alter the density of contaminating plasma membranes to improve the enrichment of rat hepatic endosomes prepared by Ficoll density centrifugation. In their study, Luzio and Stanley used primary antibodies against plasma membrane 5′-nucleotidase and an aminocellulose immunoabsorbent prepared with a secondary antibody, to perturb the plasma membrane density. The fraction that attached to the immunoabsorbent was plasma membrane, and endosomes comprised the unbound fraction. A similar approach using a high capacity cellulose immunoabsorbent was used to immunopurify cholinergic nerve terminals from contaminating mitochondrial vesicles (78).

Another example that shows the use of both negative and positive affinity selection of organelles is the isolation of functionally competent lymphocyte endosomes by Beaumelle and Hopkins (6) as illustrated in Figure 1. When they subjected the cell lysates to a discontinuous sucrose-gradient centrifugation, a crude endosome fraction was recovered from the top portion of the gradient. This endosome fraction, while contaminated with plasma membrane components, was relatively free of other cytoplasmic organelles, which have a higher density and sink lower in the gradient. They overcame this contamination by a negative affinity selection protocol based on ricin, a lectin that binds to plasma membranes by galactose or N-acetylgalactosamine residues. For this negative selection, they initially pre-incubated the cells at 4°C with a ricin–gold complex. The low temperature prevents endocytosis of the ricin–gold complex, thus limiting the label to the plasma membrane. When the cells are subsequently lysed, the ricin–gold complex that was bound to the plasma membrane changed the membrane density and caused the gold-loaded membranes to sediment to the bottom of the sucrose gradient. Similar gold conjugates prepared with wheat germ agglutinin (WGA) have been used to remove plasma membrane contaminants from an isolated Golgi-enriched fraction (37). Steck and Lavasa (83) have recently described an updated generalized protocol for density perturbation of plasma membranes using raw colloidal gold particles without a lectin or antibody attached.

There are a wide variety of general affinity techniques that exploit the known strong affinity between plasma membranes and polycationic surfaces (45). PVP-coated colloidal silica matrix (Percoll®) has long been used in self-forming density gradients during centrifugation for the isolation of cells and subcellular membrane fractions, and cationic colloidal silica is among the most popular choices for polycationic surfaces for affinity organelle isolation. In particular, it has proven to be very useful in the isolation of specific domains from plasma membranes (15). Cationic colloidal silica has been increasingly used to coat the extracellular surface of intact cells with a dense pellicle of silica particles. This pellicle greatly enhances the density of the coated plasma membrane while stabilizing it against vesiculation or lateral reorientation.

Cantrell and Ellis (13) showed that binding of colloidal silica particles to the surface of human macrophages or red blood cells increased their mobility in an electric field. The same year, Chaney and Jacobson (15) showed that coating *Dictyostelium discoideum* with colloidal silica before cell lysis, followed by centrifugation to recover the silica-coated membranes resulted in a high yield of plasma membranes with minimal cytoplasmic contamination. Cezanne et al. (14) compared the isolation of plasma membranes from CHO cells by colloidal silica treatment prior to cell disruption with more traditional differential centrifugation using self-forming Percoll gradients. Comparing these techniques, they reported that precoating the cells with colloidal silica followed by cell disruption was both a quicker procedure and resulted in a more highly enriched plasma membrane fraction. Wasserman et al. (92) used colloidal silica methods to isolate intact plasma membrane vacuoles from silica-coated beet protoplast.

In addition to providing a rapid method for plasma membrane isolation, colloidal silica has also been useful in studying macrodomains and microdomains within membranes. Typical of this approach was the study by Mason and Jacobson (64) of both the gelatin-bound and the media-exposed membrane surface from cultured HeLa cells. In their study, cells were first allowed to adhere to a gelatin surface, followed quickly by exposure to the cationic colloidal silica. Only the dorsal side of the plasma membrane, which was directly exposed to the culture media, tightly bound the silica. Following binding of the silica and cell lysis, the ventral portion of the membrane was easily recovered because it remained firmly attached to the gelatin matrix. Meanwhile, the dorsal plasma membrane fraction was also easily recovered and separated from the other cell lysate intracellular membrane fragments, including plasma membranes that had been mobilized from the surface by endocytosis, because

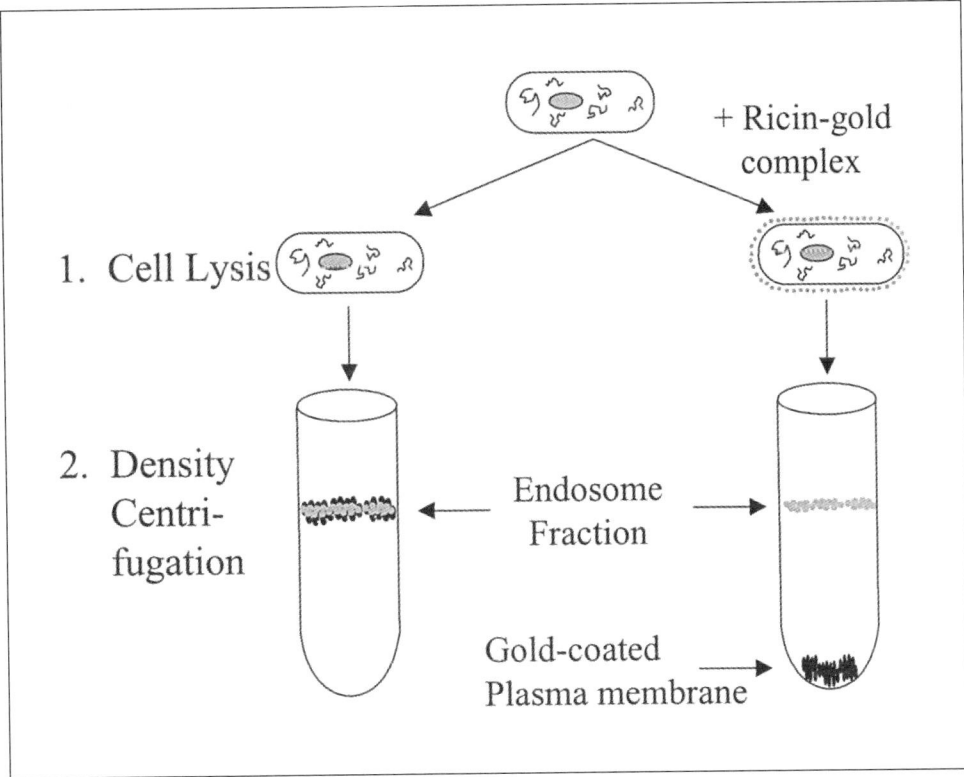

Figure 1. The density-altering protocol used to eliminate plasma membrane contaminants in an endosomal membrane preparation by density centrifugation. Typical endosomal isolation (left tube) results in an endosomal fraction at the top of the gradient contaminated with plasma membranes. Labeling the cells with a ricin-gold complex alters the density of the plasma membrane and allows the isolation of highly enriched endosomes (right tube).

of a silica-induced dramatic increase in the membrane density.

Other studies have used colloidal silica to study plasma membrane microdomains. Jacobson et al. (46) have described a method for isolating highly purified luminal plasma membrane from the rat lung microvascular endothelial cells. The endothelial surface is coated in situ by perfusion with colloidal silica followed by a polyacrylate coating to form a strongly adhering silica coat on the exposed endothelial cell surfaces. This coating withstands tissue homogenization and allows the silica-coated luminal endothelial cell plasma membranes to be isolated by centrifugation and studied free of other tissue plasma membrane sources and subcellular organelles. This approach has been used to study not only the plasma membrane macrodomain luminal surface, but also to begin probing caveolae microdomains within the plasma membrane (29,80,81). Another aspect of the colloidal silica treatment is that it can be used to glue the surface proteins in place to allow studies of changes in the lateral distribution of membrane protein components in response to environmental stimuli, as shown on con A-induced peptide capping in *D. discoideum* (68).

A variant of polycationic surface binding has been used to rapidly isolate plasma membranes that are attached to large particles. This is identical in principle to the isolation of the ventral HeLa plasma membrane surface bound to the gelatin mentioned above (64). In these protocols, the cells are generally allowed to attach to antigen- or antibody-coated magnetic particles or attach through the use of affinity ligands or even the polycationic surfaces on glass beads or other particles. After the cells are firmly attached, the surface cationic charges not involved in cell binding are neutralized, and then the attached cells are lysed. This cell lysis leaves bits of the attached plasma membrane adhering to the large particle, which can be quickly washed free of cellular debris and studied. Jacobson and Branton (44) described one such method that used 30-μm diameter glass beads coated with poly-L-lysine to produce a polycationic surface for the isolation of human erythrocyte membranes. Because erythrocytes are devoid of intracellular organelles, plasma membrane isolation could also be easily obtained by hypotonic lysis and centrifugation to recover the membrane. However, this approach leads to the formation of a mixed membrane vesicle population with the normal cytoplasmic side of the membrane exposed on the outside of some vesicles and enclosed on the inside of others. The poly-L-lysine bead approach not only allowed the isolation of the membrane, but it provided a membrane population in which all the isolated membranes were oriented with their cytoplasmic surface exposed on the beaded surface (44,49,54).

Using this same general approach, van der Meulen et al. (88) isolated plasma membranes from chromaffin cells and demonstrated which endogenous membrane proteins faced the cytoplasm surface by iodination of the isolated membrane fractions while they remained attached to the bead. The membranes were then isolated and the iodinated proteins were determined by polyacrylamide gel electrophoresis. Because the extracellular surface of the plasma membrane had been attached to the bead during the iodination step, only the cytoplasmic-exposed proteins were iodinated. Poly-L-lysine-coated beads have been applied to isolate plasma membranes from HeLa cells (16,49), human platelets (53), and *D. discoideum* (42,43). It is possible to isolate plasma membranes after first culturing the cells on the surface of a microcarrier bead, followed by cell lysis and membrane recovery directly on the growth support (35).

Stolz and Jacobson (85) combined the isolation of plasma membranes on the growth surface with the colloidal silica isolation protocol discussed earlier, to isolate two distinct populations of plasma membrane from monolayers of bovine aortic endothelial cells. The endothelial cells were initially grown in culture, and then the entire apical membrane surface was coated with cationic colloidal silica microbeads. Following cell lysis, apical and basolateral plasma membrane fractions were prepared. Angiotensin-converting enzyme was shown to be found only in the apical plasma membrane, while collagen receptors were greatly enriched with the basolateral fraction. In a companion study, this group has further studied the transcellular membrane protein distribution between the apical and basolateral surfaces (84).

PROTOCOLS

Colloidal Silica Method of Plasma Membrane Isolation from Cultured Endothelial Cells

This protocol is based upon the procedure of Schnitzer, Oh, and colleagues (67,80,81), whose publications provide additional details on how caveolae microdomains within the plasma membrane can be isolated using this general approach.

1. Grow endothelial cells in monolayers in plastic 175-cm^2 culture flasks until they reach confluence.

2. Wash the cells three times with excess Ringer's solution (114 mM NaCl, 25 mM NaHCO$_3$, 11 mM glucose, 4.5 mM KCl, 1 mM MgSO$_4$, 1 mM Na$_2$HPO$_4$, adjusted to pH 7.4 in distilled H$_2$O) supplemented with 1 mg/mL of freshly prepared sodium nitroprusside.

   Note: This solution should be filtered (0.45-μm pore size) prior to use.

3. Wash the cell surface three times with excess MES-buffered saline (20 mM 2-[*N*-morpholino]ethanesulfonic acid and 135 mM NaCl, dissolved in double-distilled H$_2$O and adjusted to pH 6.0 with NaOH).

   Note: This solution should be filtered (0.45-μm pore size) prior to use.

4. Add sufficient 1% (vol/vol) colloidal silica solution (prepared in MES-buffered saline, pH 6.0) to cover the cell monolayer and incubate for 15 min.

   Note: The pH of the colloidal solution should be checked with pH paper, not a pH electrode.

5. Remove the colloidal silica solution with a pipette, and wash the surface of the cells with excess MES-buffered saline, pH 6.0.

6. Add a 0.1% (vol/vol) solution of polyacrylic acid (prepared by dilution of a 20% stock solution of sodium polyacrylate in distilled H$_2$O) into MES-buffered saline, pH 6.0 and incubate for 15 min.

7. Remove the polyacrylic acid solution with a pipette, and wash the cells with excess MES-buffered saline, pH 6.0.

8. Add 12 mL of a sucrose–HEPES buffer prepared in KCl (250 mM sucrose, 25 mM HEPES, 20 mM KCl, pH 7.4) supplemented with protease inhibitors (0.01 mg/mL leupeptin, Pefablock SC [protease inhibitor], and pepstatin A, and 0.05 mg/mL *O*-phenanthroline).

9. Scrape the cells from the surface of the plate, and transfer them to centrifuge tubes. Centrifuge at 1000× *g* for 10 min at 4°C.

10. Resuspend the pellet in 1 mL of the sucrose–HEPES buffer with protease

inhibitors used in step 10.

Note: Three flasks of cells can be suspended in this same 1-mL volume.

11. Add an equal volume of 102% (vol/wt) Nycodenz® solution and mix by repeated gentle inversion.

12. Prepare a continuous Nycodenz–sucrose-HEPES gradient by layering 350 µL of 70, 65, 60, and 55% (wt/vol each) Nycodenz–sucrose-HEPES solution into a centrifuge tube. These solutions are prepared by adding 1 mL of a 60% sucrose solution to 7, 6.5, 6, or 5.5 mL of a 102% Nycodenz solution, adjusting the pH to 7.4, then adjusting the volume of each tube to 10 mL.

13. Carefully layer the recovered, mixed cellular material from step 11 on top of the 70%–55% Nycodenz–sucrose gradient.

14. Centrifuge at $30\,000\times g$ for 30 min at 4°C. Aspirate the supernatant, and resuspend the membrane pellet in 1 mL of MES-buffered saline.

AFFINITY PARTITIONING

Another affinity approach for the isolation of cytoplasmic membrane organelles or their membrane fragments is through the use of two-phase aqueous affinity or immunoaffinity partitioning (2,71). A historical review of this approach for membrane isolations was provided by Albertsson (1). The general principles of this technique will not be repeated here, because they have already been covered in detail as they relate to the separation of protein antigens (Chapter 4) and the separation of mixed cell populations (Chapter 8).

Often, organelle isolation or enrichment can be achieved through the use of a PEG-dextran binary mixture without the aid of any specific binding ligand. A simple application of such a two-phase system was described by Morré and Morré (65) for the isolation of highly enriched mammalian plasma membranes. After cell lysis and the preparation of a plasma membrane fraction by differential centrifugation, the concentrated plasma membrane fraction was added to a dextran–PEG solution. The solution was allowed to separate into the typical dextran-enriched lower phase and PEG-enriched upper phase. The plasma membrane preferentially partitioned into the upper phase in this system, while most of the contaminating membranes partitioned into the lower phase. Lopez-Perez et al. (59) separated rat forebrain mitochondria and synaptosomes with the same two-phase aqueous partitioning system. In their preparation, mitochondria partitioned into the dextran-rich lower phase, while synaptosomes preferred the PEG-enriched upper phase. Fristedt et al. (28) have recently used a modification of two-phase aqueous affinity partitioning to isolate primarily inside-out plasma membranes from *S. cerevisiae*.

The above studies and many others (see Reference 1) demonstrate that membrane organelles can be isolated using two-phase affinity separations without added ligands. However, as was the case for cell isolation, the affinity partitioning of cytoplasmic organelles can be greatly enhanced by the addition of specific ligands. These ligands both speed the partitioning process between the two phases and enhance the specificity for the phase partitioning. Membranes rich in cholinergic receptor molecules have been purified by affinity partitioning in a biphasic PEG–dextran system through the use of a PEG-labeled ligand that binds this receptor (27,47). The PEG-labeled receptor ligand shifts the partitioning of the cholinergic receptor-rich membrane fractions into the PEG-rich phase. A protocol for the attachment of PEG molecules to immunoaffinity ligands (antibody) was provided in Chapter 8. Other approaches that have been used for the separation of membrane organelles by the addition of affinity ligands include the addition of lectins (70,72), textile dyes (32), and antibodies (71).

PROTOCOLS

Rapid Purification of Rat Liver Plasma Membrane Isolation by Affinity Partitioning

This protocol is adapted from the work of Jergil and colleagues (25,70–72). Using this two-phase affinity partitioning system, the bulk of the membranes partition into the upper, PEG-rich phase. Unlike the ligands used in the two-phase affinity partitioning discussed in Chapter 8, the ligand in this example is added to the dextran rather than the PEG. Thus, the lectin-specific bound membrane vesicles are driven into the lower, dextran-rich phase. This protocol is highly sensitive to minor changes in the salt concentration, and improper salt concentration has been suggested as the most likely cause of failure when using this protocol for rapid plasma membrane isolation (24).

1. Dissolve 5 g of freeze-dried Dextran T-500 in 25 mL of dimethyl sulfoxide (DMSO) at room temperature and gently stir.
2. Adding one drop at a time, first add 1 mL of triethylamine to the stirring dextran solution, followed by 5 mL of dichloromethane.
3. Chill the mixture in an ice-bath, and then very slowly add 350 mg of tresyl chloride with vigorous magnetic stirring. Continue gently stirring, first for 30 min on ice, then at 4°C for 1 h, then at 20°C for approximately 16 h.
4. At room temperature, add 50 mL of dichloromethane and collect the tresyl-activated dextran precipitate following gravitational sedimentation.
5. Repeatedly wash the collected precipitate with 25 mL aliquots of dichloromethane, with stirring with a glass rod, until a firm consistency is reached.
6. Dissolve the dextran in 30 mL of double-distilled H_2O and dialyze against double-distilled H_2O overnight.
   Note: Excess prepared tresyl-activated dextran can be freeze-dried and stored at -20°C for several months for later use.
7. Dissolve 2 g of tresyl-activated dextran in 10 mL of 0.1 M NaH_2PO_4 containing 0.5 M NaCl, pH 7.5.
8. Dissolve 10 mg of WGA in 1 mL of 0.1 M NaH_2PO_4 containing 0.5 M NaCl, pH 7.5.
9. Add 1 mL of WGA (from step 8) dropwise with vigorous magnetic stirring to 10 mL of tresyl-activated dextran (from step 7), and incubate overnight at 4°C with continuous gentle mixing.
10. Add 10 mL of 400 mM Tris-HCl, pH 7.5, to the test tube from step 9 and incubate for 1 h at 4°C.
11. Remove salts and unbound ligand from the dextran by ultrafiltration (filter cut-off = 100 kDa).

12. Recover the dextran from the filter with a spatula and dissolve in 150 mL of double-distilled H_2O.

13. Determine the amount of WGA bound to the dextran by routine protein analysis (see Appendix I, *Direct Spectrophometric Technique* or *Micro-Lowry Assay*).

14. Prepare a 40% (wt/vol) PEG 3350 solution and a separate 20% (wt/vol) Dextran T-500 solution (from freeze-dried dextran), both in distilled H_2O.

    Note: Aliquots of these solutions can be stored in the freezer for later use.

15. Prepare two identical solutions, each containing a solution prepared by mixing 11.4 g of the 20% dextran, 5.7 g of the 40% PEG 3350, 15 mL of distilled H_2O, and 3 g of 0.2 M Tris-H_2SO_4, pH 7.8. Mix well and store at 4°C overnight.

    Note: These tubes should be labeled "5.7%/5.7% dextran-PEG."

16. To a pre-weighed 35-mL centrifuge tube, add 200 μg of WGA attached to dextran beads (from step 13), sufficient 20% dextran stock to raise the total added dextran mass to 3 g, 1.5 g of 40% PEG, 750 mg of 0.2 M borate buffer, pH 7.8, and sufficient distilled H_2O to adjust the total weight to 10 g. Mix well and store at 4°C overnight.

    Note: This tube should be labeled "WGA-dextan/PEG."

17. Add 3.75 g of 0.2 M borate, pH 7.8, to 23.75 mL of distilled H_2O. To this mixture, add 15 g of 20% dextran and 7.5 g of 40% PEG solutions from step 14. Mix well and store at 4°C overnight.

    Note: This tube should be labeled "6.0%/6.0% dextran-PEG."

18. Sacrifice a rat, and excise the liver. Cannulate the hepatic artery, attach it to an 8–12-cm high pressure head, and continuously perfuse the liver with ice-cold 5 mM Tris-HCl and 0.25 M sucrose, pH 8.0, until the perfusate is relatively free of blood.

19. Remove the cannula, chop the liver into several sections, and press the tissue through a garlic press.

20. Transfer 5 g of this tissue to a chilled Dounce homogenizer and add the contents from one of the two 5.7%/5.7% dextran-PEG solutions prepared in step 15. Homogenize the liver by 20–25 up-and-down strokes with a loose fitting Dounce A pestle.

21. Transfer the homogenate to a chilled 35-mL centrifuge tube and mark the meniscus on the tube with an indelible marker. Centrifuge at 150× *g* for 5 min in a swinging bucket rotor.

22. Collect the upper layer and store on ice.

23. Collect the top layer from the second tube labeled "5.7%/5.7% dextran-PEG" and use this to refill the centrifuge tube to the original meniscus mark. Cover with Parafilm, and mix the two layers thoroughly by inverting 25 times, vortex-mixing, and inverting another 25 times.

24. Centrifuge at 150× *g* for 5 min in a swinging bucket rotor. Pool the top layer with the top layer previously collected (step 22).

25. Carefully siphon off and discard the top layer from centrifuge tube that was previously labeled "WGA-dextan/PEG" (step 16) and add the pooled sample

(from step 24). Cover with Parafilm, and mix the two layers thoroughly by inverting 25 times, vortex-mixing, and inverting another 25 times. Mark the meniscus with a marker.

26. Centrifuge at 150× g for 5 min in a swinging bucket rotor. Siphon off and discard the upper layer.

27. Refill the centrifuge tube to the meniscus mark using the top layer from the tube labeled "6.0%/6.0% dextran-PEG" (step 17). Cover the centrifuge tube with Parafilm, and mix the two layers thoroughly by inverting 25 times, vortex-mixing, and inverting another 25 times.

28. Centrifuge at 150× g for 5 min in a swinging bucket rotor. Siphon off and discard the top layer.

29. Dilute the final bottom phase 10-fold in 0.1 M N-acetylglucosamine solution containing 8.5% (wt/vol) sucrose and 5 mM Tris-HCl, pH 8.0. Mix well and pellet the isolated plasma membrane by centrifugation at 100000× g for 60 min.

CHROMATOGRAPHIC TECHNIQUES

While polycationic beads and colloidal silica offer useful ways to isolate plasma membranes, these approaches are rather nonspecific and rely on coating the surface of the cells prior to cell lysis. If specific affinity or immunoaffinity ligands are available, they can be used to selectively bind and recover the membrane organelle of interest. There are typically many commercially available antibodies to extracellular antigens that are very useful for cell isolation. However, the supply of such antibodies for use in immunoaffinity separations of subcellular cytoplasmic organelles is more limited. In selecting appropriate antibodies for subcellular organelles, it is important that they be reactive with native, undenatured epitopes on the subcellular membranes. A potential source of antibodies specific for the isolation of organelle compartments includes the same antibodies used as immunoaffinity markers to identify specific organelle compartments following cell fractionation by physical techniques. For example, Fayle et al. (26) used a radiolabeled monoclonal antibody to label and track human peripheral blood monocyte plasma membranes through membrane separation by traditional isopycnic centrifugation. Based on the work of Fayle et al. (26), the same antibody would be a potential candidate for use in immunoaffinity isolation techniques for the same membrane population. Another potential source of antibodies for immunoaffinity separation techniques are ones used to label subcellular organelles in immunohistochemical studies. Obviously, such antibodies that exhibited a high degree of specificity for the desired organelle would be the most desirable as an immunoaffinity ligand.

Another potential source of immunoaffinity ligands for the isolation of cytoplasmic organelles includes the naturally occurring antibodies found in autoimmune diseases. Typically, naturally occurring antibodies are available only to extracellular antigens to which the immune system is exposed. However, there are a number of autoimmune diseases in which reactive antibodies are produced against intracellular organelles. Table 2 lists some of these self-reactive antibodies and the associated autoimmune diseases in which they are frequently produced. A number of vendors sell these self-reactive or auto-antibodies as standards for the clinical diagnosis of such diseases. When selecting auto-antibodies, it is again necessary to make sure that

Table 2. Auto-Antibodies Associated with Human Autoimmune Diseases

Auto-Antibodies	Associated Autoimmune Disease	References
Anti-mitochondrial	Idiopathic hypoparathyroidism	(8)
	Primary biliary cirrhosis	(57,58,74)
	Systemic sclerosis	(75)
Anti-nuclear and anti-DNA	Graft versus host disease	(74)
	Primary biliary cirrhosis	(17)
	Hepatitis D	(73)
	Systemic sclerosis	(75)
	Systemic lupus erythematosus	(56)
	Paraneoplastic encephalomyelitis	(21)
	Paraneoplastic opsoclonus-ataxia	(21)
Anti-smooth muscle plasma membrane	Graft versus host disease	(74)
	Viral hepatitis	(66)
Anti-microsomal membranes	With systemic lupus erythematosus	(23)
Neuromuscular junction	Myasthenia gravis	(10)

the antibody recognizes its appropriate antigen in its native, undenatured state. Antibodies can also be raised against specific subcellular organelles experimentally. Typically, this involves the crude isolation of the membrane of interest followed by the subsequent use of the isolated membrane as an immunogen. Kumar et al. (55) described a sera containing anti-mitochondrial antibodies that bound intact mitochondria, which might be suitable for the immunoaffinity isolation of mitochondria. Others have raised monoclonal antibodies to a wide variety of subcellular organelles. Typical examples include monoclonal antibodies to rat lung endothelial cell plasma membranes (30,31) and human liver mitochondrial membranes (9).

Isolation of cytoplasmic organelles by affinity or immunoaffinity chromatography is complicated by the "stickiness" of the hydrophobic membrane surfaces exposed when cells are homogenized. It is therefore important that affinity or immunoaffinity ligands are immobilized on smooth, hydrophobic surfaces to avoid nonspecific membrane binding to the surface of the chromatographic support. Brunner et al. (11) were among the first to use affinity chromatography for the isolation of specific membranes during cellular fractionation. In their study, an affinity column composed of con A-Sepharose was used to isolate the plasma membrane from rabbit thymocytes. Con A-Sepharose affinity chromatography has been used to isolate plasma membranes from many additional sources, including calf thymocytes (12,86,87) and T lymphocytes (34), mouse lymphoma cells (86), pig lymphocytes (90), and guinea pig brain synaptic vesicles (41).

In addition to isolating cytoplasmic membrane organelles from cell lysates, both affinity and immunoaffinity chromatography have been used to separate previously isolated membrane fractions into two populations based on the cytoplasmic surface either exposed on the outside of the vesicle (inside-out), or in its normal configuration on the inside of the vesicle (right-side-out). Walsh et al. (90) demonstrated the principle that membrane vesicles formed from fragmented cytoplasmic organelles are resealed into both of these orientations (Figure 2), when 30%–50% of the pig lymphocyte plasma membranes they isolated failed to bind either con A-Sepharose or specific antilymphocytic serum. Through a series of approaches, they further showed that the con A binding to the plasma membrane vesicles occurred only on right-side-out vesicles, while membrane vesicles that resealed after cell lysis in an inside-out orientation could not be bound by con A. Numerous other researchers have since used lectin-based affinity chromatography to further fractionate isolated membrane fractions into two subpopulations based on this inside-out or right-side-out orientation. Typical of the inside-out isolation of plasma membrane vesicles using con A-Sepharose affinity chromatography was the isolation of such vesicles from pig lymphocytes (90) and bovine thymocytes (76). Sepharose-bound lectins, such as *Ricinus communis* agglutinins I and II (50), have also been used as ligands for affinity chromatography to isolate normal liver lysosomes. Lotscher et al. (60) used affinity chromatography with immobilized cytochrome *c* as the ligand to isolate inside-out rat liver mitochondrial vesicles.

The separation of membrane vesicles based on their orientation is not restricted to affinity chromatography. Del Buono et al. (20) used con A-coated plates to selectively bind and recover right-side-out plasma membrane vesicles from a membrane fraction prepared by human polymorphonuclear leukocytes. D'Souza and Lindsay (22) used a more complex approach to separate right-side-out from inside-out rat liver submitochondrial membrane fragments. In this study (22), WGA was added to the mitochondrial membrane preparation. Membrane fragments that bound the WGA were then immunoprecipitated using anti-WGA IgG. The pelleted membrane was composed of right-side-out vesicles, while the unprecipitated membrane fraction contained the inside-out ones.

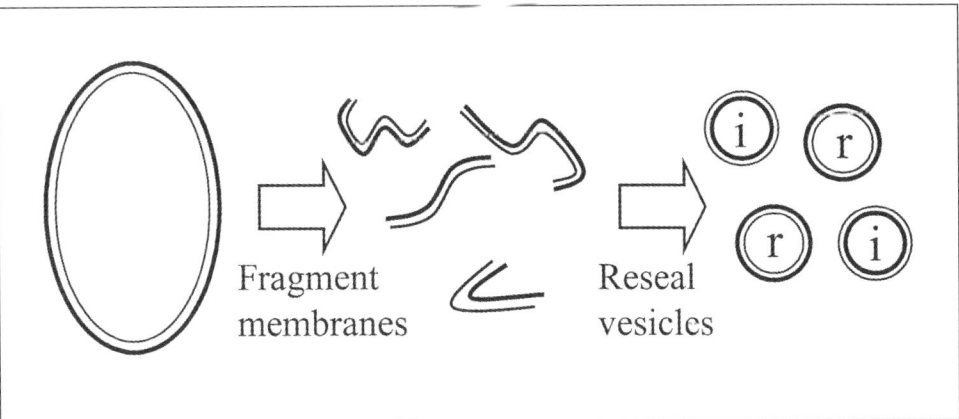

Figure 2. Membrane fragmentation results in membrane fragments that quickly reseal to form intact membrane vesicles. These vesicles can form with either half of the original membrane bilayer exposed to the external environment. Affinity or immunoaffinity techniques that recognize surface membrane components can be used to separate vesicles in the (normal) right-side-out orientation (r) from inside-out vesicles with the cytoplasmic surface exposed (i).

PROTOCOLS

Preparation of a Cytochrome *c*-Sepharose Affinity Matrix

1. Incubate 5 g of CNBr-activated Sepharose 6MB (Amersham Pharmacia Biotech) in 500 mL of double-distilled H_2O for 10 min.

   Note: Experienced chemists who are fully aware of the safety precautions required when working with CNBr can prepare their own CNBr-activated Sepharose as described in the protocol for CNBr derivatization in Chapter 2.

2. Collect the swollen Sepharose beads by filtration, and while in the filter, wash the beads sequentially with 1 L each of double-distilled H_2O, 1 mM HCl, and double-distilled H_2O again.

3. Dissolve 1 g of cytochrome *c* in 10 mL of 100 mM Na_2CO_3, pH 9.5. Add the Sepharose beads to this solution and incubate for 2 h at room temperature with gentle swirling.

4. Collect the swollen Sepharose beads by filtration, and wash the beads with 250 mL of 100 mM Na_2CO_3, pH 9.5.

5. Resuspend the beads in 50 mM ethanolamine-HCl, pH 9.5, and incubate for 1 h at room temperature.

6. Wash the beads by four complete filtration cycles through the following three sequential wash solutions in 250 mL each of 0.1 M CH_3COONa, pH 5.0; 1 M KCl containing 1% (vol/vol) Lubrol (ICN Biomedicals); and 100 mM Na_2CO_3, pH 9.5.

7. Resuspend the cytochrome *c*-labeled Sepharose beads in excess 0.25 M sucrose solution containing 1 mM EDTA and 10 mM Tris-HCl, pH 7.4. Use this preparation to pack a simple chromatographic column as described in the protocol for the construction of a simple syringe column in Chapter 7.

Affinity Purification of Right-side-out and Inside-out Rat Liver Mitochondria Using Cytochrome *c*-Sepharose Columns

This mitochondrial protocol illustrates how affinity chromatography can be used to obtain both right-side-out and inside-out membrane vesicles. The isolation protocol used for the initial mitochondrial isolation is a very generic one. For details on mitochondrial isolation from other tissue and cellular sources, as well as problems encountered in mitochondrial isolation, the reader is referred to a source such as Volume 55 of Methods in Enzymology.

1. Sacrifice a rat, and quickly excise and chill the liver in a pre-weighed beaker containing 25 mL of homogenization solution (200 mM mannitol, 50 mM sucrose, 10 mM HEPES-NaOH, 1 mM EDTA, pH 7.4). Determine the weight of the excised liver.

2. Decant the homogenization solution and, keeping the beaker on ice, finely mince the liver using dissection scissors.

3. Resuspend the minced liver in ice-cold homogenization solution at a concentration of 250 mg/mL, and homogenize on ice using a Potter-Elvehjem homogenizer for 4–5 strokes at 500 rpm.

4. Transfer the liver homogenate to a 50-mL conical centrifuge tube and centrifuge at 1000× g for 10 min at 4°C.

5. Transfer the supernatant to a fresh conical centrifuge tube, and centrifuge at 4°C at 3000× g for 10 min.

6. Carefully remove all the supernatant with a pipette, taking care to remove as much fat as possible from the surface of the supernatant and the sides of the tube during the process.

7. Add 20 mL of ice-cold homogenization buffer to the tube and gently resuspend the pellet by drawing it in-and-out of a syringe using a wide-bore metal needle.

   Note: Be careful not to resuspend any hard-packed material underneath the bulk of the mitochondrial pellet.

8. Transfer the liquid to a fresh 50-mL conical centrifuge tube, adjust the volume to approximately 45 mL, and repeat steps 5–7 twice more.

9. Resuspend the mitochondrial pellet in 10 mL of ice-cold 20 mM KH_2PO_4, pH 7.4, and determine the protein content (see Appendix I).

10. Adjust the concentration of this mitochondrial solution to 100 mg of protein/mL by adding the appropriate volume of ice-cold 20 mM KH_2PO_4, pH 7.4.

11. Place a sonication probe into a tube containing 5 mL of the mitochondrial solution on ice, assuring that the probe doesn't touch the vessel walls. Perform three cycles of sonication at 80 W for 15 s with a 1-min wait between cycles.

12. Centrifuge at 4°C at 6000× g for 10 min, and remove and store the supernatant on ice.

13. Gently resuspend the pellet in 5 mL of 20 mM KH_2PO_4, pH 7.4, by drawing it in-and-out of a syringe using a wide-bore metal needle.

14. Place a sonication probe into this tube on ice, assuring the probe doesn't touch the vessel walls. Perform three cycles of sonication at 80 W for 15 s with a 1-min wait between cycles.

15. Centrifuge at 4°C at 6000× g for 10 min, remove the supernatant, and pool it with the supernatant from step 12 on ice.

16. Centrifuge the pooled supernatants at 105 000× g for 1 h at 4°C.

17. Discard the supernatant and resuspend the pellet in ice-cold 20 mM KH_2PO_4, pH 7.4.

18. Centrifuge at 105 000× g for 1 h at 4°C.

19. Resuspend the membrane pellet by gentle hand homogenization in 5 mL of ice-cold 0.15 M KCl, 10 mM Tris-HCl, pH 7.4.

   Note: This step strips endogenous cytochrome c.

20. Centrifuge at 105 000× g for 1 h at 4°C.

21. Repeat steps 19 and 20.

22. Resuspend the membrane pellet at a concentration of 2 mg of mitochondrial protein/mL in 0.25 M sucrose containing 1 mM EDTA and 10 mM Tris-HCl, pH 7.4.

23. Apply 1 mL of membrane mixture from step 22 to the top of the cytochrome c-Sepharose column prepared in the protocol for the preparation of a

cytochrome *c*-Sepharose affinity matrix (above), and allow the solution to flow into the gel by gravity. Stop the flow by capping the column when the meniscus reaches the top of the Sepharose packing.

24. Add 20 mL of the 0.25 M sucrose containing 1 mM EDTA and 10 mM Tris-HCl, pH 7.4, and allow the column to flow by gravity, collecting 1-mL effluent fractions. Assure the column doesn't run dry, adding more of the buffer if necessary. Inside-out mitochondrial membrane vesicles can be detected in the effluent by monitoring 280 nm adsorption.

25. Once the inside-out membranes have eluted, change to an elution buffer of 0.25 M sucrose containing 1 mM EDTA, 1 M KCl, and 10 mM Tris-HCl, pH 7.4. Continue collecting 1-mL fractions, monitoring for right-side-out membrane particles by absorbance at 280 nm.

MAGNETIC SEPARATIONS

Magnetic bead technology is probably the easiest affinity or immunoaffinity separation technique to adapt for the isolation of cytoplasmic organelles. In addition, magnetic affinity and immunoaffinity isolation is both a more rapid and a gentler separation technique than those based on high speed centrifugation. The general principles of magnetic separation have been covered in Chapters 4 and 8, and thus will not be repeated here. However, it is important to realize that many magnetic particle manufacturers offer specialized particles for use specifically in subcellular isolation. One such particle is Dynal's Dynabeads M-500, which has the extremely smooth, hydrophobic surface necessary to both limit nonspecific membrane binding during organelle isolation and to prevent surface-induced injury to the potentially fragile vesicles being isolated. Kausch et al. (52) have provided a recent detailed report describing the advantages of immunoadsorption for cytoplasmic organelle isolation. In their study, they tested the effectiveness of different diameter size magnetic beads (nanoparticles [nm diameters] and microparticles [μm diameters]) for organelle isolation. In part of their study, they pre-incubated cell lysates with a biotinylated primary antibody. They then immobilized this primary antibody by the addition of commercially-supplied streptavidin-coated magnetic particles. In the novel microextraction system used, they found that nanoparticles were the most effective for the isolation of pea chloroplast and intact plant nuclei. Others have used immunoadsorption to magnetic particles as a final step after the initial physical technique-based subcellular fractionation. One such area of active study is the use of magnetic labeled beads to further purify and subsequently study Golgi complex, following an initial physical cell fractionation protocol. Jones et al. (48) described in detail the use of magnetic beads to study exocytotic budding of new transport membrane compartments from the Golgi stacks. After the physical isolation of relatively pure Golgi stacks, these authors further purified these structures by immunoadsorption onto antibody-coated Dynabeads M-500 magnetic beads. After Henley and McNiven (38) found by immunofluorescence that an antibody against a specific GTPase strongly labeled the Golgi complex in cultured cells, they then coated Dynabeads M-500 magnetic beads with the same antibody and used these beads to isolate Golgi vesicles from liver.

Wandinger-Ness et al. (91) used immunoadsorption of virally produced antigens to study the transport of membrane vesicles from the *trans* Golgi network to the plasma membrane in MDCK cells. They infected the MDCK cells with influenza or vesticular stomatitis virus and used antibodies against viral antigens to follow vesicle formation. Using antibodies directed against the respective viral antigens, these researchers showed that two distinct transport vesicles appear to exist, as influenza hemagglutinin appeared to mark vesicles targeting apical membranes while stomatitis virus G protein targeted the basolateral membrane. Saucan and Palade (79) have also used antibody-coated magnetic beads to isolate and study secretory vesicular membrane carriers that ferry material from Golgi complex to the sinusoidal plasma membrane of hepatocytes. Another carrier membrane vesicle was studied by Marquez-Sterling et al. (63), who used mouse monoclonal anti-synaptophysin antibodies immobilized on commercial magnetic beads coated with sheep anti-mouse IgG to isolate and study recycling synaptic vesicles from rat brain synaptosomal lysates.

While studies on Golgi complex and plant chloroplast have been advanced by the use of magnetic bead technology, the use of this technology for subcellular isolation is still in its infancy. As specific antibodies to cytoplasmic organelles become more readily available, we should see an explosion in the use of this technique. Already, a number of other organelles have been isolated by this technique. Lüers et al. (61) have used immunomagnetic separation to isolate peroxisomes from rat and human cells. Like most of the studies involving isolation of Golgi complex, they first prepared a peroxisome-rich fraction by routine physical subcellular fractionation. To isolate the peroxisomes, magnetic beads coated with polyclonal antibody against the cytoplasmic C terminus of a rat peroxisomal membrane protein were incubated with the peroxisome-containing light mitochondrial fraction. The peroxisomes were then recovered for subsequent study by the application of a magnetic force. Davis et al. (18) used magnetic beads coated with monoclonal antibodies against apolipoprotein B to isolate microsomes in their study of the transport and secretion of apolipoprotein B.

As noted above in the section *Affinity Density Perturbations*, researchers have used cellular iron uptake or iron overload to shift the density of lysosomes as an aid in their isolation by physical techniques (3,33,41). A marked improvement in this general technique is the use of magnetic force rather than centrifugation to recover these iron-loaded vesicles. For example, Becich and Baenziger (7) used a variation on the iron-overloading theme using affinity-labeled superparamagnetic colloidal iron dextran to load and then magnetically recover endosomes and lysosomes from a human hepatoma cell line (HepG2). In this study, the iron dextran was coated with a receptor ligand (asialofetuin) and, after receptor binding, was readily transported into lysosomes. Both lysosomal and plasma membranes were then recovered by application of a magnetic force, with the plasma membrane fraction being enriched in plasma membrane enzyme markers such as alkaline phosphodiesterase (see Table 3 for a list of some typical enzymatic markers used during subcellular membrane isolations) and containing no detectable β-galactosidase activity. Meanwhile, the reverse enzymatic marker pattern existed for the lysosomal membrane fraction isolated by this technique. In a more recent study, Perrin-Cocon et al. (69) used antigen-coated nanoparticles (10-nm diameter) to study the process of macrophage fluid-phase macropinocytosis and B-cell receptor endocytosis. Following nanoparticle internalization, the cells were lysed and cytoplasmic vesicles containing these magnetic particles were recovered in a strong magnetic field. By tracking the timing of distribution of these magnetic nanoparticles, these authors showed that macrophage pinocytosis results in the magnetic particles being initially found in endosomal vesicles, and subsequently within a lysosomal compartment. In the case of B-cell receptor uptake of the antigen-coated

Table 3. Typical Enzymatic Markers Used for Tracking Membrane Enrichments During Cytoplasmic Organelle Isolations

Organelle	Marker Enzyme
Plasma membrane	5′-nucleotidase leucine aminopeptidase alkaline phosphatase galactosyl transferase
Endoplasmic reticulum	arylsulfatase C choline phosphotransferase dolichol-P-mannosyl synthase
Golgi apparatus	galactosyl transferase
Mitochondria	monoamine oxidase
Lysosomes	β-hexosaminidase β-glucuronidase acid phosphatase
Peroxisomes	catalase

nanoparticles, immediately after exposure, the antigen-coated particles were initially found in an endocytotic compartment. After removing further extracellular antigen-coated particles and incubating for an additional 2 h, the engulfed particles were found in a cellular compartment containing a large amount of MHC Class II molecules, invariant chain, and human leukocyte antigen-DM.

PROTOCOLS

Antibody-Coated Magnetic Beads for the Isolation of Rat *Trans* Golgi Network Membranes

1. Prepare streptavidin-coated Dynabeads M-500 magnetic beads as described in the protocol for the coating of magnetic particles with streptavidin in Chapter 8.

   Note: Alternatively, any commercially available streptavidin-coated magnetic bead designed for subcellular isolation, such as the nanobeads available from Miltenyi Biotec, can be used.

2. Biotinylate anti-*trans*-Golgi Network 38 antigen (anti-TGN38; Affinity BioReagents) antibodies according to either the protocol for the hydroxysuccinimide biotinylation of antibodies or the protocol for the hydrazine biotinylation of antibodies in Chapter 5.

3. Attach the biotinylated antibody from step 2 to the streptavidin-coated magnetic bead from step 1 according to the protocol for the addition of biotinylated antibodies to streptavidin-coated magnetic particles described in Chapter 8.

4. Wash and resuspend the magnetic beads at 5 mg/mL in PBS supplemented with 5 mg/mL BSA.

5. Harvest the rat liver, and prepare and fractionate a cell homogenate as

described in steps 1–6 of the protocol for the affinity purification of right-side-out and inside-out rat liver mitochondria using cytochrome *c*-Sepharose columns, above, except transfer the supernatant from the last step to a fresh centrifuge tube.

6. Centrifuge the supernatant at $20\,000\times g$ for 30 min at 4°C, and collect the supernatant.

7. Determine the protein concentration in the supernatant (see Appendix I for routine protein assays).

8. Add 250 µg of membrane protein from the supernatant (step 6) to a test tube containing 250 µL of the anti-TGN38–coated magnetic beads prepared in step 4.

9. Incubate at 4°C for 2 h with slow end-over-end rotation.

10. Recover the Golgi-coated magnetic beads by the application of a magnetic force, and wash in PBS with BSA.

The above protocol is based on the procedure described by Jones et al. (48), which provides a recent, detailed description of the use of direct capture of organelles by immunolabeled magnetic particles. Kausch and colleagues (51,52) used an indirect capture approach for isolating chloroplast and nuclear material from lysed plant protoplast (see the protocol for chloroplast isolation by magnetic immunoabsorption, below). Their approach involved the isolation of organelle cell lysates initially exposed to biotinylated antibodies, and then they used streptavidin-coated particles to capture the biotinylated antibody-coated organelles. A variation on this theme, as discussed in Chapter 8, would be to use anti-immunoglobulin- or Protein A-coated magnetic beads to capture Fc portions of a nonbiotinylated IgG antibody.

Chloroplast Isolation by Magnetic Immunoabsorption

1. Prepare (see the protocol for the coating of magnetic particles with streptavidin in Chapter 8) or purchase streptavidin-coated magnetic particles.

   Note: Care should be taken in choosing the magnetic particle; nanosized particles (nm diameter) such as those provided by Miltenyi Biotec are reported to work best for chloroplast isolation (see Reference 52), but specialized microparticles for subcellular isolation such as the Dynabeads M-500 can also be chosen.

2. Collect 5 g of pea leave mesophyll, and wash in ice-cold H_2O and blot dry.

3. Resuspend the leaves in 500 mM sorbitol and finely mince them into 1-mm thick sections using a single-edge razor. After mincing, allow the leaves to drain.

4. Add 30 mL of digestion media [2% (wt/vol) cellulase and 0.4% (wt/vol) pectinase in 500 mM sorbitol, 5 mM MES, 1 mM $CaCl_2$, pH 6.0] to a large capacity glass container (such as a 750-mL glass beaker).

5. Place the leaves on top of the digestion media, and cover the beaker with a glass tray filled with water to act as a heat sink. Place a 60 W light bulb 20 cm above the leaves and incubate under illumination for 3 h at room temperature.

6. Decant the digestion mixture. Any protoplasts that are released into the liquid can be recovered by mild centrifugation ($100\times g$ for 5 min).

7. Wash the digested tissue twice with 25 mL each of 500 mM sorbitol, 5 mM

MES, 1 mM CaCl$_2$, pH 6.0. The washings should be poured sequentially through a 500-μm nylon mesh, then a 200-μm nylon mesh.

8. Combine the washings from step 7 with any protoplasts recovered in step 6, and centrifuge at 100× g for 5 min.

9. Gently resuspend the protoplast in 1 mL of 500 mM sucrose, 5 mM MES, 1 mM CaCl$_2$, pH 6.0, and add to a glass tube. Overlay this with 2 mL of 400 mM sucrose, 100 mM sorbitol, 5 mM MES, 1 mM CaCl$_2$, pH 6.0, followed by 1 mL of 500 mM sorbitol, 5 mM MES, 1 mM CaCl$_2$, pH 6.0. Take care not to disturb the underlying layer with each new addition.

10. Centrifuge at 100× g for 5 min. Collect intact protoplasts from the interface between the two upper layers using a Pasteur pipet and place in 500 mM sorbitol, 5 mM MES, 1 mM CaCl$_2$, pH 6.0.

11. Pellet the protoplast by centrifugation at 100× g for 5 min.

12. Discard the supernatant and resuspend the pellet in 1 mL of ice-cold 330 mM sorbitol, 50 mM HEPES, 2 mM EDTA, 1 mM MgCl$_2$, and 1 mM MnCl$_2$, pH 7.6.

13. Rupture the protoplast membranes by forcing them through a 20-μm or 35-μm mesh. Collect the filtrate and force it through the mesh again.

14. Block nonspecific immunoglobulin-binding sites by adding 100 μL of the same solution as in step 12 supplemented with 2% (wt/vol) nonfat milk, 20 mM KCl, and 0.05 mg/mL of non-immune serum.

15. Add an appropriate biotinylated antibody that recognizes pea chloroplast at a final concentration of 100 μg/mL.

   Note: The antibodies can be biotinylated according to the hydrazine procedure listed in Chapter 5 using EZ-Link™ NHS-LC-Biotin from Pierce Chemical.

16. Incubate the protoplast–antibodies solution for 10–30 min at 4°C.

17. Add the streptavidin-coated magnetic particles, and incubate with mixing for 10 min.

18. Recover the magnetic particles with attached chloroplasts by application of a magnetic force. Miltenyi Biotec makes a magnetic column suited for this recovery.

REFERENCES

1. **Albertsson, P.A.** 1986. Cell organelles and membrane vesicles, p. 147-172. *In* Partitioning of Cell Particles and Macromolecules: Separation and Purification of Biomolecules, Cell Organelles, Membranes, and Cells in Aqueous Polymer Two-Phase Systems and Their Use in Biochemical Analysis and Biotechnology. John Wiley & Sons, New York.

2. **Albertsson, P.A.** 1988. Separation of cell organelles and membrane vesicles by phase partition. Prog. Clin. Biol. Res. *270*:227-235.

3. **Arborgh, B., J.L. Ericsson and H. Glaumann.** 1973. Method for the isolation of iron-loaded lysosomes from rat liver. FEBS Lett. *32*:190-194.

4. **Bauer, H.C., E. Ferber, J.R. Golecki and G. Brunner.** 1979. Preparation and fractionation of membrane vesicles of thymocytes after osmotic cell disruption. Hoppe Seylers. Z. Physiol. Chem. *360*:1343-1350.

5. **Beaumelle, B.D., A. Gibson and C.R. Hopkins.** 1990. Isolation and preliminary characterization of the major membrane boundaries of the endocytic pathway in lymphocytes. J. Cell Biol. *111*:1811-1823.

6. **Beaumelle, B.D. and C.R. Hopkins.** 1989. High-yield isolation of functionally competent endosomes from mouse lymphocytes. Biochem. J. *264*:137-149.

7. **Becich, M.J. and J.U. Baenziger.** 1991. Ligand-specific isolation of endosomes and lysosomes using superparamagnetic colloidal iron dextran glycoconjugates and high gradient magnetic affinity chromatog-

raphy. Eur. J. Cell Biol. *55*:71-82.

8. **Betterle, C., A. Caretto, M. Zeviani, B. Pedini and C. Salviati.** 1985. Demonstration and characterization of anti-human mitochondria autoantibodies in idiopathic hypoparathyroidism and in other conditions. Clin. Exp. Immunol. *62*:353-360.

9. **Billett, E.E., B. Gunn and R.J. Mayer.** 1984. Characterization of two monoclonal antibodies obtained after immunization with human liver mitochondrial membrane preparations. Biochem. J. *221*:765-776.

10. **Boonyapisit, K., H.J. Kaminski and R.L. Ruff.** 1999. Disorders of neuromuscular junction ion channels. Am. J. Med. *106*:97-113.

11. **Brunner, G., E. Ferber and K. Resch.** 1976. Initial stages of differentiation of thymocytes in the plasma membrane. Differentiation *5*:161-164.

12. **Brunner, G., E. Ferber and K. Resch.** 1977. Fractionation of membrane vesicles. I. A separation method for different populations of membrane vesicles of thymocytes by affinity chromatography on con A-Sepharose. Anal. Biochem. *80*:420-429.

13. **Cantrell, A.C. and P. Ellis.** 1983. Reaction of colloidal silica with membranes of intact mammalian cells. Chem. Biol. Interact. *44*:169-183.

14. **Cezanne, L., L. Navarro and J.F. Tocanne.** 1992. Isolation of the plasma membrane and organelles from Chinese hamster ovary cells. Biochim. Biophys. Acta *1112*:205-214.

15. **Chaney, L.K. and B.S. Jacobson.** 1983. Coating cells with colloidal silica for high yield isolation of plasma membrane sheets and identification of transmembrane proteins. J. Biol. Chem. *258*:10062-10072.

16. **Cohen, C.M., D.I. Kalish, B.S. Jacobson and D. Branton.** 1977. Membrane isolation on polylysine-coated beads. Plasma membrane from HeLa cells. J. Cell Biol. *75*:119-134.

17. **Courvalin, J.C. and H.J. Worman.** 1997. Nuclear envelope protein autoantibodies in primary biliary cirrhosis. Semin. Liver Dis. *17*:79-90.

18. **Davis, R.A., R.N. Thrift, C.C. Wu and K.E. Howell.** 1990. Apolipoprotein B is both integrated into and translocated across the endoplasmic reticulum membrane. Evidence for two functionally distinct pools. J. Biol. Chem. *265*:10005-10011.

19. **Del Buono, B.J., F.W. Luscinskas and E.R. Simons.** 1989. Preparation and characterization of plasma membrane vesicles from human polymorphonuclear leukocytes. J. Cell Physiol. *141*:636-644.

20. **Desjardins, M. and S. Scianimanico.** 1998. Isolation of phagosomes from professional and nonprofessional phagocytes, p. 75-80. *In* J.E. Celis (Ed.), Cell Biology: A Laboratory Handbook. Academic Press, San Diego.

21. **Dropcho, E.J.** 1995. Autoimmune central nervous system paraneoplastic disorders: mechanisms, diagnosis, and therapeutic options. Ann. Neurol. (Suppl. 1) *37*:S102-S113.

22. **D'Souza, M.P. and J.G. Lindsay.** 1981. Isolation of a sealed homogeneous population of inner membrane fragments with inverted orientation from rat liver mitochondria using specific lectin immunoprecipitation. Biochim. Biophys. Acta *640*:463-472.

23. **Eggleton, P. and D.H. Llewellyn.** 1999. Pathophysiological roles of calreticulin in autoimmune disease. Scand. J. Immunol. *49*:466-473.

24. **Ekblad, L. and B. Jergil.** 1998. Purification of rat liver plasma membranes by affinity chromatography, p. 5-11. *In* J.E. Celis (Ed.), Cell Biology: A Laboratory Handbook. Academic Press, San Diego.

25. **Ekblad, L., J. Kernbichler and B. Jergil.** 1998. Aqueous two-phase affinity partitioning of biotinylated liposomes using neutral avidin as affinity ligand. J. Chromatogr. A *815*:189-195.

26. **Fayle, D.R., P.S. Sim, D.K. Irvine and W.F. Doe.** 1985. Isolation of plasma membrane from human blood monocytes. Subcellular fractionation and marker distribution. Eur. J. Biochem. *147*:409-419.

27. **Flanagan, S.D., S.H. Barondes and P. Taylor.** 1976. Affinity partitioning of membranes. Cholinergic receptor-containing membranes from Torpedo californica. J. Biol. Chem. *251*:858-865.

28. **Fristedt, U., A. Berhe, K. Ensler, Norling and B.L. Persson.** 1996. Isolation and characterization of membrane vesicles of Saccharomyces cerevisiae harboring the high-affinity phosphate transporter. Arch. Biochem. Biophys. *330*:133-141.

29. **Gafencu, A., M. Stanescu, A.M. Toderici, C. Heltianu and M. Simionescu.** 1998. Protein and fatty acid composition of caveolae from apical plasmalemma of aortic endothelial cells. Cell Tissue Res. *293*:101-110.

30. **Ghitescu, L., B.S. Jacobson and P. Crine.** 1999. A novel, 85 kDa endothelial antigen differentiates plasma membrane macrodomains in lung alveolar capillaries. Endothelium *6*:241-250.

31. **Ghitescu, L.D., P. Crine and B.S. Jacobson.** 1997. Antibodies specific to the plasma membrane of rat lung microvascular endothelium. Exp. Cell Res. *232*:47-55.

32. **Giuliano, K.A.** 1991. Aqueous two-phase protein partitioning using textile dyes as affinity ligands. Anal. Biochem. *197*:333-339.

33. **Glaumann, H., H. Jansson, B. Arborgh and J.L. Ericsson.** 1975. Isolation of liver lysosomes by iron loading. Ultrastructural characterization. J. Cell Biol. *67*:887-894.

34. **Goppelt, M., R. Eichhorn, G. Krebs and K. Resch.** 1986. Lipid composition of functional domains of the lymphocyte plasma membrane. Biochim. Biophys. Acta *854*:184-190.

35. **Gotlib, L.J.** 1982. Isolation of cell plasma membranes on microcarrier culture beads. Biochim. Biophys.

Acta *685*:21-26.

36.**Graham, J.M.** 1997. Homogenization of tissues and cells, p. 1-29. *In* J.M. Graham and D. Rickwood (Eds.), Subcellular Fractionation. A Practical Approach. Oxford University Press (IRL Press), Oxford.

37.**Gupta, D.K. and A.M. Tartakoff.** 1989. A novel lectin-gold density perturbation eliminates plasma membrane contaminants from Golgi-enriched subcellular fractions. Eur. J. Cell Biol. *48*:64-70.

38.**Henley, J.R. and M.A. McNiven.** 1996. Association of a dynamin-like protein with the Golgi apparatus in mammalian cells. J. Cell Biol. *133*:761-775.

39.**Hinton, R.H. and B.M. Mullock.** 1997. Isolation of subcellular fractions, p. 31-69. *In* J.M. Graham and D. Rickwood (Eds.), Subcellular Fractionation. A Practical Approach. Oxford University Press (IRL Press), Oxford.

40.**Howell, K.E., R. Schmid, J. Ugelstad and J. Gruenberg.** 1989. Immunoisolation using magnetic solid supports: subcellular fractionation for cell-free functional studies. Methods Cell Biol. *31*:265-292.

41.**Hultcrantz, R., J. Ahlberg and H. Glaumann.** 1984. Isolation of two lysosomal populations from iron-overloaded rat liver with different iron concentration and proteolytic activity. Virchows Arch. B Cell Pathol. Incl. Mol. Pathol. *47*:55-65.

42.**Jacobson, B.S.** 1977. Isolation of plasma membrane from eukaryotic cells on polylysine-coated polyacrylamide beads. Biochim. Biophys. Acta *471*:331-335.

43.**Jacobson, B.S.** 1980. Improved method for isolation of plasma membrane on cationic beads. Membranes from Dictyostelium discoideum. Biochim. Biophys. Acta *600*:769-780.

44.**Jacobson, B.S. and D. Branton.** 1977. Plasma membrane: rapid isolation and exposure of the cytoplasmic surface by use of positively charged beads. Science *195*:302-304.

45.**Jacobson, B.S., J. Cronin and D. Branton.** 1978. Coupling polylysine to glass beads for plasma membrane isolation. Biochim. Biophys. Acta *506*:81-96.

46.**Jacobson, B.S., J.E. Schnitzer, M. McCaffery and G.E. Palade.** 1992. Isolation and partial characterization of the luminal plasmalemma of microvascular endothelium from rat lungs. Eur. J. Cell Biol. *58*:296-306.

47.**Johansson, G., R. Gysin and S.D. Flanagan.** 1981. Affinity partitioning of membranes. Evidence for discrete membrane domains containing cholinergic receptor. J. Biol. Chem. *256*:9126-9135.

48.**Jones, S.M., R.H. Dahl, J. Ugelstad and K.E. Howell.** 1998. Immunoisolation of organelles using magnetic solid supports, p. 12-25. *In* J.E. Celis (Ed.), Cell Biology: A Laboratory Handbook. Academic Press, San Diego.

49.**Kalish, D.I., C.M. Cohen, B.S. Jacobson and D. Branton.** 1978. Membrane isolation on polylysine-coated glass beads. Asymmetry of bound membrane. Biochim. Biophys. Acta *506*:97-110.

50.**Kamrath, F.J., G. Dodt, H. Debuch and G. Uhlenbruck.** 1984. The isolation of lysosomes from normal rat liver by affinity chromatography. Hoppe Seylers. Z. Physiol. Chem. *365*:539-547.

51.**Kausch, A.P. and B.D. Bruce.** 1994. Isolation and immobilization of various plasmid subtypes by magnetic immunoabsorption. Plant J. *6*:767-779.

52.**Kausch, A.P., T.P.J. Owen, S. Narayanswami and B.D. Bruce.** 1999. Organelle isolation by magnetic immunoabsorption. BioTechniques *26*:336-343.

53.**Kinoshita, T., R.L. Nachman and R. Minick.** 1979. Isolation of human platelet plasma membranes with polylysine beads. J. Cell Biol. *82*:688-696.

54.**Kramer, R.M. and D. Branton.** 1979. Retention of lipid asymmetry in membranes on polylysine-coated polyacrylamide beads. Biochim. Biophys. Acta *556*:219-232.

55.**Kumar, V., A.N. Malviya, E.H. Beutner and W.B. Elliott.** 1979. Mitochondrial antibodies–heterogeneity and effects on mitochondrial respiration. J. Clin. Lab. Immunol. *2*:325-328.

56.**Kutsuki, H., S. Takata, K. Yamamoto and N. Tani.** 1998. Therapeutic selective adsorption of anti-DNA antibody using dextran sulfate cellulose column (Selesorb) for the treatment of systemic lupus erythematosus. Ther. Apher. *2*:18-24.

57.**Kyriatsoulis, A., M. Manns, U. Roth, G. Gerken, A. Lohse, S. Wollensak, K. Reske and K.H. Meyer zum Buschenfelde.** 1988. Strategy for the characterization of autoantigens in autoimmune diseases. Investigation of the target antigens of antimitochondrial antibodies by radioimmunoassay, immunoblotting, monoclonal antibodies and affinity chromatography. J. Immunol. Methods *109*:113-121.

58.**Leung, P.S., R.L. Coppel, A. Ansari, S. Munoz and M.E. Gershwin.** 1997. Antimitochondrial antibodies in primary biliary cirrhosis. Semin. Liver Dis. *17*:61-69.

59.**Lopez-Perez, M.J., G. Paris and C. Larsson.** 1981. Highly purified mitochondria from rat brain prepared by phase partition. Biochim. Biophys. Acta *635*:359-368.

60.**Lotscher, H.R., K. Schwerzmann and E. Carafoli.** 1979. The transport of Ca^{2+} in a purified population of inside-out vesicles from rat liver mitochondria. FEBS Lett. *99*:194-198.

61.**Lüers, G.H., R. Hartig, H. Mohr, M. Hausmann, H.D. Fahimi, C. Cremer and A. Volkl.** 1998. Immuno-isolation of highly purified peroxisomes using magnetic beads and continuous immunomagnetic sorting. Electrophoresis *19*:1205-1210.

62.**Luzio, J.P. and K.K. Stanley.** 1983. The isolation of endosome-derived vesicles from rat hepatocytes. Biochem. J. *216*:27-36.

63. **Marquez-Sterling, N.R., A.C.Y. Lo, S.S. Sisodia and E.H. Koo.** 1997. Trafficking of cell-surface beta-amyloid precursor protein: evidence that a sorting intermediate participates in synaptic vesicle recycling. J. Neurosci. *17*:140-151.

64. **Mason, P.W. and B.S. Jacobson.** 1985. Isolation of the dorsal, ventral and intracellular domains of HeLa cell plasma membranes following adhesion to a gelatin substrate. Biochim. Biophys. Acta *821*:264-276.

65. **Morré, D.J. and D.M. Morré.** 1989. Preparation of mammalian plasma membranes by aqueous two-phase partition. BioTechniques *7*:946-958.

66. **Nakamura, R.M., F.V. Chisari and T.S. Edgington.** 1975. Laboratory tests for diagnosis of autoimmune diseases. Prog. Clin. Pathol. *6*:177-203.

67. **Oh, P. and J.E. Schnitzer.** 1998. Isolation and subfractionation of plasma membranes to purify caveolae separately from glycosyl-phosphatidylinositol-anchored protein microdomains, p. 34-45. *In* J.E. Celis (Ed.), Cell Biology: A Laboratory Handbook, vol. 2. Academic Press, San Diego.

68. **Patton, W.F., M.R. Dhanak and B.S. Jacobson.** 1990. Analysis of plasma membrane protein changes in Dictyostelium discoideum during concanavalin A induced receptor redistribution using two-dimensional gel electrophoresis. Electrophoresis *11*:79-85.

69. **Perrin-Cocon, L.A., P.N. Marche and C.L. Villiers.** 1999. Purification of intracellular compartments involved in antigen processing: a new method based on magnetic sorting. Biochem. J. *338*:123-130.

70. **Persson, A. and B. Jergil.** 1992. Purification of plasma membranes by aqueous two-phase affinity partitioning. Anal. Biochem. *204*:131-136.

71. **Persson, A. and B. Jergil.** 1995. The purification of membranes by affinity partitioning. FASEB J. *9*:1304-1310.

72. **Persson, A., B. Johansson, H. Olsson and B. Jergil.** 1991. Purification of rat liver plasma membranes by wheat-germ-agglutinin affinity partitioning. Biochem. J. *273*:173-177.

73. **Philipp, T., P. Straub, M. Durazzo, R.H. Tukey and M.P. Manns.** 1995. Molecular analysis of autoantigens in hepatitis D. J. Hepatol. *22*:132-135.

74. **Quaranta, S., H. Shulman, A. Ahmed, Y. Shoenfeld, J. Peter, G.B. McDonald, J. Van de Water, R. Coppel, C. Ostlund, H.J. Worman et al.** 1999. Autoantibodies in human chronic graft-versus-host disease after hematopoietic cell transplantation. Clin. Immunol. *91*:106-116.

75. **Reimer, G.** 1990. Autoantibodies against nuclear, nucleolar, and mitochondrial antigens in systemic sclerosis (scleroderma). Rheum. Dis. Clin. North Am. *16*:169-183.

76. **Resch, K., S. Schneider and M. Szamel.** 1981. Separation of right-side-out-oriented subfractions from purified thymocyte plasma membranes by affinity chromatography on concanavalin A-sepharose. Anal. Biochem. *117*:282-292.

77. **Richardson, P.J. and J.P. Luzio.** 1988. Immunoaffinity purification of membrane fractions from mammalian cells. Subcell. Biochem. *12*:221-241.

78. **Richardson, P.J., K. Siddle and J.P. Luzio.** 1984. Immunoaffinity purification of intact, metabolically active, cholinergic nerve terminals from mammalian brain. Biochem. J. *219*:647-654.

79. **Saucan, L. and G.E. Palade.** 1994. Membrane and secretory proteins are transported from the Golgi complex to the sinusoidal plasmalemma of hepatocytes by distinct vesicular carriers. J. Cell Biol. *125*:733-741.

80. **Schnitzer, J.E., D.P. McIntosh, A.M. Dvorak, J. Liu and P. Oh.** 1995. Separation of caveolae from associated microdomains of GPI-anchored proteins. Science *269*:1435-1439.

81. **Schnitzer, J.E., P. Oh, B.S. Jacobson and A.M. Dvorak.** 1995. Caveolae from luminal plasmalemma of rat lung endothelium: microdomains enriched in caveolin, Ca(2+)-ATPase, and inositol trisphosphate receptor. Proc. Natl. Acad. Sci. USA *92*:1759-1763.

82. **Sharma, S.K. and P.P. Mahendroo.** 1980. Affinity chromatography of cells and cell membranes. J. Chromatogr. *184*:471-499.

83. **Steck, T.L. and M. Lavasa.** 1994. A general method for plasma membrane isolation by colloidal gold density shift. Anal. Biochem. *223*:47-50.

84. **Stolz, D.B., G. Bannish and B.S. Jacobson.** 1992. The role of the cytoskeleton and intercellular junctions in the transcellular membrane protein polarity of bovine aortic endothelial cells in vitro. J. Cell Sci. *103*:53-68.

85. **Stolz, D.B. and B.S. Jacobson.** 1992. Examination of transcellular membrane protein polarity of bovine aortic endothelial cells in vitro using the cationic colloidal silica microbead membrane-isolation procedure. J. Cell Sci. *103*:39-51.

86. **Szamel, M., M. Goppelt and K. Resch.** 1985. Characterization of plasma membrane domains of mouse EL4 lymphoma cells obtained by affinity chromatography on concanavalin A-Sepharose. Biochim. Biophys. Acta *821*:479-487.

87. **Szamel, M., M. Goppelt-Strube, W. Bessler, K.H. Wiesmuller and K. Resch.** 1987. Separation of plasma membrane domains of calf thymocytes by affinity chromatography on ouabain-Sepharose. Biochim. Biophys. Acta *899*:247-257.

88. **van der Meulen, J.A., D.M. Emerson and S. Grinstein.** 1981. Isolation of chromaffin cell plasma membranes on polycationic beads. Biochim. Biophys. Acta *643*:601-615.

89. **Wallach, D.F.** 1974. Affinity density perturbation. Methods Enzymol. *34*:171-177.

90. **Walsh, F.S., B.H. Barber and M.J. Crumpton.** 1976. Preparation of inside-out vesicles of pig lymphocyte plasma membrane. Biochemistry *15*:3557-3563.

91. **Wandinger-Ness, A., M.K. Bennett, C. Antony and K. Simons.** 1990. Distinct transport vesicles mediate the delivery of plasma membrane proteins to the apical and basolateral domains of MDCK cells. J. Cell Biol. *111*:987-1000.

92. **Wasserman, B.P., B.S. Jacobson, R. Schmidt, Z. Kratky and R.J. Poole.** 1984. Evaluation of the silica microbead method for isolation of red beet protoplast plasma membrane sheets. Biochim. Biophys. Acta *775*:57-63.

93. **Westwood, S.A., J.P. Luzio, D.A. Flockhart and K. Siddle.** 1979. Investigation of the subcellular distribution of cyclic-AMP phosphodiesterase in rat hepatocytes, using a rapid immunological procedure for the isolation of plasma membrane. Biochim. Biophys. Acta *583*:454-466.

94. **Wetzel, M.G. and E.D. Korn.** 1969. Phagocytosis of latex beads by Acahamoeba castellanii (Neff). 3. Isolation of the phagocytic vesicles and their membranes. J. Cell Biol. *43*:90-104.

10 Specialized and Multiple Analyte Separations

In addition to standard techniques for performing affinity and immunoaffinity separations, there are a number of useful modifications to the general procedures. The demand for ultramicroanalytical procedures has led to a growth in applications of "laboratory-on-a-chip" technologies and the development of instruments capable of analyzing samples in the nanoliter and picoliter ranges. Although certain laboratories are developing whole chromatography systems around silicon and plastic chip technology (42,82), CE (a more accessible technology) is an ideal technique for such analyses. This technology is capable of separating nanoliter samples and can be coupled to a number of sophisticated detection devices including mass spectrometers (61). The coupling of affinity and immunoaffinity separation procedures with CE further enhances the power of this exciting technology.

The culmination of much of this specialized detection is the development of biosensors, based on the coupling of biochemical or immunochemical reactions with sophisticated electronic, chemical, or optical signal detectors. These instruments are able to monitor specific biological or chemical agents in "real time." The ultimate use of these entities is the continuous monitoring of critically ill patients in a hospital setting or the monitoring of a conscious animal. Furthermore, the development of such biosensors has potential in chemistry (29), environmental monitoring (88,106), drug screening (103,114), and detection of agents of biological warfare (67).

The development of techniques for analyzing a number of different analytes in microsamples has also addressed the current need of both clinical and industrial investigative studies. Recycling techniques can range from multiple separations performed in preparative scale to the development of recycling columns capable of isolating multiple analytes from small samples. Once again, the application of affinity and immunoaffinity procedures to this area of the analytical sciences has greatly improved the scope and versatility of these techniques.

AFFINITY AND IMMUNOAFFINITY CAPILLARY ELECTROPHORESIS

Affinity capillary electrophoresis (ACE) is a relatively new approach for studying molecular interactions and biomolecular recognition (84,97). Basically, the technique

Affinity and Immunoaffinity Purification Techniques
Terry M. Phillips and Benjamin F. Dickens
© 2000 Eaton Publishing, Natick, MA

involves inclusion of a soluble ligand in the running buffer, which interacts with its target analyte, thus changing the electrophoretic properties of the analyte (100) (Figure 1). This change enables the investigator to easily isolate the molecule of interest or to measure the interaction between the ligand and the analyte. To date, this approach has been applied to chiral separation of racemic biomolecules (81,102), measurement of binding constants (19,59), estimation of kinetic on- and off-rate constants, estimation of effective charges and molecular weights of proteins, characterization of enzymatic activities, molecular binding of antigen to the MHC (45), and library screening for tight-binding drug candidates in solution. This technique demands only small amounts

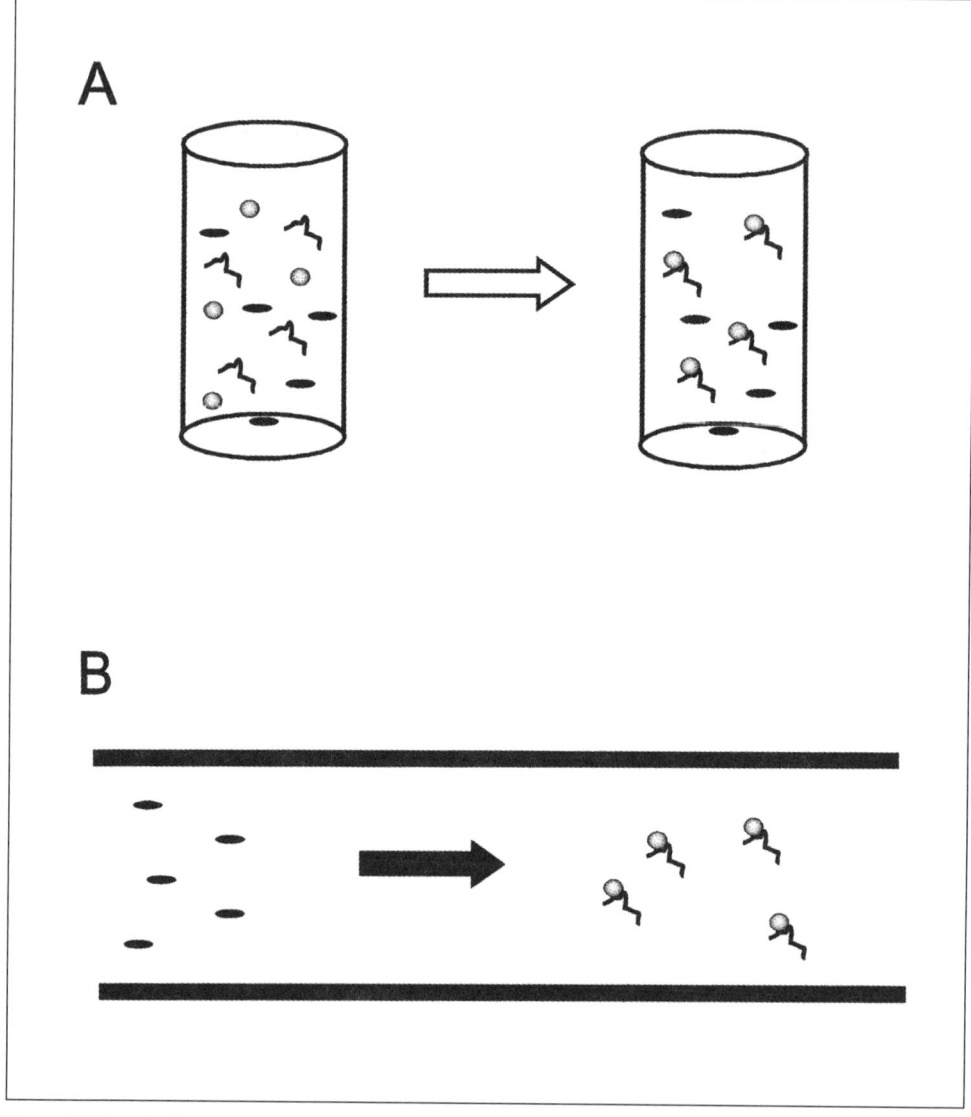

Figure 1. The principles of ACE. (A) The affinity ligand (wavy line) is mixed with the sample containing the analyte of interest (shaded ball) and another compound (black oval). The ligand binds to the analyte forming a complex. (B) The increased size and charge of the complex allow it to separate faster in the electrophoretic flow, thus enabling separation and isolation of the analyte.

of sample (nanoliter injections of picogram quantities of the proteins of interest), involves no radiolabeled materials or chemically immobilized ligands, and does not require changes in spectroscopic characteristics upon binding (18).

Affinity Capillary Electrophoresis

One of the areas in which ACE has been applied is molecular biology, where it has been used not only for the recognition of specific DNA bases and sequences but also for the investigation of the binding properties of DNA-based drugs. In such studies, the ligand is usually incorporated into the separation gel and the running buffer (3). Guttman and Cooke (37) added ethidium bromide to a linear polyacrylamide gel to optimize the separation of DNA restriction fragments. Using this approach, they estimated that high resolution separations are possible with separation efficiencies up to 10^7 theoretical plates per meter of capillary.

The addition of simple ions to the CE running buffer has been reported to aid in ACE separations. The addition of an ion pairing agent (100 mM heptanesulfonic acid in 40 mM sodium phosphate buffer, pH 2.5) was used to characterize the glycopeptide map of recombinant human erythropoietin (87). The inclusion of this agent aided peptide resolution by its affinity for asialoglycopeptides. In a similar vein, the inclusion of the metal chelate iminodiacetate-Cu(II) coupled to PEG has been employed in studies of the interactions of model proteins with metal chelates (41). Using ribonuclease, cytochrome c, chymotrypsin, and kallikrein as the model proteins, the authors of these studies were able to obtain data suitable for calculating fast on/off kinetics of the interactions.

The addition of cyclodextrins and crown ethers to buffer systems in CE for the affinity separation of racemic forms of a number of drugs is well documented (108). Affinity separations can also be achieved by the addition of suitable protein or peptide ligands, which has led to the application of ACE for the separation of racemic forms of a number of different drugs (Figure 2). Barker et al. (4) employed BSA as an additive for separation of the (6R)- and (6S)-stereoisomers of leucovorin. The application of protein ligands for the affinity separation of chiral entities using CE has been recently reviewed by Hage (38). A useful modification of ACE is vacancy affinity capillary electrophoresis (VACE), which can be used to study the displacement of a target drug from a protein by the simultaneous administration of the drugs under investigation. Erim and Kraak (28) studied the displacement of the target drug warfarin from BSA by furosemide and phenylbutazone, by monitoring the shift in the mobility of warfarin in the presence of these displacer drugs.

Affinity ligands such as lectins and bacterial proteins have also been used to enhance specific analyte separations. Shimura and Kasai (95) studied the dissociation constants between the lectin con A and unlabeled neutral monosaccharides using a modification they called affinity probe capillary electrophoresis (APCE). The probe was synthesized by labeling p-aminophenyl α-D-mannopyranoside with a fluorochrome, using glutathione as a negatively charged linker. In these studies, the authors describe a decrease in the mobility of the probe in the presence of con A, which could be analyzed as a function of the concentration of the lectin. Examination of the migration patterns also enabled the investigators to determine the dissociation constant of the lectin–monosaccharide complex. Similar findings were reported using a mannoside probe coupled to the free thiol group of N-succinylated glutathione to investigate the binding of pea lectin by ACE (96). Using this approach, the authors state that the procedure requires as little as 0.14 ng of protein to perform the studies and is therefore applicable to a number of microscale studies. Taga et al. (101) reported the separation of 1-phenyl-3-methyl-5-pyrazolone derivatives of simple disaccharides using a number of lectin ligands. The addition of *Lens culinaris* agglutinin lectin to solutions of maltose, cellobiose, gentiobiose, lactose, and melibiose demonstrated a selective retardation in the migration patterns of glucobioses, depending on the amount of lectin present. The addition of another lectin, *R. communis*, gave a different migration profile, favoring retardation of derivatives of galactosyl glucoses. However, neither lectin gave complete separation of its target analyte.

The incorporation of bacterial proteins such as protein A has also been reported. Lausch et al. (58) reported a procedure wherein the determination of IgG in culture medium was achieved using fluorochrome-labeled protein A as the affinity ligand. Detection of the affinity complex was achieved by laser-induced fluorescence. Use of

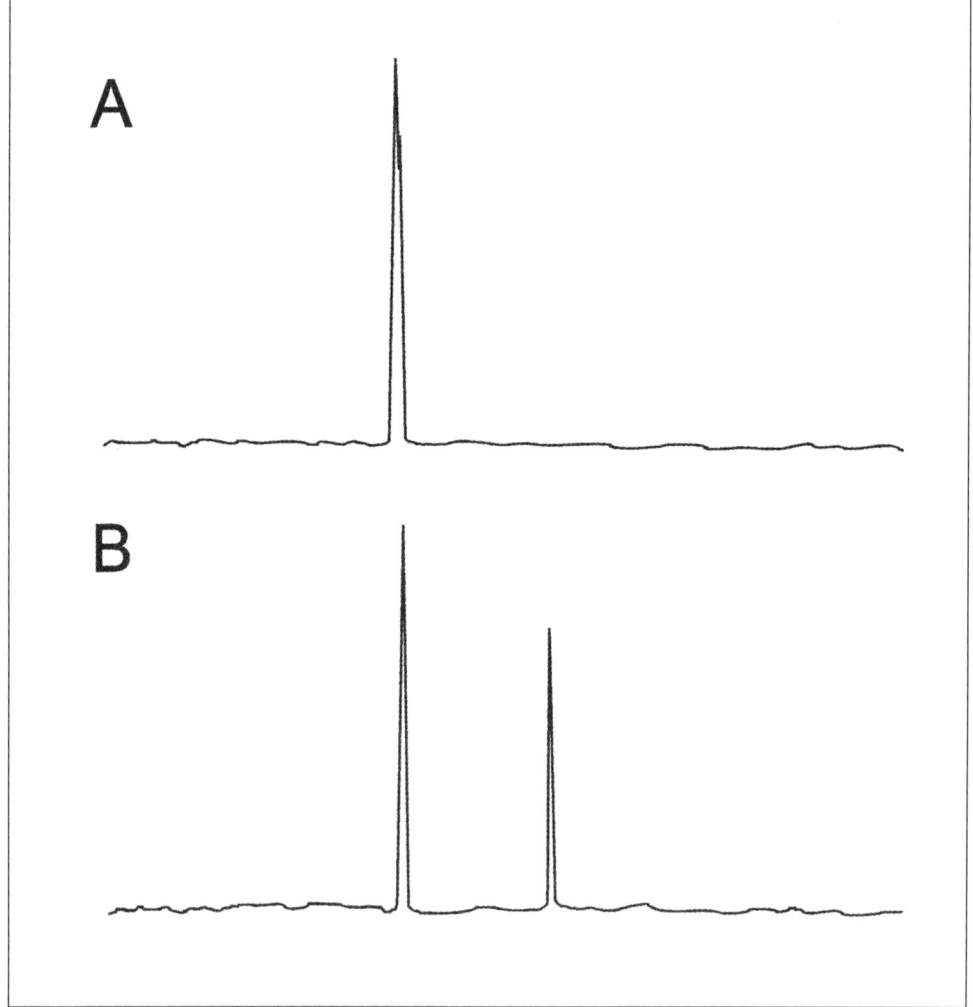

Figure 2. CE chiral separation of racemic forms of a test drug solution by incorporation of a β-cyclodextrin into the running buffer. (A) Untreated sample illustrating the poor resolution of the two enantiomers of the drug. (B) Successful separation of the enantiomers following combination of one form with the cyclodextrin.

the complete protein A molecule appeared to cause interference with the assay, which could be rectified by substituting recombinant monovalent fragments for the whole molecule.

Heparin has a number of interesting properties that make it an ideal affinity ligand. These properties have been used by a number of workers to develop useful ACE techniques for studying the binding interactions between several molecules of biological importance and the heparin ligand. Heegaard et al. (46) used ACE procedures to study the linear heparin-binding sites in the serum protein amyloid P (SAP). In these studies, the authors employed various amounts of heparin in solution to study the migration shift patterns of the different fragments. In later studies, Heegaard (44) further characterized the binding of heparin and a tryptic digest fragment (14–38 amino acids designated as T3) of SAP. The inclusion of heparin and heparin fragments as ligands allowed screening for various glycosaminoglycans. Plots based on quantitative measurements of the resolved analyte peaks (free and heparin-complexed T3) demonstrated a dissociation constant of 1.5 µM for its interaction with heparin. VanderNoot et al. (105) employed an immobilized glycosaminoglycan as a ligand to study the affinity resolution of synthetic heparin-binding peptides using ACE. Heparin and heparin sulfate were immobilized onto fused silica capillaries by the biotin–avidin complex and were shown to be able to distinguish peptides based on the heparin-binding domain of acidic fibroblast growth factor (residues 125–144). Wu and Lindhardt (112) also used immobilized heparin to study its interaction with antithrombin III and secretory leukocyte proteinase inhibitor. Table 1 gives selected examples of ligands used in ACE.

Immunoaffinity Capillary Electrophoresis

Antibody-based immunoassays performed by CE or immunoaffinity capillary electrophoresis (ICE) are becoming a popular new addition to CE applications. This approach has proven useful for studying interaction kinetics and antibody–analyte interactions by studying either electrophoretic migration changes or mass changes. Migration shifts caused by the formation of antibody–antigen complexes have been used to study interaction kinetics between a specific antibody and a model protein (43). Likewise, Qian and Tomer (79) used mass spectrometry to investigate the binding of a monoclonal antibody to different epitopes on the HIV core protein p24. Shimura and Karger (94) employed a fluorochrome-labeled FAb fragment of a monoclonal antibody to determine the concentrations of methionyl recombinant human growth hormone. This technique could detect the analyte down to concentrations of approximately 5×10^{-12} M and could distinguish between deaminated and nondeaminated variants of the analyte of interest. Chiem and Harrison (16) described a microchip variation of CE designed for the rapid determination of affinity constants. Using a fluorescent-labeled model protein (BSA), they employed a microfluidic chip to perform ICE analyses in less than one minute.

Approaches using immobilized antibodies as capture agents prior to electrophoretic analysis have been described (Figure 3). Phillips and Chmielinska (74) described the application of immobilized anti-cyclosporin antibodies coupled to a capillary zone electrophoresis system for the monitoring of both the native molecule and its clinically important metabolites in patients receiving topic cyclosporin following corneal transplantation (Figure 4). This approach has also been applied to the isolation of neuropeptides in tissue samples and to the monitoring of recombinant

Table 1. Ligands Used in ACE

Ligand	Application and Reference
Ethidium bromide	DNA restriction fragments (37)
Heptanesulfonic acid	Glycoproteins (87)
Iminodiacetate -Cu(II) – PEG	Model protein interaction with metal chelates (41)
Cyclodextrin	Racemic forms of drugs (108)
Con A	Monosaccharides (95)
Pea lectin	Glutathione (96)
L. culinaris agglutinin lectin	Glucobioses (101)
R. communis lectin	Galactosyl glucoses (101)
BSA	Isomers of leucovorin (4) Warfarin (28,38)
Protein A	IgG (58)
Heparin	Amyloid P (44,45) Acidic fibroblast growth factor (105) Antithrombin III (112) Secretory leukocyte proteinase inhibitor (112)

cytokines in various human body fluids (73,75). Dalluge and Sander (22) employed a pre-analytical immunoaffinity bed to detect the presence of human cardiac troponin I in samples obtained from patients with myocardial infarction. The technique used a 5-mm section of capillary packed with antibody-coated porous silica (Figure 5). This procedure was able to detect the specific analyte at 2 nM/L in serum samples. Ljungberg and colleagues (60) have developed a system based on weak bioaffinity recognition for the separation and detection of carbohydrates possessing similar structural characteristics. They polymerized monoclonal antibodies into a gel within a silica capillary for the separation of the carbohydrate molecules. Although promising, these findings do not appear to demonstrate a major improvement in ICE technology. In their excellent review of ACE, Heegaard et al. (47) report not only the application of pre-analytical affinity concentrators and traps but also an exciting new approach to ICE. They describe a variation they call chemicoaffinity capillary electrophoresis/capillary electrochromatography in which molecular imprinting techniques are used to create "plastic antibodies" suitable for immunoaffinity selection.

New Technology

Several investigators have proposed new designs for CE, based on chip technology. Effenhauser et al. (27) proposed an integrated CE system built around a glass or polymeric chip. They also discussed important issues such as electrokinetic fluid handling, detection systems, and chip designs for a number of applications including immunoassays. Colyer et al. (20) described a microchip CE system for the analysis of serum proteins in human samples. However, this work was performed on model systems and had not been performed on natural biological samples. Later, Mangru and Harrison (62) described a chemiluminescence detection system based on the horseradish peroxidase-catalyzed reaction of luminol with peroxide for postseparation detection in microchip-based ACE. In this study, the entire system was fabricated on planar glass wafers.

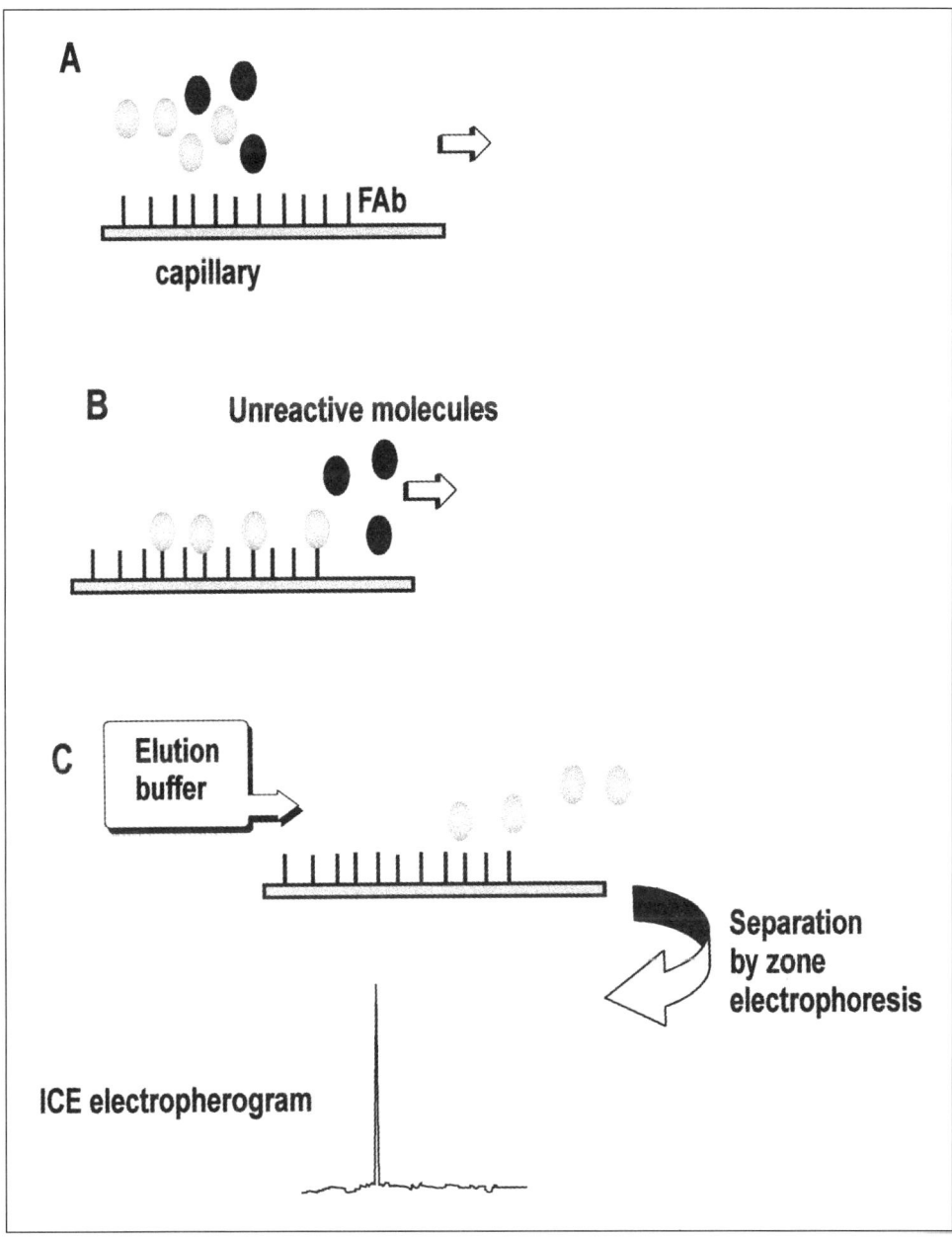

Figure 3. Immunoaffinity capillary electrophoresis. (A) The sample is injected into the capillary and the analyte of interest (shaded balls) comes into contact with the FAb antibody fragments immobilized to the internal wall of the capillary. (B) The immobilized FAbs capture the analyte, allowing the unreactive molecules (filled balls) to be flushed through the capillary. (C) The bound analyte is released from the FAb following the introduction of the elution buffer and the released molecules are analyzed by zone electrophoresis, producing the single peak electropherogram shown at the bottom of the figure.

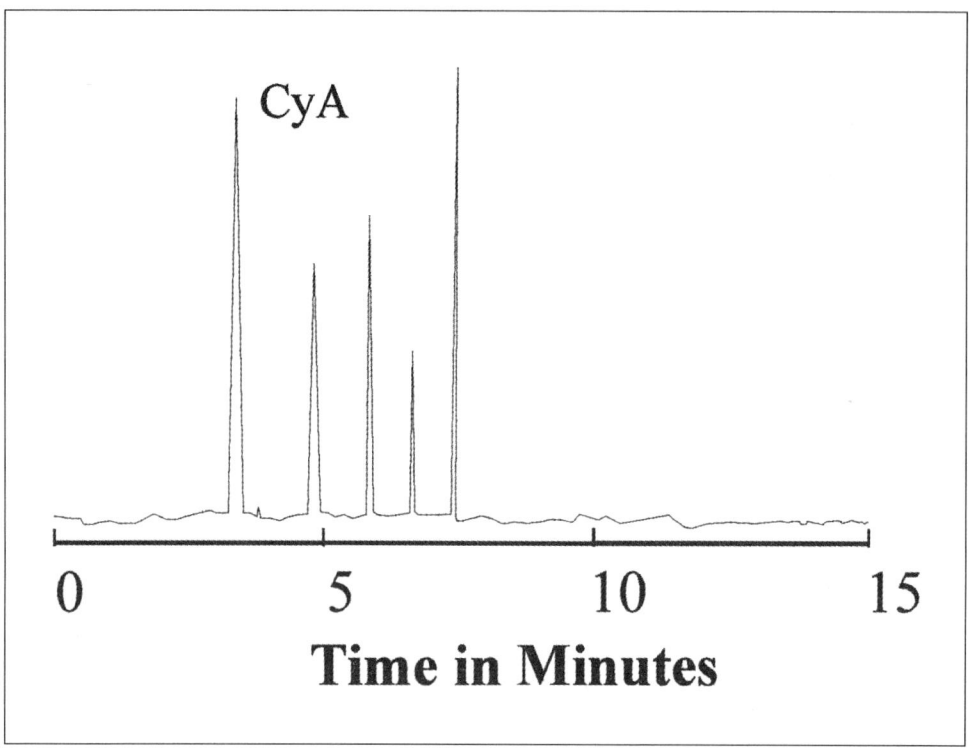

Figure 4. ICE electropherogram of the immunosuppressive drug cyclosporin and its major metabolites. The native drug is found in peak 1 (labeled CyA) and the major metabolites in the four peaks.

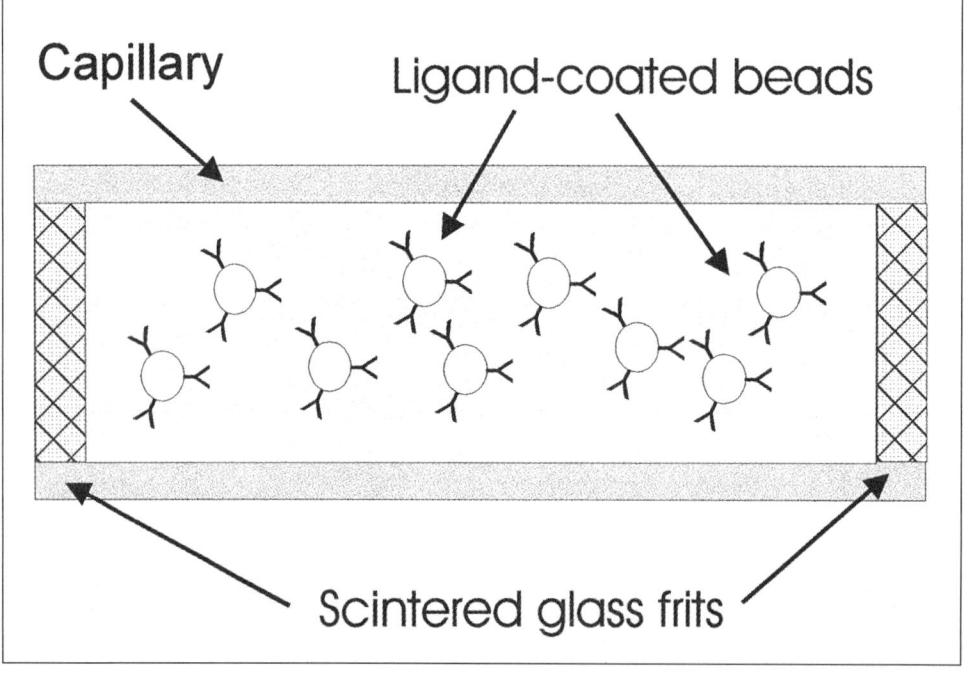

Figure 5. A pre-column concentrator used in ACE and ICE.

294

CE is well suited to analyze samples obtained by microdialysis. Dawson (23) reviewed the application of CE to the analysis of microdialysis samples and compared its effectiveness with other chromatographic techniques. We have applied ICE for the analysis of cytokines secreted from cells in culture using a microdialysis sampling technique (76). In our system, four cytokines were simultaneously measured over time in microdialysis samples obtained from 5-cell cultures (Figure 6).

PROTOCOLS

Separation of Racemic Forms of Warfarin by Chiral CE Using β-Cyclodextrin

1. Prepare 50 mL of running buffer (20 mM NaH_2PO_4 and 14 mM sulfobutyl ether [4]-β-cyclodextrin), pII 7.0, and filter through a 0.22-μm syringe filter.
2. Fill both the anodal and cathodal reservoirs of the instrument with the running buffer. Place the inlet of the capillary into the anodal reservoir and the outlet

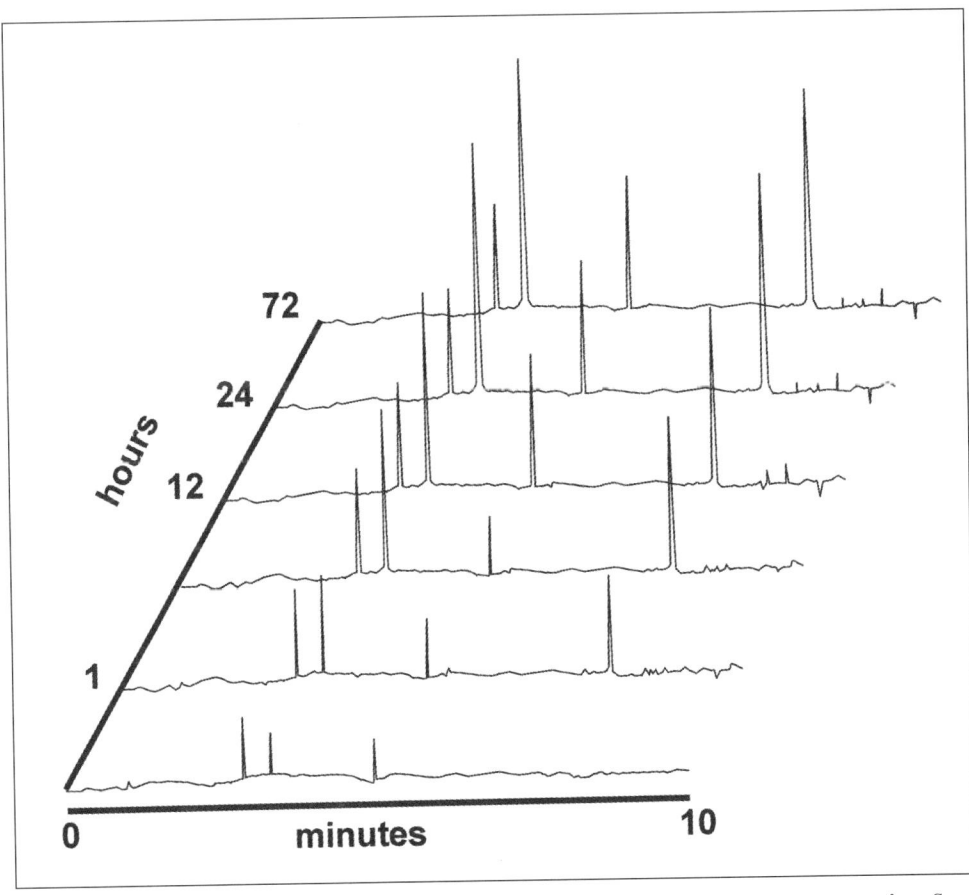

Figure 6. A series of ICE electropherograms showing the detection of four different cytokines over time. Samples taken by microdialysis from 5-cell cultures following stimulation with the neuropeptide substance P. The x-axis represents the duration of each ICE electropherogram and the y-axis represents time in hours. Only selected samples are shown for clarity. The four peaks (from left to right) represent the concentrations of interleukin (IL)-2, IL-4, interferon-gamma, and IL-10, respectively.

into the cathodal reservoir.

3. Attach an uncoated 50-μm capillary (60-cm in length) with a flow-cell positioned at 40-cm from one end.

4. Fill the capillary with the running buffer, then inject 5–10 nL of a 10-μg/mL sample.

5. Run for 5 min at 25 kV, 30–35 μA, continuously monitoring the electropherogram until two distinct peaks representing the racemic forms of the drug are resolved.

For other molecules, it may be necessary to vary the concentration of the cyclodextrin, the composition of the running buffer, and the pH of the running buffer.

ACE Isolation of a Glycoprotein Using Immobilized Peanut Lectin

1. Prepare an 100-cm CDI-coated capillary as described in the protocol for CDI-derivatized capillaries in Chapter 2.

2. Flush the capillary with distilled H_2O.

3. Pipet 500 nL of a 10-μg/mL solution of peanut lectin (dissolved in 0.1 M NaH_2PO_4, pH 7.2) onto a Parafilm sheet.

4. Allow the capillary to take up the liquid by capillary action, then seal the capillary ends with tape.

5. Incubate at room temperature for 6–8 h or overnight at 4°C.

6. Flush the capillary with 1 mL of 0.1 M NaH_2PO_4, pH 7.2.

7. Make a flow cell 65-cm from the inlet by burning off a 0.5-cm section of the polyimide coating of the capillary.

8. Insert the capillary into a CE system, making sure that the flow cell is aligned with the detector port.

9. Fill the capillary with 0.1 M NaH_2PO_4, pH 7.4.

10. Inject a 30–50-nL sample containing an immobilized lectin into the capillary.

11. Incubate the sample for 5–10 min at room temperature.

12. Purge the capillary with 200 μL of a solution of 0.1 M NaH_2PO_4, 0.01% (vol/vol) Brij 35, pH 7.4.

13. Fill the anodal and cathodal buffer reservoirs with a solution of 0.1 M NaH_2PO_4, 0.01% (vol/vol) Brij 35 to which 0.25 M α-mannose has been added.

14. Place the inlet of the capillary into the anodal reservoir and the outlet into the cathodal reservoir.

15. Run the elution-separation at 75 μA constant current for 30–50 min at room temperature or until the single peak is detected on an electropherogram.

ICE Using a Pre-Analytical Concentrator

1. Activate glass beads (10 μm in diameter; Supelco) with CDI as described in the protocol for CDI activation of silica and glass in Chapter 2.

2. Coat the CDI-activated beads with antibodies directed against the desired analyte as described in the protocol for the direct ligand attachment for derivatized

supports in Chapter 5.

3. Fuse a glass frit onto one end of a capillary.

4. Pack the coated beads into the capillary.

5. Fuse a glass frit onto the open end of the capillary.

6. Attach the concentrator to the anodal end of a 60-cm fused silica analytical capillary using polyimide sleeves (Supelco).

7. Attach the capillary to the CE system and inject the sample into the concentrator.

8. Allow the sample to react with the immobilized antibodies for 10 min at room temperature.

9. Fill the anodal and cathodal reservoirs of the CE with 0.1 M NaH_2PO_4, 0.01% (vol/vol) Brij 35, pH 1.5.

10. Place the free end of the concentrator into the anodal reservoir.

11. Place the other end of the analytical capillary into the cathodal reservoir.

12. Electro-elute the bound analyte into the analytical capillary.

13. Run the system at 8–10 kV for 30–40 min, continuously monitoring the developing electropherogram either by UV at 200-nm or by fluorescence.

ICE Using an Antibody Immobilized Directly to the Internal Walls of the Capillary

1. Prepare a 60-cm antibody-coated capillary as described in the protocol for the attachment of ligands to thiol-activated capillaries in Chapter 4, substituting the antibody of choice for the "ligand" in step 3 of that protocol.

2. Make a flow cell 40-cm from the inlet by burning off a 0.5-cm section of the polyimide coating of the capillary.

3. Insert the capillary into the CE system, making sure that the flow cell is aligned with the detector port.

4. Fill the capillary with 0.1 M NaH_2PO_4, pH 7.4.

5. Inject a 30–50 nL sample with the immobilized antibody coating into the capillary.

6. Incubate the sample for 2–5 min at room temperature.

7. Purge the capillary with 200 µL of a solution of 0.1 M NaH_2PO_4, 0.01% (vol/vol) Brij 35, pH 7.4.

8. Fill the anodal and cathodal buffer reservoirs with a solution of 0.1 M NaH_2PO_4, 0.01% (vol/vol) Brij 35, pH 1.5.

9. Place the inlet of the capillary into the anodal reservoir and the outlet into the cathodal reservoir.

10. Run the elution-separation phase at 100-µA constant current for 10–20 min at room temperature or until the single peak is detected on an electropherogram.

IMMUNOSENSORS

Biosensors are basically hybrid instruments coupling biological selectivity with electronic and/or optical sensitivity. A biosensor is composed of a bioaffinity ligand immobilized onto a surface that is capable of being interfaced with a detector and/or measuring device (Figure 7). Signal transduction can be accomplished in many ways, by the generation of heat, light, density, and electrical changes. Several specialized approaches to immobilizing ligands to suitable surfaces for the development of immunosensors have been reported. Kossek et al. (55) employed the streptavidin–biotin system to attach estradiol to the surface of a silicon chip. They also note that immobilized streptavidin can be used to attach any biotinylated ligand to the sensor. Nakanishi and colleagues (65) immobilized antibodies onto the surface of quartz crystals using an ethylenediamine plasma-polymerized film matrix, incorporating an abundance of amino groups. These antibody-coated crystals were used to manufacture quartz crystal microbalance-based immunosensors. Antibodies can be easily attached to thioctic acid following self-assembly of the acid on gold electrodes, to manufacture capacitive immunosensors. Such sensors have been reported to specifically detect human chorionic gonadotropin hormone at a range of 1–1000 pg/mL (7).

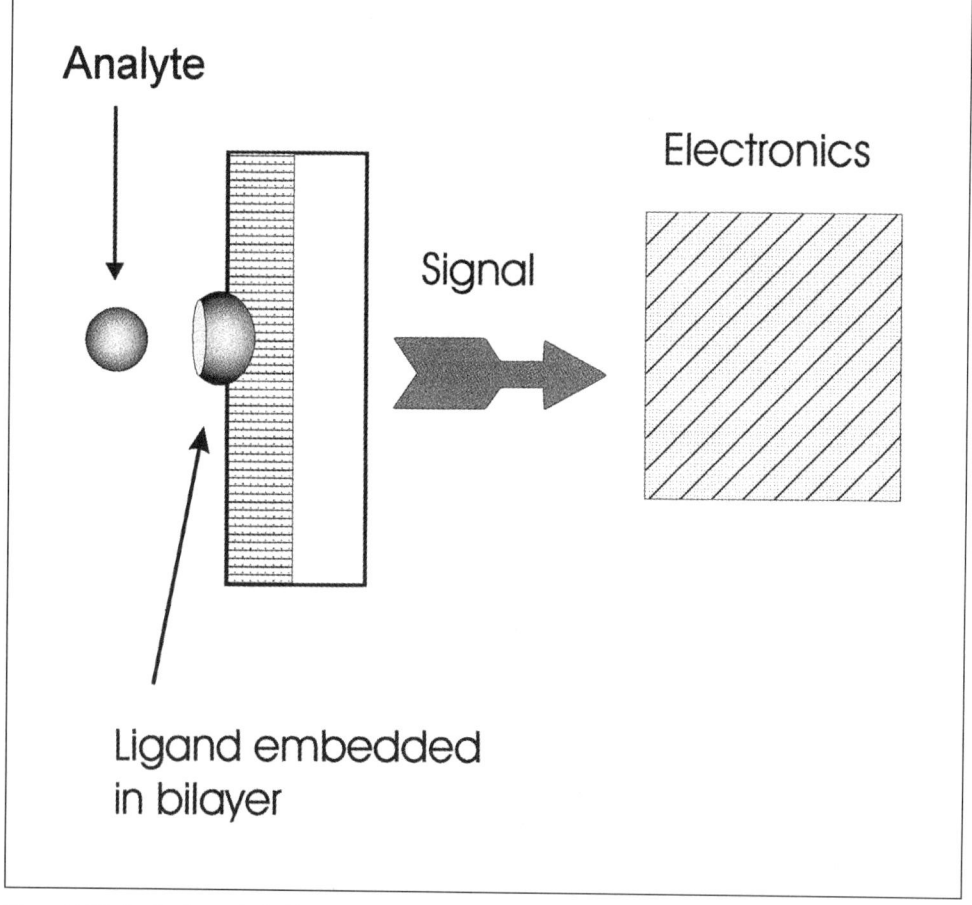

Figure 7. The basic element of a biosensor.

The bioaffinity ligand can be any of the materials routinely used for affinity or immunoaffinity separations (see Chapter 4). The use of immobilized enzymes and receptors has gained popularity and provided the selectivity required for producing a selective biosensor. However, the increasing variety of antibodies commercially available today has made immunologically based biosensors (immunosensors) popular and the design focus for many improved sensors (32,34). Immobilized antibodies possess the necessary selectivity and reversibility required for biosensors. Additionally, under optimal conditions, immobilized antibodies can be regenerated many times, making them ideal as capture ligands (90). Biosensor construction based on immunological reactions has been reported since the mid-1980s using a number of different approaches. These techniques range from immunoassays with enzyme or electrical impulse generation through to the development of immunoassays performed on the tips of fiber optics. In 1986, Karube and Suzuki (51) described a number of immunoassay-based sensors using enzyme reactors as the reporter systems. A year later, the co-immobilization of glucose oxidase and anti-T4 antibodies was employed to build an immunosensor for thyroxine measurements (113). This sensor was able to selectively detect T4 to a detection limit of 15 nM. Paek and Schramm (68) suggested that mathematical modeling could be applied to improve the selectivity and sensitivity of enzyme-based immunosensors. They made a significant contribution by reviewing the potential problems arising from the performance of such immunosensors under non-equilibrium conditions. Furthermore, they developed mathematical solutions that could be used to model optimal conditions for immunosensors operating under non-equilibrium conditions.

Immunosensors based on short columns containing immobilized antibodies or receptors have also been reported. Bouvrette and Luong (8) demonstrated that antibodies immobilized onto aminopropyl-activated glass beads could be used as an effective immunoreactor for the detection of *E. coli* in artificially contaminated foods. This sensor relied on the production of the fluorescent by-product 4-methylumbelliferone, following interaction of the captured bacteria with the substrate 4-methylumbelliferyl β-D-glucuronide. The authors report that the sensor could detect approximately 5×10^7 organisms per mL and that the detection could be performed in less than 30 min. Additionally, the sensor was reusable for greater than 300 assays and remained stable for 3 months under refrigerated conditions. Ghindilis et al. (33) developed a flow-through immunosensor based on a high-surface-area carbon immuno-electrode. The immunosensor element of this apparatus consisted of a disposable column containing antibody-coated carbon particles, which acted as a working electrode. Using this sensor to detect rabbit IgG in solution and human IgM in plasma, the authors report an average analytical time of 22 min. The detection limit for this sensor was in the picogram range for rabbit IgG and nanogram range for human IgM. We have employed recombinant receptors immobilized on glass beads and placed into a post-affinity column immunoreactor to measure bioactive neuropeptides (72) (Figure 8). This system employed a series of enzyme-labeled antibodies to detect the captured analytes, measuring the different enzyme–substrate by-products produced following injection of a series of enzyme-specific substrates.

Enzyme signal-generating reporter systems can also be used to produce electrochemical signals that can be measured by electrochemical detectors or ion-selective electrodes. Duan and Meyerhoff (26) described a novel separation-free sandwich enzyme immunoassay for proteins in which electrochemical detection was achieved by the production of aminophenol and oxidation at a gold electrode. This sensor was based on the diffusion of analyte and substrate through a gold-coated microporous membrane and used to measure human chorionic gonadotropin. Bauer et al. (5) employed a dual enzyme system coupled to a flow injection system for the measurement of alkaline phosphatase in competitive immunoassays for the pesticide 2,4-dichlorophenoxyacetic acid. A Clark-type electrode was used to detect the phenol end product produced by the enzyme using a membrane containing entrapped tyrosinase and quinoprotein glucose dehydrogenase. Phenol is oxidized in the sensor membrane by the tyrosinase to *o*-quinone, which is reconverted to catecholamine by the glucose dehydrogenase. The authors reported that this sensor was able to detect 320 zM (zeptomole)/100 µL of alkaline phosphatase; however, this translates to a sensor sensitivity of approximately 100 ng/mL of the pesticide. Sanden et al. (89) used an amperometric enzyme-linked immunoassay for specific detection of *Nitrobacter* species. The sensor used for this detection was based on the principle that the immunological reaction takes place at the tip of a glassy carbon electrode. The capture immunoassay used an immobilized primary antibody and an alkaline phosphatase-labeled secondary antibody. Detection was achieved by measuring the electrochemical end product produced by the enzyme–substrate interaction. Using this sensor, the authors report that they were able to detect approximately 3×10^6 *Nitrobacter* cells/mL, which is comparable to conventional enzyme-linked immunoassays.

Cook (21) described an electrochemical immunosensor for real-time determination of corticosteroids in conscious animals. This sensor was based on competitive binding of endogenous corticosteroid and a corticosteroid–peroxidase conjugate with antibodies immobilized on a platinum electrode. Additionally, the probe was encased in a dialysis membrane, thus protecting the electrode from direct contact with the surrounding biological fluids. This design plus the small size (350 µm outer diameter) enabled the sensor to be placed either directly into tissues or into the circulation of living animals. The lifetime (200–400 sequential measurements) and sensitivity (2–6 ng/mL) of this immunosensor opens the possibility for "real-time" monitoring of hormones in vivo. An amperometric immunosensor for the detection of the lactate

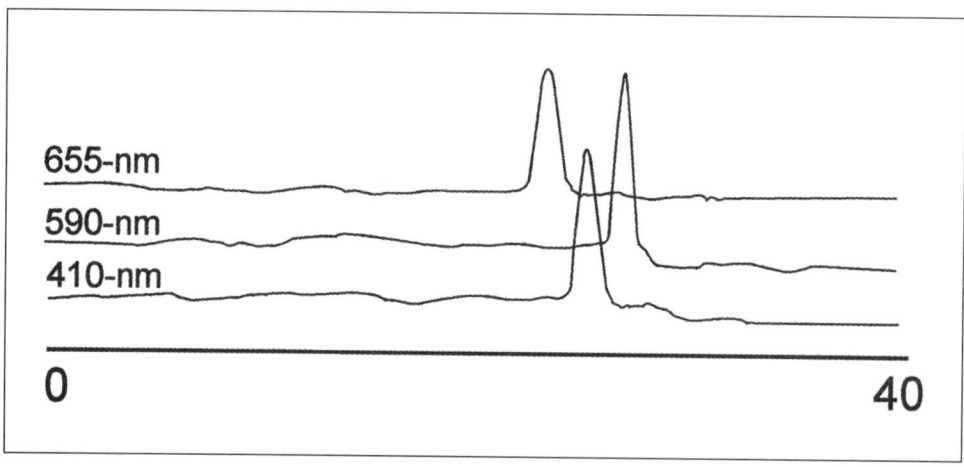

Figure 8. Detection of three neuropeptides from human biopsy samples using a receptor-based biosensor with enzyme-immunoassay detection. Three biologically active neuropeptides were captured by their respective immobilized recombinant receptors and, following reaction with three different enzyme-antibody reagents, the by-products of each enzyme reaction were detected by a diode-array detector. The chromatogram represents the analysis of these by-products: substance P (410-nm), calcitonin gene-related peptide (590-nm), and vasointestinal peptide (655-nm).

dehydrogenase isoenzyme LD_1 has been described (53). This sensor utilizes antibodies immobilized onto a membrane; the end product is recorded by a platinum-working electrode polarized at +600 mV vs. an Ag/AgCl electrode. The isoenzyme activity was measured by detection of the oxidation of nicotinamide adenine dinucleotide (reduced form) (NADH) at the electrode surface. The immunosensor exhibited reasonable sensitivity, detecting a calibration curve for lactate dehydrogenase (LDH) in the 0.005–0.12 U range with 3% repeatability and 10% reproducibility.

There have also been reports of the development of disposable electrochemical immunosensors. Wang et al. (109) developed a disposable electrochemical immunosensor based on potentiometric stripping analysis of a metal tracer for HSA. The sensor was based on an "on-chip" assay format using immobilized antibodies on a thick-film electrode. The assay was performed in a competition style between bismuth-labeled analyte and the analyte of interest. The immunosensor exhibited a dynamic range of 0.3–30 µg/mL and a detection limit of 0.2 ng/mL. Likewise, Baumner and Schmid (6) developed a disposable amperometric immunosensor for the detection of triazine pesticides in environmental samples, based on a thick-film electrode printed on polyvinylchloride (PVC) as the strip-type transducer. Monoclonal antibodies against atrazine and terbutylazine served as biorecognition elements. Hapten-tagged liposomes entrapping ascorbic acid as a marker molecule were chosen for generation and amplification of the signal. A capillary gap and a wicking filter membrane strip served as the migration zone. For signal detection on a graphite electrode, liposomes were lysed by Triton X-100, and the released ascorbic acid was quantified at a potential of +300 mV vs. a printed Ag/AgCl electrode. Signal response time was 1–3 min, and sensitivity of measurements in tap water was below 1 µg/100 mL for atrazine and terbutylazine. Atrazine determinations in soil extracts correlated well with standard procedures based on ELISA and HPLC.

Although there are a growing number of reports on novel designs for immunosensors, detection of the reaction product appears to be one of the stumbling blocks of this technology. Many sensors are based on optical detection and measurement of a specific product, and within this area, there are several schools of thought on the optimal design. Place et al. (78) reviewed a number of optical techniques that were used to measure immunological reactions on synthetic surfaces (i.e., Langmuir-Blodgett films) and concluded that internal reflectance spectroscopy and ellipsometry were the techniques offering the most versatility for immunosensor detection. They also concluded that ellipsometry was the most sensitive, although internal reflectance spectroscopy probably provided more specificity. Many detection systems employed in immunosensors have focused on reflective light systems. Schipper et al. (91) developed a planar waveguide immunosensor that measured changes in the refractive index following an immunological reaction at the waveguide surface. The authors report that the lower limit of detection achieved by this system was equivalent to 12 pg/mm^2 of analyte coverage. This sensitivity is comparable to that achieved by surface plasmon resonance (SPR) or grating coupler sensors. Planar waveguide technology has also been applied to the development of a fluorescence-based immunosensor for the simultaneous analysis of multiple samples. A patterned array of antibodies immobilized on the surface of a planar waveguide was used as the analyte capture ligand coupled to a diode laser laser-induced fluorescence (LIF) detector. A CCD camera was used to detect the pattern of fluorescent emissions on the sensor surface and this pattern was further analyzed using an image analysis software package. This immunosensor was used to detect physiologically relevant concentrations of a number of clinically relevant markers in spiked human samples (86).

SPR is becoming a popular approach for detection of bound analytes to immobilized ligands, mainly due to its sensitivity (2), its ability to study the dynamic aspects of macromolecular interactions (92), and the availability of commercial instruments such as the BIACORE® X (Biacore Instruments). SPR-based immunosensors have been developed for the detection of the mycotoxin fumonisin B1 (64). This sensor utilized antibodies adsorbed onto gold film coupled to a glass prism in the Kretschmann configuration. Detection of bound analyte was achieved by recording changes in the resonance angle (caused by shifts in the angular profile of reflected light resulting from the addition of the bound analyte). A detection limit of 6 µg/mL in serum samples was reported for this immunosensor. SPR immunosensors have also been developed for measuring human IgG using anti-IgG antibodies immobilized to silver surfaces (25) and for measuring racemic forms of α-amino acids (49). Schuck et al. (93) developed a closed-loop sampling device for the BIACORE X that enabled the continuous circulation of a sample over two detection points (a ligand binding area and a blank reference area). This modification enabled these investigators to calculate binding isotherms by a stepwise titration, thus allowing for studies of high affinity intramolecular interactions. Another approach is the use of surface acoustic wave devices (SAWs) based on horizontally polarized shear waves. These instruments can be used to produce mass-sensitive immunosensors. Wessa et al. (110) used CNBr-activation of a polyimide-coated aluminum transducer to immobilize anti-glucose oxidase antibodies.

A piezoelectric immunosensor for the determination of human IgM has been reported (17). This sensor utilized immobilized antibodies directed against human IgM as the ligand. These antibodies were bound to crystals with a basic resonance frequency of 9 MHz and were able to detect IgM in the range of 5–93 µg/mL. Additionally, following washing with tetrahydrofuran, the crystal could be reused 20 times without detectable loss of sensitivity. A direct piezoelectric flow injection analysis immunoassay for the detection of both African swine fever virus and anti-virus antibodies was developed by Uttenthaler et al. (104). The sensor used in this immunoassay was based on immobilization of either the virus protein or virus-specific antibodies onto a quartz crystal microbalance. Accumulation of the analyte on the surface of this mass-sensitive biosensor resulted in a shift of the resonant frequency, which could be correlated with the amount of bound analyte. The sensor was able to perform one measurement cycle every 30 minutes and was stable for approximately 30 days. The sensor could detect 0.3–1.0 µg/mL of viral protein and 0.1–0.2 µg/mL of antibody in either saline or serum samples.

Reactions based on fluorescence are popular and, according to a number of authors, reasonably sensitive (8,9,24,57,85). Kusterbeck et al. (57) developed a continuous flow system capable of detecting either a radiolabeled or fluorochrome-labeled model antigen (2,4-dinitrophenol-lysine or DNP-lysine) to a detection limit of 143 nM. Other investigators have used fiber-optic sensors for the detection of cocaine and its derivatives (24). These workers demonstrated that the fluorescence-based unit was selective and had a lower limit of 5 ng/mL for cocaine, 5 ng/mL cocaethylene, a 10 ng/mL limit for tropacocaine, and a 29 ng/mL limit for norcocaine. The specificity of the unit was based on the selectivity of an immobilized monoclonal antibody. Rabbany et al. (80) reviewed the potential of immunosensors and the mechanisms used to develop these entities. They presented an excellent overview of the diversity of techniques employed, as well as some of the potential problems facing the future development of biosensors in general and immunosensors in particular. Hanbury et al. (40) described a self-contained fiber-optic immunosensor for the measurement of myoglobin. The sensing element was constructed by entrapment of Cascade Blue-labeled antibody within a polyacrylamide gel at the distal face of an optical fiber. This allowed diffusion of the analyte but excluded larger interfering proteins such as hemoglobin from the sensing area. Detection was achieved by measuring the amount of fluorescence-quenching that occurred when the myoglobin bound to the antibody, thus inhibiting the fluorescent signal produced by fluorescence energy transfer between the Cascade Blue and the heme group of myoglobin. The sensor was reported to be able to detect approximately 83 µg/L with response times of 15–130 min.

The current trends and engineering aspects of immunosensor design and construction have recently been reviewed (32). The authors of this review give a brief description of the main principles and limitations of conventional immunoassays together with alternative approaches to overcome the shortcomings of previous work. The review covers the application of flow-injection techniques to the development of immunological sensing systems and gives a classification of immunosensors based on the detection principle employed. This review covers electrochemical immunosensors, piezoelectric immunosensors, and sensors based on optical detection of the immune reactions. The discussion, however, heavily focuses on electrochemical immunosensors.

PROTOCOLS

Building a Basic Fluorescence Fiber-Optic Immunosensor

1. Prepare a 5-cm–long thiol-derivatized capillary as described in the protocol for the attachment of ligands to thiol-activated capillaries in Chapter 4. Substitute a FAb fragment of anti-human serum albumin for the "ligand" in step 3 of that protocol.

    Note: Other antibodies can be used depending on the analyte to be measured.

2. Remove the polyimide coating from the capillary and attach the ends to 0.2-mm id PEEK tubing.

3. Assemble a fiber-optic LIF detector as described in the protocol for building a fiber-optic fluorescence detector for increased sensitivity in Appendix II and use the antibody-coated capillary as the flow cell.

4. Attach a syringe pump to one end of the PEEK tubing (inlet). This will enable loading of the flow cell.

5. Flush the flow cell with 1 mL of 0.01 M NaH_2PO_4, pH 7.4.

Measurement of Serum Albumin Using a Basic Fluorescence Fiber-Optic Immunosensor

1. Make a set of albumin standards as described in the protocol for the preparation of standards for a standard curve in Appendix I.

2. Label each standard and the test samples with a fluorochrome by any of the protocols described in section *Fluorescence* of Appendix II.

3. Turn on the diode laser, inject the lowest standard, and record the fluorescence intensity (this is recorded automatically by the spectrometer and usually stored in Microsoft Excel format).

4. Flush the flow cell with 500 µL of 0.01 M NaH_2PO_4, pH 7.4.

5. Inject the second standard and record the fluorescence intensity.

6. Repeat steps 4 and 5 until all of the standards and samples have been read.

7. Using the readings obtained from the different standards, plot a calibration curve and use this curve to calculate the concentration of albumin in each test sample.

Measurement of Bioactive Molecules Using an Immobilized Receptor Cartridge

1. Prepare CDI-activated glass beads as described in the protocol for CDI activation of silica and glass in Chapter 2.

2. Suspend 0.2 g of activated beads in 500 µL of 50 mM Na_2CO_3, pH 9.0.

3. Suspend 200 ng of recombinant receptor in 500 µL of 50 mM Na_2CO_3.

4. Combine the beads and the receptors in a microcentrifuge tube and place on an overhead mixer for 18 h at 4°C.

5. Wash the beads three times in 0.01 M NaH_2PO_4, pH 7.2.

6. Slurry-pack the coated beads into a 2-mm × 50-mm PEEK chromatography column.

7. Attach the column to a pump and flush with 4 mL of 0.01 M NaH_2PO_4, pH 7.2.

8. Label the sample with Cy5 as described in the protocol for labeling with Cy5 in Appendix II.

9. Inject the sample into the column and allow the sample to reside in the column for 2–5 min.

10. Flush the column with 1 mL of 0.01 M NaH_2PO_4, pH 7.2.

11. Pump 4 mL of 2.5 M NaSCN through the column at a flow-rate of 0.25 mL/min and continuously monitor the column effluent with a LIF or fluorescence detector.

12. Record the resolved peaks.

Multi-Analyte Batch Techniques and Related Techniques

Multi-analyte separations using batch techniques are theoretically simple to perform and are capable of being modified to extract more than one analyte from complex matrices such as spent tissue culture media or bulk culture feedstock. However, this approach has not gained great popularity. Miller et al. (63) used a batch technique incorporating con A-coated Sepharose as the affinity matrix to isolate rod and cone pigments from chicken photoreceptor extracts. They reported separating the extract into four distinct fractions, three of which contained highly enriched visual pigments. Batch separations of mixed yeast RNA have been reported using m-aminophenylboronic acid immobilized to agarose (98). This approach was successful in separating several RNAs when barium ions were applied rather than the more conventional magnesium ions. Busch et al. (10) described an immunoaffinity batch technique for the isolation of two different analytes. They reported the successful and rapid isolation of human cholesteryl ester and triglyceride exchange protein from plasma using a single-step batch extraction technique. Likewise, Petrescu et al. (71) immobilized two different antibodies onto nitrocellulose powder to simultaneously isolate *Datura innoxia* lectin and murine tyrosinase. The authors compared batch and column techniques and reported greater than 75% recoveries of both analytes.

Affinity phase separations have been used to isolate two different analytes simultaneously. Albertsson and Birkenmeier (1) described a three-phase system for the isolation of serum albumin and pre-albumin from human serum. Combinations of the textile dyes Cibacron Blue F36A and Remazol Yellow GCL that were bound to either dextran or PEG resulted in successful partitioning of the two proteins first from other serum proteins, followed by separation from each other. Persson and colleagues (69,70) used wheat germ lectin attached to dextran as an affinity ligand for the separation of plasma membranes from other cellular membranes in homogenates of rat liver. They reported that this approach was comparable to sucrose density gradient separations. Another variation on this approach included the use of affinity microspheres incorporated into the phase systems to isolate *E. coli* fusion proteins of IgG and albumin-binding domains (54). In this system, the fusion protein–microspheres partitioned into the PEG-rich zone, thus purifying them from contaminating cells and cell debris. Jaschke et al. (50) employed oligonucleotide-coated PEG to successfully partition complementary nucleic acid strands into the PEG zone. In this manner, they were able to use hybridization-based affinity partitioning to isolate sequence-specific nucleic acids.

An interesting new approach to batch or preparative separations of proteins is affinity isolation using expanded beds (13). This technique is readily applicable to the separation of analytes from feedstocks and other large volume samples. The method is based on pumping of the mobile phase in an ascending manner through the column bed, thus causing the bed to expand. This produces a similar effect to plug flow. The main advantage of this form of separation is that it enables protein recovery directly from the sample, even in the presence of particulate matter such as cells or cell debris (31). Although not widely used to date for the affinity or immunoaffinity separation of multiple analytes, Chase and Draeger (14) report the use of expanded beds to selectively isolate polyclonal IgG using protein A-coated Sepharose and phosphofructokinase using Cibacron Blue Sepharose.

PROTOCOLS

Isolating Multiple Analytes Using a Series of Affinity Membranes

1. Prepare a contactor containing a number of different lectin affinity membranes (10 for each lectin) as outlined in the protocol for the isolation of a specific glycoprotein from a bioreactor feedstock using a lectin membrane unit in Chapter 6.

2. Connect the inlet of the contactor to the analyte source.

3. Connect the outlet of the contactor to a suitable collection vessel and run the sample through as described in steps 7 through 10 of the bioreactor feedstock protocol from Chapter 6.

4. Make 0.5 M solutions of the sugars that act as substrates for the different

lectins (see Table 1 in Chapter 4 for details on the different lectins and their reactive sugar groups). These sugars can be used as elution agents.

5. Pump 5 vol (calculated from the capacity of the contactor; small laboratory contactors require between 1 and 5 mL, while preparative contactors may require 100 mL or more) of one of the sugar solutions through the contactor and collect the effluent.

6. Repeat step 5 for each of the other sugar solutions.

7. Remove the elution sugar from each fraction by dialysis against 0.01 M sodium phosphate, pH 7.4, for 5–10 h.

Immunoaffinity Partitioning Using a Panel of Antibodies

1. Prepare a two-phase system as described in the protocol for the immunoaffinity partitioning of proteins in Chapter 6; however, make the PEG phase with PEG coated with two or more different antibodies.

2. Mix the sample with the phase mixture and allow the phases to separate at 4°C.

3. Centrifuge at 1000× g for 10 min at 4°C to ensure complete phase separation.

4. Recover the top phase with a pipette and mix 1:1 (vol:vol) with 1 M glycine.

5. Incubate at 4°C for 60 min.

6. Centrifuge at 30 000× g for 20 min at 4°C to sediment the PEG.

7. Recover the supernatant with a pipette and dialyze against 0.01 M NaH_2PO_4, pH 7.4, overnight at 4°C.

The efficiency of this protocol can be greatly improved by selecting antibodies with different affinities. In this way, the bound analytes can be differentially eluted by replacing the 1 M solution of glycine in step 4 with a series of glycine solutions of increasing molarity and varying pH. This would differentially recover the bound analytes according to their affinity binding to the immobilized antibodies.

Recycling Affinity and Immunoaffinity Techniques

The idea of using a series of different chromatographic ligands to separate analytes has been in existence for some years. Early work involved the recycling of the analyte through a series of different separation processes, such as ion-exchange followed by reverse-phase or size-exclusion chromatography (11,12,48). Fonseca et al. (30) employed a pseudo-affinity–based continuous system for the recovery and purification of penicillin acylase from crude extracts. Separation was achieved by recycling the sample through three vessels containing phenyl-Sepharose gel. Using this approach, the authors recovered the analyte of interest at approximately 74% purity. Regnier and Huang (83) proposed a multidimensional chromatographic system in which the analyte is switched from one dimension to another during the separation process. They indicated that the employment of affinity-based techniques would help to eliminate peak drift during the valve-switching process. They illustrated their position by citing a number of cases in which analytes isolated by immunoaffinity were cycled into analytical reverse-phase columns.

Recycling techniques for the extraction of specific analytes is becoming popular, especially procedures employing continuous extraction (35). Additionally, recycling techniques in which multiple analytes are separated by a series of affinity or immunoaffinity columns is also gaining popularity. In both the biological and bio-medical sciences, there is a growing need for the development of techniques designed to measure a variety of different analytes within the same sample (15,56). Similar situations occur when analyzing valuable archival materials, low-density cell cultures, or samples from small experimental animals. Measurement of multiple analytes in clinically relevant samples is of great interest in the biomedical sciences. This interest stems partially from the need to examine and monitor pediatric and elderly patients, in whom multiple sampling is problematic. Such analyses are also of interest when examining large archival materials collected during epidemiological surveys. Although most reports use such antibodies for single analyte analysis, mixtures of antibodies have been used to analyze a number of analytes in the same sample (77,99). These procedures may involve techniques such as sequentially incubating the sample with a number of different immobilized or labeled antibodies to isolate and measure a number of different analytes. Using a sequential immunoassay system, Steffen and Ebersole (99) measured five cytokines in a 50-µL sample of gingival crevicular fluid. In recent studies, we have been able to isolate and measure 10 cytokines from 25-µL samples of a number of different biological fluids using immunoaffinity columns linked in series by microswitching valves (77) (Figure 9). In our hands, this system provided reliable and reproducible results in whole dried blood spot eluates, plasma, and urine. This technique was applied to the analysis of a series of archival dried human blood spots. In this study, a total of 53 analytes (Table 2) were analyzed in approximately one-third of a blood spot using recycling immunoaffinity chromatography (36,66). In both studies, samples were recycled through sets of 10 micro-immunoaffinity columns, set in series. Recycling affinity or immunoaffinity chromatography is a versatile technique and can easily be adapted to automated or miniaturized systems. Using the system developed in our laboratory,

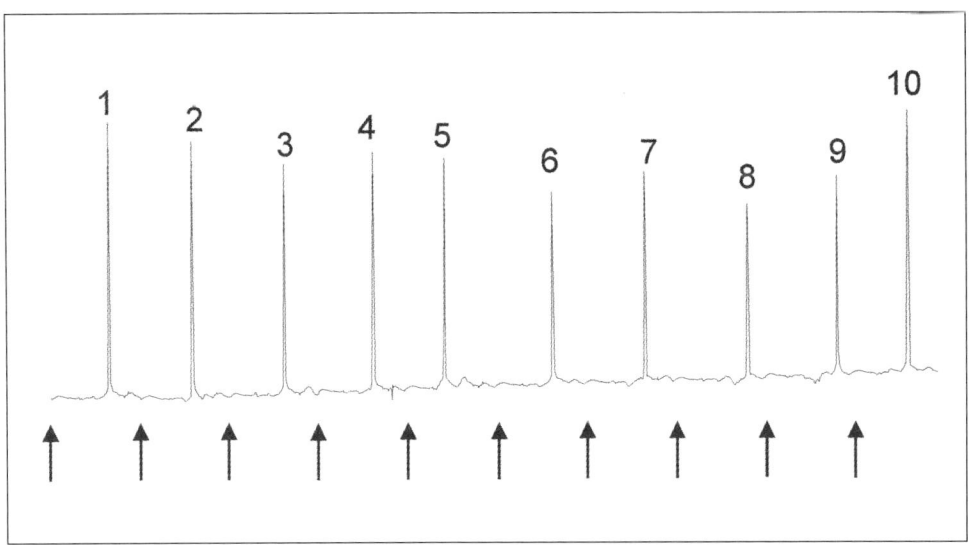

Figure 9. Chromatogram of a recycling immunoaffinity separation of 10 cytokines from a single sample. The peaks represent the recovery of each cytokine as they elute from the recycling system. The arrows indicate the points in the chromatogram where individual column elution was started. Peaks represent 1. IL-1; 2. IL-2; 3. IL-4; 4. IL-5; 5. IL-6; 6. IL-10; 7. IL-12; 8. IL-13; 9. Tumor necrosis factor alpha; 10. Interferon gamma.

Table 2. Analytes Recovered from a Single Blood Spot by Recycling Immunoaffinity
Chromatography

Number of analytes	Type of analyte
14	Cytokines
6	Chemokines
4	Growth factors
3	Neuropeptides
4	Immunoglobulin classes
12	Complement components
4	Coagulation factors
4	Thyroid hormones
2	Auto-antibodies

we have found that the nonreactive fraction from the initial battery of columns can be re-analyzed by a different battery of columns. This procedure appears to be possible for at least 10 cycles.

Other uses for multi-analyte immunoaffinity techniques have been described. van Ginkel (107) reviewed the applications of multi-analyte immunoaffinity chromatography for drug residue analysis. This author discusses the use of several permutations of the immunoaffinity technique including single-antibody–multi-analyte procedures and multi-antibody–multi-analyte procedures. Likewise, Katz and Siewierski (52) discussed the use of immunoaffinity techniques for the analysis of a number of important drugs in a variety of different tissues. Wheatley (111) described a procedure wherein a mixture of antibodies specific for two separate analytes was employed for the immuno-extraction of these analytes from IgG preparations. This author also discusses the use of other techniques, such as cationic exchange chromatography coupled with immunoaffinity, as alternative approaches to purification using subtraction rather than positive isolation. In his recent review of the applications of immunoaffinity chromatography, Hage (39) describes the applications of immunoaffinity in a number of fields including cleanup procedures prior to further analysis and multi-analyte analysis.

PROTOCOLS

Recycling Analytes Through a Lectin Affinity Column Followed by a Reverse-Phase Column

1. Prepare CDI-activated glass beads as described in the protocol for CDI activation of silica and glass in Chapter 2.
2. Coat the beads with con A as described in the protocol for the direct ligand attachment to derivatized supports in Chapter 5.
3. Wash the beads three times in 0.01 M NaH_2PO_4, pH 7.2.
4. Slurry-pack the coated beads into a 2-mm × 50-mm PEEK chromatography column.

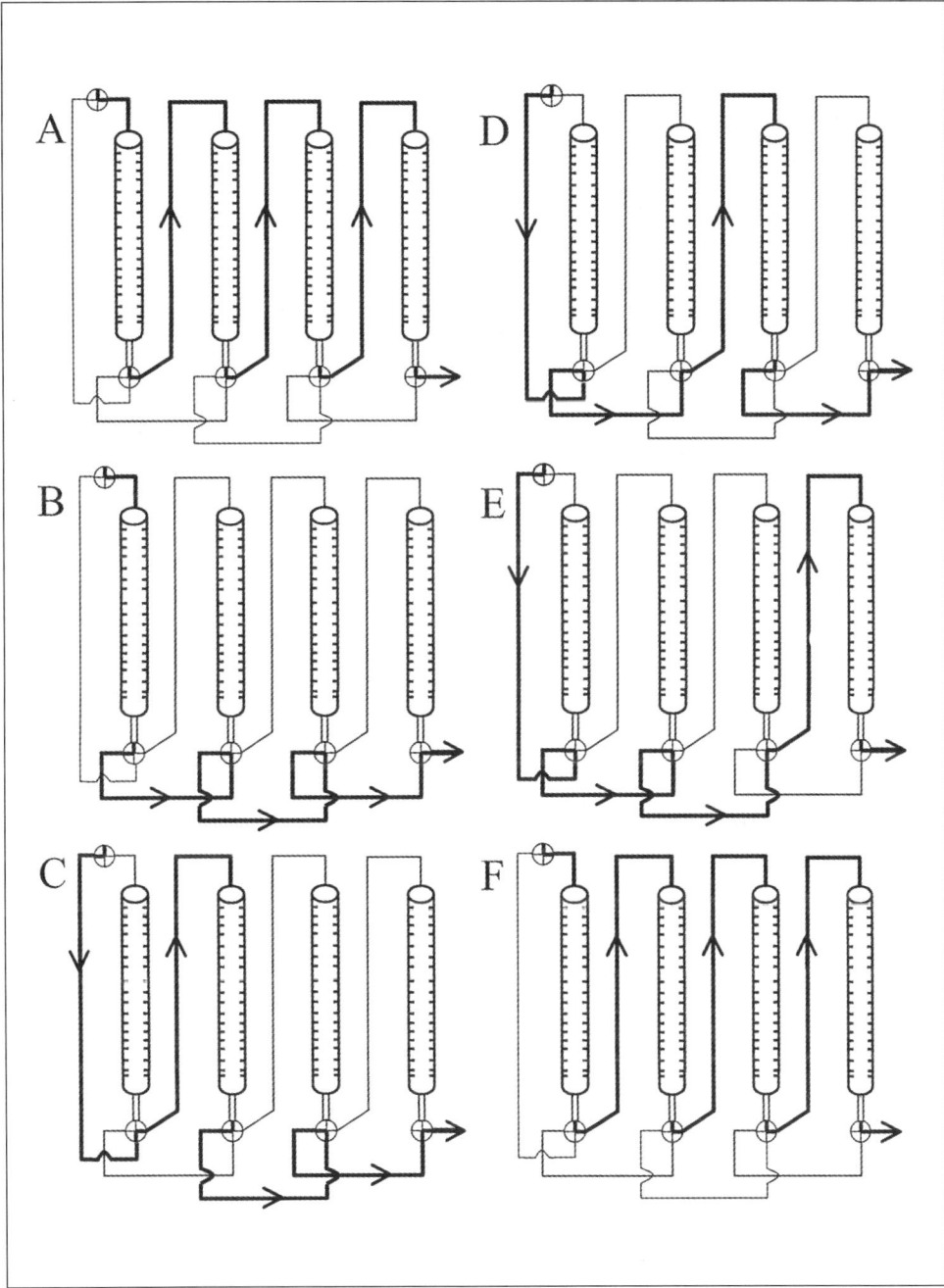

Figure 10. Diagrammatic representation of a four-column recycling immunoaffinity system. The four different immunoaffinity columns are inter-connected via a series of 4-way switching valves placed at the outlet of each column and a single 4-way valve placed at the inlet of the first column. The bolded line indicates the flow pattern through the columns. (A) The sample is loaded onto the first column and passes through all of the columns in a serpentine manner, thus allowing the capture of a single analyte within each column. (B) Following a wash through, the valves are readjusted so that the flow passes only through the first column. An elution gradient is started and the bound analyte is released and recorded by the detector. (C) The valves are again readjusted in order to elute the second column and the elution gradient repeated. (D) and (E) The valves are readjusted in a manner similar to step C in order to elute the third and fourth columns. (F) Finally, the valves are readjusted back to the original positions and the system regenerated. In our labaoratory, up to 10 different immunoaffinity columns have been used in this system.

5. Attach the inlet of the affinity column to the injection port of an HPLC system and the outlet of the column to port 1 of a 3-way valve.

6. Attach the inlet of a 4.6-cm × 150-cm C_{18} column to port 3 of the 3-way valve. Attach the outlet of the C_{18} column to the inlet of the detector flow cell.

7. Set the 3-way valve to connect ports 1 and 2.

8. Inject the sample and run the system at 0.5 mL/min in 0.01 M NaH_2PO_4, pH 7.4.

9. Run the sample through the affinity column and collect the effluent emerging from port 2 of the 3-way valve.

10. Stop the flow and readjust the 3-way valve to connect ports 1 and 3.

11. Change the running buffer to 25 mM α-mannose dissolved in 25% acetonitrile:80% distilled water (vol/vol).

12. Set the pump to 0.5 mL/min and run for 60 min.

13. Elute the bound analyte from the affinity column and analyze on the reverse-phase (C_{18}) column.

14. Continuously monitor the developing chromatogram with a UV or refractive index detector.

Multi-Analyte Analysis Using Multicolumn Recycling Immunoaffinity Chromatography

1. Prepare a panel of 4–10 antibody-coated glass beads as described in the protocol for ICE using a pre-analytical concentrator (above), using one antibody per batch of beads.

2. Pack each batch of beads into a separate column as described in the protocol for slurry-packing a small affinity or immunoaffinity column by hand in Chapter 7.

3. Assemble the panel of immunoaffinity columns as shown in Figure 10 and connect to a conventional HPLC or FPLC system.

4. Prior to analysis, set 4-way valves (at the bottom of each column and one at the top of the first column) to allow the sample to pass from one column to the next in a serpentine fashion (ports 1–2 connection).

5. Inject 25-µL of sample into the system via the injection port on the HPLC system.

6. Start the pump and set to 0.5 mL/min and run the system in an isocratic mode for 20 min using 0.01 M NaH_2PO_4, pH 7.0, plus 0.01% (vol/vol) Brij 35 as the running buffer.

7. Program (or manually turn) the bottom valve on column 1 to the 1–4 position (leaving the top valve in the 1–2 position). Adjust the valves on the other columns to the 3–4 position, except for the last column, which should be adjusted to the 2–3 position.

8. Change the running buffer to 2.5 M NaSCN and continue to run this buffer through column 1 for 3 min. This will elute the bound analyte.

9. Change the top valve to the 1–4 position and change the bottom valve on column 1 to the 3–2 position. Change the bottom valve on column 2 to the 1–4 position. Do not adjust any other valves.

10. Leave the top valve in the 1–4 position and change the lower valve on column 1 to the 3–4 position. Change the valve on column 2 to the 3–2 position and the valve on column 3 to the 1–4 position.

11. Repeat the sequence of valve changes outlined in step 10 for each column in the series. Remember to set the valve on the column just eluted to the 3–2 position and the valve on the column to be eluted to the 1–4 position. When eluting the last column, keep the valve in the 1–2 position.

12. Constantly monitor the column effluent throughout the procedure and record the chromatogram (Figure 9).

REFERENCES

1. **Albertsson, P.A. and G. Birkenmeier.** 1988. Affinity separation of proteins in aqueous three-phase systems. Anal. Biochem. *175*:154-161.
2. **Attridge, J.W., P.B. Daniels, J.K. Deacon, G.A. Robinson and G.P. Davidson.** 1991. Sensitivity enhancement of optical immunosensors by the use of a surface plasmon resonance fluoroimmunoassay. Biosens. Bioelectron. *6*:201-214.
3. **Baba, Y.** 1996. Capillary affinity gel electrophoresis. New tool for detection of the mutation on DNA. Mol. Biotechnol. *6*:143-153.
4. **Barker, G.E., P. Russo and R.A. Hartwick.** 1992. Chiral separation of leucovorin with bovine serum albumin using affinity capillary electrophoresis. Anal. Chem. *64*:3024-3028.
5. **Bauer, C.G., A.V. Eremenko, E. Ehrentreich-Forster, F.F. Bier, A. Makower, H.B. Halsall, W.R. Heineman and F.W. Scheller.** 1996. Zeptomole-detecting biosensor for alkaline phosphatase in an electrochemical immunoassay for 2,4-dichloro-phenoxyacetic acid. Anal. Chem. *68*:2453-2458.
6. **Baumner, A.J. and R.D. Schmid.** 1998. Development of a new immunosensor for pesticide detection: a disposable system with liposome-enhancement and amperometric detection. Biosens. Bioelectron. *13*:519-529.
7. **Berggren, C. and G. Johansson.** 1997. Capacitance measurements of antibody-antigen interactions in a flow system. Anal. Chem. *69*:3651-3657.
8. **Bouvrette, P. and J.H. Luong.** 1995. Development of a flow injection analysis (FIA) immunosensor for the detection of *Escherichia coli*. Int. J. Food Microbiol. *27*:129-137.
9. **Bright, F.V., T.A. Betts and K.S. Litwiler.** 1990. Regenerable fiber-optic-based immunosensor. Anal. Chem. *62*:1065-1069.
10. **Busch, S.J., C.R. Duvic, J.L. Ellsworth, J. Ihm and J.A. Harmony.** 1986. Immunoaffinity purification of the lipid transfer protein complex directly from human plasma. Anal. Biochem. *153*:178-188.
11. **Cai, J. and J. Henion.** 1997. Quantitative multi-residue determination of beta-agonists in bovine urine using on-line immunoaffinity extraction coupled column packed capillary liquid chromatography-tandem mass spectrometry. J. Chromatogr. B Biomed. Sci. App. *691*:357-370.
12. **Campins-Falco, P., R. Herraez-Hernandez and A. Sevillano-Cabeza.** 1993. Column-switching techniques for high-performance liquid chromatography of drugs in biological samples. J. Chromatogr. *619*:177-190.
13. **Chase, H.A.** 1994. Purification of proteins by adsorption chromatography in expanded beds. Trends Biotechnol. *12*:296-303.
14. **Chase, H.A. and N.M. Draeger.** 1992. Affinity purification of proteins using expanded beds. J. Chromatogr. *597*:129-145.
15. **Chen, G. and A.G. Ewing.** 1997. Chemical analysis of single cells and exocytosis. Crit. Rev. Neurobiol. *11*:59-90.
16. **Chiem, N.H. and D.J. Harrison.** 1998. Monoclonal antibody binding affinity determined by microchip-based capillary electrophoresis. Electrophoresis *19*:3040-3044.
17. **Chu, X., Z.H. Lin, G.L. Shen and R.Q. Yu.** 1995. Piezoelectric immunosensor for the detection of immunoglobulin M. Analyst *120*:2829-2832.
18. **Chu, Y.H. and C.C. Cheng.** 1998. Affinity capillary electrophoresis in biomolecular recognition. Cell. Mol. Life Sci. *54*:663-683.
19. **Colton, I.J., J.D. Carbeck, J. Rao and G.M. Whitesides.** 1998. Affinity capillary electrophoresis: a physical-organic tool for studying interactions in biomolecular recognition. Electrophoresis *19*:367-382.
20. **Colyer, C.L., S.D. Mangru and D.J. Harrison.** 1997. Microchip-based capillary electrophoresis of human serum proteins. J. Chromatogr. A *781*:271-276.
21. **Cook, C.J.** 1997. Real-time measurements of corticosteroids in conscious animals using an antibody-

based electrode. Nat. Biotechnol. *15*:467-471.

22.**Dalluge, J.J. and L.C. Sander.** 1998. Precolumn affinity capillary electrophoresis for the identification of clinically relevant proteins in human serum: application to human cardiac troponin I. Anal. Chem. *70*:5339-5343.

23.**Dawson, L.A.** 1997. Capillary electrophoresis and microdialysis: current technology and applications. J. Chromatogr. B Biomed. Sci. App. *697*:89-99.

24.**Devine, P.J., N.A. Anis, J. Wright, S. Kim, A.T. Eldefrawi and M.E. Eldefrawi.** 1995. A fiber-optic cocaine biosensor. Anal. Biochem. *227*:216-224.

25.**Disley, D.M., D.C. Cullen, H.X. You and C.R. Lowe.** 1998. Covalent coupling of immunoglobulin G to self-assembled monolayers as a method for immobilizing the interfacial-recognition layer of a surface plasmon resonance immunosensor. Biosens. Bioelectron. *13*:1213-1225.

26.**Duan, C. and M.E. Meyerhoff.** 1994. Separation-free sandwich enzyme immunoassays using microporous gold electrodes and self-assembled monolayer/immobilized capture antibodies. Anal. Chem. *66*:1369-1377.

27.**Effenhauser, C.S., G.J. Bruin and A. Paulus.** 1997. Integrated chip-based capillary electrophoresis. Electrophoresis *18*:2203-2213.

28.**Erim, F.B. and J.C. Kraak.** 1998. Vacancy affinity capillary electrophoresis to study competitive protein-drug binding. J. Chromatogr. B Biomed. Sci. App. *710*:205-210.

29.**Fishman, H.A., D.R. Greenwald and R.N. Zare.** 1998. Biosensors in chemical separations. Annu. Rev. Biophys. Biomol. Struct. *27*:165-198.

30.**Fonseca, L.J., I.M. Patricio and J.M. Cabral.** 1998. Preliminary studies on continuous recovery and purification of the penicillin acylase under pseudo-affinity conditions using phenyl-Sepharose gel. J. Mol. Recognit. *11*:252-254.

31.**Galaev, I.Y.** 1998. New methods of protein purification. Expanded bed chromatography. Biochemistry *63*:619-624.

32.**Ghindilis, A.L., P. Atanasov, M. Wilkins and E. Wilkins.** 1998. Immunosensors: electrochemical sensing and other engineering approaches. Biosens. Bioelectron. *13*:113-131.

33.**Ghindilis, A.L., R. Krishnan, P. Atanasov and E. Wilkins.** 1997. Flow-through amperometric immunosensor: fast "sandwich" scheme immunoassay. Biosens. Bioelectron. *12*:415-423.

34.**Gizeli, E. and C.R. Lowe.** 1996. Immunosensors. Curr. Opin. Biotechnol. *7*:66-71.

35.**Gordon, N.F., H. Tsujimura and C.L. Cooney.** 1990. Optimization and simulation of continuous affinity-recycle extraction (care). Bioseparation *1*:9-21.

36.**Grether, J.K., K.B. Nelson, J.M. Dambrosia and T.M. Phillips.** 1999. Interferons and cerebral palsy. J. Pediatr. *134*:324-332.

37.**Guttman, A. and N. Cooke.** 1991. Capillary gel affinity electrophoresis of DNA fragments. Anal. Chem. *63*:2038-2042.

38.**Hage, D.S.** 1997. Chiral separations in capillary electrophoresis using proteins as stereoselective binding agents. Electrophoresis *18*:2311-2321.

39.**Hage, D.S.** 1998. Survey of recent advances in analytical applications of immunoaffinity chromatography. J. Chromatogr. B Biomed. Sci. App. *715*:3-28.

40.**Hanbury, C.M., W.G. Miller and R.B. Harris.** 1997. Fiber-optic immunosensor for measurement of myoglobin. Clin. Chem. *43*:2128-2136.

41.**Haupt, K., F. Roy and M.A. Vijayalakshmi.** 1996. Immobilized metal ion affinity capillary electrophoresis of proteins – a model for affinity capillary electrophoresis using soluble polymer-supported ligands. Anal. Biochem. *234*:149-154.

42.**He, B., N. Tait and F.E. Regnier.** 1998. Fabrication of nanocolumns for liquid chromatography. Anal. Chem. *70*:3790-3797.

43.**Heegaard, N.H.** 1994. Determination of antigen-antibody affinity by immunocapillary electrophoresis. J. Chromatogr. *680*:405-412.

44.**Heegaard, N.H.** 1998. A heparin-binding peptide from human serum amyloid P component characterized by affinity capillary electrophoresis. Electrophoresis *19*:442-447.

45.**Heegaard, N.H., B.E. Hansen, A. Svejgaard and L.H. Fugger.** 1997. Interactions of the human class II major histocompatibility complex protein HLA-DR4 with a peptide ligand demonstrated by affinity capillary electrophoresis. J. Chromatogr. *781*:91-97.

46.**Heegaard, N.H., H.D. Mortensen and P. Roepstorff.** 1995. Demonstration of a heparin-binding site in serum amyloid P (SAP) component using affinity capillary electrophoresis as an adjunct technique. J. Chromatogr. *717*:83-90.

47.**Heegaard, N.H., S. Nilsson and N.A. Guzman.** 1998. Affinity capillary electrophoresis: an overview of some recent developments. J. Chromatogr. B Biomed. Sci. App. *715*:29-54.

48.**Hendrickson, T.L. and G.S. Wilson.** 1994. Improved clean-up method for the enkephalins in plasma using immunoaffinity chromatography. J. Chromatogr. B Biomed. Appl. *653*:147-154.

49.**Hofstetter, O., H. Hofstetter, M. Wilchek, V. Schurig and B.S. Green.** 1999. Chiral discrimination using an immunosensor. Nat. Biotechnol. *17*:371-374.

50. **Jaschke, A., J.P. Furste, V.A. Erdmann and D. Cech.** 1994. Hybridization-based affinity partitioning of nucleic acids using PEG-coupled oligonucleotides. Nucleic Acids Res. *22*:1880-1884.

51. **Karube, I. and M. Suzuki.** 1986. Novel immunosensors. Biosensors *2*:343-362.

52. **Katz, S.E. and M. Siewierski.** 1992. Drug residue analysis using immunoaffinity chromatography. J. Chromatogr. *624*:403-409.

53. **Kelly, S., D. Compagnone and G. Guilbault.** 1998. Amperometric immunosensor for lactate dehydrogenase LD-1. Biosens. Bioelectron. *13*:173-179.

54. **Kondo, A., T. Kaneko and K. Higashitani.** 1993. Purification of fusion proteins using affinity microspheres in aqueous two-phase systems. Appl. Microbiol. Biotechnol. *40*:365-369.

55. **Kossek, S., C. Padeste and L. Tiefenauer.** 1996. Immobilization of streptavidin for immunosensors on nanostructured surfaces. J. Mol. Recognit. *9*:485-487.

56. **Kricka, L.J.** 1992. Multianalyte testing. Clin. Chem. *38*:327-328.

57. **Kusterbeck, A.W., G.A. Wemhoff, P.T. Charles, D.A. Yeager, R. Bredehorst, C.W. Vogel and F.S. Ligler.** 1990. A continuous flow immunoassay for rapid and sensitive detection of small molecules. J. Immunol. Methods *135*:191-197.

58. **Lausch, R., O.W. Reif, P. Riechel and T. Scheper.** 1995. Analysis of immunoglobulin G using a capillary electrophoretic affinity assay with protein A and laser-induced fluorescence detection. Electrophoresis *16*.636-641.

59. **Liu, J., S. Abid, M.F. Hail, M.S. Lee, J. Hangeland and N. Zein.** 1998. Use of affinity capillary electrophoresis for the study of protein and drug interactions. Analyst *123*:1455-1459.

60. **Ljungberg, H., S. Ohlson and S. Nilsson.** 1998. Exploitation of a monoclonal antibody for weak affinity-based separation in capillary gel electrophoresis. Electrophoresis *19*:461-464.

61. **Lyubarskaya, Y.V., Y.M. Dunayevskiy, P. Vouros and B.L. Karger.** 1997. Microscale epitope mapping by affinity capillary electrophoresis-mass spectrometry. Anal. Chem. *69*:3008-3014.

62. **Mangru, S.D. and D.J. Harrison.** 1998. Chemiluminescence detection in integrated post-separation reactors for microchip-based capillary electrophoresis and affinity electrophoresis. Electrophoresis *19*:2301-2307.

63. **Miller, J.L., R.W. Morton and R.S. Fager.** 1985. Simplified concanavalin-A Sepharose adsorption method for separation of cone visual pigments from rhodopsin. Exp. Eye Res. *40*:471-476.

64. **Mullett ,W., E.P. Lai and J.M. Yeung.** 1998. Immunoassay of fumonisins by a surface plasmon resonance biosensor. Anal. Biochem. *258*:161-167.

65. **Nakanishi, K., H. Muguruma and I. Karube.** 1996. A novel method of immobilizing antibodies on a quartz crystal microbalance using plasma-polymerized films for immunosensors. Anal. Chem. *68*:1695-1700.

66. **Nelson, K.B., J.M. Dambrosia, J.K. Grether and T.M. Phillips.** 1998. Neonatal cytokines and coagulation factors in children with cerebral palsy. Ann. Neurol. *44*:665-675.

67. **Paddle, B.M.** 1996. Biosensors for chemical and biological agents of defence interest. Biosens. Bioelectron. *11*:1079-1113.

68. **Paek, S.H. and W. Schramm.** 1991. Modeling of immunosensors under nonequilibrium conditions. I. Mathematic modeling of performance characteristics. Anal. Biochem. *196*:319-325.

69. **Persson, A. and B. Jergil.** 1992. Purification of plasma membranes by aqueous two-phase affinity partitioning. Anal. Biochem. *204*:131-136.

70. **Persson, A., B. Johansson, H. Olsson and B. Jergil.** 1991. Purification of rat liver plasma membranes by wheat-germ-agglutinin affinity partitioning. Biochem. J. *273*:173-177.

71. **Petrescu, S., N. Branza-Nichita, M. Nita-Lazar, A.J. Petrescu and C. Motas.** 1995. Immunoaffinity chromatography on antibodies immobilized on nitrocellulose powder. Anal. Biochem. *229*:299 303.

72. **Phillips, T.M.** 1996. Measurement of bioactive neuropeptide using a chromatographic immunosensor cartridge. Biomed. Chromatogr. *10*:145-150.

73. **Phillips, T.M.** 1998. Determination of in-situ tissue neuropeptides by capillary electrophoresis. Anal. Chim. Acta *372*:209-218.

74. **Phillips, T.M. and J.J. Chmielinska.** 1994. Immunoaffinity capillary electrophoretic analysis of cyclosporin in tears. Biomed. Chromatogr. *8*:242-246.

75. **Phillips, T.M. and B.F. Dickens.** 1998. Analysis of recombinant cytokines in human body fluids by immunoaffinity capillary electrophoresis. Electrophoresis *19*:2991-2996.

76. **Phillips, T.M., L.M. Kennedy and E.C. De Fabo.** 1997. Measurement of secretion products from neuropeptide-stimulated lymphocytes using immunoaffinity capillary electrophoresis coupled with microdialysis sampling. J. Chromatogr. B *697*:101-109.

77. **Phillips, T.M. and J.M. Krum.** 1998. Recycling immunoaffinity chromatography for multiple analyte analysis in biological samples. J. Chromatogr. B *715*:55-63.

78. **Place, J.F., R.M. Sutherland and C. Dahne.** 1985. Opto-electronic immunosensors: a review of optical immunoassay at continuous surfaces. Biosensors *1*:321-353.

79. **Qian, X.H. and K.B. Tomer.** 1998. Affinity capillary electrophoresis investigation of an epitope on human immunodeficiency virus recognized by a monoclonal antibody. Electrophoresis *19*:415-419.

80. **Rabbany, S.Y., B.L. Donner and F.S. Ligler.** 1994. Optical immunosensors. Crit. Rev. Biomed. Eng. *22*:307-346.
81. **Radaz, S., J.L. Veuthey, C. Desiderio and S. Fanali.** 1998. Use of cyclodextrins in capillary electrophoresis: resolution of tramadol enantiomers. Electrophoresis *19*:2883-2889.
82. **Regnier, F.E., B. He, S. Lin and J. Busse.** 1999. Chromatography and electrophoresis on chips: critical elements of future integrated, microfluidic analytical systems for life science. Trends Biotechnol. *17*:101-106.
83. **Regnier, F.E. and G. Huang.** 1996. Future potential of targeted component analysis by multidimensional liquid chromatography-mass spectrometry. J. Chromatogr. *750*:3-10.
84. **Rippel, G., H. Corstjens, H.A. Billiet and J. Frank.** 1997. Affinity capillary electrophoresis. Electrophoresis *18*:2175-2183.
85. **Robinson, G.A.** 1991. Optical immunosensing systems – meeting the market needs. Biosens. Bioelectron. *6*:183-191.
86. **Rowe, C.A., S.B. Scruggs, M.J. Feldstein, J.P. Golden and F.S. Ligler.** 1999. An array immunosensor for simultaneous detection of clinical analytes. Anal. Chem. *71*:433-439.
87. **Rush, R.S., P.L. Derby, T.W. Strickland and M.F. Rohde.** 1993. Peptide mapping and evaluation of glycopeptide microheterogeneity derived from endoproteinase digestion of erythropoietin by affinity high-performance capillary electrophoresis. Anal. Chem. *65*:1834-1842.
88. **Sadik, O.A. and J.M. Van Emon.** 1996. Applications of electrochemical immunosensors to environmental monitoring. Biosens. Bioelectron. *11*:i-xi.
89. **Sanden, B., L.H. Eng and G. Dalhammar.** 1998. An amperometric enzyme-linked immunosensor for Nitrobacter. Appl. Microbiol. Biotechnol. *50*:710-716.
90. **Scheller, F. and F. Schubert.** 1992. Biosensors. Techniques and Instrumentation in Analytical Chemistry, Vol. 11. Elsevier, Amsterdam.
91. **Schipper, E.F., R.P. Kooyman, A. Borreman and J. Greve.** 1996. The critical sensor: a new type of evanescent wave immunosensor. Biosens. Bioelectron. *11*:295-304.
92. **Schuck, P.** 1997. Use of plasmon resonance to probe the equilibrium and dynamic aspects of interactions between biological macromolecules. Annu. Rev. Biophys. Biomol. Struct. *26*:541-566.
93. **Schuck, P., D.B. Millar and A.A. Kortt.** 1998. Determination of binding constants by equilibrium titration with circulating sample in a surface plasmon resonance biosensor. Anal. Biochem. *265*:79-91.
94. **Shimura, K. and B.L. Karger.** 1994. Affinity probe capillary electrophoresis: analysis of recombinant human growth hormone with a fluorescent labeled antibody fragment. Anal. Chem. *66*:9-15.
95. **Shimura, K. and K. Kasai.** 1995. Determination of the affinity constants of concanavalin A for monosaccharides by fluorescence affinity probe capillary electrophoresis. Anal. Biochem. *227*:186-194.
96. **Shimura, K. and K. Kasai.** 1996. Determination of the affinity constants of pea lectin for neutral sugars by capillary affinophoresis with a monoligand affinophore. J. Biochem. *120*:1146-1152.
97. **Shimura, K. and K. Kasai.** 1998. Capillary affinophoresis as a versatile tool for the study of biomolecular interactions: a mini-review. J. Mol. Recognit. *11*:134-140.
98. **Singh, N. and R.C. Willson.** 1999. Boronate affinity adsorption of RNA: possible role of conformational changes. J. Chromatogr. *840*:205-213.
99. **Steffen, M.J. and J.L. Ebersole.** 1996. Sequential ELISA for cytokine levels in limited volumes of biological fluids. BioTechniques *21*:504-509.
100. **Sun, P., A. Hoops and R.A. Hartwick.** 1994. Enhanced albumin protein separations and protein-drug binding constant measurements using anti-inflammatory drugs as run buffer additives in affinity capillary electrophoresis. J. Chromatogr. B *661*:335-340.
101. **Taga, A., Y. Yabusako, A. Kitano and S. Honda.** 1998. Separation of disaccharides by affinity capillary electrophoresis in lectin-containing electrophoretic solutions. Electrophoresis *19*:2645-2649.
102. **Thorman, W. and J. Caslavska.** 1998. Capillary electrophoresis in drug analysis. Electrophoresis *19*:2691-2694.
103. **Toppozada, A.R., J. Wright, A.T. Eldefrawi, M.E. Eldefrawi, E.L. Johnson, S.D. Emche and C.S. Helling.** 1997. Evaluation of a fiber optic immunosensor for quantitating cocaine in coca leaf extracts. Biosens. Bioelectron. *12*:113-124.
104. **Uttenthaler, E., C. Kosslinger and S. Drost.** 1998. Characterization of immobilization methods for African swine fever virus protein and antibodies with a piezoelectric immunosensor. Biosens. Bioelectron. *13*:1279-1286.
105. **VanderNoot, V.A., R.E. Hileman, J.S. Dordick and R.J. Linhardt.** 1998. Affinity capillary electrophoresis employing immobilized glycosaminoglycan to resolve heparin-binding peptides. Electrophoresis *19*:437-441.
106. **Van Emon, J.M., C.L. Gerlach and K. Bowman.** 1998. Bioseparation and bioanalytical techniques in environmental monitoring. J. Chromatogr. B *715*:211-228.
107. **van Ginkel, L.A.** 1991. Immunoaffinity chromatography, its applicability and limitations in multi-residue analysis of anabolizing and doping agents. J. Chromatogr. *564*:363-384.
108. **Vespalec, R. and P. Bocek.** 1994. Chiral separations by capillary electrophoresis: present state of the art.

Electrophoresis *15*:755-762.

109. **Wang, J., B. Tian and K.R. Rogers.** 1998. Thick-film electrochemical immunosensor based on stripping potentiometric detection of a metal ion label. Anal. Chem. *70*:1682-1685.

110. **Wessa, T., M. Rapp and H.J. Ache.** 1999. New immobilization method for SAW-biosensors: covalent attachment of antibodies via CNBr. Biosens. Bioelectron. *14*:93-98.

111. **Wheatley, J.B.** 1992. Multiple ligand applications in high-performance immunoaffinity chromatography. J. Chromatogr. *603*:273-278.

112. **Wu, X. and R.J. Linhardt.** 1998. Capillary affinity chromatography and affinity capillary electrophoresis of heparin binding proteins. Electrophoresis *19*:2650-2653.

113. **Yao, T. and G.A. Rechnitz.** 1987. Amperometric enzyme-immunosensor based on ferrocene-mediated amplification. Biosensors *3*:307-312.

114. **Zahn, M. and S. Seeger.** 1998. Optical tweezers in pharmacology. Cell. Mol. Biol. *44*:747-761.

Appendix: Protein Measurement

I

INTRODUCTION

Measurement of total protein is common in basic science and clinical applications, and its routine measurement is required for virtually all of the immunochemical techniques described in this book. This appendix describes several of the most common techniques used in measuring protein, including the direct spectrophotometric, biuret (4), Bradford (2), Lowry (5), bicinchoninic acid (BCA) (6), fluorescamine (3,7) and dotMETRIC™ techniques. These procedures vary from the quick and relatively insensitive (range 0.1–3 mg) direct spectrophotometric measurement of the UV ratio at 280 nm vs. 260 nm to highly sensitive fluorescamine (0.1–50 µg) or dotMETRIC assays (0.5 ng–50 µg). Table 1 compares the sensitivity for all of the protein measurement protocols given, and provides a list of disadvantages for each method. Finally, we provide a short description of how to prepare a working protein standard curve for these assays.

PROTEIN ASSAYS

Direct Spectrophotometric Technique

Direct spectrophotometric estimation of protein content is the quickest and most inexpensive method of protein determination. The only equipment required is a UV spectrophotometer and a centrifuge, and reagents do not have to be purchased or prepared. The disadvantages include the inability to measure proteins in very dilute samples, the serious problems caused by turbidity, and interference caused by nucleic acids and lipids.

1. Remove particulate material in the sample by centrifugation at 10 000–15 000× *g* for 5 min.

Affinity and Immunoaffinity Purification Techniques
Terry M. Phillips and Benjamin F. Dickens
© 2000 Eaton Publishing, Natick, MA

Table 1. Comparison of Protein Measurement Assays

Assay	Sensitivity	Comments
Direct Spectrophotometry Technique	0.1–3 mg	Cannot measure low protein concentrations. Sample turbidity can be a serious problem. Interference can be caused by lipids and nucleic acids.
Biuret Method	1–10 mg	Cannot measure low protein concentrations. Nonlinear at high protein concentrations. Some variation with protein concentration.
Bradford Assay	1–100 µg	Some variation with the protein composition of the sample. Interference can be caused by detergents.
Micro-Lowry Assay	1–50 µg	Some variation with the protein composition of the sample. Interference can be caused by detergents, but this can be cleared with SDS.
BCA Assay	0.5–50 µg	Sensitive, easy to perform, and free of interference.
Fluorescamine Assay	0.1–50 µg	Very sensitive and free of interference. Rapid and easy to perform.
dotMETRIC Assay	0.5 ng–50 µg	Very sensitive, rapid assay that uses very little sample material (as low as 1 µL). Measuring protein solutions with less than 5 µg/mL involves serial dilutions and rapidly consumes expensive test kit materials.

2. Transfer an aliquot of the clarified sample protein solution to a quartz cuvette with a 1-cm light path.

3. Fill the reference cuvette of the spectrophotometer with the same buffer used to dissolve the sample protein.

4. Use the reference cuvette to zero the spectrophotometer at 280 nm.

5. Determine the sample absorbance at 280 nm.

   Note: If the sample reading is off of the scale (generally above 2.00), then dilute the sample 1:10 and remeasure.

6. Adjust the wavelength to 260 nm, and zero the spectrophotometer using the reference cuvette.

7. Determine the sample absorbance at 260 nm.

8. Check the purity of the protein sample by calculating the 280:260 ratio. (Ratios above 0.6 indicate possible contamination with nucleic acids.)

9. If the extension coefficient of the sample protein is not known, calculate the protein concentration from the following formula:

$$\text{Protein concentration (mg/mL)} = (1.55 \times \text{Abs}_{280\,nm}) - (0.77 \times \text{Abs}_{260\,nm})$$

   where Abs_x = absorbance at wavelength x.

10. If the extension coefficient of the sample at 280 nm is known, then calculate the protein concentration by the more accurate formula:

Protein concentration (mg/mL) = ($Abs_{280\ nm}$/extinction coefficient at 280 nm) × 10 mg/mL

If it was necessary to dilute the sample to get a reading in step 5 above, this dilution must be adjusted for when calculating the protein concentration.

Biuret Method

The biuret method was one of the first chemical tests designed to estimate protein content (4). This method depends on a direct complex between the peptide bond in proteins and copper ions. The sensitivity of this reaction is very poor, as it measures proteins in the range of 1–10 mg, and the standard curve becomes flattened at higher protein concentrations. Proteins containing a high percentage of aromatic amino acids will also give an abnormally high reading. Because of its poor sensitivity, the biuret method is generally unsuitable for protein assays for immunochemical techniques.

1. To prepare the biuret reagent, dissolve 1.5 g of cupric sulfate pentahyrate and 6 g of potassium tartrate tetrahydrate in 500 mL of distilled H_2O. Add 300 mL of a 10% (wt/vol) solution of NaOH and adjust the total volume to 1 L. Store this biuret stock solution in plastic and protect it from light exposure.
2. Add 1 mL of test sample to a test tube.
3. Prepare a series of protein standards (as described in the protocol for the preparation of a standard curve, below). Add 1 mL of each standard protein concentration to a separate tube.
4. Add 0.1 mL of 10% (wt/vol) deoxycholate and sufficient distilled H_2O to each tube to obtain a final volume of 1.5 mL.
5. Add 1.5 mL of biuret reagent to each tube and mix well.
6. Incubate the tubes for 30 min in a 30°C water bath.
7. Read the absorbance of each tube at 540 nm against a reagent blank.
8. Generate a standard curve by plotting the concentration of the protein standards (x axis) vs. the absorbance (y axis).
9. Determine the concentrations of unknown samples by reading the values off the standard curve prepared in step 8.

Micro-Lowry Assay

The Lowry assay (5) is perhaps the most frequently used protein assay. This assay is based on the reduction of peptide–divalent copper ions by Folin-Ciocalteu reagent under alkaline conditions. Three specific amino acids (tyrosine, cysteine, and tryptophan) react with the divalent copper and Folin-Ciocalteu reagent to produce a blue molybdenum–tungsten product that can be spectrophotometrically measured. Because proteins rich in these three amino acids give darker color and proteins poor in these amino acids produce a lighter color, some variations result based on specific protein concentrations. In addition, detergents can cause interference with this assay, but this interference can usually be cleared by SDS.

1. Dissolve 2 g of sodium bicarbonate in 100 mL of 0.1 *N* sodium hydroxide.
2. Add 1 g of copper sulfate and mix well.

   Note: Make this reagent fresh before assaying the protein.
3. Place 200 µL of each test sample into a separate, labeled glass tube.
4. Prepare a series of protein standards in the range required (as described in the protocol for the preparation of a standard curve, below). Add 1 mL of each standard protein concentration to a separate tube.
5. Add 1 mL of the copper sulfate solution and 1 mL of 2% (wt/vol) sodium potassium tartrate to each tube. Mix well and allow to stand at room temperature for 10 min.
6. Add 200 µL of 10% (wt/vol) SDS and 100 µL of a 1:2 dilution of Folin-Ciocalteu reagent.
7. Mix well and read in a spectrophotometer after 30 min. All the tubes should be read within 10 min of each other.
8. Plot the reading against the known concentration of the standards and calculate the concentrations of the unknowns from the standard curve.

Bradford Assay

The Bradford protein assay (2) is based on the principle of a chromatographic dye binding to the protein of interest, then measuring the absorbance of the dye. While this procedure can be performed rapidly and is relatively sensitive, it has a number of limitations. This assay is subject to aggregate formation, leading to a loss of absorbance over time. There are also variations in results based on protein sample composition, and the detergents often used in protein isolation can cause interference with this assay. The Bradford reagent is commercially available from Bio-Rad and Pierce Chemical, or can be prepared as given in the protocol below.

1. Dissolve 100 g of Coomassie Brilliant Blue G250 in 50 mL of 95% ethanol.
2. Prepare the Bradford reagent by adding 100 mL of 85% (vol/vol) phosphoric acid to the ethanolic solution of Coomassie Brilliant Blue G250, and adjust the volume to 200 mL by adding distilled H_2O.
3. Filter the Bradford reagent through Whatman No. 1 filter paper.

   Note: The solution can be stored at room temperature until ready to use, and if a precipitate forms, the solution can be refiltered.
4. Prepare a series of protein standards in the range required (as described in the protocol for the preparation of a standard curve, below).
5. Add 500 µL of test sample to a test tube.
6. Add 500 µL of each standard protein concentration to be used to prepare a standard curve to additional separate tubes.
7. Add 5 mL of the Bradford reagent to each tube, mix well, and read the absorbance at 590 nm in a spectrophotometer.
8. Generate a standard curve by plotting the concentration of the protein standards (x axis) vs. the absorbance (y axis).
9. Determine the concentrations of unknown samples by reading the values off the standard curve prepared in step 8.

Bicinchoninic Acid Assay

This assay uses BCA to reduce divalent copper ions bound to peptide bonds under alkaline conditions. As with the micro-Lowry procedure, the molybdenum–tungsten product produces a blue color that can be spectophotometrically measured. The BCA assay requires only one incubation step (compared to two for the micro-Lowry assay), but it requires a 30-min incubation at 60°C while the Lowry assay is performed at room temperature. The reagents are available from Pierce Chemical.

1. Prepare the BCA working stock solution by mixing 100 parts of Pierce Reagent A with 2 parts of Pierce Reagent B.
2. Add 10 µL of each unknown protein sample to a separate well of a microplate.
3. Add 10 µL of distilled H_2O to a separate well as a reagent blank.
4. Prepare a series of protein standards in the range required (as described in the protocol for the preparation of a standard curve, below). Add 10 µL of each standard protein concentration to additional wells in the microplate.
5. Add 200 µL of the BCA working solution to each well and mix well by gently tapping the sides of the plate.
6. Cover the plate with Parafilm and incubate at 37°C for 30 min.
7. Read the plate in a microplate reader at 540 nm or in a spectrophotometer at 560 nm.
8. Prepare a standard curve by plotting the absorbance of the known protein standards vs. their concentration. Read the concentration values for each unknown from this standard curve.

Fluorescamine Assay

The use of fluorescence-based assays has greatly enhanced the sensitivity of total protein determination. The most popular of these is the fluorescamine assay originally described by Udenfriend et al. (7) and modified by Castell et al. (3). Fluorescamine (4-phenylspiro[furan-2(3H), 1'-phthalan]-3,3'-dione) rapidly reacts with primary amines within proteins to give a highly fluorescent product. The resulting fluorescence can then be readily determined using microplates and a fluorescence plate reader, with a lower sensitivity limit of approximately 0.1 µg of protein. The advantages of the fluorescamine assay are its sensitivity and the fact that if the fluorescent product is protected from light, it is stable for hours. The disadvantage is that fluorescamine reacts with the free primary amine at the N terminus of all proteins and with the primary amine found on each lysine moiety in the protein. Because the number of lysines vary, some variations result, based on specific protein composition. Other fluorescent molecules that react with primary amines have also been used successfully to measure total protein, including o-phthalaldehyde (1) and 3-(4-carboxy-benzoyl)quinoline-2-carboxaldehyde (8).

1. Add to a test tube 1 mL of 0.2 M sodium borate, pH 9.5, to 250 µL of each test sample and standard.
2. Prepare a series of protein standards as described in the protocol for the preparation of a standard curve, below.
3. Dissolve 2 mg of fluorescamine in 10 mL of acetone and add 250 µL to each

test sample and standard. Vortex-mix well.

4. Incubate at room temperature for 10–15 min.

5. Read in a fluorometer at excitation and emission wavelengths of 390 and 480 nm, respectively.

6. Plot the reading against the known concentration of the standards and calculate the concentrations of the unknowns from the standard curve.

dotMETRIC Assay

Geno Technology sells its dotMETRIC microprotein assay kit for the measurement of protein in a 1-µL volume using their own proprietary test strips and reagents. The advantages of this method are the extremely small volume required and the speed of the assay (8–10 min). The primary disadvantage is the cost of the individual kits. Each kit comes with test strips and four solutions (dilution buffer, fixer, sensitizer, and developer).

1. Add 1 µL of unknown sample to 5–20 µL of dilution buffer supplied with the dotMETRIC kit.

2. Apply 1–5 µL of each protein dilution to the test strip.

3. Prepare working fixer by diluting 0.8 mL of stock fixer with 9 mL of distilled H_2O in a 15-mL tube.

4. Incubate the test strip in working fixer for 2 min at room temperature.

5. Add 0.8 mL of sensitizer to the tube containing the working fixer and test strip, and incubate for an additional 2 min.

6. Prepare a working developer by diluting 0.8 mL of stock developer with 9 mL of distilled H_2O.

7. Transfer the test strip to the working developer solution and incubate for 3–5 min in the dark.

8. Use the dotMETRIC card to match the diameter of each spot to its corresponding protein concentration.

It should be noted that the diameter of the spot is proportional to protein concentrations at 5 µg/mL or greater. To determine the protein content of samples with less than 5 µg of protein/mL, a more laborious procedure requiring serial dilution of the sample is required. Basically, the sample is serially diluted and repeatedly spotted onto test strips until such time as no protein spot can be detected after the spots are developed. In general, the limit of protein detection by this procedure is 0.3–0.6 ng. By calculating back from the dilution required to reach the limit of detection, the sample protein concentration can be reasonably estimated.

Geno Technology has also developed an additional new microprotein assay called the Non-Interfering Protein Assay™. This assay is based on an interesting variation of copper ions binding peptide groups under alkaline conditions (similar to the Lowry and BCA assays). However, instead of measuring the reduction of copper ions, they colorimetrically measure the amount of free copper (that is, copper not bound to proteins). Such an assay is independent of the variability caused by different amino acid side chains, which hampers many of the protein assays described above. They also use a proprietary precipitating agent to remove the test protein from potentially interfering agents used in the isolation of the protein. While the authors have not yet tested this protein assay, the manufacturer reports a linear sensitivity between 0 and 50 ng of protein per assay tube.

PREPARATION OF STANDARDS FOR A STANDARD CURVE

The validity of any of the given protein measurement techniques depends on the development of an accurate protein standard curve. Therefore, great care should be taken when preparing a standard curve mixture. Protein standards should be prepared fresh weekly, and stored at 4°C when not in use.

1. Accurately weigh out 1 mg of BSA.
2. Prepare a 100-µg/mL standard by dissolving the 1 mg of BSA in 10 mL of 0.01 M sodium phosphate.
3. Prepare a serial dilution of this standard by labeling 6 tubes and place the following amounts of phosphate buffer and stock protein standard in each tube:

Tube Label (µg/mL Protein)	Volume of Phosphate Buffer (in mL)	Volume of 100 µg/mL Protein Standard Stock (in mL)
10 µg/mL	0.9	0.1
20 µg/mL	0.8	0.2
40 µg/mL	0.6	0.4
60 µg/mL	0.4	0.6
80 µg/mL	0.2	0.8
100 µg/mL	0.0	1.0

REFERENCES

1. **Benson, J.R. and P.E. Hare.** 1975. o-Phthalaldehyde: fluorogenic detection of primary amines in the picomole range. Comparison with fluorescamine and ninhydrin. Proc. Natl. Acad. Sci. USA 72:619-622.
2. **Bradford, M.** 1976. A rapid and sensitive method for the quantitation of microgram quantities of protein utilizing the principle of protein-dye binding. Anal. Biochem. 72:248-254.
3. **Castell, J.V., M. Cervera and R. Marco.** 1979. A convenient micromethod for the assay of primary amines and protein with fluorescamine. A reexamination of the conditions of reaction. Anal. Biochem. 99:379-391.
4. **Gornall, A.G., C.J. Bardawill and M.M. David.** 1949. Determination of serum proteins by means of the biuret reaction. J. Biol. Chem. 177:751-766.
5. **Lowry, O.H., N.J. Rosebrough, A.L. Farr and R.J. Randall.** 1951. Protein measurement with the folin phenol reagent. J. Biol. Chem. 193:265-275.
6. **Smith, P.K., R.I. Krohn, G.T. Hermanson, A.K. Mallia, F.H. Gartner, M.D. Provenzano, E.K. Fujimoto, N.M. Goeke, B.J. Olson and D.C. Klenk.** 1985. Measurement of protein using bicinchoninic acid. Anal. Biochem. 150:76-85.
7. **Udenfriend, S., S. Stein, P. Böhlen, W. Dairman, W. Leimgruber and M. Weigele.** 1972. Fluorescamine: a reagent for assay of amino acids, peptides, proteins, and primary amines in the picomole range. Science 178:871-872.
8. **You, W.W., R.P. Haugland, D.K. Ryan and R.P. Haugland.** 1997. 3-(4-Carboxybenzoyl)quinoline-2-carboxaldehyde, a reagent with broad dynamic range for the assay of proteins and lipoproteins in solution. Anal. Biochem. 244:277-282.

II Appendix: Analyte Labeling Techniques

O ne of the major obstacles in bioanalytical separations is the ability to detect and measure the isolated analyte in either dilute solutions or complex biological matrices. This problem can often be overcome by direct measurement approaches such as UV or refractive index spectrometry. In these cases, on-line detectors can be calibrated to determine the concentrations of the isolated peaks and to record these findings to an integrator. However, in many situations, the analyte of interest is present at dilute concentrations and often below the detection limit of direct measurement detectors. In such situations, other techniques have to be employed to detect and measure the analyte. These techniques include pre- and post-analytical labeling of the analyte with agents such as isotopes, fluorochromes, or enzymes. Once labeled, the analytes can then be detected by the appropriate detection system (radiometry, fluorescence, chemiluminescence, immunoassay, or mass spectrometry). The following appendix will outline the different approaches to these labeling techniques from a practical standpoint.

IODINATION

One of the most popular techniques for labeling analytes prior to separation and analysis is iodination or labeling with radioactive iodine (7,23). There are numerous techniques for performing this type of labeling, the most common of which are the Bolton and Hunter, chloramine T, and the IODO-GEN® techniques (14,20,24,39). Although it has been reported that there is little or no difference between these techniques (24), it has also been reported that the Bolton and Hunter technique is better for some native proteins (14,48). However, IODO-GEN (1,3,4,6-tetrachloro-3 α, 6 α-diphenylglycoluril), especially in the solid-phase bead form commercially available from Pierce Chemical, represents the most straightforward approach to radiolabeling labile biological analytes.

Affinity and Immunoaffinity Purification Techniques
Terry M. Phillips and Benjamin F. Dickens
© 2000 Eaton Publishing, Natick, MA

PROTOCOLS

Protein Iodination Using the Bolton and Hunter Technique

1. Dissolve 1 µg of protein in 10 µL of 0.1 M $Na_2B_4O_7$, pH 8.5.
2. Place 1 mCi of Bolton and Hunter reagent [N-succinidimyl 3-(4-hydroxy, 5-[^{125}I] iodophenyl)propionate] in a microcentrifuge tube.
3. Remove the solvent using a gentle jet of nitrogen.
4. Resuspend the Bolton and Hunter reagent in 10 µL of 0.1 M $Na_2B_4O_7$, pH 8.5.
5. Add the protein solution and gently agitate for 15 min at 0°C.
6. Add 100 µL of a solution of 0.1 M $Na_2B_4O_7$ and 0.2 M glycine, pH 8.5, and agitate for 5 min at 0°C.
7. Dialyze for 3–6 h against 0.01 M sodium phosphate, pH 7.4.
8. Inject into the affinity or immunoaffinity system.

Chloramine-T Iodination of Proteins

1. Dissolve 1–5 µg of protein in 10 µL of 0.05 M sodium phosphate, pH 7.4.
2. Add 2 mCi of carrier-free $Na^{125}I$ in 10 µL of 0.05 M NaH_2PO_4.
3. Add 10 µg of chloramine T freshly dissolved in 10 µL of 0.05 M sodium phosphate, pH 7.4.
4. Quickly mix for 15 s by flicking the tube with your finger.
5. Immediately add 20 µL of 0.05 M sodium phosphate, pH 7.4, containing 20 µg of sodium metabisulfite.
6. Incubate at 4°C for 5 min.
7. Add 500 µL of 0.05 M sodium phosphate, pH 7.4, containing 1 µg/mL of BSA.
8. Dialyze for 3–6 h against 0.01 M sodium phosphate, pH 7.4.
9. Inject into the affinity or immunoaffinity system.

IODO-GEN Solid-Phase Iodination

1. Place an IODO-GEN bead (Pierce Chemical) into a 12-mm × 75-mm glass tube.
2. Dissolve 200–500 µCi of $Na^{125}I$ in 500 µL of 0.1 M Tris, pH 7.0.
3. Add the iodine solution to the bead and incubate at room temperature for 5 min.
4. Dissolve 10 µg of protein in 200 µL of 0.1 M Tris, pH 7.0.
5. Add the protein solution to the bead and incubate for 10–15 min at room temperature.
6. Dialyze for 3–6 h against 0.01 M sodium phosphate, pH 7.4.
7. Inject into the affinity or immunoaffinity system.

FLUORESCENCE

One of the most widely used derivatization agents for fluorescence detection is ophthalaldehyde (OPA), which forms a fluorescent product upon reaction with a primary amino group. Additionally, OPA forms a stable product in a relatively short period of time (30 seconds) and is reactive at room temperature. This reagent has been applied to both pre- and post-separation derivatization procedures for a number of different analytes, including proteins and peptides (51), enantiomeric species of drugs (47), histidine (43), and biotin (33). The technique is relatively simple to perform and is reasonably cost-effective. Tang et al. (42) used this reagent to develop an automated system for the fluorescent determination of gabapentin [1-(amino-methyl)cyclohexaneacetic acid] in serum samples by HPLC. Derivatization was so simple that it could be easily controlled during robotic manipulation of the process.

Basic fluorochromes such as fluorescamine (51) and dansyl chloride (5-[dimethyl-amino]naphthalene-1-sulfonyl chloride) have been employed for amino acid detection for a number of years, especially in amino acid sequencing (27) and selected drug screening (41). Although dansyl chloride is a reasonable reagent with good fluorescent characteristics, another reagent (FITC) (21) is easier to couple to peptides and proteins. This reagent has gained popularity as a simple procedure for labeling a number of different molecules including the N terminus of peptides (50) and DNA (2).

The improved detection afforded by the use of fluorescent agents has led to the development of an increasing number of unique reagents that have been applied to the detection of a wide variety of biological molecules including glycans (5) and fatty acids or their derivatives (53). Table 1 gives a selection of novel agents used for the fluorescent detection of analytes.

Laser-Induced Fluorescence Reagents

Although the simplest approach to detection is the employment of an on-line fluorescence detector based on a spectral lamp output (such as a xenon or mercury arc lamp), the introduction of CE has led to the development of laser-induced fluorescence (LIF) detectors. Based on this development, O'Keefe (34) employed the thiol-reactive reagent monobromobimane to selectively label protein cysteine residues prior to CE analysis. This approach allowed detection at the 10-pM level of concentration, but the fluorescence intensity was dependent on the number of cysteine residues present in the analyte. Fadden and Haystead (13) labeled phosphoserine residues in different proteins and peptides with 1,2-ethanedithiol and found that they could quantitate the phosphoserine content of the analytes down to approximately 75 attomole (10^{-18} M). Baars and Patonay (3) employed a near infrared dye (NN382; LI-COR) to label angiotensin-related peptides prior to analysis by CE. This unique dye was activated by a near infrared diode laser system and enabled investigators to detect analytes within the 100–300 zM range. LIF detection has also been used for the study of the binding characteristics of photoaffinity labeled receptors (9) and for profiling complex carbohydrate profiles following enzymatic degradation of biologically important glycoproteins (17).

One of the major advances in LIF detection was the introduction of the cyanine dyes. These dyes were first introduced as isothiocyanate derivatives and used to detect cell types by immunological staining and flow cytometry (31). Later, the same dyes became available as succinimidyl derivatives – an advance in protein labeling (30). Although several species of these dyes exist, Cy5 (Research Organics) is the most popular for analytical work and has been used to study antigen–antibody interactions by CE (12), bioanalysis of drugs (11,32), tissue neuropeptides (35), recombinant cytokines in human body fluids (36) and DNA sequencing (26). Cy5 in combi-

Table 1. Selected Reagents Used for Fluorescence Detection

Fluorescent Reagent	Excitation Wavelength	Emission Wavelength	Reference
4-(5,6-dimethoxy-2-phthalimidinyl) -2-methoxyphenylsulfonyl chloride	318	406	(45)
4-(4,5-diphenyl-1H-imidazol-2-yl) benzoyl chloride	330	440	(1,49)
cyano[f]benzoisoindole	420	470	(10)
4-(2-phthalimidyl) benzoyl chloride	312	420	(16)
9-fluorenylmethyl-oxycarbonyl chloride	255	315	(40)
1,2-diphenylethylenediamine	330	445	(22)
Tris(2,2'-bipyridyl)ruthenium	454	607	(52)
7-fluorobenzo-2-oxa-1,3-diazole-4- sulfonate	385	515	(15)
2,3-diaminonaphthalene	350	540	(46)
7-fluoro-4-nitrobenzo-2-oxa-1,3-diazole	495	533	(19)

nation with one or more other dyes in the Cy series (Cy3, Cy5.5, and Cy7) has been applied to multi-analyte screening (8) and in situ tumor imaging (4). Recently, Moody et al. (29) reported on the use of a tricarbocyanine dye, indocyanine green, as a fluorescent label for the noncovalent labeling of proteins.

One of the problems with fluorescent labeling that is becoming a recurrent issue is the question of how to label minute biological samples. This problem can be overcome by a number of different techniques. Pinto et al. (38) employed a membrane to capture and concentrate a model protein (insulin B chain) prior to labeling with 3-(2-furoyl)quinoline-2-carboxaldehyde in situ. The authors report that this technique was able to handle as little as 5 pg of the protein and afford detection at 2.4×10^{-21} M. In another report by the same group (37), protocols for handling extremely small volume protein samples are given, using a number of model proteins including ovalbumin. For readers wishing to further investigate the problems of small sample labeling, Krull et al. (25) have produced an excellent review on this subject.

PROTOCOLS

o-Phthalaldehyde–thiol Technique for Labeling Primary Amines

1. Prepare a stock solution by dissolving 27 mg of o-phthalaldehyde–thiol (OPT; Sigma Chemical) in 500 μL of absolute ethanol alcohol.
2. Add 5 mL of 0.1 M $Na_2B_4O_7$, pH 8.0.
3. Mix thoroughly and add 50 μL of 2-mercaptoethanol.
4. Mix thoroughly and store in the dark.
5. Add 200 μL of the OPT stock solution to 50 μL of protein sample.

6. Incubate in the dark for 2 min at room temperature.

7. Inject into the affinity or immunoaffinity system.

Labeling Proteins and Peptides with Dansyl Chloride

1. Dissolve 1 mg of dansyl chloride in 200 μL of acetone.

2. Dissolve 1 μg of protein in 200 μL of 0.5 M Na_2CO_3, pH 9.0, and place in a microcentrifuge tube. Cover the tube with aluminum foil to protect it from light.

3. Slowly add 50 μL of the dansyl chloride solution and mix by gentle vortex-mixing while protecting from light.

4. Cap the tube and incubate on an overhead mixer for 2–3 h at 4°C. Protect from light.

5. Inject into the affinity or immunoaffinity system.

Fluorescamine Labeling

1. Dissolve 1 μg of fluorescamine in 5 μL of acetone.

2. Dissolve 1 μg of protein in 25 μL of 0.1 M $Na_2B_4O_7$, pH 8.5.

3. Slowly add the fluorescamine to the protein solution.

4. Incubate the mixture at 37°C for 5 min.

5. Readjust the pH to 7.4 with 0.1 N HCl.

6. Inject into the affinity or immunoaffinity system.

Labeling Proteins with FITC

1. Dissolve 10 μg of FITC in 1 mL of dimethylformamide.

2. Add 10 μl of the FITC solution to 100 μL of 0.5 M Na_2CO_3, pH 9.5.

3. Dissolve 1 μg of protein in 100 μL of 0.5 M Na_2CO_3, pH 9.5, and place in a microcentrifuge tube. Cover the tube with aluminum foil to protect it from light.

4. Slowly add the FITC solution (drop-wise) to the protein solution with continuous gentle vortex-mixing.

5. Cap the tube and mix on an overhead mixer for 30–60 min at room temperature. Protect from light.

6. Inject into the affinity or immunoaffinity system.

Labeling with Cy5

1. Dissolve 50 μg of Cy5 in 1 mL of 0.5 M Na_2CO_3, pH 9.5.

2. Dissolve 1 μg of protein in 20 μL of 0.5 M Na_2CO_3, pH 9.5.

3. Slowly add 20 μL of the Cy5 solution.

4. Place on an overhead mixer for 10 min at room temperature.

5. Clarify the solution by centrifugation at $12\,000\times g$ for 1 min.

6. Inject into the affinity or immunoaffinity system.

Building a Fiber-Optic Fluorescence Detector for Increased Sensitivity

The fiber-optic detector shown in Figure 1 can be created by following the steps in the protocol below.

1. Secure a 4-mW miniature diode laser module (Edmund Scientific) into a laser mount on a fiber launch module (Thorlabs) and attach the module to a light bench. Select the wavelength of the laser to match the absorbance maximum of the fluorescent label.

2. Attach a 10–15-cm, 200-µm fiber optic to the laser output by a fiber connector (Thorlabs).

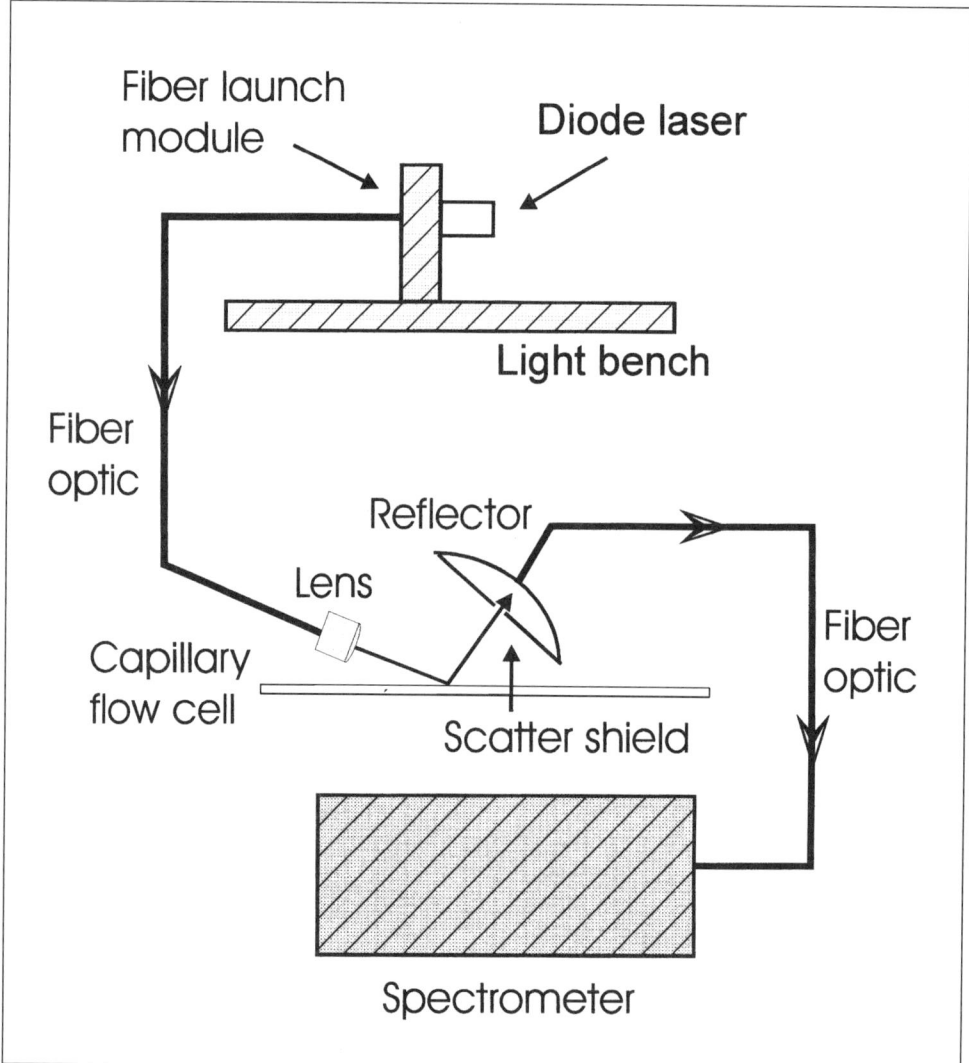

Figure 1. A fiber-optic LIF detector. A 4-mW diode laser is attached to a fiber launch module on a light bench. A fiber optic delivers the laser light by a focusing lens to the fused silica capillary flow cell. The fluorescence generated when the fluorochrome-labeled analyte is excited by the laser light is focused onto the reflector by the scatter shield, where it is collected and passed to a spectrometer by another fiber optic.

3. Using a 50-mm convex lens, focus the laser beam onto the capillary flow cell at a 15° angle.

4. Set up a 10-mm radius parabolic reflector equipped with a scatter shield to cover the flow cell.

5. Arrange the scatter shield so that it positions the flow cell at the reflector's point of focus.

6. Place a 671 ± 10-nm interference filter in front of another optical fiber and attach to the focal point of the reflector.

7. Attach the output end of the detection fiber to a miniature fiber-optic spectrometer (Ocean Optics).

CHEMILUMINESCENCE

The introduction of chemiluminescence heralds a potential new era in detection sensitivity, although this has still to be proven. The generation of light through either a chemical or enzymatic reagent can be used to sufficiently amplify a signal for an on-line detector to make adequate measurements. This situation allows for the detection of analytes at extremely dilute concentrations. Reaction systems based on enzyme catalysts such as hydrogen peroxide and alkaline phosphatase are commercially available from a number of companies, including Tropix and Amersham Pharmacia Biotech. Tsunoda et al. (44) used post-column oxidation coupled with ethylenediamine derivatization and peroxyoxalate chemiluminescence to detect catecholamines in rat plasma by HLPC. Using this system, they reported detection limits of 3–10 fM/50-μL sample. Likewise, Hamachi et al. (18) used pre-separation derivatization with 4-(N,N-dimethylaminosulfonyl)-7-hydrazino-2,1,3-benzoxadiazole to determine by HPLC the concentration of propentofylline in microdialysis samples taken from living rat brains. This adaptation of the peroxyoxalate chemiluminescence technique resulted in detection of 0.031–1.25 ng/sample. Bolden and Danielson (6) employed a different approach. They used UV irradiation-enhanced tris(2,2′-bipyridyl)ruthenium(III) [Ru(bpy)3(3+)] chemiluminescence in the detection of a number of aromatic amines by HPLC. This approach allowed the investigators to detect these analytes in the 2–20-pM range.

Mangru and Harrison (28) used an enzyme-catalyzed luminol–peroxide reaction in a post-analytical reactor to generate the signal required for chemiluminescence detection. These reactions were performed following CE separation of the analytes of interest. Zhu and Kok (54) have written a good review of the instrumentation and procedures involved in performing post-separation fluorescence and chemiluminescence measurements.

PROTOCOLS

Post-Analysis Peroxyoxalate Chemiluminescence

1. Prelabel the sample with dansyl chloride as described in the protocol for labeling proteins and peptides with dansyl chloride, above.

2. Pack a 5-cm × 2-mm id PEEK column with 75-μm acid-washed glass beads

(Supelco). (This will act as the reactor column.)

3. Set up a post-analysis system as shown in Figure 2.

4. Load the post-column reservoir with a solution of 1 mM bis(2,4,6-trichlorophenyl) oxalate and 100 mM H_2O_2 in 80% acetonitrile:20% H_2O (vol/vol).

5. Inject the labeled sample into the affinity or immunoaffinity column and isolate the analyte of interest.

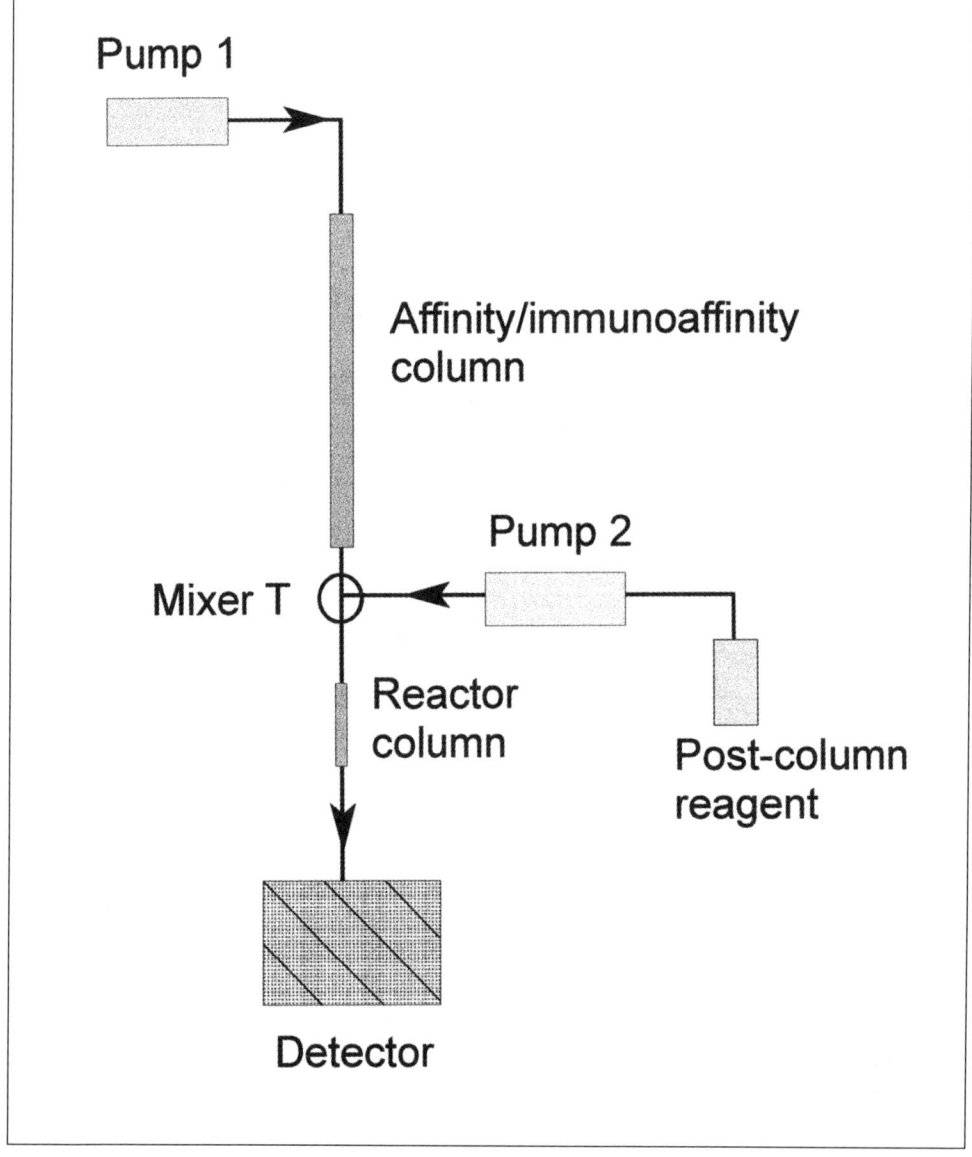

Figure 2. A post-separation chemiluminescence system. The sample is pumped through the affinity or immunoaffinity column and onto the detector by pump 1. The post-column chemiluminescence reagent is pumped into the mixer T connector by pump 2. Reactions between the resolved peaks and the post-column reagent take place in the glass bead-filled reactor column. Finally, the chemiluminescent end product is passed on to the detector where it is measured.

6. During the elution or recovery phase of the separation, continuously pump the post-column reagent into the mixer T at a flow rate of 0.5 mL/min (the flow rate of the post-column reagent should match the flow rate of the main column).

7. Measure the chemiluminescence in an on-line detector.

Alkaline Phosphatase-Catalyzed Post-Separation Chemiluminescence

1. Immobilize the desired antibody to the bottom of a suitable well (i.e., the bottom of a 96-well microplate) by any of the protocols given in Chapter 4.

2. Collect the eluted peaks from an affinity or immunoaffinity column.

3. Add 25 μL of each peak to a separate antibody-coated well.

4. Add 25 μL of suitable controls to a further series of antibody-coated wells.

5. Incubate for 1–2 h at room temperature.

6. Wash each well three times with 0.1 M sodium phosphate, 0.05% Tween 20, pH 7.4.

7. Add 25 μL of a 1:5000 dilution of alkaline phosphatase-labeled antibody (the same antibody as in step 1 or an antibody reactive to another epitope on the captured analyte) in 0.1 M sodium phosphate, pH 7.4 (without Tween 20) to each well.

8. Incubate for 60 min at room temperature.

9. Wash three times in 0.1 M sodium phosphate, 0.05% Tween 20, pH 7.4.

10. Prepare a solution of 25 mM disodium 3-(4-methoxyspiro [1,2-dioxetane-3,2′-(5′-chloro)-tricyclo-decan]-4-yl)phenyl phosphate (CSPD; Tropix) in 0.1 M diethyleneamine, 1 mM $MgCl_2$. Adjust the pH to 10.0 with HCl.

11. Add 200 μL to each well and incubate for 2–5 min at room temperature.

12. Measure the chemiluminescence in a luminometer.

5-Bromo-4-chloro-indolyl Phosphate as a Chemiluminescence Substrate for Alkaline Phosphatase

1. Follow steps 1 through 9 in the protocol for alkaline phosphatase-catalyzed post-separation chemiluminescence, above.

2. Prepare a solution of 0.19 mM 5-bromo-4-chloro-indolyl phosphate, 25 μM isoluminol, 5 U/mL horseradish peroxidase, 0.1% BSA, and 0.1 M $MgCl_2$ in 0.1 M Tris-HCl, pH 9.5.

3. Add 200 μL to each well and incubate for 2–5 min at room temperature.

4. Measure the chemiluminescence in a luminometer.

These last two procedures are best used to amplify a capture immunoassay detection system. The enzyme-labeled reagent can also be replaced with enzyme-labeled avidin or streptavidin, provided that the analyte has previously been biotinylated.

REFERENCES

1. **Al-Dirbash, O., J. Qvarnstrom, K. Irgum and K. Nakashima.** 1998. Simple and sensitive high-performance liquid chromatographic determination of methamphetamines in human urine as derivatives of 4-

(4,5-diphenyl-1H-imidazol-2-yl) benzoyl chloride, a new fluorescence derivatization reagent. J. Chromatogr. B *712*:105-112.

2. **Azadnia, A., R. Campbell and M. Sharma.** 1994. The scope of dansyl vs. fluorescein label in fluorescence postlabeling assay for DNA damage. Anal. Biochem. *218*:444-448.

3. **Baars, M.J. and G. Patonay.** 1999. Ultrasensitive detection of closely related angiotensin I peptides using capillary electrophoresis with near-infrared laser-induced fluorescence detection. Anal. Chem. *71*:667-671.

4. **Ballou, B., G.W. Fisher, J.S. Deng, T.R. Hakala, M. Srivastava and D.L. Farkas.** 1998. Cyanine fluorochrome-labeled antibodies in vivo: assessment of tumor imaging using Cy3, Cy5, Cy5.5, and Cy7. Cancer Detect. Prev. *22*:251-257.

5. **Bigge, J.C., T.P. Patel, J.A. Bruce, P.N. Goulding, S.M. Charles and R.B. Parekh.** 1995. Nonselective and efficient fluorescent labeling of glycans using 2-amino benzamide and anthranilic acid. Anal. Biochem. *230*:229-238.

6. **Bolden, M.E. and N.D. Danielson.** 1998. Liquid chromatography of aromatic amines with photochemical derivatization and tris(bipyridine)ruthenium(III) chemiluminescence detection. J. Chromatogr. *828*:421-430.

7. **Breslav, M., A. McKinney, J.M. Becker and F. Naider.** 1996. Preparation of radiolabeled peptides via an iodine exchange reaction. Anal. Biochem. *239*:213-237.

8. **Chen, F.T. and R.A. Evangelista.** 1994. Feasibility studies for simultaneous immunochemical multianalyte drug assay by capillary electrophoresis with laser-induced fluorescence. Clin. Chem. *40*:1819-1822.

9. **Dong, M., W.Q. Ding, D.I. Pinon, E.M. Hadac, R.P. Oda, J.P. Landers and L.J. Miller.** 1999. Structurally related peptide agonist, partial agonist, and antagonist occupy a similar binding pocket within the cholecystokinin receptor. Rapid analysis using fluorescent photoaffinity labeling probes and capillary electrophoresis. J. Biol. Chem. *274*:4778-4785.

10. **Eisenberg, E.J. and K.C. Cundy.** 1998. High-performance liquid chromatographic determination of GS4071, a potent inhibitor of influenza neuraminidase, in plasma by precolumn fluorescence derivatization with naphthalenedialdehyde. J. Chromatogr. B Biomed. Sci. App. *716*:267-273.

11. **Ellingson, A. and H.T. Karnes.** 1998. Investigation of far red dyes for use in peroxyoxalate chemiluminescence detection and analysis of the Cy5 derivative of amantadine hydrochloride in human plasma. Biomed. Chromatogr. *12*:8-12.

12. **Evangelista, R.A. and F.T. Chen.** 1994. Analysis of structural specificity in antibody-antigen reactions by capillary electrophoresis with laser-induced fluorescence detection. J. Chromatogr. *680*:587-591.

13. **Fadden, P. and T.A. Haystead.** 1995. Quantitative and selective fluorophore labeling of phosphoserine on peptides and proteins: characterization at the attomole level by capillary electrophoresis and laser-induced fluorescence. Anal. Biochem. *225*:81-88.

14. **Frantzen, F., D.E. Heggli and E. Sundrehagen.** 1995. Radiolabeling of human haemoglobin using the ^{125}I-Bolton-Hunter reagent is superior to oxidative iodination for conservation of the native structure of the labelled protein. Biotechnol. Appl. Biochem. *22*:161-167.

15. **Gautier, F.C., C. Berneron and P.J. Douce.** 1999. Determination of homocysteine in plasma by liquid chromatography with fluorescence detection. Biomed. Chromatogr. *13*:239-243.

16. **Gong, A. and C. Ye.** 1998. Analysis of trace atrazine and simazine in environmental samples by liquid chromatography-fluorescence detection with pre-column derivatization reaction. J. Chromatogr. *827*:57-63.

17. **Guttman, A. and C. Starr.** 1995. Capillary and slab gel electrophoresis profiling of oligosaccharides. Electrophoresis *16*:993-997.

18. **Hamachi, Y., M.N. Nakashima and K. Nakashima.** 1999. High-performance liquid chromatography with peroxyoxalate chemiluminescence determination of propentofylline concentrations in rat brain microdialysate. J. Chromatogr. B Biomed. Sci. App. *724*:189-194.

19. **Hui, Y.H., S. Carroll and K.C. Marsh.** 1997. Development of a sensitive method for quantitation of ABT-089 in plasma using fluorescence labeling with 7-fluoro-4-nitrobenzo-2-oxa-1,3-diazole. J. Chromatogr. B *695*:337-347.

20. **Hussain, A.A., J.A. Jona, A. Yamada and L.W. Dittert.** 1995. Chloramine-T in radiolabeling techniques. II. A nondestructive method for radiolabeling biomolecules by halogenation. Anal. Biochem. *224*:221-226.

21. **Johnson, D.A. and R. Cushman.** 1988. Purification and characterization of four monofluorescein cobra alpha-toxin derivatives. J. Biol. Chem. *263*:2802-2807.

22. **Kai, M., H. Iida, H. Nohta, M.K. Lee and K. Ohta.** 1998. Fluorescence derivatizing procedure for 5-hydroxytryptamine and 5-hydroxyindoleacetic acid using 1,2-diphenylethylenediamine reagent and their sensitive liquid chromatographic determination. J. Chromatogr. B *720*:25-31.

23. **Karonen, S.L.** 1990. Developments in techniques for radioiodination of peptide hormones and other proteins. Scand. J. Clin. Lab. Invest. Suppl. *201*:135-138.

24. **Kienhuis, C.B., J.J. Heuvel, H.A. Ross, J.A. Foekens and T.J. Benraad.** 1992. High-performance liquid chromatography of ^{125}I-labeled mouse epidermal growth factor radioiodinated by six different methods. Clin. Chem. *38*:681-686.

25. **Krull, I.S., R. Strong, Z. Sosic, B.Y. Cho, S.C. Beale, C.C. Wang and S. Cohen.** 1997. Labeling reac-

tions applicable to chromatography and electrophoresis of minute amounts of proteins. J. Chromatogr. B *699*:173-208.

26. **Lieberwirth, U., J. Arden-Jacob, K.H. Drexhage, D.P. Herten, R. Muller, M. Neumann, A. Schulz, S. Siebert, G. Sagner, S. Klingel et al.** 1998. Multiplex dye DNA sequencing in capillary gel electrophoresis by diode laser-based time-resolved fluorescence detection. Anal. Chem. *70*:4771-4779.

27. **Lundblad, R.L.** 1995. Methods for sequence determination, p. 35-50. Techniques in Protein Modification. CRC Press, Boca Raton.

28. **Mangru, S.D. and D.J. Harrison.** 1998. Chemiluminescence detection in integrated post-separation reactors for microchip-based capillary electrophoresis and affinity electrophoresis. Electrophoresis *19*:2301-2307.

29. **Moody, E.D., P.J. Viskari and C.L. Colyer.** 1999. Non-covalent labeling of human serum albumin with indocyanine green: a study by capillary electrophoresis with diode laser-induced fluorescence detection. J. Chromatogr. B *729*:55-64.

30. **Mujumdar, R.B., L.A. Ernst, S.R. Mujumdar, C.J. Lewis and A.S. Waggoner.** 1993. Cyanine dye labeling reagents: sulfoindocyanine succinimidyl esters. Bioconjug. Chem. *4*:105-111.

31. **Mujumdar, R.B., L.A. Ernst, S.R. Mujumdar and A.S. Waggoner.** 1989. Cyanine dye labeling reagents containing isothiocyanate groups. Cytometry *10*:11-19.

32. **Nagaraj, S., S.V. Rahavendran and H.T. Karnes.** 1998. Visible diode laser induced fluorescence detection for capillary electrophoretic analysis of amantadine in human plasma following precolumn derivatization with Cy5.29.OSu. J. Pharm. Biomed. Anal. *18*:411-420.

33. **Nojiri, S., K. Kamata and M. Nishijima.** 1998. Fluorescence detection of biotin using post-column derivatization with OPA in high performance liquid chromatography. J. Pharm. Biomed. Anal. *16*:1357-1362.

34. **O'Keefe, D.O.** 1994. Quantitative electrophoretic analysis of proteins labeled with monobromobimane. Anal. Biochem. *222*:86-94.

35. **Phillips, T.M.** 1998. Determination of in-situ tissue neuropeptides by capillary electrophoresis. Anal. Chim. Acta. *372*:209-218.

36. **Phillips, T.M. and B.F. Dickens.** 1998. Analysis of recombinant cytokines in human body fluids by immunoaffinity capillary electrophoresis. Electrophoresis *19*:2991-2996.

37. **Pinto, D.M., E.A. Arriaga, D. Craig, J. Angelova, N. Sharma, H. Ahmadzadeh and N.J. Dovichi.** 1997. Picomolar assay of native proteins by capillary electrophoresis precolumn labeling, submicellar separation, and laser-induced fluorescence detection. Anal. Chem. *69*:3015-3021.

38. **Pinto, D.M., E.A. Arriaga, S. Sia, Z. Li and N.J. Dovichi.** 1995. Solid-phase fluorescent labeling reaction of picomole amounts of insulin in very dilute solutions and their analysis by capillary electrophoresis. Electrophoresis *16*:534-540.

39. **Reynolds, A.J. and I.A. Hendry.** 1999. A technique for ^{125}I-labeling of neurotrophins and the use of retrograde axonal transport as a bioassay. Brain Res. Brain Res. Protocol *3*:308-312.

40. **Sastre Torano, J. and H.J. Guchelaar.** 1998. Quantitative determination of the macrolide antibiotics erythromycin, roxithromycin, azithromycin and clarithromycin in human serum by high-performance liquid chromatography using pre-column derivatization with 9-fluorenylmethyloxycarbonyl chloride and fluorescence detection. J. Chromatogr. B *720*:89-97.

41. **Suckow, R.F., M.F. Zhang, E.D. Collins, M.W. Fischman and T.B. Cooper.** 1999. Sensitive and selective liquid chromatographic assay of memantine in plasma with fluorescence detection after pre-column derivatization. J. Chromatogr. B *729*:217-224.

42. **Tang, P.H., M.V. Miles, T.A. Glauser and T. DeGrauw.** 1999. Automated microanalysis of gabapentin in human serum by high-performance liquid chromatography with fluorometric detection. J. Chromatogr. B *727*:125-129.

43. **Tateda, N., K. Matsuhisa, K. Hasebe, N. Kitajima and T. Miura.** 1998. High-performance liquid chromatographic method for rapid and highly sensitive determination of histidine using postcolumn fluorescence detection with o-phthaldialdehyde. J. Chromatogr. B *718*:235-241.

44. **Tsunoda, M., K. Takezawa, T. Santa and K. Imai.** 1999. Simultaneous automatic determination of catecholamines and their 3-O-methyl metabolites in rat plasma by high-performance liquid chromatography using peroxyoxalate chemiluminescence reaction. Anal. Biochem. *269*:386-392.

45. **Tsuruta, Y. and H. Inoue.** 1998. 4-(5,6-dimethoxy-2-phthalimidinyl)-2-methoxyphenylsulfonyl chloride as a fluorescent labeling reagent for determination of amino acids in high-performance liquid chromatography and its application for determination of urinary free hydroxyproline. Anal. Biochem. *265*:15-21.

46. **Ummus, R.E., J. Onuki, D. Dornemann, M.H. Medeiros and P. Di Mascio.** 1999. Measurement of 4,5-dioxovaleric acid by high-performance liquid chromatography and fluorescence detection. J. Chromatogr. B *729*:237-243.

47. **Vermeij, T.A. and P.M. Edelbroek.** 1998. High-performance liquid chromatographic analysis of vigabatrin enantiomers in human serum by precolumn derivatization with o-phthaldialdehyde-N-acetyl-L-cysteine and fluorescence detection. J. Chromatogr. B *716*:233-238.

48. **Vigorito, E., A. Robles, H. Balter, A. Nappa and F. Goni.** 1995. [125I]IgM (KAU) human monoclonal

cold agglutinin: labelling and studies on its biological activity. Appl. Radiat. Isot. *46*:975-979.

49. **Wada, M., S. Kinoshita, Y. Itayama, N. Kuroda and K. Nakashima.** 1999. Sensitive high-performance liquid chromatographic determination with fluorescence detection of phenol and chlorophenols with 4-(4,5-diphenyl-1H-imidazol-2-yl)benzoyl chloride as a labeling reagent. J. Chromatogr. B *721*:179-186.

50. **Weber, P.J., J.E. Bader, G. Folkers and A.G. Beck-Sickinger.** 1998. A fast and inexpensive method for N-terminal fluorescein-labeling of peptides. Bioorg. Med. Chem. Lett. *8*:597-600.

51. **Wehr, T.C.** 1992. Post column reaction systems for fluorescence detection of polypeptides, p. 579-586. *In* C.T. Mant and R.S. Hodges (Eds.), High-Performance Liquid Chromatography of Peptides and Proteins. CRC Press, Boca Raton.

52. **Woltman, S.J., W.R. Even and S.G. Weber.** 1999. Chromatographic detection using tris(2,2'-bipyridyl)ruthenium(III) as a fluorogenic electron-transfer reagent. Anal. Chem. *71*:1504-1512.

53. **Yoshida, T., A. Uetake, H. Yamaguchi, N. Nimura and T. Kinoshita.** 1988. New preparation method for 9-anthryldiazomethane (ADAM) as a fluorescent labeling reagent for fatty acids and derivatives. Anal. Biochem. *173*:70-74.

54. **Zhu, R. and W.T. Kok.** 1998. Post-column derivatization for fluorescence and chemiluminescence detection in capillary electrophoresis. J. Pharm. Biomed. Anal. *17*:985-999.

 # Appendix: Selected Suppliers

Contact information for all suppliers mentioned in this book is provided below, but this list is not meant to be exhaustive.

Accurate Chemical & Scientific Corp.
Westbury, NY, USA
www.accuratechemical.com
800-645-6264

Advanced Biotechnologies, Ltd.
Surrey, UK
www.adbio.co.uk
44-(0)1372-723-456

Affinity Biologicals Inc.
Hamilton, ON, Canada
www.affinitybiologicals.com
800-903-6020

Affinity BioReagents, Inc.
Golden, CO, USA
www.bioreagents.com
800-527-4535

Agilent Technologies
Hewlett-Packard Chemical Analysis
 Group
Little Falls, DE, USA
www.chem.agilent.com
800-227-9770

Aldrich
Milwaukee, WI, USA
www.sigma-aldrich.com
800-558-9160

Alexis Corp.
San Diego, CA, USA
www.alexis-corp.com
800-900-0065

Alltech Associates, Inc.
Deerfield, IL, USA
www.alltechweb.com
800-ALLTECH

American Bioanalytical
Natick, MA, USA
www.americanbio.com
800-443-0600

American Peptide Co., Inc.
Sunnyvale, CA, USA
www.americanpeptide.com
800-926-8272

Affinity and Immunoaffinity Purification Techniques
Terry M. Phillips and Benjamin F. Dickens
© 2000 Eaton Publishing, Natick, MA

American QUALEX
San Clemente, CA, USA
www.aqsp.com
800-341-2235

Amersham Pharmacia Biotech, Inc.
Piscataway, NJ, USA
www.apbiotech.com
800-526-3593

Amicon
Beverly, MA, USA
www.amicon.com
800-426-4266

AMS Biotechnology (Europe) Ltd.
Witney, Oxon, UK
www.immunok.com
44-(0)1993-706500

Ana-Gen Technologies, Inc.
Palo Alto, CA, USA
www.ana-gen.com
800-654-4671

Ancell Corp.
Bayport, MN, USA
www.ancell.com
800-374-9523

Antibodies Incorporated
Davis, CA, USA
www.antibodiesinc.com
800-824-8540

Autogen Bioclear UK Ltd.
Calne, Wiltshire, UK
www.autogen-bioclear.co.uk
44-(0)1249-819008

Bachem Bioscience Inc.
King of Prussia, PA, USA
www.bachem.com
800-634-3183

Bangs Laboratories, Inc.
Fishers, IN, USA
www.bangslabs.com
800-387-0672

Beckman Coulter, Inc.
Fullerton, CA, USA
www.beckman.com
800-233-4685

Berkeley Antibody Co.
Richmond, CA, USA
www.babco.com
800-92BABCO

Biacore Inc.
Piscataway, NJ, USA
www.biacore.com
800-242-2599

BIO 101, Inc.
Carlsbad, CA, USA
www.bio101.com
800-424-6101

Bioprocessing Inc.
Princeton, NJ, USA
www.bioprocessing.co.uk
609-951-2243

Bioprocessing Ltd.
Consett, Durham, UK
www.bioprocessing.co.uk
44-(0)1207-581555

Bio-Rad Laboratories
Hercules, CA, USA
www.bio-rad.com
800-4-BIORAD

BioSeparations, Inc.
Tucson, AZ, USA
www.bioseparations.com
520-622-5882

BioSepra Inc.
Marlborough, MA, USA
www.biosepra.com
800-752-5277

BioSource International Inc.
Camarillo, CA, USA
www.biosource.com
800-242-0607

BioSpectra, Inc.
Sciota, PA, USA
www.biospectrausa.com
570-992-1243

Bodman Industries
Aston, PA, USA
www.bodman.com
800-241-8774

Calbiochem-Novabiochem Corp.
San Diego, CA, USA
www.calbiochem.com
800-854-3417

Cambio Ltd.
Cambridge, UK
www.cambio.co.uk
44-(0)1223-366500

CEDARLANE Laboratories Ltd.
Hornby, ON, Canada
www.cedarlanelabs.com
800-268-5058

Chemicon International Inc.
Temecula, CA, USA
www.chemicon.com
800-437-7500

Chromatochem Inc.
Missoula, MT, USA
406-728-5897

CLONTECH Laboratories Inc.
Palo Alto, CA, USA
www.clontech.com
800-662-2566

Commonwealth Biotechnologies, Inc.
Richmond, VA, USA
www.cbi-biotech.com
800-735-9224

Cortex Biochem, Inc.
San Leandro, CA, USA
www.cortex-biochem.com
800-888-7713

CPG, Inc.
Lincoln Park, NJ, USA
www.cpg-biotech.com
800-362-2740

Crescent Chemical Co.
Hauppauge, NY, USA
www.creschem.com
800-645-3412

Dynal Inc.
Lake Success, NY, USA
www.dynal.no
800 638-9416

Edge BioSystems
Gaithersburg, MD, USA
www.edgebio.com
800-326-2685

Edmund Scientific Co.
Barrington, NJ, USA
www.edsci.com
800-728-6999

EM Industries, Inc.
Hawthorne, NY, USA
www.emindustries.com
914-592-4660

Fisher Scientific
Pittsburgh, PA, USA
www.fishersci.com
800-766-7000

Fluka
Milwaukee, WI, USA
www.sigma-aldrich.com
800-358-5287

FMC BioProducts
Rockland, ME, USA
www.bioproducts.com
207-594-3400

Geno Technology, Inc.
Maplewood, MO, USA
www.genotech.com
800-628-7730

ICN Biomedicals, Inc.
Costa Mesa, CA, USA
www.icnbiomed.com
800-854-0530

Immunicon Corporation
Huntingdon Valley, PA, USA
www.immunicon.com
215-938-0100

Immunochemistry Technologies, LLC
Bloomington, MN, USA
www.mm.com/ichem
800-829-3194

ImmunoKontact
Witney, Oxon, UK
www.immunok.com
44-(0)1993-706500

INDOFINE Chemical Co., Inc.
Somerville, NJ, USA
www.indofinechemical.com
888-463-6346

InnoGenex
San Ramon, CA, USA
www.innogenex.com
877-449-4636

Irvine Scientific
Santa Ana, CA, USA
www.irvinesci.com
800-577-6097

Jackson ImmunoResearch Laboratories, Inc.
West Grove, PA, USA
www.jacksonimmuno.com
800-367-5296

Kimble/Kontes
Vineland, NJ, USA
www.kimble-kontes.com
888-546-2531

Kirkegaard & Perry Laboratories, Inc.
Gaithersburg, MD, USA
www.kpl.com
800-638-3167

Leipziger Arzneimittelwerk GmbH
Leipzig, Germany
49-(0)341-2582-0

Lemargo Inc.
Mississauga, ON, Canada
www.lemargo.com
800-469-3932

LI-COR Biotechnology Division
Lincoln, NE, USA
www.licor.com
800-645-4267

Life Technologies, Inc. (GIBCO BRL)
Rockville, MD, USA
www.lifetech.com
800-828-6686

Macherey-Nagel GmbH & Co. KG
Düren, Germany
www.macherey-nagel.com
49-(0)2421-969-0

Macherey-Nagel Inc.
Easton, PA, USA
www.macherey-nagel.com
610-559-9848

Marsh Biomedical Products, Inc.
Rochester, NY, USA
www.biomar.com
800-445-2812

MBL International Corp.
Watertown, MA, USA
www.mblintl.com
800-200-5459

Merck KGaA
Darmstadt, Germany
www.merck.de
49-6151-72-0

Merck Ltd.
Poole, Dorset, UK
www.merck-ltd.co.uk
44-1202-665599

Midwest Scientific
St. Louis, MO, USA
www.midsci.com
800-227-9997

Millipore Corporation
Bedford, MA, USA
www.millipore.com
800-MILLIPORE

Miltenyi Biotec Inc.
Auburn, CA, USA
www.miltenyibiotec.com
800-367-6227

Molecular Probes, Inc.
Eugene, OR, USA
www.probes.com
800-438-2209

NEN Life Science Products
Boston, MA, USA
www.nenlifesci.com
800-551-2121

Novagen Inc.
Madison, WI, USA
www.novagen.com
800-526-7319

Ocean Optics, Inc.
Dunedin, FL, USA
www.oceanoptics.com
727-733-2447

PE Biosystems
Foster City, CA, USA
www.pebio.com/ab
800-345-5224

Pegasus Scientific Inc.
Burtonsville, MD, USA
www.pegasusscientific.com
800-734-0078

PeproTech, Inc.
Rocky Hill, NJ, USA
www.peprotech.com
800-436-9910

Pfaltz & Bauer
Waterbury, CT, USA
www.pfaltzandbauer.com
800-225-5172

PGC Scientifics Corp.
Frederick, MD, USA
www.pgcscientifics.com
800-424-3300

BD PharMingen
San Diego, CA, USA
www.pharmingen.com
800-848-6227

Pall Gelman Laboratory
Ann Arbor, MI, USA
www.pall.com/gelman
800-521-1520

Pierce Chemical Co.
Rockford, IL, USA
www.piercenet.com
800-874-3723

Polysciences, Inc.
Warrington, PA, USA
www.polysciences.com
800-523-2575

Promega Corp.
Madison, WI, USA
www.promega.com
800-356-9526

Prosep Filter Systems Ltd.
Huddersfield, West Yorkshire, UK
44-1484-432343

Prozyme
San Leandro, CA, USA
www.prozyme.com
800-457-9444

QED Bioscience Inc.
San Diego, CA, USA
www.qedbio.com
800-929-2114

QIAGEN Inc.
Valencia, CA, USA
www.qiagen.com
800-426-8157

Quantum Biotechnologies Inc.
Montreal, QC, Canada
www.quantumbiotech.com
888-DNA-KITS

R&D Systems
Minneapolis, MN, USA
www.rndsystems.com
800-343-7475

Rainin Instrument Company, Inc.
Woburn, MA, USA
www.rainin.com
800-4-RAININ

Repligen Corp.
Needham, MA, USA
www.repligen.com
800-622-2259

Research Diagnostics, Inc.
Flanders, NJ, USA
www.researchd.com
800-631-9384

Research Organics, Inc.
Cleveland, OH, USA
www.resorg.com
800-321-0570

Roche Molecular Biochemicals
Indianapolis, IN, USA
biochem.roche.com
800-262-1640

Rockland Immunochemicals, Inc.
Gilbertsville, PA, USA
www.rockland-inc.com
800-656-ROCK

Santa Cruz Biotechnology, Inc.
Santa Cruz, CA, USA
www.scbt.com
800-457-3801

Sartorius North America Inc.
Edgewood, NY, USA
www.sartorius.com
800-635-2906

Schleicher & Schuell Inc.
Keene, NH, USA
www.s-and-s.com
800-245-4024

Schott Glass Technologies Inc.
Duryea, PA, USA
www.schottglasstech.com
570-457-7485

Scytek Laboratories, Inc.
Logan, UT, USA
www.scytek.com
800-729-8350

Seikagaku America, Inc.
Falmouth, MA, USA
www.seikagaku.com
800-237-4512

Sepracor Inc.
Marlborough, MA, USA
www.sepracor.com
877-SEPRACOR

Seradyn Inc.
Indianapolis, IN, USA
www.seradyn.com
800-428-4007

Serotec Inc.
Raleigh, NC, USA
www.serotec.co.uk
800-265-7376

SERVA Electrophoresis GmbH
Heidelberg, Germany
www.serva.de
49-(0)6221-13840-0

Sigma Chemical
St. Louis, MO, USA
www.sigma-aldrich.com
800-325-3010

Sigma RBI
Natick, MA, USA
www.callrbi.com
800-736-3690

Spectrum Laboratories, Inc.
Laguna Hills, CA, USA
www.spectrumlabs.com
800-634-3300

Spherotech, Inc.
Libertyville, IL, USA
www.spherotech.com
800-368-0822

Stratagene
La Jolla, CA, USA
www.stratagene.com
800-424-5444

Supelco
Bellefonte, PA, USA
www.sigma-aldrich.com
800-247-6628

Synergy Scientific
Bethesda, MD, USA
www.edvotek.com/synergy
877-467-9637

Technikrom
Evanston, IL, USA
www.technikrom.com
800-865-4100

The Binding Site
San Diego, CA, USA
www.bindingsite.co.uk
800-633-4484

Thomas Scientific
Swedesboro, NJ, USA
www.thomassci.com
800-345-2100

Thorlabs, Inc.
Newton, NJ, USA
www.thorlabs.com
973-579-7227

Tropix, Inc.
Bedford, MA, USA
www.tropix.com
800-542-2369

United States Biological
Swampscott, MA, USA
www.usbio.net
800-520-3011

Upchurch Scientific
Oak Harbor, WA, USA
www.upchurch.com
800-426-0191

Vector Laboratories, Inc.
Burlingame, CA, USA
www.vectorlabs.com
800-227-6666

VWR Scientific Products
West Chester, PA, USA
www.vwrsp.com
800-932-5000

Waters Corporation
Milford, MA, USA
www.waters.com
800-252-4752

Worthington Biochemical Corp.
Lakewood, NJ, USA
www.worthington-biochem.com
800-445-9603

XC Corporation
Lowell, MA, USA
www.xcbeads.com
978-323-4450

Zymed Laboratories Inc.
South San Francisco, CA, USA
www.zymed.com
800-874-4494

Terry M. Phillips, PhD, DSc

Terry M. Phillips has been an immunochemist for the past 32 years, working on a variety of topics ranging from immune regulation to cancer immunology, autoimmunity, immune parasitology, and comparative immunology. He came to the United States in 1977, first holding the position of Assistant Professor of Pathology at Georgetown University Medical Center and later joining the staff of the George Washington University Medical Center in 1980 as an Associate Professor of Medicine and Director of the newly formed Immunochemistry Laboratory. Currently, he holds the position of Professor of Microbiology and Immunology at that institution. He received his PhD from the University of London and later a DSc from the same institution, both in immunology. Although formally trained as an immunochemist, he developed an early interest in the development of analytical procedures based on immunochemical techniques, particularly immunoaffinity chromatography. He has served as a scientific reviewer for the National Institutes of Health, National Science Foundation, Howard Hughes Foundation, the Wellcome Foundation, the Environmental Protection Agency, the Food and Drug Administration, and the Congressional Office for Technology Assessment. Additionally, he has acted as a technical consultant for the US Navy, the Smithsonian Institute, and the Canadian Medical Research Council. During his academic career, he has authored or co-authored 165 peer-reviewed papers, 25 book chapters, and three books, all focused on immunology, immunochemistry, and bioanalytical techniques. He is a member of five editorial boards and extensively reviews for a number of scientific journals and educational book publishers. He has experience not only in analytical techniques such as high-performance liquid chromatography (HPLC), capillary electrophoresis, gel electrophoresis, and immunoassays but also has developed a number of procedures and instruments for performing these analyses. Recent developments focus on the manufacture and refinement of miniaturized instruments for performing multiple analyses on biological fluids and tissues.

Benjamin F. Dickens, PhD

Benjamin F. Dickens has been a faculty member at the George Washington University Medical Center, Washington, DC, since 1985, and is currently an Associate Research Professor of Microbiology and Immunology, and Chief of the Ultramicro Analytical Section of its Immunochemistry Laboratory. He received his PhD from the Department of Microbiology and Cell Science at the University of Florida, Gainesville, FL, and is the author or co-author of over 75 articles, book chapters, and abstracts. His training included a three-year service as a NIH Carcinogenesis Trainee at the University of Texas and an Associate Cardiovascular Research Scientist at the Oklahoma Medical Research Foundation from 1982 to 1985. He regularly reviews for a number of scientific organizations such as the US Department of Agriculture and has 26 years of experience as a chromatographer, including gas chromatography with Mass spectrometry and flame ionizing detection (FID), high-performance liquid chromatography (HPLC), capillary electrophoresis, and affinity and immunoaffinity chromatography.

345

INDEX